# ENVIRONMENTAL ASSESSMENT OF SOCIOECONOMIC SYSTEMS

# NATO CONFERENCE SERIES

I    Ecology
II   Systems Science
III  Human Factors
IV   Marine Sciences
V    Air—Sea Interactions

## II SYSTEMS SCIENCE

*Volume 1*  Transportation Planning for a Better Environment
          Edited by Peter Stringer and H. Wenzel

*Volume 2*  Arctic Systems
          Edited by P. J. Amaria, A. A. Bruneau, and P. A. Lapp

*Volume 3*  Environmental Assessment of Socioeconomic Systems
          Edited by Dietrich F. Burkhardt and William H. Ittelson

*Volume 4*  Earth Observation Systems for Resource Management and Environmental Control
          Edited by Donald J. Clough and Lawrence W. Morley

*Volume 5*  Applied General Systems Research: Recent Developments and Trends
          Edited by George J. Klir

# ENVIRONMENTAL ASSESSMENT OF SOCIOECONOMIC SYSTEMS

Edited by

## Dietrich F. Burkhardt

*Industrieanlagen-Betriebsgesellschaft mbH*
*Ottobrunn, West Germany*

and

## William H. Ittelson

*University of Arizona*
*Tucson, Arizona*

Published in coordination with NATO Scientific Affairs Division by

**PLENUM PRESS · NEW YORK AND LONDON**

Library of Congress Cataloging in Publication Data

Main entry under title:

Environmental assessment of socioeconomic systems.

    (NATO conference series: II, Systems science; v. 3)
    Proceedings of a conference held in Istanbul, Turkey, Oct. 4-8, 1976, sponsored by Sys-
the NATO Special Program Panels on Systems Science and Human Factors.
    Includes index.
    1. Human ecology—Congresses. 2. Environmental policy—Congresses. 3. Technolo-
gy—Social aspects—Congresses. 4. Planning—Congresses. I. Burkhardt, Dietrich. II.
Ittelson, William H. III. NATO Special Program Panel on Systems Science. IV. NATO
Special Program Panel on Human Factors. V. Series.
GF3.E58                          301.3                          77-23528

ISBN-13: 978-1-4684-2522-2         e-ISBN-13: 978-1-4684-2520-8
DOI: 10.1007/978-1-4684-2520-8

Proceedings of a conference on Environmental Assessment of Socioeconomic Systems held in Istanbul, Turkey, October 4—8, 1976, sponsored by the NATO Special Program Panels on Systems Science and Human Factors

# ACKNOWLEDGEMENTS

The editors would like to thank Dr. B.A. Bayraktar for his assistance and the administrative director of the conference, Professor Dr. I.I. Karayalçin and his staff, whose activities guaranteed the success of the conference, also the chairmen of the sessions: M. R. Crucn, Professor Dr. Dogrusöz, Professor Dr. J. Klaus, Mr. P.C. Roberts, Dr. G.H. Winkel and Dr. Ch.P. Wolf, whose efforts substantially enriched the scientific level of the discussion. Last but not least we would like to thank Mrs. P. Bydlinski for her patience in typing and correcting the manuscripts and Mrs. B. Kane for all her help in preparing the conference.

CONTENTS

Introduction . . . . . . . . . . . . . . . . . . . . . . .  xi

SESSION I
METHODOLOGY: PLANNING

Systematic Distortion in Planning and Assessment . . . . . . .  3
    J. Davoll

Planning Sociotechnical Systems . . . . . . . . . . . . . .  17
    E. Rupp

Demo-Economic Policy Simulations for Belgium . . . . . . . . .  29
    R. Lesthaeghe, S. Wijewickrema, M. Despontin,
    and A. De Kerpel

Socio-Economic Systems, Urban Growth and
        Settlement Patterns (The Case
        of Turkey) . . . . . . . . . . . . . . . . . . . . .  49
    R. Keles

SESSION II
METHODOLOGY: PREDICTION

Some Future Uses of Analytic Forecasting . . . . . . . . . . .  71
    E. Solem

The Development of Policy-Sensitive Models . . . . . . . . . .  83
    H.-P. Weber

Intuitive Prediction of Growth . . . . . . . . . . . . . . . 103
    W. A. Wagenaar and H. Timmers

Models for Prediction under Conditions
        of Interaction . . . . . . . . . . . . . . . . . . . 123
    P. C. Roberts

SESSION III
METHODOLOGY: ASSESSMENT

Universal Checklists in the Concept
        of "Impact Trees" . . . . . . . . . . . . . . . . . 133
  U. Liebermeister

Problems with People, Decision, Conflict,
        Language and Measurement . . . . . . . . . . . . . . 159
  K. C. Bowen and J. I. Harris

Long-Term Policy Assessment of Energy/Environment
        Systems: A Conceptual and Methodological
        Framework . . . . . . . . . . . . . . . . . . . . . 183
  W. K. Foell

Economic and Environmental Assessment of
        a Water Quality Management System
        (River Neckar) . . . . . . . . . . . . . . . . . . . 203
  J. Klaus

Behavior Settings: The Building Blocks of
        Social Systems . . . . . . . . . . . . . . . . . . . 225
  M. Gürkaynak and W. A. LeCompte

SESSION IV
CASE STUDIES: ASSESSMENT OF TECHNOLOGICAL SYSTEMS

Government Regulation and Rail Safety . . . . . . . . . . . . 241
  H. P. Johri, J. D. Milne, and R. E. Wright

Towards a Social Psychology of the Traffic Environment . . . 263
  C. K. Knapper and A. J. Cropley

Control Methodology of the U.K. Road Traffic System . . . . . 279
  M. R. C. McDowell and D. F. Cooper

Traffic Policing, Effectiveness Measurement
        and Resource Allocation . . . . . . . . . . . . . . 299
  S. Freckleton, N. Ferguson, and M. E. Moncaster

A Suggested Framework for Evaluating the
        Assessment of a Technological System . . . . . . . . 317
  D. W. Fischer and R. W. Keith

Assessment of Alternative Energy/Environment Futures:
        A Comparative Case Study of Wisconsin (USA), the
        German Democratic Republic, and Rhone-Alpes
        (France) . . . . . . . . . . . . . . . . . . . . . . 335
  W. K. Foell, F. Buehring, W. Buehring, R. Dennis,
  K. Ito, R. Keeny and B. Lapillonne

SESSION V
CASE STUDIES: ASSESSMENT OF SOCIAL SYSTEMS

The Distribution of Environmental Quality:
      Some Canadian Evidence . . . . . . . . . . . . . . . 361
  B. A. Emmett, L. E. Perron and P. F. Ricci

Social Stability Analysis - Can Carrying
      Capacity Provide an Answer to Public
      Policy in Canada? . . . . . . . . . . . . . . . . . 377
  P. F. Ricci

Environmental Quality Change in Several
      Socio-Economic Contexts in France:
      1970 - 1995 . . . . . . . . . . . . . . . . . . . . 395
  J. Theys

Environmental Control: A Methodology for Planning
      Change in Public Housing . . . . . . . . . . . . . . 433
  R. M. Fenker

Environment and Public Policy in the United
      States: A Diffusion Analysis . . . . . . . . . . . . 441
  V. Mahajan and M. Agarwal

Responses to Changing Urban Systems: An
      Analysis of Public Education . . . . . . . . . . . . 469
  B. Anderson and J. Mark

SESSION VI
CASE STUDIES: ASSESSMENT OF SOCIAL IMPACT

Environment and Energy on the Texas Gulf
      Coast: An Economic Evaluation Model of
      Alternative Policies . . . . . . . . . . . . . . . . 493
  K. E. Haynes, J. E. Hazleton and W. T. Kleeman

Environmental Pollution and Its Social Impact . . . . . . . . 525
  H. J. Karpe and D. Scholz

Social Perception of Industrial Odors . . . . . . . . . . . . 535
  D. Agrafiotis, A. Baumerder, J. Brenot,
  and F. de Lavergne

Impact Assessment and Participation: Case Studies
      on Nuclear Power Siting in West Germany . . . . . . 551
  V. J. Hartje and M. Dierkes

Environmental Planning and Social Response
        at the Strategy Level . . . . . . . . . . . . . . . 569
  F. E. Joyce and C. W. Sinclair

Register of Names . . . . . . . . . . . . . . . . . . . . 581

Index . . . . . . . . . . . . . . . . . . . . . . . . . . 587

# INTRODUCTION

Neglect of the relation between the socio-economic system and its natural environment has had detrimental consequences in the past, for example

- the pollution of the natural environment (water, air and soil) by producing, using and consuming the products of our industrialized economy,

- the forseeable exhaustion of natural resources by continuing the increase of industrial production.

Most of the recent activities, both in research and in administration, against these impacts have been technically oriented, with the aim of stimulating and introducing new technologies of production and new products to diminish the environmental pollution.

But these efforts, which are of course necessary, cannot be successful in approaching the aim - which should and must in the long-term view be defined as the development of society in balance with the natural environment. Therefore, in addition to an assessment of technologies, emphasis should be put on an assessment of socio-economic systems. On different levels, individual and social preferences determine quantities and qualities of production and consumption using economical values, e.g., market prices as regulators. Following this argumentation, an environment assessment of activities against pollution has especially to consider the social response to environmental impacts.

Of course, this topic must be a subject of interdisciplinary research. The challenge concerned in this context is to increase the knowledge of the relationship between social, economic and technical subsets.

The authors of the following papers are all experts on the various scientific disciplines. The necessity for an interdisciplinary treatment of the complex topic could therefore be fulfilled in a very comprehensive way.

The underlying problem can by no means be solved by an integration of applicable scientific methods. It is also necessary to stimulate communication between scientists and members of the public administration who are dealing with the present problems in environmental planning, and who should apply methods and models successfully. Therefore, the editors tried to balance the position of both groups in this book by giving methods on the one hand, and case studies on the other hand the same priority. The "consultants" for public affairs are the third group; they should try to undertake the task of combining possible different positions between the two groups. This task can perhaps be defined in the following way: to transfer scientific methods into applicable models for public decision makers.

The underlying topic is too complex and rather new. Therefore, it is impossible to expect that in the following papers the problems which are involved could be completely clarified or, even more, already solved. One of the main objects of this book is, therefore, to compile different concepts and methods in order to initiate and to carry forward the discussion about this topic. The results of the conference have obviously shown that

. more research, especially interdisciplinary research, is necessary

. the communication between the groups should be made more efficient

. there is a need for models on a higher level of application for analyzing, planning and evaluating social and economic structures and individual behavior under an environmental point of view.

In the following pages some aspects which arise in this context will be described in a very rough way, in order to give the reader a possible guide through the book. The topic involves explicitly the assessment problem. An assessment of systems or components of systems, be it a technical, economic or social problem, always needs an answer to the following questions:

- What is the subject that has to be evaluated?

- Who should evaluate?

- Which is the underlying objective-function?

- Which scale and dimension should be taken?

The answer to the first question not only requires an operational definition of the subject, but also enforces a search for alternatives, because every evaluation will be relative; that means it consists of a comparison between at least two possibilities.

According to the main topic, most of the papers are dealing with social and/or economical effects of activities which are environmental biased, and with the social response to the impact of these activities. The authors try to illuminate these problems from different points of view.

The question of who should evaluate raises especially institutional problems; for example, which individuals or which groups have to be considered? In which way should the weights be divided among the groups and/or individuals? Obviously evaluation is based on preferences, and preferences are subjective and often change in time. Therefore, scientists can only try to make the process of evaluation more transparent. It is almost impossible to give inflexible rules for an interpersonal comparison of preferences. The challenge will remain a political one, closely connected to the more or less democratic institutions of a society.

The question concerning the objective-function raises, for example, the problem of the multi-dimensionality. Although the topic has been concentrated on the environmental goal, it is nevertheless necessary for a realistic assessment to consider further aims, such as economic or cultural ones. The reader will notice that some authors focus interest on objects other than the environmental one. But this deviation from the main goal has to be seen as an enrichment of the discussion.

Considering the scale and the dimension on which the evaluation should be based, it is necessary to decide whether to take an abstract utility function or a monetary value. Both involve advantages and disadvantages. It depends on the particular circumstances of research and the methodological treatment of the subject as to which way the author has preferred.

No less important than the assessment problem is the decision about the planning horizon which underlies a special plan, model, etc. According to whether the planning horizon is long or short, fundamental changes in the assessment may arise. The decision about the time factor raises in a very obvious way the so-called "generation conflict". This conflict may perhaps become more serious in the future because decisions, based on present values, will probably diminish, more dramatically than in the past or the present, the possibility for forthcoming generations to arrange their own life.

It is particularly on this question that the points of view of the scientists and the politicians as well as the members of the public administration, differ greatly. The activities of the public sector are usually decision-oriented and implicit rules in most political systems force the politician to act under a rather short-time horizon. Therefore, they are from a practical point of view more interested in short than in long range planning. In contrast to this the scientific research has generally the aim of recognition and knowledge about hitherto unknown relations.  Especially because most of the social processes are long range, research in this field needs long-range planning and forecasting. Obviously an approach of both aspects is necessary to realize a more effective assistance for both sides by each other.

This leads directly to the problems which are connected with the application of the results of scientific research in more realistic models for a better planning, prediction, and assessment.  A summary of all papers and especially of the general discussion at the conference under this aspect would rather establish a sceptical than an optimistic position.  Without doubt the so-called implementation gap, which is mentioned so often by all professionals, has many different and highly complex reasons.  Some of them are:

. the difficulty of building a model which represents reality in such a way that on the one hand it is not too confusing because of too many details and, on the other, it is not too abstract or too highly aggregated,

. neglecting personal and institutional conditions and circumstances which complicate or even obstruct implementation and successful application,

. the quality and availability of data on a sufficient level of disaggregation,

. last but not least the cost effectiveness of implementing and using the model itself.

In other words, the implementation gap can be explained by technical, economical and psychological reasons.  The "model-builder" will therefore be well advised to regard the results and activities of the sociological and psychological research in a more professional way.

The implementation of models dealing with environmental problems often fails to succeed because the feedback between the model and the men outside the administrative system has been neglected. It is necessary to consider and to analyse the response to special activities which have environmental as well as social and economic impacts.  As environmental planning has tremendous effects on the living conditions, their success depends greatly on the reactions of groups and of individuals.  In this context social structures and individual patterns of behavior should be a major topic of research.

In conclusion it can be postulated that the human factors should be explicitly considered in those models which are regarded by the politicians as instruments to realize better environmental planning.

Considering the above-mentioned, briefly discussed aspects, it seems to be reasonable to arrange this book in the same way as the conference, namely

- a methodological part
  and
- a part dealing with case-studies

The methodological part is divided in the following three sessions:

    I   Planning
   II   Prediction
  III   Assessment

In this part the instruments are discussed and analyzed in a more exemplary way. According to the topic the interdisciplinary concept will be especially regarded.

In the second part attention has been focused on the application of the different methods and models. The case studies are sub-divided under session IV - VI:

   IV   Assessment of technology systems
    V   Assessment of social systems
   VI   Assessment of social impacts

A further reason for organizing this book in such a way has been to point out the necessities and expectations of the "practice" of planning on the one hand and the possibilities and research strategies of different sciences on the other hand. The editors hope that the papers stimulate new and fruitful discussion about the possibility of approaching the methods of systems science and human factors in order to realize a better concept of environmental planning.

                                                                            D.F.B.
                                                                            W.H.I.

# Session I

# METHODOLOGY: PLANNING

# SYSTEMATIC DISTORTION IN PLANNING AND ASSESSMENT

John Davoll

The Conservation Society

Chertsey, Surrey, U.K.

The publication of "The Limits to Growth" by Meadows et al. in 1972, with its claims that continued growth of human numbers or material consumption would terminate in catastrophic collapse, made a substantial impact on public opinion. This was speedily followed by a counter-reaction, described by one commentator as "the conventional wisdom strikes back". Economists were first into the fray, propelled by a fear that acceptance of limits to growth would render much of their stock in trade valueless, and uninhibited by any real knowledge of how the physical world actually works.

More serious criticisms came later as other teams began their own studies of global modelling and found flaws both in the basic data and the feed-back loops of the model used by Meadows. Indeed, a further study under the auspices of the Club of Rome, carried out by M.Mesarovic and E.Pestel and published as "Mankind at the Turning Point" (1974), came to markedly different conclusions from its predecessor's. Recognition of the difficulties and limitations of modelling appears to have led to some waning of interest, certainly among the public who bought "The Limits to Growth" in such numbers. Although many of those involved in these studies acknowledge that the present and anticipated impact of human activities on the biosphere is of profound importance, the relative failure of available techniques to elicit scientifically respectable conclusions seems to have discouraged many workers from saying anything prescriptive.

This ought to cause surprise, as the subject matter is not merely an academic discipline, but is arguably a question of life or death for billions of people and perhaps the whole human species. Even tentative recommendations would be useful, and men in the past have not proved wholly incapable of wise judgement even without

help from infallible advisers. Why then are those professionally
engaged in these studies so hesitant in proffering advice on the
most important problems, preferring instead to concentrate on small
parts of the structure rather than on the stability of the whole
edifice, and to work within a near time horizon at a time when
human activities can profoundly affect the remote future?

The central problem concerns the stability of the total terres-
trial system, which now comprises two distinct but interacting sub-
systems. The first of these is the legacy of some 3,000 million
years of evolution, during which the relationship of living things
to their habitat was conditioned by feed-back mechanisms that con-
ferred a high degree of stability on the system. This, at least,
seems to follow from the unbroken continuity of life from its earliest known origins. Long term changes have occurred in the composition of the atmosphere and oceans, but have been accompanied
by mutual adjustment with the complex of living organisms. Little
is known of the mechanisms by which this very large scale homeostasis
operates, but work by Lovelock is beginning to suggest subtle processes which might act as regulators (J.Lovelock and S.Epton, "The
quest for Gaia", New Scientist, 65, 304-306 (1975) ).

Whatever these prove to be, it seems likely that but for one
crucial evolutionary step, and excluding cosmic accidents, the ebb
and flow of species would have persisted during the remainder of
the 8,000 million years during which a star such as the sun maintains
a reasonably constant output of radiation. The break occurred when
intelligence appeared, although some time elapsed before the full
effects began to be revealed. Without attempting to define intelligence precisely, it certainly includes a novel ability to make
mental projections of the consequences of possible courses of action;
an examination of these projections then enables the organism to
exercise a conscious choice from among its possible actions in the
present in the light of their more distant probable results. This
bringing of the future into the present has the most profound consequences. Most obviously, an intelligent organism's competitive
position among other species will be improved so greatly as to
ensure its rapid rise to a dominant position. Secondly, the introduction of "choice" among alternative possibilities breaks the direct
controls that previously ensured adjustment to the requirements for
overall stability of the environment. Thirdly, the imaginative
powers that come with intelligence and speech must eventually enable
their possessors to understand the physical world and harness the
enormous flows and reservoirs of physical energy in their surroundings; these can then be applied to produce rates of environmental
change comparable to or greater than those produced by natural forces.

Fourthly, although the ability to choose among conceivable
goals makes possible a degree of planning against future contingen-
cies, including possible distant adverse consequences of actions

that are advantageous in the near future, there is no a priori reason why consequences that lie more than (at most) a few generations
ahead should have much influence on decisions in the present. Writers
such as Olaf Stapledon have speculated that elsewhere in the cosmos
intelligence may have arisen in potentially immortal collectivities
of individual units which are replaced as necessary without loss of
the overall pattern.  However, for life forms more akin to terres-
trial ones the span of a few generations - the "significant" future
for most planning - is likely to be infinitesimal in comparison with
geological periods; much longer generation times would not provide
sufficient evolutionary adaptability for a species.  It also seems
likely than an exclusive application of intelligence within a near
time-horizon (a generation or less) will be advantageous to the
individual in intra-specific competition.  Hence this type of character may well be selected for against one more oriented towards
long term wisdom, so that qualities for short term dominance may be
favoured in a way that precludes long term survival of the species
once sufficiently potent weapons are developed.

        Finally, a more philosophical dilemma emerges: although intel-
ligence makes possible the intellectual exploration and comprehension of the physical universe, there is as yet no sign that it reveals any meaning or purpose, and consequently it may be that as
the old biological drives to survival are perceived as purely mechanical, the species - particularly one such as man that consists of
short-lived individuals - will lose its collective interest in long
term survival, which will appear pointless.  There is some evidence
suggesting that this may already be contributing to the general
malaise of our own time.

        This very broad survey indicates that within the geologically
short period of about 10,000 years the application of human intelligence,especially through increasing mastery of technology, has led
to a radical change in the interactive components of the total terres-
trial system.  There are therefore legitimate grounds for concern
about the future behaviour of this system, although in the absence
of more specific information it is not clear whether the present
situation is one that demands urgent corrective action.

        Governments and economic planners demonstrate in practice that
they either consider that urgent action is not necessary or that
they are indifferent to the more remote consequences of their actions.
The latter attitude may arise from a belief, at least superficially
reasonable, that the distant consequences of present actions are so
uncertain that they cannot be given any great weight against immediate and more predictable results.  Two further arguments are used to
reinforce the case for business as usual; the first holds that the
physical magnitude of human intervention, now and anticipated, is
negligible in comparison with natural homeostatic mechanisms.  According to the second argument, if dangerous limits should be approached

in future, some combination of physical, economic and social mechanisms will provide adequate feed-back controls on human activities.

I shall argue that these reassurances are misleading, and are
based on artificially isolating parts of the total system for study.
An examination of the behaviour of the human institutions that determine the course of technology, and the nature of their interactions with changing physical circumstances, shows that the prospects
for man are bleaker and the need for action more urgent than the
attitudes of most planners and forecasters suggest. The reasons for
this discrepancy can be found in the personal interests and preferred methodologies used by professionals in this field.

The first question to consider is the significance of human
activities in relation to natural ones. It is true that there is no
unequivocal evidence that global equilibrium is as yet seriously endangered, but the relaxation time of the biosphere is probably so
long in comparison with the time scale of planning that by the time
significant general effects were observed the damage would already
be irrevocable. Accurate quantitative predictions seem unattainable,
but this does not rule out informed judgement and prudent recommendations about the future envisaged in most conventional planning.
This assumes a further doubling of world population (at a conservative estimate) and the provision of industrial development for
the resulting 8,000 million people. Since decrease of population
size by non-catastrophic means is inevitably a slow process, and
the voluntary abandonment of industrial civilization an improbable
one, the implication is that a rate of human impact on the environment at least 10 times the present one will have to be achieved and
then sustained for several centuries. It must be remembered that
economic growth in material terms is not a velocity but an accelera-
tion, and even if material growth is eventually halted, rapid, and
cumulative, man-made change will continue.

To assume that the massive re-working of the whole planet implicit in this scenario would not seriously affect its homeostatic
capabilities seems foolhardy in the extreme. Emergencies of resource
depletion, pollution and environmental degradation would be inevitable, but it cannot be taken for granted (as the second argument
for the pursuit of growth assumes) that these emergencies would con-
strain human technology to a more prudent course in the same way
that the numbers and range of non-human species are governed by environmental limitations. In order to envisage what the response
might be one must examine the dynamics of human societies and institutions.

One of the most obvious characteristics of human societies is
a propensity for growth; in former times this was manifested mainly
by an increase in numbers, but more recently there has been an additional expansion in technological impact. Unchecked expansion is

not compatible with dynamic equilibrium, as Istock pointed out ("A corollary to the dismal theorem", BioScience, 19, 1079-1081 (1969) ): "Thus there arises a contradiction between the practice of economics in industrial states and the requirements for stable ecosystems. The two are probably at odds at every point of contact, with human population increase a concomitant of almost every difficulty. So general is the conflict between human economics and the ecology of this planet, and so pervasive, powerful and unswerving the dynamic of the industrial state that it is impossible to image a suitable ecology under prevailing economic theory."

A facile analogy is sometimes drawn between the growth of human numbers and technological impact on the one hand and the processes of growth found in nature on the other. Growth, it is said, is "natural", conveying a reassuring implication that human expansion is no more a threat to the environment than is a temporary increase on the part of any species. Nothing could be further from the truth; whereas biological evolution is controlled by the necessity of adaptation to the physical environment, cultural evolution, as Rapoport pointed out ("Conflict in Man-Made Environment"; Penguin Books, 1974, p. 76), generates a man-made environment, and individual adaptations are made to this: "... there is no guarantee that these adaptations enhance the survival potential of the culture." In a culture such as the one we enjoy, or endure, the promotion of material growth is the effective measure of success and reward, and it is difficult to see how this can be changed without changing the culture itself almost beyond recognition.

If technological activities were directed to some general goal, in the shape of an envisaged and desired state of society, then it might be expected that as the goal was approached negative feed-back mechanisms that drive growth also ensure that cessation of growth leads not to stability but to decline and slump, so that governments anxious to survive always strive for continuous growth; if they fail they are likely to be voted out or forcibly overthrown, according to the custom of the country.

Nor is this the only powerful mechanism promoting growth. The promise of steadily increasing general wealth eases social tensions and facilitates generosity towards the poorer sections of the community; once the total amount available ceases to grow, arguments over allocation become sharper. These can be even more intractable between than within nations, since within nations there do at least exist more or less effective regulatory agencies and political structures for resolving differences peacefully. On the international scene the necessary conditions for regulating the use of common resources are rudimentary or non-existent. Very large differences of consumption prevail between countries; there is no general consensus on the apportionment of limited resources, and no mechanisms exist to enforce regulation even if it could be agreed upon. It is

perhaps not surprising that, regardless of evidence, a belief that
universal affluence will eventually sidestep these problems is widespread and compulsive.

Returning now to the conventional forecast of world economic
development, a curious contradiction can be discerned. On the one
hand, it seems physically implausible that anything like this can
be accomplished, and if one has the long term interests of the
species in mind it must be hoped that the enterprise will be halted
by mundane rate and magnitude problems before irrevocable damage
has been done to the planet. On the other hand, it is also difficult
to envisage how ingrained and institutionalized habits of growth can
be abandoned. Unlike some economists, I believe that physical impos-
sibilities have the last word when they clash with political ones,
and I conclude that the next century will be one of enforced and
traumatic adaptation.

It will also be a period of acute danger. An animal species
whose numbers have exceeded the carrying capacity of the environment
experiences a high death rate, or population crash, until equilibrium
is restored. Human beings do not just die peaceably, however, and in
an emergency of widespread starvation or economic collapse one or
both of two reactions seems likely. First, any prudence about exploiting the environment will be abandoned in the interests of feed-
ing the starving or getting production going; a very minor manifestation of this reaction occurred in Britain when the implementation
of the Control of Pollution Act was deferred until economic growth
resumed. Second, competition between rich and poor countries, and
among rich countries, will become more acute, perhaps leading to
military conflict. When the United States recently felt threatened
by an oil embargo that might have reduced its per capita oil consumption to only four times the global average, Dr. Kissinger made
ominous noises about possible military seizure of the production
facilities. Again, would it be surprising if poor countries, facing
starvation, used the plutonium from the reactors we are assiduously
exporting to them for a little blackmail?

On present form, a future rather like this seems all too probable. The question that must now be asked is whether there are
alternatives, and whether the professional activities of the forecasters and planners are likely to increase public understanding of
the dangers and offer ways of reducing them.

Futurology is not, and cannot be, a science in the sense that
physics is a science, and even forecasts dressed up in the language
of the exact sciences are heavily influenced by subjective factors.
As Martin Shubik remarked ("Alternative Futures: Expectations";
Annals of the New York Academy of Sciences, 261, 20-24 (1975) ),
"... given enough degrees of freedom in a system, one can prove virtually anything. This being the case, it is possible for a decision-

maker to say to a futures-planner or simulator, 'This is the way I want the future to look.' If the model builder is given a sufficient number of degrees of freedom he can make the future look that way and still be consistent with virtually any given set of empirical information. The large scale simulation, if used without care, opens up new dimensions for 'rationalizing man'." Nevertheless, a debate between proponents of widely differing views might be effective as a way of stimulating public discussion and broadening perceptions of the possible, provided that all reasonable views were represented. I shall argue that this is not the case, and that there is a systematic bias towards a particular set of basic assumptions among those professionally concerned with the future of society.

It is important to emphasize that we are concerned here with a much larger class than futurologists and model builders in the strict sense. Much economic theory, and most ordinary forward planning by governments, is based on unspoken assumptions about the world. Examples are Professor Beckerman's opinion that economic growth can probably continue for a million years or so ("Controversy"; BBC-TV, July 1972), and the well-known statement by Barnett and Morse ("Scarcity and growth" (1963), Baltimore, John Hopkins Press; page 10): "Few components of the earth's crust, including farm land, are so specific as to defy economic replacement, or so resistant to technological advance as to be incapable of eventually yielding extractive products at constant or declining costs." It would be hard to find more sweeping and implausible generalizations in any speculative work.

Most government and industrial planning works to a rather short time scale, and at first sight seems distinct from the wider-ranging speculations of futurology. This is not the case, however, since the incremental and short term decisions combine into a trajectory largely determined by concealed assumptions on the part of the planners about the nature of the physical world, the aims of human society, and the possible range of interactions between them.

A few years ago it appeared that a potentially fruitful debate was opening up on the proper aims of humanity, and beliefs long held to be self-evident were being challenged. However, when the smoke cleared away it was seen that nothing had really changed, and the character of industrial society continued to be largely determined by a particular set of attitudes entrenched in the minds of the decision-makers and their advisers. Although these would hold, often sincerely, that their actions and plans were rationally and objectively defensible, it is more likely that what they believe to be possible is circumscribed by what they personally regard as acceptable, because it does not cause them material disadvantage and because they can handle it in a way that enhances their status in their jobs. Their approach tends to be scientific, reductive,

rational, and analytic, and they are uneasy with strong emotional
and political commitments.  Their interests include a desire for
generous remuneration and high professional status, and intellectuals
and academics are usually important figures in the future societies
they envisage (or implicitly assume), and in the programmes of action they recommend as planners and advisers.  The influence exerted
on society by these individuals is profound and ramifying, not least
because it effectively limits the choices that can be made.  Nevertheless, because it operates under the disguise of objective rationality, it is rarely apparent as the personal bias it actually represents.  The examples which follow are illustrative rather than exhaustive.

One of the most striking paradoxes of modern industrial societies is that in spite of vast expenditure on control systems they
appear to be out of control.  In Hazel Henderson's phrase, they
appear to be examples of "the entropy state", in which " ... complexity and interdependence have reached such unmanageable proportions that the transaction costs generated equal or exceed its productive capacities" ("The End of Economics"; The Ecologist, May 1976,
pp. 137-146).  Certainly there is a widespread feeling among the
public that they have little influence on remote decisions by governments, although in truth it is difficult to name any society much
larger than a tribal unit in which things were much better in this
respect.  Far more alarming is the pervasive impression that no-one
is in control, and that the institutional machine we have created
is now autonomous.  Of course, no human organisation is strictly
autonomous; collective decisions by its members can always, in theory,
remodel or dismantle it.  It may, however, prove difficult to do
this if influential members of the organization feel that their contribution is validated by their skills and training in managing complexity.  Hence they will resist suggestions that the organization
be simplified until it is comprehensible to a majority of citizens.

Complexity not only puts many vitally important decisions beyond
the capacity of all but experts, whose very concentration on specific areas may blind them to the repercussions of their recommendations on adjacent areas.  Even more insidiously, the need to keep
the increasingly complex organization of a nation state functioning
leads to a progressive reduction of personal autonomy.  In the high
technology societies anticipated, with approval, by the conventional
wisdom, material affluence would be accompanied by total subjection
to the supposed imperatives of technology.  The results might well
be as forecast long ago by Dostoevski:

"Moreover, even if man were the keyboard of a piano, and could
be convinced that the laws of nature and mathematics had made
him so, he would still decline to change.  On the contrary, he
would once more, out of sheer ingratitude, attempt the perpe-
tration of something which would enable him to insist upon

himself; and if he could not effect this, he would then proceed to introduce chaos and disruption into everything, and to devise enormities of all kinds, for the sole purpose, as before, of asserting his personality. But if you were to tell me that all this could be set down in tables - I mean the chaos, and the confusion, and the curses, and all the rest of it - so that the possibility of computing everything might remain, and reason to continue to rule the roost - well, in that case, I believe, man would become a lunatic, in order to become devoid of reason, and therefore able to insist upon himself." (F. Dostoevski, "Letters from the Underworld", cited by Robert S. Morison, "Science and Social Attitudes"; Science, <u>165</u>, 150-156 (1969) ).

Modern society offers no easy trade-offs between more material goods and more personal autonomy; such a choice is not allowed by the guardians, or slaves, of the system.

Because planners are only happy with what appears predictable, they prefer to envisage the future as a smooth continuation of the past, and since the recent past has been characterized by rapid technological expansion, this too is expected to continue. If it were accepted that a deliberate decision should be made to reduce the pace of change on the grounds that its costs now exceed its benefits, major discontinuities in planning would be inevitable, or, even worse, fewer planners might be needed ("Planning in an age of stagnation"; David Eversley, Built Environment, January 1975, 14-16). In order to avoid such disruptions in the smooth tenor of their lives, planners find it convenient to believe that advancing technology will cope with the increasingly serious problems that past and present technology leaves as a legacy for the future. At the point when the debts come up for settlement, the acutely dangerous confrontations mentioned earlier will arise. Reliance on massive deployment of nuclear fission power to keep industrial societies going provides a good example of this attitude.

It is true, of course, that the optimists assure us that feedback mechanisms, usually operating through economic forces, will correct these dangerous tendencies in time. Proponents of this belief rarely seem to appreciate the difficulties of monitoring and the long lead times involved. One is reminded of Zeno's Paradox of Achilles, in which it is supposed that Achilles never overtakes the tortoise, since as he covers each successively smaller increment of distance towards it, the tortoise advances a set fraction of this distance. In real life, industrial expansion overtakes the feeble attempts to determine and correct its adverse environmental effects, just as in real life Achilles would soon overtake the tortoise.

It would not be unfair to summarize the dominant attitude as:

"We realise that we are depleting and damaging the environment, but
we hope that our descendants will be smart enough to find ways of
getting by. If not, that will be their bad luck." In the conveniently distancing language more commonly used in this type of discussion, "Technological change is the only possible solution to this
problem of intergenerational equity, since the world is not finite
as long as technological change is possible." (David A.Huettner,
"Net Energy Analysis: An Economic Assessment"; Science, 192, 101-
104 (1976) ). It does indeed require an impressive confidence in
technology to believe that it can provide,in time, a replacement
for the natural service functions of the biosphere as it is pro-
gressively debilitated.

Because planners seek professional respectability, they crave
value-free and preferably mathematical approaches to the problems
they are supposed to deal with. If there is some way, however distorting, of reducing the judgement of an issue to the comparison of
numbers, it will be used. If there is no way of doing this, the
question, however important, will be neglected.

Consider, for example, the concept of "progress" in human
affairs, and the measures by which a nation judges its achievements.
It is a commonplace that the amount of a person's material possess-
ions is not the only, or even (beyond a certain modest limit) the
main factor making for happiness. However, it is quite easy to find
sets of numbers relating to the national economy, such as the consumption of energy in gigajoules or of steel in tonnes, and the
econometricians can then derive endless pleasure by manipulating
these (using, of course, the latest equipment) and then pronouncing
on the state of health of the economy. At least one national economy was and is widely praised and envied for its growth rate at a
time when the poorest sector of its population was becoming poorer
and the use of torture was widespread. Such matters are, of course,
difficult or impossible to quantify.

It has already been noted that planners and forecasters are
uneasy about issues, such as the distribution of benefit and cost,
that have a political and ethical content. Once again, the reasons
can be found in the personal interests of the planners; with incomes
well above the average they are disinclined to investigate solutions
that involve transfers from the rich to the poor without concomitant
overall growth. In addition, recourse to allegedly objective methods
of assessment, such as cost and benefit analysis, is welcome as a
means for concealing sectional interests; by costing a business
man's time at more than a bus passenger's the existing distribution
of incomes is not only apparently legitimated but continues to be
built into the planning process.

A further corollary of this concentration on quantifiable
changes in consumption and production and the neglect of social and

ethical factors in judging a society is that there is little ability
to meet, or prepare to meet, problems of scarcity, except when it
can be presented as a temporary state - a mere hiccup in the upward
curve of growth.  A society that has no idea how to make the best
of physical scarcity, and even to find in it a stimulus to ethical
progress, will be compelled to believe that any scarcity can be re-
lieved by technical virtuosity, in the pursuit of which it will take
great environmental risks.  Those who like to affect left-wing political attitudes in spite of having achieved some measure of affluence
are especially resistant to suggestions that redistribution without
overall growth may be necessary.

Finally, planners and forecasters will tend to produce plans
that please their clients, whether governments or private industry,
or perhaps their clients will only employ and support forecasters
who can be relied upon not to say anything very disturbing.  Thus,
forecasts prepared for governments tend to be bland, to avoid
"political" issues, and to convey a discreet impression that whatever happens will be capable of resolution by the civil service. If
anything further is needed, the market mechanism will be relied on,
in spite of evidence that it is currently failing to solve problems
far less severe than those anticipated.

Such "official" forecasts are in any case effectively sanitized
by restrictive terms of reference, especially in regard to time horizons.  For reasons perhaps distantly connected with millenarian expectations, 2000 AD is a favourite boundary, now less than a quarter
of a century away.  It is this choice of some date beyond which the
forecaster disclaims all responsibility that allows forecasts to be
accepted even when the conditions anticipated at the time horizon
are almost certainly highly unstable in physical and social terms.
This acceptance is doubtless connected with the present orientation
of industrial society towards maximizing a process - physical growth -
rather than seeking to achieve a goal.  In its turn, the absence of
a comprehensible goal is a plausible contributor to the malaise and
disorientation that often co-exists with material plenty.

In summary, then, as far as achieving a satisfactory and dur-
able world order is concerned, planners appear to be more of a hin-
drance than a help.  This is not, of course, to blame everything on
them; they are subject to all the pressures of their man-made environment and the ultimate argument that "if I don't do it, someone
else will."

Nevertheless, some tentative recommendations can be made; even
if they do not appear realistic now, they may be more applicable in
a future of enforced and traumatic change, not merely in the technological mode as in the past, but in the deeper levels of social
relationships and man's view of himself and his purposes.

The most immediately applicable suggestions relate to advice given to governments and industry. My impression is that quite absurdly optimistic assumptions about the availability of energy, industrial raw materials, and food underlie official planning about, for example, private car usage and the use of agricultural land. A greater determination to press home the reality of rate and magnitude constraints, even at the cost of some unpopularity, would undoubtedly contribute to a more realistic approach, and even to a dawning realization that until more global equality has been obtained, the obsession among the rich with growing ever richer is somewhat obscene. The consequences of a failure of the more optimistic forecasts to be realized after they have been used as a basis for planning need to be strongly emphasized, and, finally, much more effort should be put into analysis of who gains and who loses from possible alternative developments, so that political and ethical issues can be separated from technical ones for judgement and public discussion.

Adoption of these suggestions would no doubt be useful, but they seem inadequate against the magnitude of the challenge of adaptation we face. One might have expected, or hoped, that the simultaneous realization that man depends on the functional integrity of the natural world, and that he is thoughtlessly mangling it for immediate gain, would have called forth some response of the spirit. Although there have been many signs at an individual level of a growing perception of this new relationship, there is little evidence that it has had any effect on mainstream national policies. Indeed, official spokesmen and advisers make every effort to convince their audiences that nothing has really changed, so that they themselves need make no uncomfortable changes in their thinking and methods of work. Bland reassurances smother even the recognition of the tragic contradiction between human potentialities and the sordid downfall that threatens us, so discouraging the possibility of a new movement of human consciousness towards steadying the race for a long future.

The most one can do may be to try to foster the conditions for such a transformation, and surely one condition is that society should be simple enough for most people to understand and (if they wish) participate in decisions that affect their lives. Another condition is that technological change should not be so rapid as to destroy the sense of meaningful continuity between past, present and future. There is, it must be admitted, no guarantee that even if these conditions were met the majority of people would readily accept the labour of fashioning a new way of life, and the even more arduous task of discarding old assumptions; one can only claim that movement in the direction of a comprehensible society may well be a necessary, but not sufficient, pre-condition for a successful transition from material growth to dynamic equilibrium.

For planners and others who enjoy and make their living from the large scale manipulation of people and things such an equilibrium society might be unexciting. Many of them are skilled, or at least academically qualified, in the social sciences, and at first sight it might appear that they will have an increasing role in managing change; Lord Ashby has suggested (Nature, 259, 170 (1976) ) that in the face of a supposed danger of social breakdown we should concentrate more on the social sciences than on the physical ones. There is, however, an essential difference: the physical sciences give men power over things, but the social sciences (in so far as they work) tend rather to give some men power to manipulate other men, because of their greater technical knowledge. Moreover, from a desire to ape physical scientists, social scientists move towards generalization, simplification, and abstraction, whereas one of the important attributes of a successful society beyond the turmoil of present change must surely be a respect for the particularity of things, even if this makes them more recalcitrant to the efforts of the planners.

Whether planners and those in power can nevertheless reconcile themselves to the prospect, and to becoming participants rather than external manipulators, seems to me a test of whether they are part of the solution or part of the problem.

# PLANNING SOCIOTECHNICAL SYSTEMS

E. Rupp

Program "Applied Systems Analysis" in Arbeitsgemeinschaft
der Großforschungseinrichtungen (AGF)
Köln, West Germany

## TECHNICAL PROGRESS IS IMPEDED BY THE METASTABILITY OF SOCIAL SYSTEMS

Technical innovations, equipment, and systems are most easily
and most rapidly introduced into our daily way of living and practical activities subsequent to disasters, political revolutions,
and wars. Whenever standards and customary patterns of behaviour
are undermined, whenever beaten tracks are blurred, and whenever
there is a shortage of goods and services, we are quite ready to
welcome new ideas and procedures, to accept new goods and to learn
how to control and use them.

This does not only apply to the individual, but even more so
to institutions and to social systems in general. This fact is
demonstrated by a number of well-known examples provided by the
history of innovations. The most illustrative example so far has
been the massive application of ideas, technical and sociological
innovations and procedures in the early stage of the European industrialization. In deliberate allusion to the previous bourgeois
revolution, Friedrich ENGELS called this epochal change in the production and distribution systems taking place in the 19th century,
the "revolution industrielle", with "industry" meaning a new organization of the use of production factors, which could only be realized through the existence of new forms of production machinery.

Further examples are to be found in the novel technologies pene-
trating the international market after the first and second World
Wars; in the vigorous industrialization, based on technology and
technology transfer, of the feudal-agrarian Russia after the bolshe-
vistic revolution of 1917; and today in the enormously progressing

modernization of the national economies of some developing countries.

Besides, there are other, less incidental and less spectacular, forms of introducing new technologies, viz. selective and long-term measures to influence and enhance Technical Progress. Only think of the elaboration and funding of long-term technological programs by governmental agencies and multinational enterprises, where it is not always clear from the beginning to whom these programs will ultimately be beneficial, and whether the risk involved can still be borne in the future. For the majority of such and similar measures it is true that the cost of innovation - i.e. the cost required for marketing and distributing at a profit a new technology - frequently is ten and a hundred times greater than the cost of research and inventive activity. In some sectors, this discrepancy has become so marked that governmental and also non-governmental agencies have started to drastically increase the funds intended for studying the innovation barriers. However, such studies have not yet shown any remarkable results.

Difficulties of a similar extent are encountered when new technologies are introduced by way of the free play of the market forces. Those who want to be successful, must either "be in command of the free play of these forces" or they must try to "organize chance".

In general, the introduction of novel technologies destroys a number of customary and well-established rules of conduct, and we are forced to set up new rules of conduct and to reassess the old ones. As a result, we oppose the introduction of novel technologies, especially when we are compelled to perform new tasks whose outcome is rather uncertain and whose success cannot be assured by futurologists, i.e. when we are compelled to take a risk. Frequently, the risk only consists of the fact that, by means of a novel technology, we become aware of the risk of a customary technology, which is reason enough for us to offer resistance to the novel technology. Groups of people or institutions show a similar behaviour when faced with such situations. There is, however, one important difference: the resistance offered to novel technologies by social systems, i.e. societies, business enterprises, bureaucracies, political parties, trade unions, insurance companies, associations etc., is in general much more effective than that offered by individuals.

To summarize, our living habits tend to oppose, and guard themselves against, the introduction of novel technologies and, in general, this tendency is displayed even more distinctly in social systems, i.e. in social organizations and institutional associations; there are a number of sociological laws opposing the introduction of certain novel technologies and Technical Progress in general.

I am including the notion Technical Progress deliberately, since the Technical Progress still is a measure and a category of

the productive capability of our societies, a measure also of our
organizational capability - not "in spite of", but "because" of its
vague definition in the modern economic theory. Technical Progress
means to us something like the socioeconomic principle of evolution;
a principle which permanently impels us to reorient our way of thinking, to readapt ourselves, to overcome risks, and which exposes us
to unknown strains and liberations, and that in opposition to our behavior, which we wish to be determined by reason, i.e. by economical
thinking, by control, clarity, and inherent strength. In contrast
to this, there is the tendency of social systems to always use or
remodel all capabilities and also all resistance only to consolidate
their own stability. In terms of sociology, social organizations
still tend towards metastability after they have long lost their
social or economic importance.

    There are technologies and also novel technologies which inten-
sify these tendencies. The attempt e.g. to confine production only
to products and resources which are oriented towards the previous
traditional profits, may favour anti-evolutionary technologies; concentration of our organisational and reflective capabilities on the
prevention of hazardous changes and surprises possibly produces
technologies which will hinder us from using risks and sudden changes
in a productive and rapid way to improve our living conditions. It
is not improbable that technologies of this nature increase as the
Technical Progress is stopped. One essential reason for this stop-
ping of the Technical Progress is the law of the metastability of
social systems, according to the first thesis. This law states that
social systems generally only permit those innovations which contribute to their own stabilization. This does, of course, not exclude that social systems may be forced to accept innovations by
means of appropriate external measures, which does, however, not
alter the basic tendency of social systems to be metastable and to
reject or remodel any technological intervention which would force
them sooner or later into hazardous changes.


   II. THE METASTABILITY OF SOCIAL SYSTEMS DOES NOT FIND SUFFICIENT
     CONSIDERATION IN MODERN TECHNOLOGY PLANNING AND INNOVATION
                          ENHANCEMENT

    The second thesis states that today this anti-progress stabili-
zation tendency of social systems is not sufficiently taken into
consideration in technology planning and innovation enhancement.

    Of the different reasons for this deficiency, I will mention
two:
    The first reason is a certain prejudice held by the scientists
and engineers responsible for technology planning and innovation enhancement. This prejudice implies that Technical Progress is necessitated by a quasi-natural law, and that it is the task of the plan-

ners and scientists to identify this law and help it to be successful, as the one and best way, in the economic and political reality. Planning is composed of an analysis of the predetermined developments and of the measures required to realize the development, which has been projected by systems analysis, in the most effective way. Transformation of social systems is - if at all - regarded as a consequence of the introduction of novel technologies, not as an instrument and precondition of the introduction of novel technologies. One of the most popular mistakes of modern technology planning, resulting from the above prejudice, is the so-called Technology Assessment. The supporters of Technology Assessment intend to include the transformation of social systems in technology planning. Their intention is based on the opinion that transformation is primarily an effect and result of the introduction of novel technologies and that technology itself is a kind of deus ex machina for the transformation of social systems. In addition, they hold the opinion that this impact of technologies on social systems follows such a regular course that it is possible to project and assess the effects also for those technologies to be introduced in the near and more distant future. It is possible that this approach has induced the one or the other technology planner to include the role of social changes in his activities. In my opinion, however, it is all the more important to realize that this approach of Technology Assessment is not even useful as fig leaf to conceal technological failures, since it is contrary to all historical experience made with technological innovations and reveals a state of sociological understanding corresponding to that prevailing at the beginning of our time: then, too, much effort and time was used to assess the individual, social or even global effects of certain planetary conjunctions or the appearance of new stars or comets; and who could prove that changes in the planetary space around the earth would have no impact on our social conduct? What I want to say is that Technology Assessment is worthless for technology planning and innovation enhancement and that it even keeps us from judging the value of social changes to be positive for the Technical Progress.

The second reason for disregarding the metastability of social systems in technology planning and innovation enhancement is related to the first reason: Today, the development and the reforms of social systems are scarcely dealt with by scientific research and development. This is partly due to the lack of scientific methodology and capacities, but also to the general prejudice against any scientific, thus systematic, control of social changes and social systems. This prejudice is largely based on the historical experience which we have made, especially in this century, in liberalistic/pluralistic and with socialistic/centralistic social systems. Both, the heirs of the bourgeois revolution and the bolshevist revolution each had, in the beginning, claimed the possibility of a scientifically controlled transformation and advancement of society or their individual social systems.

Both had to realize that the scientifically controlled development and transformation of social systems led to hostility and be-
ligerent destruction, whenever they were used as a means of uncontrolled and elite power.  This experience has had a lasting effect
and each envisaged transformation of social systems easily comes under suspicion of elite abuse of power, also and just there where it
is combined with the use of novel technologies.

III. THE STUDY AND THE DEVELOPMENT OF SOCIAL SYSTEMS MAY BECOME AN
INSTRUMENT OF TECHNICAL PROGRESS

The third thesis states that the study and the scientifically
controlled development of social systems may in future be increasingly used as an instrument of Technical Progress.  This means that
Technological Systems can be regarded as a part of social systems and
that planning and innovation of technological systems also includes
a transformation of social systems, if unintentional, intentional
or even controlled.  This is meant in the following by the term
Planning Socio-technical Systems.  The special relationship between
social and technological systems, which is the primary subject of
our considerations, is no longer that of Technology and Technology
Assessment, neither that of Technology and Social Environment.  The
sociotechnical approach is rather based on a criticism of the basic
subordination of social systems to the functional definitions and
performance requirements of technological systems.

The sociotechnical approach replaces the simple subordination
of social systems to the performance requirements of technological
systems by the positive and instrumental relationship between social
and technological systems.

This means,
- first, that technological systems are an integral part of social
systems and that technological systems are only introduced, changed
or replaced by novel technologies in accordance with the methodology
of social systems; that there exists thus no quasi-natural compul-
sion for the introduction of new technologies, unless certain development conditions of social systems themselves.

- Secondly, the positive and instrumental relationship between technological and social systems indicates that roughly since the beginning of the Industrial Revolution in the past century new technologies have been developed and used by social systems primarily to
solve social conflicts and to overcome system crises, that consequently technological systems represent an important means of modern social systems to solve conflicts.

As an explanation, especially for natural scientists, I should
like to add that social conflicts - such as the birth of triplets
within a family; a strike or sudden reduction in orders within a

small enterprise; an inundation, which has been realized too late, of an army with university graduate recruits – and the solution strategies of most of the social systems have little in common with the way the natural scientist identifies and solves problems.  Social systems, on the contrary, tend to solve problems by shifting the problems between the various units and sectors of the social system or by pushing the problem off to a different social system or possibly even to a social system which has been solely created for this problem.  Through such transposition new social and other conflicts may emerge.  Experience has shown that new technologies can be used as an excellent instrument and means to adapt and solve problems in accordance with the method peculiar to social systems.

– Thirdly, the sociotechnical approach is based on the observation that social systems do not only invent, introduce or replace novel technologies, when social conflicts break out and an immediate solution is required.  Modern social systems more and more commence to develop and study organisations, behavioral patterns and procedures, which can be used to identify or simulate future potential conflicts before they break out.  Thus, social systems will be able to tailor their technological potential to expected or projected conflicts at an early stage.  I need not mention that the simulation or projection of a possible conflict and the appropriate orientation of the technological potential towards this simulation can already prevent the conflict – a classical form of sociological conflict transposition.

The most popular example of this process of controlled conflict transposition of social systems by the use of most advanced technologies, a process which is in contradiction to all natural scientists' methodology, is the system of strategic balance between the East and the West and the role of weapons and warfare technology, in particular of nuclear technology, in this system.

Summarizing our previous observations, we get the following picture:
Periods of socio-economic crises provide a favorable basis for the readiness of social systems to accept innovations – this has been demonstrated by the history of innovations since the first Industrial Revolution.

But also outside periods of crises and critical situations, the development of novel technologies represents a significant means of modern social systems to solve conflicts.  Such novel technologies are not the actual solution of social conflicts; sociological findings rather show that the problem-solving behavior typical of social systems cannot be replaced, but decisively improved, by the development and the use of new technologies.

Furthermore, we can observe today that, with the application

of systems analysis and other innovation measures, social systems
endeavour to use the planning and introduction of novel technologies
also as a means of preventing crises.

I have tried to outline, by means of three short theses, a problem at which, as far as the exact scientific definition is concerned,
we are looking somewhat helplessly. Both in economic innovations research and empirical sociology there is a lack of exactly elaborated
problem tasks, by means of which sociotechnical systems analysis
could be enhanced as part of a long-term technology planning. For
this reason I have confined myself to depicting, in the form of
theses, a relationship which is a challenge to present sociological
proficiency and which will in future also permit a systematic empirical consideration. It is the relationship between the characteristic behavior of social systems in the identification and solution of problems and the invention, development, and introduction
of novel technologies. As we have seen, the social problems, for
whose identification and solution novel technologies are introduced,
are primarily traditional problems, which always occur within and
between social systems and which are in general produced within and
between groups of society through the change in the natural resources
and in the performance spectrum. Secondarily, these are problems
generated by the introduction of technological performance units
within the social system - an absolutely usual process of problem
generation for social systems, as it is suitable for social systems
to solve problems also by transposing it to one or a number of different problems (cf. III).

This basic restriction of the sociotechnical field of study to
the problem-adapting and problem-solving behavior of social systems
opens up new aspects and new perspectives for future activities in
the field of sociotechnical systems analysis. I will briefly point
out some of these:

IV. SELECTION OF SOCIOTECHNICAL SYSTEMS ANALYSIS STUDIES SHOULD
MAINLY INCLUDE SOCIAL AND SOCIOECONOMIC FIELDS CHARACTERIZED BY
A HIGH CONFLICT PROBABILITY

According to the results of a sociological analysis, which we
carried out in the Federal Republic of Germany, large suburban
housing areas e.g. show an extremely high social conflict potential,
due to their unsatisfactory social and economic infrastructure and
due to the unbalanced social structure of the people living there.
This potential, which is already active and will continue to be vir-
ulent, is the basis for a relatively great readiness to accept innovations of technological systems, which lead to a visible and desirable improvement of the existing services or even to the presentation of completely new types of services. The sociological analysis
had as its objective to investigate new fields of application of

cable TV technology. We have learned from this analysis that the
technical possibilities of cable TV technology can only be fully
utilized if they are tailored to the specific services structure and/
or deficits of - in our case - large suburban housing areas. It was
shown that completely new requirements had to be imposed on individual system parts of cable TV technology. We are reviewing these
requirements for their technical and economic realizability. In addition, this example has demonstrated that the selection of an innovation sector oriented towards the conflicts and deficits of social systems differs to a great extent from the previous market types of novel technologies and that, in future, an entirely new, so-
ciologically experienced type of promoter and consultation company
is required.

## V. SELECTION OF SOCIOTECHNICAL SYSTEMS ANALYSIS STUDIES SHOULD ALSO INCLUDE SOCIAL FIELDS CHARACTERIZED BY A HIGH CONFLICT CAPABILITY

This requirement has always been fulfilled throughout the history of novel technology, which is also the history of the bourgeois
society. Novel technologies have always found acceptance most easily
and most rapidly by the power elites of the bourgeois society, as
these groups, due to their assured financial position, could in general bear the risks of novel technologies more easily than the finan-
cially dependent groups of society. The shift in the power elite
of the modern industrial societies has entailed changes which had a
largely negative effect on the innovation capability of their elites.

Since in the aspiring developing countries of today the power
elites are still identical with the plutocratic elites, a considerable demand for most advanced technologies must be expected to develop in the years to come in these countries - the social conflicts
in these countries are an extraordinary stimulus for such a demand.
The assumption that the industrialized countries could sell to the
aspiring developing countries their outdated technologies of yester-
day, is based on a disrespect and a misestimation of the positive
role which the dynamics of social systems plays in their innovation
capability. It seems more realistic to suppose that these nations
will sooner or later order such new technologies and inventions from
the developed industrialized nations, which could there only be introduced by the utmost effort. Hence, we should direct our searching for social fields characterized by a high conflict capability
not to the traditional plutocratic elites such as banks and large-
scale enterprises, but look for new innovators who combine in a new
form liquidity and social power. Insurance companies, organized in
the manner of Lloyds, consumer companies or even new and potent religious communities, for example, will perhaps play a more important
role in the field of innovation in the next 30 years than e.g. AEG-
Telefunken or British Petroleum.

VI. EMPIRICAL SOCIOLOGY SHOWS REMARKABLE BEGINNINGS OF SOCIO-
TECHNICAL FIELD RESEARCH REPRESENTING ALREADY TODAY IMPORTANT
COMPONENTS OF SOCIOTECHNICAL SYSTEMS ANALYSIS

The socio technical analyses carried out today by sociologists
are almost exclusively confined to clearly arranged fields of study,
where the usual relationship between technologies and the social system directly surrounding these technologies is disturbed and where
this disturbance and/or the arising conflicts entail, or are likely
to entail, economic, social or even legal changes of a major extent.

Now would you please direct your attention to a table presenting
the different sectors and questions handled today by technology-
oriented, empirical sociologists.  The information given is based on
an enquiry recently carried out in the Federal Republic of Germany.

The first and classical conflict sector of sociotechnical research is the relationship between technology, labor and business
organization in the field of industrial production.

The studies refer among other things to economic and socioeconomic interferences between labor market and technical change, to technological interferences between labor organization and technical optimization or between labor performance and automation, to sociological interferences between qualified labor organization and flexible
production and automation.

Social science still is "relatively helpless" as regards definition and analysis of the sociotechnical interferences in many fields
of the services sector.

The sociological fields of study included in this sector are
Education and Schooling, Media, Office, Research and Development,
Medicine, and Traffic.

The first field of study "Media" investigates interferences and
conflicts associated with the application of electronically supported information and communication technologies.  In addition to market research activities in this field, which had been promising in
the beginning, but as a whole were unsuccessful - the most recent
example being the video disc - sociology started to deal with some
selected areas, e.g. organization problems, functional problems, and
user problems of technical media in various fields of governmental
administration.

Interferences and problems connected with the introduction and
application of novel data processing and communication systems in
the offices of the big industrial and services enterprises are still
relatively easy to identify.

FIELDS OF STUDY OF TECHNOLOGY-ORIENTED SOCIOLOGY

| Secondary sector (Industrial production of goods) | Tertiary sector (Services) | | | | | |
|---|---|---|---|---|---|---|
| | Education | Media | Office | Research & Development | Medicine | Traffic |
| Production organization | Tech.-supported teaching and learning procedures | Market analyses | Organizational structure | Mobility | Market prognoses | Minimization of traffic volume |
| Labor market | Mobility research | User behavior | Working groups | Advanced training | Change of services spectra | Risk behavior of road users |
| Business Organization | Learning group analysis | Changes of infrastructure | Management | Management | Epidemiology | Market and consumer research (public relations) |
| Labor performance, Labor organization | | | | | | |

Source: AGF/ASA, Intermediate Evaluation of an Inquiry on Technology-Oriented Sociology, Cologne, February 1976.

Other studies deal with the influence exerted by data processing on the organizational structure of insurance companies in particular, with working methods in groups, applications of smaller and medium-size data technology and office technology as well as with the role of data processing experts in the management sector.

Sociological investigations of the sector "Research and Development" are primarily concerned with the relationship between scientific-technical productivity and various forms of labor organization, with special consideration being given to the (lack of) mobility of the scientists.

The sociotechnical problems in the public health sector are of a wide extent and more serious than had been anticipated by recent market prognoses e.g. for data processing systems and medical technologies. Medical sociology – a relatively young sector of social science – only gains gradual importance also for the preparation and control of measures and projects of research and technology policy.

The increased use of EDP and CATV technologies in the public health system seems only justified if it is not only connected with an improvement but also with an extensive and gradual shift of the services offered by regular public health institutions towards prophylactic and hygienic education services. This seems, however, only possible by means of organizational changes in individual areas of the medical care system, as without such changes even the most advanced technology will not be able to improve, and reduce the cost of, medical services.

In the sector of "Traffic" a distinction can be made between three groups of sociotechnical field research:

1. Attempts to determine economic and sociological conditions and possibilities of minimizing and/or redistributing and optimizing traffic volume;

2. Study of the sociotechnical factors of the risk behavior of road users as well as single fields of accident research;

3. Investigation of the users' behavior, in particular with respect to the implementation of new transportation technologies as well as image-building in the field of market research.

In the course of many of these studies, which to a large extent have been and are being carried out by scientists working independently from each other, methods have been developed and empirical results been achieved, which may be of great benefit for technology planning and specifically for sociotechnical systems analysis. This presupposes a single-minded understanding between sociologists, economists, technicians, and engineers, to which this Congress supplies a valuable contribution.

DEMO-ECONOMIC POLICY SIMULATIONS FOR BELGIUM

R. Lesthaeghe, S. Wijewickrema, M. Despontin, A.De Kerpel[*]

Vrije Universiteit Brussel

Belgium

## 1. INTRODUCTION

All countries that have gone through what is known in the relevant literature as the "demographic transition" have experienced the fertility decline which belongs to the very essence of this transition. Such a decline in fertility which needs about 100 to 150 years from beginning to end is part of the history of all the industrialized countries of the world. This major fertility decline was accompanied by short term rises and falls in fertility. Thus a fertility decline is no new phenomenon. However, the fertility decline experienced in Western Europe, North America and Australia since the 1960's seems to get into a class apart because of the following reasons:

1. The rapidity and degree of fall has hardly ever before been witnessed outside times of war or major economic depression (see Table 1).

2. The discussion and debate concerning the decline under consideration have been markedly characterized by a great deal of prudence. Social scientists for their part have been less categoric and more nuanced in what they have said and consequently less pessimistic than ever before. On the other hand governmental authorities have evidenced greater hesitation in adopting the same attitudes that

[*] We thank the Centrum voor Bevolkings- en Gezinsstudièn, Brussels (Belgian Ministry of Public Health and the Family) for its financial support of the project presented in this paper.

characterized policy decisions in the past. This overall prudence
can be attributed to the prevailing awareness of the complexity of
the situation: - involving as it does considerations both qualitative and quantitative which engage the attention of disciplines
other than that of pure demography - carrying with it problems related to all levels of society.

3. While part of the explanation of the rapid fall in fertility
resides in the fact that cross-sectional (or transversal) fertility
indicators for the 1950's and early 1960's took on inflated values
arising from changes in the tempo of reproductive behavior a good
portion of the decline has to be attributed to the advent of a new
mentality regarding desired family size. The changes in reproductive cohort timing referred to is the result of the lowering of the
mean age at marriage accompanied by the concentration of marital
fertility at the lower end of the marriage duration scale. The new
mentality referred to seems to be of such a nature that its action
would be effective in lowering fertility even in the absence of efficient contraceptive techniques ( 5), ( 29).

Table I

Values of the Gross Reproduction Rates in a Number of Western
Countries, 1963 - 74

| Country | 1963 | 1972 | last estimate 1973 or 1974 | index last estimate relative to 1963 ( 1963 = 100) |
|---|---|---|---|---|
| Canada | 1.82 | 0.98 | 0.94 | 52 |
| United States | 1.69 | 0.93 | 0.90 | 53 |
| Australia | 1.64 | 1.34 | 1.21 | 74 |
| Netherlands | 1.53 | 1.04 | 0.91 | 59 |
| France | 1.38 | 1.15 | 1.03 | 75 |
| England & Wales | 1.37 | 1.05 | 0.97 | 71 |
| Belgium | 1.28 | 1.07 | 0.95 | 74 |
| Federal Rep. Germany | 1.21 | 0.83 | 0.75 | 62 |
| Sweden | 1.12 | 0.92 | 0.90 | 80 |

Arising from the above considerations we were led to confine our attention to the quantitative aspects of the economic consequences of the Belgian demographic situation past, present and future, and were inclined to think that the demographic future of Belgium is likely to be characterized by a low level of fertility as long as the above-mentioned new mentality prevails. Since guessing about the future is necessarily hazardous it is thus impossible to estimate how long this new mentality would last, we decided that after an initial concession to this new mentality we could not do better than simulate alternative patterns of fertility behavior (+ accompanying population projections)around and about the central theme of a zero growth population. The resulting projections have the following characteristics.

1. Starting point of projections = the Belgian situation on the 31st. December, 1970.

2. Initial concession to the above-mentioned new mentality consists of a 9 year (1970 to 1979) extrapolation of the fertility decline witnessed immediately anterior to 1970. The value taken for the net reproduction rate (NRR) in 1979 equals 0.71. (NRR = transversal measure of the potential of a population to reproduce itself. NRR = 1 implies that the population exactly replaces itself).

3. Alternative population models: those leading to the stationary state
those leading to the stable state (constant age structure + positive or negative growth rate; stationary state = stable state with zero growth)
those leading to the state of stable population cycles (referred to as "pseudo-stationary" state).

4. Conditions of projection: mortality constant and absence of migration.

## 2. ALTERNATIVE POPULATION MODELS

A stationary population is characterized by a constant age structure and a zero population growth rate. In our simulations, the stationary state is reached in each of three ways.

1 – Hypothesis H1. The annual number of births is kept constant from 1979 onwards. The stationary state is reached in 100 years (100 years was taken as the upper age limit in our population).

2 – Hypothesis H2. The stationary state is reached by keeping the population size constant from 1979 onwards.

3 – Hypothesis H3 (1.0). The NRR is made to rise from its 1979

value (of 0.71) to unity in 1987, the value of one being retained thereafter.

A stable population is characterized by a constant age structure and a constant positive or negative population growth rate. (A stationary population is a special case of a stable population). The stable state is reached via:

1 - H3 (1.2). The NRR rises from its 1979 value to 1.2 in 1987 and is maintained constant thereafter. The resulting stable population has a positive constant growth rate.

2 - H3 (0.8). The constant value of the NRR attained in 1987 and kept constant thereafter equals 0.8. The terminal stable population is a constantly decreasing one.

The state of stable population cycles is characterized by the periodic repetition of the series of age structure of population witnessed during any one cycle, together with the presence of a birth function which can be decomposed to an oscillatory part accompanied by an exponential part. This state was reached through an NRR which, having attained unity in 1987 as in H3(1.0) was made to oscillate sinusoidally about the mean value of unity with an amplitude equal to 0.2. Simulations following three hypothesis are worked out:

1 - H4(26). A period of oscillation approximately equal to the mean age of childbearing (here equal to 26 years) gives rise, through a resonance-effect to large swings in the birth cycle. This is merely the result of a terminal state in which large cohorts of mothers experience the higher portion of the fertility cycle while small cohorts are subject to the lower levels of the fertility cycle as the cohorts move through their reproductive age span.

2 - A periodicity of 52 years produces, through anti-resonance small amplitude birth cycles. In this case, large cohorts experience a minimal fertility and vice versa.

3 - A periodicity of 13 years produces an intermediate case.

Figure 1A shows the marked differences between the various population models as regards the ratios between specific large age groups. Clearly models H3(0.8) and H1 produce a considerable "ageing effect" of the population in the non-terminal state. Among the H4-type models, the evolution witnessed in H4(13) very quickly approaches the evolution of the stationary models H2 and H3(1.0) while H4(26) leads to great fluctuations in the age structure ratios. As regards the ageing index (65+/0 - 14) model H4(52) also approaches the level of the terminal stationary state but then only in the very long run (since it takes a very long time before the initial perturbations in age structure are completely levelled off).

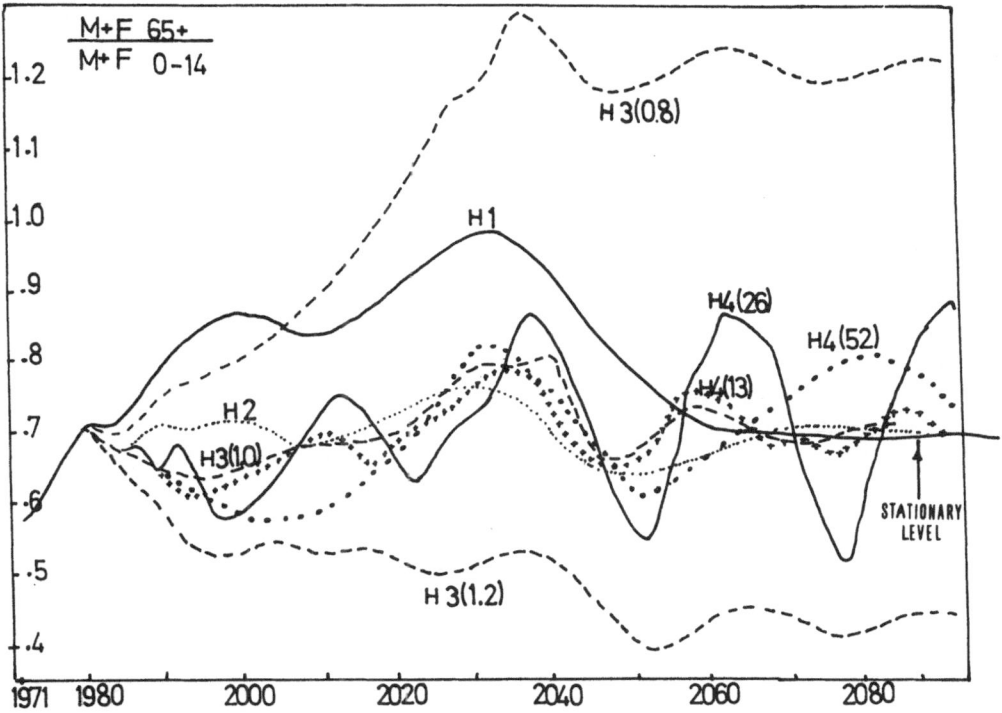

Figure I.   Evolution of the index of ageing in the eight population
models.

## 3. LINKING THE POPULATION MODELS TO A DEMO-ECONOMIC MODEL OF THE BELGIAN ECONOMY: SIMULATION RESULTS

The use and construction of demo-economic models goes back to 1958 when COALE and HOOVER ( 9) presented a simple model specifically designed for evaluating the effects of demographic policies in a developing country.  Since then the use of such models as well for developing as for developed countries has become rather widespread ( 6), ( 11), ( 14),( 15).  Most models applied to industrialized nations are of the neoclassical type, i.e. they mainly consist of a production function and a relation that determines the accumulation of capital.  As was shown elsewhere ( 13), ( 14) the economic outcome of such types of models depends heavily on the equations that determine the capital stock.  If these equations are such that they imply a very sensitive reaction of investment to demographic change then every demographic variation gives rise to striking repercussions witnessed in the economic growth rate.  Moreover, the neoclassical demo-economic models are mostly capacity type models in that they measure economic growth solely as growth in production capacity.  We therefore thought it interesting to estimate a demo-economic model for Belgium based on the following principles.

1 - The model remains basically neoclassical in that a Cobb-Douglas production function underlies the employment and investment decision but this production function itself is not used in the model as a means of computing total output (10). The process of capital accumulation is now determined by an estimated relationship based on economic mechanisms present in the Belgian economy for the period of estimation (1948 - 1973).

2 - Elements of the demand side of the economy are introduced in the model so that the simulation results are no longer exclusively dependent on the evolution of the production capacity of the economy.

3 - Government variables traditionally enter an econometric model as exogenous factors.  In that way they also allow for the application of multi criteria analysis (see section 4).  However, in order to establish a standard simulation run these variables are treated in our model as endogenous and are then dependent on the evolution of the GNP and the unemployment rate.  In doing so only basic economic principles of government intervention are adopted.  Thus, governmental action is presumed to prevent the unemployment rate from rising above a specified high level through the raising of government expenditure.  Introducing such a relationship in the model implies that our simulation results are also determined by the offer of labor (and no longer solely by the demand side of the economy) because of the relation : greater demographic growth→ greater labor force→ more unemployment→ more government intervention.

4 - The demographic inputs in the model are:-the total labor force
                                      :-a coefficient of labor
   efficiency which is age and sex dependent,
                                      :-the total amount of con-
   sumer units (1 consumer unit = imaginary coefficient equal to 1
   for a male adult consumer)
                                      :-total population.

    Let us now take a closer look at the simulation results, rep-
resented in Figure IIA and IIB. At first, it should be noted that
the various simulations differ considerably with respect to economic
growth only from 1990 - 2000 onwards. This stresses the long term
nature of the problem under study. All population models start from
1970 with a nine year decline of the net reproduction rate (see
section 2). Moreover the diversive effects of alternative evolutions
in NRR only become economically visible after 15 to 20 years, i.e.
after new small or great cohorts reach economically significant age
groups (total labor force, number of adult consumers). Hence: 1970
+ 9 + 15 = 1994.

    As can be seen on Figure IIA all stationary models generate
the same economic growth rate of about 2 per cent in the terminal
state. Some markedly different evolutions occur during the transition to this state. Model H1 (annual number of births constant from
1979 onwards) leads to a 30 year period of low, but still positive
growth during which the economy is of course more vulnerable to ex-
ternally generated catastrophies. It should also be kept in mind
that model H1 implies a reduction in total population of the present
9.6 million to about 7.6 million.

    Therefore it could be stated (and this could serve as a guide-
line to policy making) that the transition to a stationary population for Belgium does not generate an economic recession unless the
envisaged final population size (and consequently also total labor
force and total number of adult consumers) is considerably lower than
the present 9.6 million.

    The evolution of the GNP growth rate in the models leading to
a stable population (see Figure 2B) is as expected: 1 - permanently
high in the case of H3(1.2) because of the stimulus stemming from
increased employment and consumption in such a growing population,
2 - continuously low in the case of H3(0.8).

    The pseudo-stationary population models generate particularly
intriguing results as regards the economic growth rate. In model
H4(26) an exploding sequence of fluctuations in the growth rate of
the GNP appears. This is due to the resonance-effect in the birth
cycle (explained earlier), a resonance-effect that persists in the
age groups with economic significance. On the other hand, model

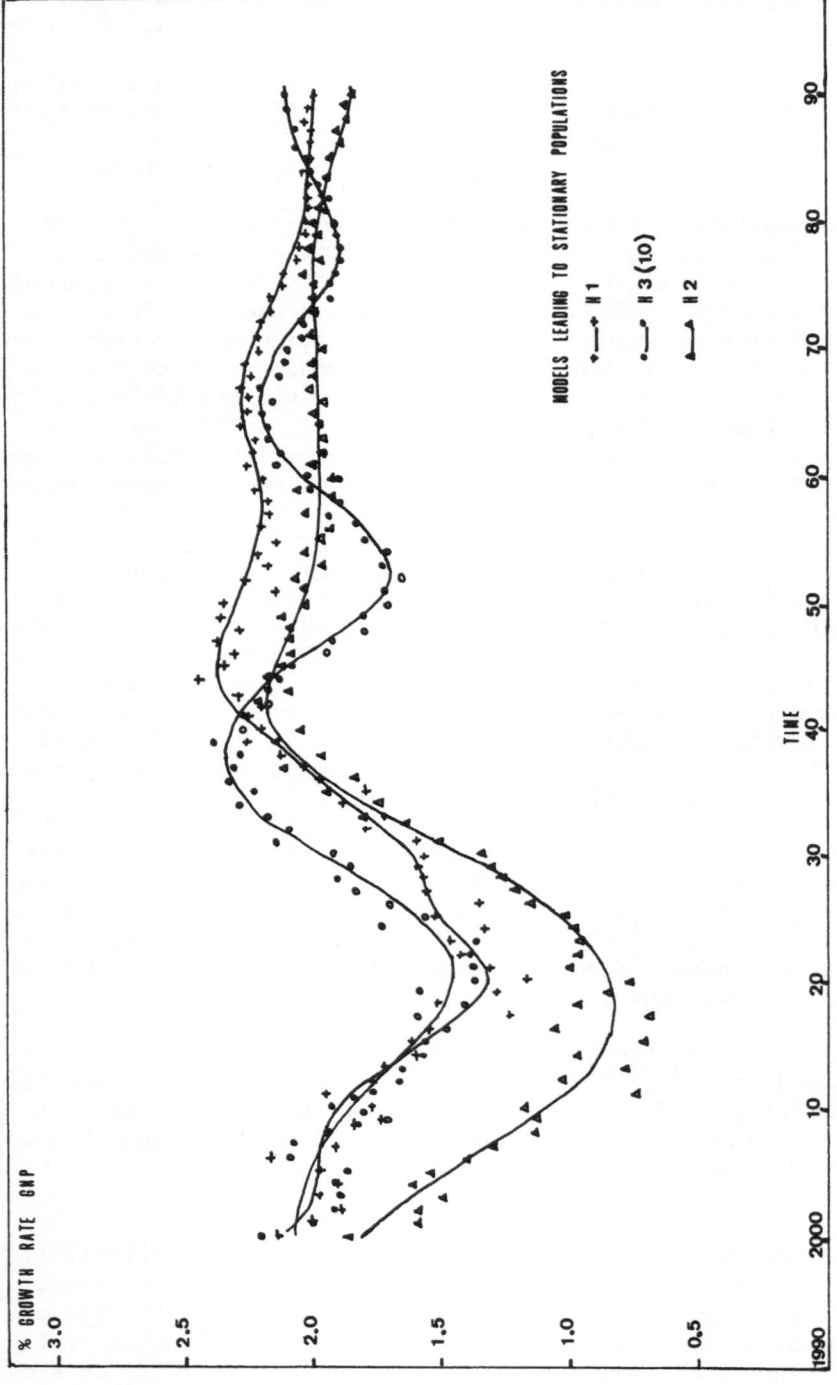

Figure IIA. Evolution of the GNP growth rate in the models leading
to stationary populations.

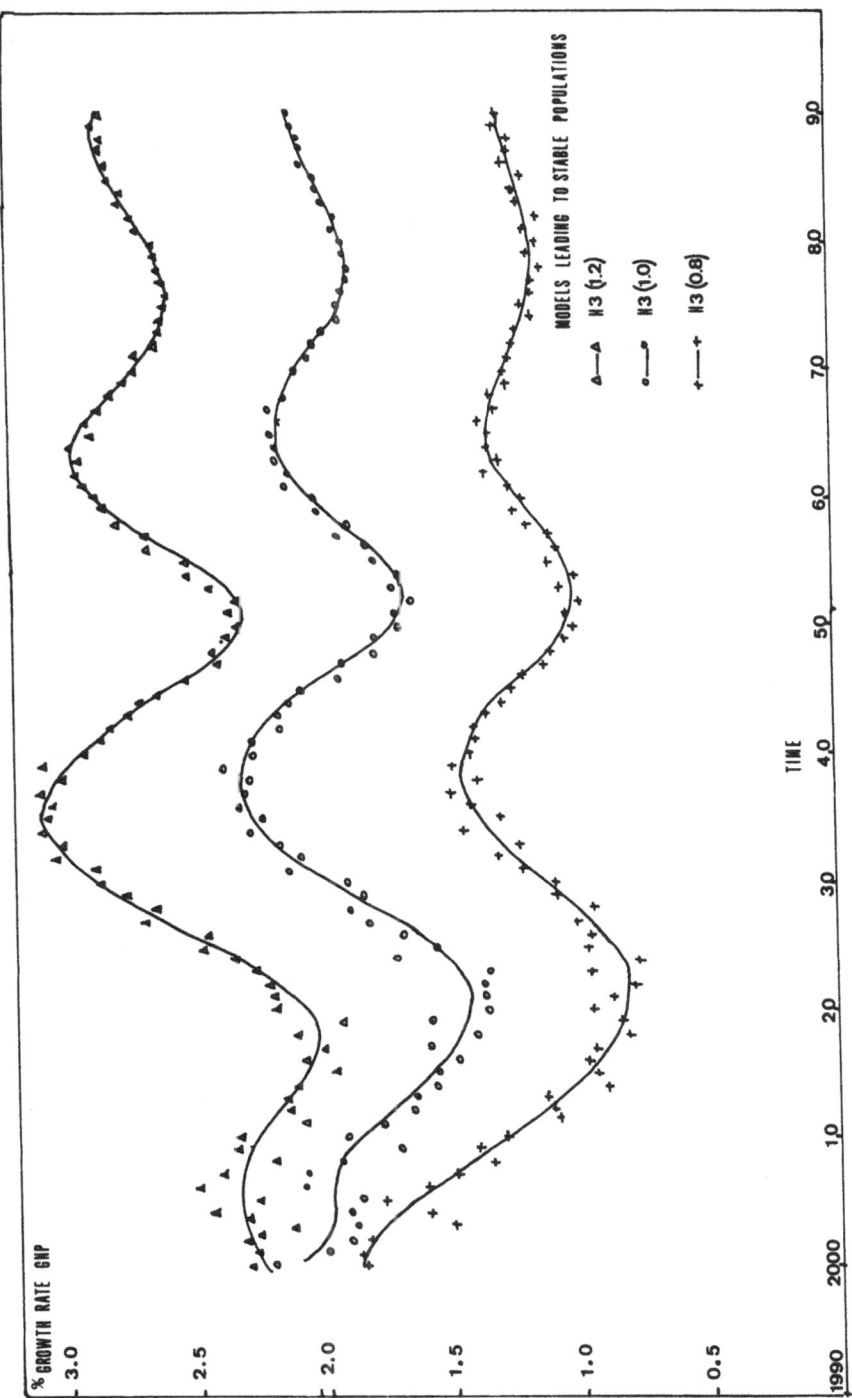

Figure IIB. Evolution of the GNP growth rate in the models leading
to stationary populations.

H4(13) produces only fluctuations with small and even slightly diminishing amplitude. The economic outcome of this model resembles
very much the outcome of a pure stationary model (H2 or H3(1.0) ).
We stated earlier that model H4(52) produces a birth cycle with anti-
resonance. However, the anti-resonance effect is not fully translated into correspondingly mitigated variations of the GNP growth rate
(see Figure IIC). This is due to the fact that in this model generations are subject to high or low fertility during a very long span
of time which allows for significant increases or decreases in the
age groups of economic significance i.e. age groups in which are
found total labor force and number of consumer units.

A commonly used technique of comparing the economic advantages
of one demographic evolution as against another is by comparing the
evolution of the per capita GNP proper to the different cases. It
appears that in our simulations after a 100 to 150 years the greatest differences in per capita GNP lie only within a range of 5 to
6 percent – the "best" performances occurring in the case of models
that imply a decrease in population (H1, H3(0.8) ). Since these differences are so small a slight change in one of the estimated coefficients of the econometric model might possibly change the magnitude or direction of the differences involved. A sensitivity analysis on the coefficients of our model now seems to be appropriate.
Such an analysis has not yet been carried out. What we can state
for the moment is that the differences in the evolution of per capita GNP in the various population models – if any – are so small
that in themselves they do not constitute a sound basis for opting
for one population model as against another.

In short our simulation results lead to the following conclusions.

1 – No major decline in GNP growth rate occurs in a population
tending to the stationary state. This statement needs to be nuanced
in the case of evolutions involving a strong population reduction on
their way to stationarity.

2 – The evolution of the growth rate of the gross national product witnessed in some of the pseudo-stationary models is very similar to that occurring in the case of the models which usher in a
pure stationary population. The crucial factor determining the economic outcome of a pseudo-stationary model is the period of fluctuation in the net reproduction rate and not so much the amplitude of
this fluctuation.

3 – The various population models do not differ substantially
in the evolution of the per capita GNP.

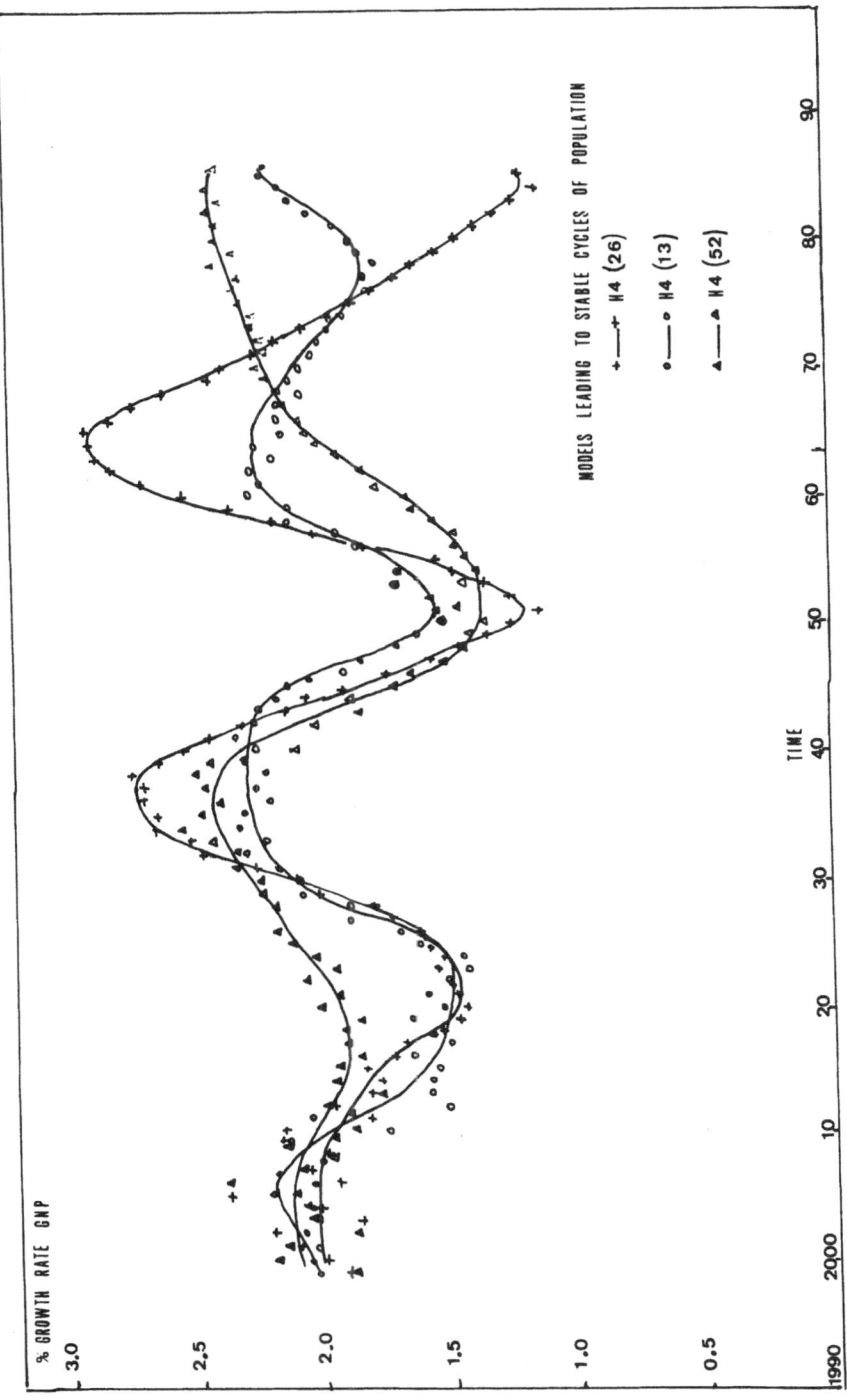

Figure IIC.   Evolution of the GNP growth rate in the models leading
to stable cycles of population.

## 4. MULTIPLE CRITERIA DECISION AID IN ECONOMIC POLICY

How could we use the Janus-model as instrument for the formulation of the "optimal" economic strategy?

Formally a linear econometric model can be stated as:

(4.1)     $\underline{y}_t = \underline{A}\ \underline{x}_t + \underline{b}_t + \underline{y}_t$

where $\underline{y}_t$, $\underline{A}_t$, $\underline{x}_t$, $\underline{b}_t$ and $\underline{y}_t$ are respectively matrices of dimension mx1, mxn, nx1, mx1 and mx1.

$\underline{y}_t$ are the endogenous variables, $\underline{x}_t$ the exogenous variables, $\underline{y}_t$ the stochastic disturbances and $\underline{b}_t$ the vector of constants.

$\underline{A}$ is a matrix of estimated and definitional coefficients, and t the time-subscript. The value of $\underline{y}_t$ being unknown, let us assume for the sake of simplicity $\underline{y}_t = 0$.

(4.1) can then be written as:
(4.2) $\underline{y}_t = \underline{A}\ \underline{x}_t + \underline{b}_t$

Alternative approaches can be chosen: $\underline{y}_t$ can for example be drawn from a stochastic distribution.

Some of the endogenous variables will be targets and some of the exogenous variables can be used as instruments. Let us denote the vectors of instruments and target respectively $x_t$ and $y_t$ which are of dimension (kx1) and (1x1). When we enter the value of the exogenous non-instrument variables and delete the superfluous equations we obtain:
(4.3) $y_t = Ax_t + b_t$
where A is a matrix (1xk) of coefficients and $b_t$ a vector (1x1) of constants.

A first solution was given by TINBERGEN (28) when the number of targets is equal to the number of instruments (k=1). He proposed to specify desired values for the targets and determine the values for the instruments by solving the system of equations (4.3).

THEIL (26) proposed to optimize a welfare function subject to the restrictions of the econometric model. In a first step we have to determine desired values for the targets and for the instruments. Let us denote them respectively $y_t^*$ and $x_t^*$. In order to obtain a solution we minimize the deviations between the realized and desired values for both the targets and the instruments. If the quadratic criterion is chosen we have to solve the following program:

(4.4)     $\text{Min}\ (y_t - y_t^*)'\ Q(y_t - y_t^*) + (x_t - x_t^*)'\ R(x_t - x_t^*)$

          subject to: $y_t = Ax_t + b_t$

Where Q and R are a priori specified matrices of dimension (1x1) and (kxk) of "welfare weights".

RUSSELL and SMITH (24) suggested that it was not necessary to define a welfare function when we have more instruments than targets. They proposed to specify desired values for the targets and to minimize the norm of the vector of instruments subject to the system of equations (4.3) using the generalized inverse. This is more a mathematical than an economic solution.

Primarily the specification of an optimal strategy is a multiple criteria problem. We try to optimize several objectives of society under restriction of technical and behavioural relations. Although the problem was stated more than twenty years ago by KUHN and TUCKER (18) little attention has been paid until recent years to the simultaneous optimization of several objectives.

As ZELENY and COCHRANE (32) recalled, in view of application to economic policy we can formulate the problem as:

(4.5)          $\text{Max } Cz_t$

subject to:  $y_t = Ax_t + b_t$

where $z_t = \begin{pmatrix} y_t \\ x_t \end{pmatrix}$ is a vector of policy variables with $r = k+1$ components and C is a matrix (pxr) of coefficients of the p objective functions we want to optimize.

$C = \begin{pmatrix} c_1 \\ c_2 \\ \vdots \\ c_p \end{pmatrix}$ where $c_i$ is a vector of dimension (1xr); i=1,2,...p.

Due to political, economic and social constraints the admissibile interval of the policy variables will normally be restrained and the program can more generally be stated as:

(4.6)          $\text{Max } Cz_t$

subject to:  $y_t - Ax_t = b_t$

$z_t \leq b_t^u$

$z_t \geq b_t^l$

where $b_t^u$ is an (rx1) vector of upper bounds and $b_t^l$ a vector of lower bounds of same dimension on the policy variables. Other linear constraints of any kind can be added.

A lot of methods have been published to solve the multiple criteria linear programming problem, among others by AUBIN and NASLUND (1), BELENSON and KAPUR (3), BENAYOUN, TERGNY, DE MONTGOLFIER and LARICHEV (4), BOYD (7), BRAGARD (8), GEOFFRION, DYER and FEINBERG (17), MONARCHI (19), NIJKAMP and RIETVELD (20), VINCKE (30), ZELENY (32) and, ZIONTS and WALLENIUS (34).

Although this approach to economic policy does not imply the choice of any particular multiple criteria algorithm recent experiments on KLEIN's model for the United States in the nineteen-thirties (12) showed the usefulness of the algorithm of VINCKE. This algorithm reaches the solution in an interactive man-computer procedure. We think this is very important in view of practical application. Indeed, this algorithm does not suppose any specification of the implicit welfare function and allows to reach all efficient or Pareto-optimal solutions. The optimum is attained in an iterative way in which the algorithm proposes certain perturbations in the so far obtained solution and the consequences of possible decisions. On the basis of this information the decision-maker can decide whether the proposed perturbation should be carried through.

The interactive way we suggest for reaching the optimal solution tries to face principally three problems in quantitative economic policy: (1) Some objectives cannot be quantified; (2) An econometric model remains a model and cannot be more than an approximation of reality; (3) Society has an unknown utility function.

In order to apply this algorithm which makes use of the simplex-method both the variables and the value of the objective functions should be non-negative. It is clear that this can be achieved without difficulty and will be assumed here. Our policy problem can then be formulated as:

$$(4.7) \qquad \text{Max } Cz_t$$
$$\text{subject to: } y_t - Ax_t = b_t$$
$$z_t \leq b_t^w$$
$$z_t \geq b_t^i$$
$$z_t \geq 0$$
$$c_i z_t \geq 0 \qquad i=1,2,\ldots,p$$

Solving the $p$ unicriterion programming problems separately we obtain $p$ unicriterion optimal solutions $\tilde{z}_t^i$ $(i=1,2\ldots,p)$.

Let us define

$$N = \begin{pmatrix} N_1 \\ N_2 \\ \cdot \\ \cdot \\ \cdot \\ N_p \end{pmatrix} = \begin{pmatrix} c_1 \tilde{z}_t^1 \\ c_2 \tilde{z}_t^2 \\ \cdot \\ \cdot \\ \cdot \\ c_p \tilde{z}_t^p \end{pmatrix}$$

The vector N is called the ideal point, i.e. it is the vector in the space of the objectives which corresponds to a vector of policy variables in the decision space and optimizes all unicriterion problems simultaneously. If this point exists the multiple criteria problem is actually solved. Generally speaking this vector of policy variables does not exist and N will be an unattainable target.

Following the first step which is identical to that of the STEM-method we solve the following program which is a minimax solution to the problem of finding the feasible solution nearest to the ideal point.

(4.8)  Min $\alpha$

subject to:  $f_i = c_i z_t$  $\qquad$ $i=1,2,\ldots,p$

$\qquad$ $[N_i - c_i z_t] \pi_i \leq \alpha$  $\qquad$ $i=1,2,\ldots,p$

$\qquad$ $y_t - Ax_t = b_t$

$\qquad\qquad$ $z_t \leq b_t^u$

$\qquad\qquad$ $z_t \geq b_t^l$

$\qquad$ $\alpha, z_t \geq 0$

$\qquad\qquad$ $f_i \geq 0$  $\qquad$ $i=1,2,\ldots,p$

The $\pi_i$ are "sensitivity weights" which measure the sensitivity of the different objectives on perturbations of the solution vector. For example, those proposed by the STEM-method can be used.

Introducing the slack variables $t_1, t_2, \ldots, t_p, s_1^u, s_2^u, \ldots, s_r^u, s_1^l, s_2^l, \ldots, s_r^l$, the artificial variables $v_1, v_2, \ldots, v_p, w_1, w_2, \ldots, w_r$ and defining M as an arbitrarily large number we obtain:

(4.9)  Min $\alpha + M [\sum_{i=1}^{p} v_i + \sum_{i=1}^{r} w_i ]$  $\qquad\qquad$ (4.9.1)

subject to:  $f_i - c_i z_t = 0$  $\qquad$ $i=1,2\ldots,p$  (4.9.2)

$\qquad$ $\pi_i c_i z_t + \alpha - t_i + v_i = \pi_i N_i$  $\quad$ $i=1,2\ldots,p$  (4.9.3)

$\qquad$ $y_t - Ax_t = b_t$  $\qquad\qquad$ (4.9.4)

$\qquad$ $z_t + s^u = b_t^u$  $\qquad\qquad$ (4.9.5)

$\qquad$ $z_t - s^l + w = b_t^l$  $\qquad\qquad$ (4.9.6)

$\qquad$ $\alpha, z_t, s^u, s^l, w \geq 0$  $\qquad\qquad$ (4.9.7)

$\qquad$ $f_i, t_i, v_i \geq 0$  $\qquad$ $i=1,2\ldots,p$  (4.9.8)

$$\text{where } s^u = \begin{pmatrix} s_1^u \\ s_2^u \\ \cdot \\ \cdot \\ s_r^u \end{pmatrix} \; ; \; s^l = \begin{pmatrix} s_1^l \\ s_2^l \\ \cdot \\ \cdot \\ s_r^l \end{pmatrix} \; ; \; w = \begin{pmatrix} w_1 \\ w_2 \\ \cdot \\ \cdot \\ w_r \end{pmatrix}$$

This program yields a first compromise solution which is presented to the decision-maker. At the same time the consequences of possible perturbations are given. This is quite easy due to the properties of the simplex-tableaus. The increase (or decrease) of the constants of equations (4.9.5) and (4.9.6) represent a relaxation or contraction of the domain of admissible values for the policy variables.

The consequence on the different policy variables and the value of the objective functions of increasing (or decreasing) one of these constants by unity can immediately be read in the column of the slack variable corresponding to the equation of the implied constant in the final simplex-tableau of program (4.9.). Moreover, we can immediately and easily compute the maximal change in this constant which does not alter the base. Indeed, it is the maximal change which allows all constants in the adapted simplex-tableau to remain non-negative.

We can follow the same procedure in changing the constants of equations (4.9.3). Changing these constants by one unity implies a change of component of the ideal point by $1/\pi_i$. This means our target is altered.

If it seems preferable to the decision-maker to change any of these constants the simplex-tableau is adapted and the new compromise-solution and information on possible new perturbations are given. The procedure ends when no perturbation is not attractive anymore to the decision-maker. Then he has chosen this Pareto-optimal solution which maximizes utility.

Although no welfare function is assumed it is clear that for example an implicit THEIL-type welfare function can be treated in this manner, in which the desired values for both targets and instruments can be decided in an interactive way with a knowledge of the consequences of the determination of these "desired values a posteriori".

BIBLIOGRAPHY

(1)  Aubin, J.P., Naslund, B., "An Exterior Branching Algorithm".
     European Institute for Advanced Studies in Management, Working
     paper 72-42, 1972.

(2)  Bauer, V., Wegener, M., "A Community Information Feedback Sys-
     tem with Multiattribute Utilities", paper presented at the
     IIASA Workshop on Decision-Making with Multiple Conflicting
     Objectives, Laxenburg, October 20-24, 1975.

(3)  Belenson, S.M., Kapur, K.C., "An Algorithm for Solving Multi-
     criterion Linear Programming Problems with Examples", Opera-
     tional Research Quarterly, 24, No. 1, 1973.

(4)  Benayoun, R., Tergny, J., de Montgolfier, J., Larichev, O.,
     "Linear Programming with multiple objective function: step
     method (STEM)", Mathematical Programming, 1, No. 3, 1971.

(5)  Blake, J. & P. Das Gupta, "Reproductive Motivation versus
     Contraceptive Technology", Population and Development Review,
     December 1975.

(6)  Blandy, R., Wery, R., "Croissance Démographique et Emploi-
     Bacchue 1". Revue Internationale du Travail, No. 5, May 1973,
     p. 479-489.

(7)  Boyd, D., "A Methodology for Analyzing Decision Problems In-
     volving Complex Preference Assignments." Decision Analysis
     Group, Stanford Research Institute, Menls Park, California 1970.

(8)  Bragard, L., "La Programmation à Objectifs Multiples," Revue
     des Sciences Economiques de l'Université de Liège, 1974.

(9)  Coale, A.J., Hoover, E.M.,"Population Growth and Economic De-
     velopment in Low-Income Countries", Princeton University Press,
     Princeton, 1958.

(10) Coen, R.M., Hickman, B.G., "Constrained Joint Estimation of
     Factor Demand and Production Functions," Review of Economics
     and Statistics, No. 3, 1970, p. 287-300.

(11) Denton, F.J., Spencer, B.G., "A Simulation Analysis of the
     Effects of Population Change on a Neoclassical Economy",
     Journal of Political Science, No. 2, Part I, March/April 1973.

(12) Despontin, M., "Multicriteria Economische Politiek" (in Dutch),
     Centrum voor Statistiek en Operationeel Onderzoek, Vrije Uni-
     versiteit Brussel, 1976.

(13) Despontin, M., De Kerpel, A., Lerou, A.M., Lesthaeghe, R., van
     Maldeghem, R., Wijewickrema, S., "Janus 1: Van brutoreproduk-
     tiecijfer naar bruto binnenlands produkt. 1. De alternatieve
     bevolkingsmodellen," (in Dutch), Bevolking en Gezin, 1975, 2,
     153-186.

(14)  Despontin, M., de Kerpel, A., Lerou, A.M., Lesthaeghe, R.,
      van Malderghem, R., Wijewickrema, S., "Janus 1: Van bruto-
      reproduktiecijfer naar bruto binnenlands produkt. 2. De macro-
      economische simulaties," (in Dutch), Bevolking en Gezin, 1975,
      3, 367-400.

(15)  Enke, S., "Economic Consequences of Rapid Population Growth",
      Economic Journal, London, December 1971, 800-811.

(16)  Fox, K.A., Sengupta, J.K., Thorbecke, E., "The Theory of
      Quantitative Economic Policy", North-Holland, 1973.

(17)  Geoffrion, A.M., Dyer, J.S., Feinberg, A., "An Interactive
      Approach for Multi-Criterion Optimization with an Application
      to the Operation of an Academic Department," Management
      Science, 19, No. 4, 1972.

(18)  Kuhn, H.W., Tucker, A.W., "Non Linear Programming" in: Pro-
      ceedings of the Second Berkeley Symposium on Mathematical
      Statistics and Probability, J. Neyman, editor, University of
      California Press, Berkeley, 1951.

(19)  Monarchi, D.E., "An Interactive Algorithm for Multiobjective
      Decision-Making," University of Arizona, Tucson, Arizona,
      Report No. 6, 1972.

(20)  Nijkamp, P., Rietveld, P., "Multi-Objective Programming Mo-
      dels: New Ways in Regional Decision-Making," Free University
      Amsterdam, Research Memorandum No. 43, 1976.

(21)  Pitchford, J.P., "Population in Economic Growth",North-Holland/
      American Elsevier, Amsterdam/New York, 1974, 280 p.

(22)  Roy, B., "Management Scientifique et Aide à la Décision,"
      SEMA, Direction Scientifique, rapport de synthèse No. 86, 1974.

(23)  Roy, Brans, Vincke, "Aide à la Decision Multicritère," Revue
      Belge de Statistique, d'Informatique et de Recherche Opération-
      nelle, Vol. 15, No. 4, 1975.

(24)  Russell, C.S., Smith, V.K., "Targets, Instruments and Genera-
      lized Inverses," European Economic Review, 1975.

(25)  Theil, H., "Optimal Decision Rules for Government and Industry"
      North-Holland, 1964.

(26)  Theil, H., "Economic Forecasts and Policy", North-Holland 1961.

(27)  Thiriez, H. Zionts, S., eds.. "Multiple Criteria Decision
      Making," Lecture Notes in Economics and Mathematical Systems,
      Springer-Verlag, 1976.

(28)  Tinbergen, J., "On the Theory of Economic Policy", North-
      Holland, 1952.

(29)  Van Maldergheim, R., "De evolutie van de gezinsgrootte bij de
      Nederlandse cultuurgemeenschap in België, 1966-1971,"

Bevolking en Gezin, No. 1, 1975, 49-73 (in Dutch).

(30)  Vincke, Ph., "Une méthode interactive en programmation linéaire
      à plusieurs fonctions économiques," Revue Francaise d'Automa-
      tique et de Recherche Opérationnelle, Vol. 10, No. 6, June 1976.

(31)  Wallenius, J., "Interactive Multiple Criteria Decision Methods:
      An Investigation and an Approach," The Helsinki School of Eco-
      nomics, 1975.

(32)  Zeleny, M., "Compromise Programming," in: Multiple Criteria
      Decision Making, J.L. Cochrane and M. Zeleny, eds., Univer-
      sity of South Carolina Press, Columbia, 1973.

(33)  Zeleny, M., Cochrane, J.L., "A Priori and A Posteriori Goals
      in Macroeconomic Policy Making," in: J.L. Cochrane and M.
      Zeleny, eds., "Multiple Criteria Decision Making," University
      of South Carolina Press, Columbia, 1973.

(34)  Zionts, S., Wallenius, J., "An Interactive Programming Method
      for Solving the Multiple Criteria Problems," European Insti-
      tute for Advanced Studies in Management," Working paper 74-
      10, 1974.

# SOCIO-ECONOMIC SYSTEMS, URBAN GROWTH AND SETTLEMENT PATTERNS

## (The Case of Turkey)

Ruşen Keleş

Ankara University

Turkey

Villages, local communities, neighbourhoods, towns, cities and metropolises are all territorial entities that possess the characteristics of highly complex socio-economic systems. Their spatially defined boundaries do not necessarily delimit the operational range of other systems, whose people, goods, messages and decisions move back and forth over much larger areas (B.Gross, 1966:74). The aim of this paper is not to deal with these territorial entities in all their relationships, but rather to see the main approaches of the principal economic systems to the problems of spatial systems, urbanization and environment. This subject was not studied much until recently. This lack of information was mainly due to the scarcity of studies on urbanization and settlement patterns in countries with different economic and social structures, political regimes as well as economic and social development policies. Particularly in the socialist countries and in the countries of the Third World, the phenomena of city and urban growth have been subject to only a limited number of scholarly studies.

Within the framework of this presentation, only the ideologies of capitalism and socialism, as the main economic systems, will be reviewed. It is assumed that general principles of either capitalism or socialism prevail in the mixed economy systems of which some are more close to capitalism and others range all the way to democratic socialism. Similarly, the Third World countries may be considered under either the capitalist or the socialist groups by looking at the prevailing aspects of their developmental policies[1].

## CITY IN THE MARKET ECONOMIES

Capitalism may be defined as an economic system where all production factors are owned by individuals and the profit motive is the principal guide for all kinds of productive activity (G.Grossman, 1967:26). It is characterized by the freedom for separate entities to buy and sell competitively in a climate ruled by free price formation and profit maximization. The emergence of capitalism was accompanied by the egalitarian doctrine that property and freedom were universal and inalienable rights. As a general rule, the possession of these rights imposed no obligation to support the society which guaranteed them. Consequently, the protection of the individual members of the society was to be ensured by the "self-regulating market" or by the "hidden hand" of Adam Smith.

### City Structure

In market economies, there are striking class differences in the internal structure of the cities. Poor, middle-income and wealthy districts are definitely separated from each other. Models of urban ecological structure formulated by American sociologists during the 1930's are based on similar observations (S.F.Chapin, 1965: 10-21). In fact, this differentiated structure is due to the identity of the city with the whole society of which it is part. As it was observed by G.Sjoberg, pre-industrial, transitional and industrial cities are sub-systems of the larger socio-economic systems. Each of the sub-systems mentioned above corresponds to pre-industrial or feudal, to developing or modernizing, and to developed and modern societies, respectively (G. Sjoberg, 1965: 213-263). The fact that each city has the identical characteristics of a society at the same level of development is also a corollary of the distinction made by Karl Marx between the primitive, slave-owning, feudal, capitalist and socialist societies.

---

[1] The Third World countries are generally underdeveloped economies. Another common characteristic which unites them is their firm attitude against colonialism and imperialism. However, they differ greatly in their economic and international policies. B. Berry regards these countries and social welfare states as separate socio-political systems in addition to capitalism and socialism. (1974).

## Center-Periphery Relations

In the industrialized western countries, urbanization is a process historically parallel to the industrial revolution. Those countries had a higher level of industrialization when they had reached the present urbanization level of the developing countries. One important reason for this is the "dependent urbanization" which is the product of the last century colonialism and which depends on the asymmetric relations between the developed and the developing countries. The developing commercial capitalism together with the industrial and financial imperialism has developed large business, commercial and administrative centers in the so-called Third World countries. The identity of interest between the developed metropolises of imperialist powers and the large urban centers of developed countries is called the "Structural Theory of Imperialism" (J. Galtung, 1971: 84).

Contemporary metropolitanism is embedded in a global economy of great complexity. That economy is hierarchically ordered with local centers dominating local hinterlands, more important metropolitan centers dominating lesser centers, and all centers outside the socialist countries being ultimately subordinate to the central metropolitan areas of North America and Western Europe (D. Harvey, 1973: 262). In other words, at each step along the way of this exploitation relation which in chain-like fashion extends the capitalist link between the capitalist world and the great metropolises to the rest of the world, the relatively few capitalists above exercise monopoly power over the many below. (A.G.Frank, 1969: 31-32).

The model of urbanization of the capitalist societies is generally disorderly, unplanned and uncontrolled. As the economic and social development is not a planned one in these countries, there is no possibility of handling the rural-urban migration within the context of a development plan. As a result, one or several cities in the country overdevelop against the rest. Consequently, countries such as Great Britain and France were obliged to take measures to limit the growth of London and Paris. The concept of "the optimal city size" was invented by the planners and economists of these countries as a solution to the overgrowth of cities (H.W.Richardson, 1973; E.V.Böventer, 1973: 145-151).

In market economies, state intervention in urban life takes place in the reorganization of the central business districts, in the relocation of low income groups to higher standard residential areas, in the prevention of social conflicts stemming from the ethnic segregation and in the ensuring of the low-income citizens to benefit equally not only from health, education, housing and similar public services but also from the right of participating in the urban planning process (P. Davidoff, 1964; B.Frieden and R. Morris, 1968).

## Regional Inequalities

In the capitalist countries with no state intervention, not only the growth of cities but also the development of the backward regions is left to the free functioning of the market forces.  Regional disparities are considered an expected outcome of development process of these countries.  Because, as G.Myrdal has pointed out, the market mechanism causes most of the more productive forms of economic activity to group themselves in certain localities, thus establishing a system of haves and have-nots within a country. This tendency towards regional inequality is a self-reinforcing process and it is an increasing function of the poverty concerned. One would expect regional disparities to be much wider in poorer countries.  Myrdal suggests that this type of inequality will retard expansion on the national level (G. Myrdal, 1957: 39).  The existence of large cities stabilizes the inequalities among regions.  According to one student of urban affairs, "The big city monopolizes most of the country's activity and the pick of its resources and is therefore responsible for the backwardness of the other towns and of the rural areas which it exploits for its own benefits" (M. Santos, 1975: 351).

On the other hand Bela Balassa contrasted Myrdal's position with that held by the classical economists who felt that the free play of the market forces would eliminate inter-regional income differences (B. Balassa, 1965: 23).  Other writers sharing his views claim that the market economy itself is not the source of the inbalances between the various city sizes and regions (E. Mills, 1974).

Capitalist societies are mainly concerned with the increase of real income rather than its distribution among social classes and geographical regions.  Therefore, the income distribution in urban centers seems to be moving towards a state of greater inequality and greater injustice.  Unless this trend can be reversed, an intensifying conflict within the urban system may become unavoidable. Both in the U.S. and in Great Britain, there is enough evidence to indicate that an open conflict is beginning (T.L.Blair, 1973; R.Goodman, 1972).  Therefore, one can conclude that it will be disastrous for the future of the social system to plan ahead simply to facilitate existing trends (D. Harvey, 1973: 93-94).

## Housing Conditions

To improve housing conditions of low income families is a major problem in the cities of capitalist countries.  Because housing is viewed as a consumption good and its price is fixed according to the market forces.  Most of the investment in housing in these countries is made by the private sector.  Housing cooperatives and the public sector exceptionally assume any responsibility for providing dwellings for the poor. As it was observed by F. Engels, a hundred years ago, "as long as the capitalist mode of production continues to exist it is folly to hope for an isolated solution of the housing question." (F. Engels, 1872: 74-77). The experience gained from implementing urban housing policies in contemporary American cities indicates some similarities to Engels' account, and it is difficult to avoid concluding that the inherent contradiction in the capitalist market mechanism contributes to it. (D. Harvey, 1973: 143). Devices like urban renewal and slum clearance are not suitable to solve the main problem which is consisted of wiping out the poverty altogether, by reducing the income differences among social classes, but not of moving it around. (Castells, 1975: 366).

## Urban Land Speculation

In capitalist societies, because of private land ownership, landowners and consequently high income groups are effective in urban development and planning. So much that in this system, some economists encourage the high increases in land prices as a result of speculation. Because they claim that the inflated land prices create an income increase from which ultimately the members of the whole society benefit (W. Lean, 1969: 77; G. Vickers, 1968: 2-10). They underestimate the effects of speculation such as distorting the income distribution, channeling the scarce resources into nonproductive sectors and preventing the planned development of cities. As Harvey has pointed out, they ignore the fact that "the maximization of exchange values by diverse actors produces disproportionate benefits to some groups and diminishes the opportunities for others" (D. Harvey, 1973: 175).

In fact, the diversity of actors operating in a land-use system and the monopolistic nature of the absolute ownership rights render the micro-economic theories of urban land use entirely inadequate. It is so because of the imperfections of the market economy and the dysfunctional character of the land market. It is not without reason that a recent U.N. report described the rampant land speculation as "one of the most serious barriers to the rational development of urban areas" (U.N., 1975a: 234).

Although we tend to consider the system of providing the poor families with dwellings on a pure profit basis as one of the deplorable phenomena of the last century, private ownership in land is still accepted everywhere in non-socialist Europe. And even the democratic socialist political parties do not seem to be much interested in the question of private land ownership. They are quite willing to restrict the rights of the landowner when they consider it necessary for public purposes, but they do not make an important point of question of land ownership in principle. One can hardly decline to share the prediction that in order to solve the housing

crises of industrialized capitalist countries, private property in
land will gradually fade away, if not officially abolished in the
future (E.W.Hofstee, 1967: 73). As Kenneth Galbraith rightly noted:
"Like many others, I see the key to this order in the control of
land. I continue to be puzzled as to why, when socialists gather
to celebrate their faith, they give so little attention to the so-
cialisation of urban land. For no other form of property is the
case so strong. It is commending height on which, more than anything else, the quality of urban development depends" (J.K. Galbraith,
1974: 13).


                            Environment

     Industrial development, urbanization and the progress of tech-
nology create various kinds of externalities in the form of air,
water, land and noise pollution. These are the effects of production and consumption activities on other producers and consumers
without corresponding payments. The externalities are usually over-
taken by the society.

     It is interesting to see whether the pace of environmental de-
terioration has been historically more rapid in the capitalist economies. One view is that differences exist between countries of
different levels of development, with different population densities,
different degrees of urbanization, different geographical and climatic conditions, but there is no evidence whatsoever that the economic systems, the political organization and the ideology of the
governments have been of any importance in this connection. This
is based on the observation that the state-owned companies have been
no different from the private ones in the extent to which they have
damaged the environment (E. Dahmén, 1971: 44-45). In addition, it
is asserted that a comparatively low value has been placed on environmental conditions in both economic systems as compared with
other sectors of the economy on the ground of economic efficiency.
(W. Beckerman, 1972: 19).

     The other view is that the nature of environmental deteriora-
tion is different in the market and non-market economies. The major reason for this difference is that consumption is not as much
as an important drive behind the economic activities in the non-
market systems. Besides, the fact that the differences between economic systems regarding the cause of environmental pollution is not
so clear does not exclude the possibility that different systems
might have different means of pursuing a successful environmental
policy. It would be wrong to assume that the problem has nothing
to do with the kind of economic system, since only under the market
conditions polluters do not take into account the externalities that
they create for the society.

Finally, general economic philosophy of capitalism is not suitable to find solutions for many of the urban growth problems, because it does not allow to plan for the future and because, as it was underlined in the Cocoyoc Declaration, the classical market put the resources at the disposal of those who are able to buy them rather than those who need them (Cocoyoc Declaration, 1975). Antiplanning together with the lack of control over the urban land market plus a _laissez-faire_ attitude towards regional inequalities make again the capitalist system inadequately equipped for urban problems. The word "profitopolis" put forward by the German architect Lehmbrock is nothing but a symbol for a model of a typical capitalist city which maximizes the interests of the individuals as against the public interest, and which alienates the human beings from their environment. As such, profitopolis identifies itself with the great metropolises of the capitalist countries. To conclude, one may refer, as a witness, to a talented economist of a capitalist country:

"There is first the fact that capitalism is seriously incompetent in providing the things and services that cities most require. And it is quite competent in providing those things that cause problems for the city. Capitalism has never in any country been efficient for building houses, which it seems necessary to stress, are a rather important urban artifact. Nor does it provide good health services ... Nor does capitalism provide efficient transportation for people - something that also makes vital the life of the metropolis ... In Western Europe and Japan ... these failures of capitalism are at least partly accepted, though they are still regarded as adventitious rather than inherent. Perhaps, the most urgent need for improved urban existence is to accept that the modern city is, by its nature, a socialist enterprise. Without a large area of successful socialism, urban life will be defective in the most immediate essentials: housing, health care and transportation" (J.K. Galbraith, 1974: 12).

Similarly, H. Stretton, in a recent book about the economic systems and environment, defines the disorder of the contemporary cities as a crisis of capitalism (H. Stretton, 1976).


THE RISE OF INTERVENTIONISM

Beginning from the years preceding the Second World War, the capitalist system was obliged to give up the strict _laissez-faire_, attitude so far as the problems of urbanization are concerned, even though it was contrary to its ideology. In these countries, social reform laws were passed to prevent the pressure of land speculation on low-income families, to influence the location of new towns and industries and to safeguard the tenants from the effects of chronic inflation during political and economic crises. To try to find

solutions to the housing problem of the poor has become one of the
important duties of the welfare states (P. Wendt, 1962; A.A.Nevitt,
1967). Because, otherwise, as was pointed out by Karl Polanyi, "To
allow the market mechanism to be the sole director of the fate of
human beings and their environment ... would result in the demoli-
tion of society" (K. Polanyi, 1944).

The solution of the problems stemming from the capitalist system can be eased by trying to provide territorial social justice.
To reach this aim, in these systems, the state interferes in the
social and economic life in order to improve the income distribution,
to remove disparities within the cities, between the cities of various sizes, between urban and rural areas and to charge the ones who
cause externalities. There are significant achievements in this
respect, in social democracies where capitalism is rationalized as
neo-capitalism and where the principle of social justice is given
priority in urban life. For example, in some of these countries,
the share of public ownership in urban land and the shares of the
non-profit housing organizations such as cooperatives in the volume
of house building is very high as compared with other capitalist
countries (K. Aström, 1967; E.W. Hofstee, 1967: 73-74). The city
of Amsterdam owns two thirds of the land within its boundaries and
the State of Israel owns 90% of all the national land. The rational
urban land policies of the Scandinavian countries are well known.
Similarly, the U.S., Great Britain and France were obliged to take
measures to make up the relative backwardness and stagnation of
regions such as Appalachia and the areas outside London and Paris.

The concepts such as new and satellite towns and the optimum
city size are the tools used by the planners of the capitalist countries to create a more rational settlement pattern. But it should
be pointed out immediately that the capitalist countries are searching for solutions both for their housing and unbalanced settlement
patterns within the framework of micro-analyses of the liberal system
(W. Thompson, 1965; H.O. Nourse, 1968; H. Siebert, 1969 and D. Netzer, 1970). Utopian socialists like Owen, Proudhon and Fourier and
some reformists like E. Howard have had much influence on the new
policies followed up by the neo-capitalist countries (M. Bookchin,
1974: 112-141; F. Choay, 1965). It can be argued that the rise of
neo-capitalism has been parallel to the birth of the City Planning
movement.

To summarize, one could say that even the capitalist world has
observed that pure capitalism is incapable of meeting the needs of
the contemporary cities and that "the continued pursuit of economic
growth by Western societies is more likely on balance to reduce
rather than increase social welfare" (A.J. Mishan, 1967: 219).
Therefore, they have begun to develop new definitions of economic
policy and new concepts of ownership (D.P. Denham, 1967: 89-101;
W. Kapp, 1971: 228-262 and J.P. Gillie, 1975). All these reformu-

lations are based in the refusal of the principle regarding each
man as the best judge of his own welfare disregarding what he does
for the society.


## THE THIRD WORLD

The rules of urban planning and the measures of social welfare
are implemented as they should be in the industrialized capitalist
countries, regardless of the social classes they actually prefer to
favor.  It is only the underdeveloped capitalist countries, in other
words, the pre-capitalist countries that face everyday the negative
effects of land speculation and of business-like functioning of the
real estate market as well as the non-implementation of master plans
due to individual, class and party interests.

There is no correlation whatsoever between the pseudo-urbanization of these countries and the increase in their industrial population.  Some of the urban sociologists explain the existence of
the "parasitic cities" in the Third World by the dependent nature
of the urbanization process in these societies.  In fact, in the
pre-capitalist Third World countries, the over-growth of one or two
metropolises has important inconveniences both for fulfilling their
own functions and for causing the continuation of regional disparities (M. Castells, 1975: 62-88).

Observations on the cities of the Third World countries reveal
that pseudo-urbanization transforms rural poverty into urban poverty.
And it does not play a role of modernizing these countries.  One
such study made on South Asian cities demonstrated that "what appear
to be workers in secondary and tertiary sectors are in fact urban
peasants performing manual or very low skilled tasks with no other
assets than the strength of their backs or arms" (M.A. Qadeer, 1974:
274).  It also concluded that Indian and Pakistani cities can be
described as "enlarged villages" (279).  There are numerous other
examples from Arabic countries supporting the same observations
(S.E.M.Ibrahim, 1975).  This kind of urbanization that retards the
integration of the country may finally result in social and political upheavals.

Pre-capitalist countries are dissimilar to the capitalist ones
not only with regard to rapid urbanization, regional disparities and
land speculation but also with respect to environmental pollution.
The pollution of the developed countries which stems from the increase of production and consumption is called "pollution of affluence" whereas the pollution in the less developed countries is
caused by hunger, poverty, unemployment, unsatisfactory health conditions and the like, and it is called "pollution of poverty" (de
Araujo Castro, 1972a: 408).  In other words, pollution number one
in these countries is neither air pollution nor water and noise

pollution but the underdevelopment itself (de Araujo Castro, 1972b).

There is no doubt that for the large masses in the metropolitan centers of the Third World countries, to clean the human environment from several cancers like urban poverty, ignorance, hunger, disease and crime is as important as to control the pollution of the physical environment. The people of these countries cannot be satisfied to go on living in endless poverty even though they are provided with a clean environment not polluted by the effects of urbanization and industrialization. They, too, should have the privilege of having enough industries to pollute the environment. That is why the "zero growth" proposal of the M.I.T. experts' report prepared upon the request of the Club of Rome was not acceptable to the poor nations (D.H. Meadoes, 1972).

It is a well accepted method in the capitalist countries to charge the ones causing the pollution by the cost of solving the problems of environment. But it is expected in the countries with unbalanced income distribution that the cost which is charged will be an additional burden for the poorer families as it is reflected on the prices.

In regard to housing, social inequalities in developing nations inhibit the provision of standard housing for the whole population while simultaneously fostering the land consumptive patterns of the middle and upper income classes. In this case, the price system does not recognize that people do not start from the same point and that the existing distribution of wealth determines who gets what (U.N., 1975b: 95). Since a capitalist system of ownership of urban land culminates in an urban structure which is dysfunctional in terms of meeting the collective requirements of an urbanizing society, developing nations should not pattern their set of ownership rights directly after those of the already developed nations. As was suggested by an Expert Group of the U.N., a concept of ownership which recognizes the role of land as a natural resource must be accepted. Such a role "would function in the context of land use policies that relate to the social functions of land and the obligation of the individual landowners to the society" (U.N., 1975b: 116). That would put an end to the speculation in urban land that renders the solution of the housing question in the Third World countries almost impossible.


## URBAN SYSTEMS IN SOCIALIST ECONOMIES

Socialist theory assumes that the relation between the society and the space or urbanization is a function of a certain production mode that evolves during the course of history. The fundamental principle of this system is the elimination of the private ownership of the production units and the relative balance in the distribution of wealth and income (G. Grossman, 1967).

## City Structure

It is difficult to distinguish socialist cities from the ones of the capitalist countries simply by using their physical appearance as a criterion. Because, the differences between the visible features of their internal structure are not quite clear. They rather differ in the principles of the planned urban and regional development that are rigidly implemented in the former. There are, however, no contradictions among various districts of the socialist cities. For the social structure of these classless societies is highly homogenous. Residential areas are of such standards that every family can easily meet its immediate needs for housing and urban public services at least at an adequate minimum level. Luxurious districts in sharp contrast with the slum-like residences of the working classes and a "reserved workers army" peculiar to large cities of the capitalist world, are foreign to the cities of socialist countries.

## Balanced Regional Development

To handle the agriculture and industry as a whole, to decentralize both the population and manufacturing industries over the whole country in a balanced way are among the major principles of the Communist Manifesto (L.S. Feuer, 1957: 27-38). Marx and Engels had given a clearer picture of rural-urban contradictions in their German Ideology and they had pointed out that division of labor separates the agricultural and non-agricultural ways of production in any country, thus giving rise to the antagonisms between towns and villages (K. Marx and F. Engels, 1879: 26-27). In order to justify the necessity of eliminating the antagonisms between town and country, the classics of Marxism-Leninism suggest the harmonious development of productive forces according to a single overall development plan, a greater equalization of the distribution of large-scale industry and population over the country and overcoming the excessive concentration of population in large cities. (D.G. Khodzhaev and B.S. Khorev, 1973: 44).

The roots of Owen's ideal town and Fourier's phalanstère can be found in these reformist ideas that favored a balanced economic and social decentralization. Both the schemes of these utopian socialists and the garden-city invention of Ebenezer Howard, the founding father of the modern town planning, aimed to unite the advantages of towns and villages in a period where town planning did not exist.

As an application of these thoughts, in the USSR, in the East

European countries and in China, large amount of non-utilized row
lands were transformed into big industrial centers to settle the
rapidly growing working classes (A. Kopp, 1975; N.V. Baranov, 1967;
J.V. Lewis, 1971: 13). The accomplishments of socialist countries
with respect to reducing regional disparities were analyzed by several economists (Koropecky, 1972: 83-84).

## URBAN LAND IN THE SERVICE OF THE SOCIETY

In socialist countries, as a rule, the private ownership of
land is abolished and the nationalized land rent is simply used for
the purposes of social welfare. This socialist principle provides
a means for the State and for the planners to ensure the development
of cities in a planned way. The land rent that is a product of the
private ownership not only has negative effects on the agricultural
production and the pattern of urbanization, but at the same time it
paralyzes qualitatively the development process and creates suburbs
that are neither villages nor cities. Similarly, slums and metropolises as centers of widening contradictions are the outcome of the
deviations from the principle of public land ownership. In capitalist societies, as a result of the integration of pre-capitalist
social relations with capitalism, real estate speculation widespreads
on one hand, and the rural-urban antagonisms are sharpened on the
other (H. Lefebvre, 1972: 149-153). Even the Western scholars agree
that by abolishing the private ownership rights in land, the Soviets
have realized unquestionable accomplishments (B.J.L. Berry, 1974:
72).

Because of the injustices created by the individual ownership
rights that in recent years there has been a growing concern that
the concept of ownership should emphasize the role of land as a
natural resource and that its use should benefit the whole of society (U.N., 1975b: 98). On the other hand, several countries, like
Cuba, went further and have made an effort to change entirely their
traditional land ownership patterns. For example, the National Land
Reform Law in Cuba has tried to a) equalize the urban society, b)
to prevent the evil aspects of the social forces which guide urbanization, c) to suppress speculation and d) to minimize the present
unjust social differences (U.N., 1975b: 104). Bolivia and Brazil
have taken similar measures.

### Environmental Deterioration

As to the approach of the Marxist system towards environment,
Lenin noted that the creation of great capitalist production and the
capitalist competition accompany the "dilapidation of the productive
forces of the world" (Lenin, 353). According to this view, history
shows that the capitalist mode of production and the congenital

sicknesses of the capitalist society cause intensifying of the negative influences over the natural environment and prevent their
elimination. On the other hand, new attitude towards the environment which is the socialist mode of production makes is possible to
multiply in a systematic way richnesses of the globe (I. Guérassimov,
1975: 30-31).

In other words, it is not the industrialization itself that de-
teriorates environment. Because there is no notion of abstract vice
of "modern, industrial or mechanical development". On the contrary,
the development of industries and machines can only contribute to
free men. However, under the capitalist system, it puts men under
pressure and it fabricates the unemployed instead of reducing working
hours (Biolat, 1973: 118). In the capitalism, the prime objective
of the power is to increase the profit rate of the private monopolies
which is openly in contradiction with a policy of environment in the
service of the community (G. Biolat, 1973: 128).


THE CASE OF TURKEY

Turkey is one of the most rapidly urbanizing countries with her
average rate of urbanization of 7%, although she cannot be regarded
as an urbanized one. The degree of urbanization as measured in terms
of the share of total population living in cities is only 37.3 (x)
in 1975. This percentage was 18.4 in 1945, 22.5 in 1955 and 29.8
in 1965. The number of cities having 100,000 and more inhabitants
increased from 4 in 1945, to 5 in 1955, to 14 in 1965 and to 23 in
1975. The share of population of these large cities in the urban
population has risen from 40.3% in 1945, to 50.7% in 1965 and to
59.5% in 1975. It is estimated that it will reach 65.2% in one decade.

In other words, during the last quarter of century, the annual
average increase of the total population has been 2.7%,while it was
1.6% for rural population, 7-8% for urban population and 10% for
the cities over 100,000 inhabitants. As an example, the capital
city of Ankara has become a metropolis of 2 million in 50 years
when it was a small town of 70 thousand at the day where it was proclaimed as the capital of the new State. Its population doubled in
every ten years.

This rapid urbanization was made possible especially since the
early 50's by the changes in the structure of the economy, partly
because of the earlier efforts to industrialize, partly because of
the introduction of a large scale road system all over the country
and certainly by the mechanization of agriculture. However it is
generally accepted that, on the balance, rural-urban migration in
Turkey is happening under the push factors prevailing in the agricultural sector, rather than the pull factors of the urban industries.

The rate of urbanization not only differs among the city (x).
Localities having 10,000 and more population are regarded as urban
size groups, but at the same time among the geographical regions.
The concentration of the largest cities in the Western Anatolia creates an unbalanced urban system.  Not only in terms of urban population, but also in terms of all indicators of socio-economic progress, the West of the country is a well developed region as compared with the East and Southeastern regions with their semi-feudal
socio-economic structures.

It is almost a common belief among the scholars today that most
of the rural migrants are absorbed by the informal or marginal sector and not by manufacturing industries.  And thus, the process of
urbanization carries rural underemployment into urban centers creating a reserved army of unemployed.

Each district in the large metropolises possesses its own mix
of pre-industrial, transitional and industrial characteristics.
There are obvious contrasts between the ways of living, world views
and incomes of the families living in different parts of the rapidly
growing urban centers.  It is interesting to note that one fourth
of the urban population lives in the gecekondus, while this figure
is as high as 65% for Ankara, 45% for Istanbul and 35% for Izmir.
Speculation in urban land within the centers and in outlying areas,
urban transportation, water supply, electricity, pollution of the
environment are the major bottleneck areas.  One cannot talk about
one single urban poverty but large segments of the cities that are
in sharp contrast with small islands of welfare.

Until 1950 there were no conscious efforts to reduce disparities
between regions and between rural and urban areas and to ensure a
balanced settlement pattern, but the State had established some fac-
tories in small towns and cities along the railways.  From 1950 to
1963, the underdeveloped Eastern regions did not get any special attention, although a great many investments were made in settlements
other than big cities, socio-economic disparities still favored the
West.  The first two of the Five Year Development Plans covering the
period of 1963-1973 emphasized on the growth center concept and on
the development of backward regions without much improvement. However, the last Development Plan (1973-1977) committed to rapid industrialization did not favor any sacrifice from the economic growth
targets in favor of accelerated growth of backward regions.  Neither
in the pre-plan period nor during the planned development period,
conscious policies have been implemented to reduce the size of rapidly growing metropolises and to provide a balanced urban hierarchy.

On the other hand, private initiative in the housing sector,
lack of an efficient mass transportation, inadequacies of local revenues, rapid inflation of land prices due to uncontrolled speculation, haphazard and unplanned developments in the outskirts of the

municipalities and non-conservation of natural beauties seem all to be the outcome of capitalist, in other words, pre-capitalist ways of production and consumption. The economic system which is adopted in this country and it is called "mixed economy" does not either have the rigidity of socialism or the rationalism of developed capitalism to bring some solutions to the problems of urban poverty, the disorderly urban development, regional imbalances and the housing crises.

## CONCLUDING REMARKS

Since the beginning of the Twentieth Century, the problems of urban growth required the intervention of the states in social and economic life. Land speculation, housing crises, unplanned growth of metropolitan regions, sanitation, transportation and the quality of urban environment have become major concerns not only of the policy makers but also of all people both in developed and in developing societies. City planning movement was born as a result of the neo-capitalism which was largely strengthened by the World Economic Crisis of the late 20's. For more than fifty years, the techniques of urban planning have really helped many nations to overcome the difficulties they faced during their process of urbanization. Developing countries of Asia, Africa and Latin America have tried to import these planning techniques from the highly industrialized capitalist countries of the West. It was assumed that the stock of experience of capitalist countries would protect the newly emerging nations of the Third World from committing the mistakes of the past.

However, the failure of the so-called planning efforts in most of the less developed countries suggests that the major responsibility for urban crisis is not the inadequacy of the planning approaches, but the socio-economic systems themselves. For instance, Turkey, who has been trying to solve her urban growth problems by a pre-capitalistic economic system with negligible public intervention since the early 20's, is quite far from being successful in this respect. In fact, the socialist system has several obvious advantages over the capitalist, especially as far as the impact of public land ownership upon urban development is concerned. The Third World countries cannot shape their settlement patterns according to the principles of pure capitalism. Because they cannot afford the social injustices created by the price mechanism and individual profit maximization, disregarding the long-term interests of the society. At least with respect to meeting the essential needs of rapidly growing urban centers of the Third World countries, the principles of the socialist system seem highly promising.

REFERENCES

Aström, K. (1968), City Planning in Sweden, The Swedish Institute, Stockholm.

Balassa, Bela (1965), Economic Development and Integration, Centro de Estudios Monetarios Lateno Americanas, Mexico.

Baranov, N.V. (1967), "Building New Towns", U.N., Planning of Metropolitan Areas and New Towns, 209-215.

Beckerman, Wilfred (1972), "Environmental Policy Issues: Real and Fictitious", O.E.C.D., Problems of Environmental Economics, Paris, 19-38.

Berry, Brian J.L. (1974), "Comparative Urbanisation Strategies," Harry Swain and Ross D. MacKinnon (eds.), Issues in the Management of Urban Systems, Schloss Laxenburg, 66-79.

Biolat, Guy (1973), Marxisme et Environment, Editions Sociales, Paris.

Blair, Thomas L. (1973), The Poverty of Planning: Crisis in the Urban Environment, MacDonald, London.

Bookchin, Murray (1974), The Limits of the City, Harper, Colophon Books, New York.

Castells, Manuel (1975), La Question Urbaine, Maspero, Paris.

Choay, Francoise (1965), L'Urbanisme: Utopies et Réalités, Ed. du Seuil, Paris.

Chapin, Stuart F. Jr. (1965), Urban Land Use Planning (2nd. edition), University of Illinois Press, Urbana.

Cocoyoc Declaration, (1975), International Organization, Vol. 29, No. 3, 893-902.

de Araujo Castro, J. Augusto (1972a), "Environment and Development: The Case of the Less Developed Countries," Internation Organization, Vol. 26, No. 2, 401-416.

de Araujo CASTRO, J. Augusto (1972b), "Pollution Problem No.1: Underdevelopment", Unesco Courier, Jan., 20-23.

Dahmen, Erik (1971), "Environmental Control and Economic Systems," Peter Böhm and Allen V. Kneese (eds.), The Economics of Environment, MacMillan, New York, 44-52.

Davidoff, Paul (1964), "Advocacy and Pluralism in Planning," Journal of the American Institute of Planners, Vol. 31, No. 4.

Denman, D.R. (1967), "Towards a Modern Theory of Property", Howick and Others (eds.), Land and People, Leonard Hill, London, 89-100.

Engels, Fredrich (1872), The Housing Question (1935 edition), New York.

Echols, John M. (1975), "Politics, Budgets, and Regional Equality in Communist and Capitalist Systems", Comparative Political Studies, Vol. 8, No. 3, Oct., 259-292.

Feuer, Lewis S. (ed.) (1959), Marx and Engels: Basic Writings on Politics and Philosophy, Doubleday, New York.

Frank, André Gunder (1969), Capitalism and Underdevelopment in Latin America, Modern Reader, New York.

Frieden, Bernard and Morris, Robert (eds.)(1968), Urban Planning and Social Policy, Basic Books, New York.

Galbraith, John Kenneth (197L), "Modern City or History as the Future", Journal of the Royal Institute of British Architects, Oct., 9-13.

Galtung, Hohan (1971), "A Structural Theory of Imperialism", The Journal of Peace Research, Vol. 8, No. 2, 81-117.

Gillie, J.P. (1975), Redéfinir le Droit de Propriété, Centre de Recherche d'Urbanisme, Paris.

Goodman, Robert (1972), After the Planners, Penguin, London.

Gross, Bertram (1966), "The State of Nation: Social Systems Accounting", Raymond Bauer (ed.), Social Indicators, The M.I.T. Press, Cambridge, 154-271.

Grossman, Gregory (1967), Economic Systems, Princeton University Press, New Jersey.

Guerassimov, I., (ed.) (1975), Homme, Société et Environnement, Académie des Sciences de l'URSS, Institut de Géographie, Moscou.

Harvey, David (1973), Social Justice and the City, Edward Arnold, London.

Hofstee, E.W. (1967), "Land Ownership in Densely Populated and Industrialized Countries," Howick and Others (eds.), Land and People, Leonard Hill, London, 55-74.

Ibrahim, Saad E.M. (1973), "Over-Urbanization and Under-Urbanism: The Case of the Arab World," International Journal of the

_Middle Eastern Studies_, Vol. 6, 29-45.

Kapp, William (1971), _The Social Costs of Private Enterprise_, Shocken Books, New York.

Khodzhaev, David G. and Khorev, Boris S. (1973), "The Concept of Unified Settlement System and the Planned Control of the Growth of Towns in the USSR", _Geographica Polonica_, Vol. 27, 43-51.

Kopp, Anatole (1975), _Changer la Vie, Changer la Ville_, Union Générale d'Editions, 10-18, Paris.

Koropecky, I.S. (1972), "Equalization of Regional Development in Socialist Countries: An Empirical Study", _Economic Development and Cultural Change_, Vol. 21, No. 1, Oct., 68-86.

Lean, William (1969), _Economics of Land Use Planning_, The Estates Gazette, London.

Lefebvre, Henri (1972), _La Pensée Marxiste et la Ville_, Casterman, Paris.

Lenin, V., _Oevres_, Tome: 6, Editions du Progrés, Moscou.

Lewis, John Wilson (ed.) (1971), _The City in Communist China_, Stanford University Press, Stanford.

Marx, Karl and Engels, Fredrich (1970), _The German Ideology_, International Publishers Edition, New York.

Meadows, Donella H and Others, (1972), _The Limits to Growth_, A Patomac Associates Book, Pan Books Ltd., London.

Mills, Edwin (1974), "Do Market Economy Distort City Sizes?", Harry S. Swain and Ross MacKinnon (eds.), _Issues in the Management of Urban Systems_, International Institute of Applied Systems Analysis, Schloss Laxenburg, 80-99.

Mishan, A.J. (1967), _The Costs of Economic Growth_, Penguin, London.

Myrdal, Gunnar (1957), _Rich Lands and Poor_, Harper, New York.

Netzer, Dick (1970), _Economics and Urban Problems_, Basic Books, New York.

Nevitt, Adam Adela (ed.) (1967), _The Economic Problems of Housing_, MacMillan, London.

Nourse, Hugh O. (1968), _Regional Economics_, McGraw-Hill, New York.

Polanyi, Karl (1944) The Great Transformation, Boston.

Qadeer, M.A. (1974), "Do Cities Modernize the Developing Countries? An Examination of the South Asian Experience", Comparative Studies in Society and History, Vol. 16, No. 3, June, 266-283.

Richardson, Harry W. (1973), The Economics of Urban Size, Saxson House, Lexington.

Santos, Milton (1975), "Space and Domination: A Marxist Approach", International Social Science Journal, Vol. 26, No. 2, 346-363.

Siebert, Hans (1969), Regional Economic Growth: Theory and Policy, International Textbook Co., New York.

Sjoberg, Gideon (1965), "Cities in Developing and in Industrial Societies: A Cross-Cultural Analysis", Philip Hauser and Leo F. Schnore (eds.), The Study of Urbanization, John Wiley, New York, 213-263.

Slater, David (1975), "Capitalisme Sous-Développé et Amémagement de l'Espace: Le Pérou", Revue Tiers-Monde, Tome: 16, No. 64, (oct.-Dec.), 707-734.

Stretton, Hugh (1976), Capitalism, Socialism and the Environment, Cambridge University Press, London.

Thompson, Wilbur (1965), A Preface to Urban Economics, The John Hopkin Press, Baltimore.

U.N., (1975a), Report on the World Social Situation, 1974, New York.

U.N., (1975b), Urban Land Policies and Land Use Control Measures, Vol. VII, Global Review, New York.

Vickers, Geoffrey (1968), "The Uses of Speculation", Journal of the American Institute of Planners, Vol. 34, No. 1, (Jan.), 2-10.

Von Böventer, E., (1973), "City Size Systems: Theoretical Issues, Empirical Regularities and Planning Guides," Urban Studies, Vol. 10, No. 2, (June), 145-162.

Wendt, Paul (1962), Housing Policy: The Search for Solutions, University of California Press, Berkeley.

# Session II

# METHODOLOGY: PREDICTION

# SOME FUTURE USES OF ANALYTIC FORECASTING

E. Solem

Department of National Defence

Ottawa, Ontario, Canada

## INTRODUCTION

As yet we do not possess an adequate, in the sense of fully comprehensive statement of Futures problems, let alone ways of resolving them. However, recent progress in different fields of forecasting and analysis suggests that we are now in a position to attempt this. The present paper seeks to identify the remaining problems facing futures research while suggesting possible ways of solving these.

## FUTURES PROBLEMS

The relationships between social, economic and technical actions and their direct and indirect impacts on the environment, as well as the response of social systems to these impacts will take an increasing importance in the years to come. What is needed, first and foremost, is the ability to <u>confirm</u> knowledge of these relationships, and to <u>monitor</u> information and knowledge in such a manner as to obtain the maximum influence on such action in due time prior to the impacts, if necessary to avert those which are dysfunctional in nature to the systems within which they exist.

So far, there are only limited sections of human and social interactions where short or middle-range <u>predictive</u> forecasting can be made in some respect with scientific justification. By scientific justification I mean an explicit demonstration of the <u>premises</u> (suppositions, facts, data etc.) and the <u>methods</u> used in arriving at the given prediction, preferably also an assessment of its <u>probability</u> of becoming true. The typical example of predictive forecas-

ting of this kind may be the modern meteorological prognosis which
does not predict simply rain for Sunday, but specifies "75 percent
chance of rain on Sunday" -- a much more useful statement for many
practical purposes.  Often they are accompanied by an overview, or
a map of already existing weather conditions.  Such a map would include not only data about the existing situation, it would also demonstrate the "fronts" and "flows" which may appear and thus tend
to determine tomorrow's weather.

Something comparable to this sort of predictive forecasting has
become available to us within the realm of social sciences, but
merely in those few branches of human and societal activities where
present resources, inputs and productive capacities are known, and
limit to a considerable extent the range of possible future developments within determined time-spans, and where major trends can be
fairly well mapped out and quantified.  The rather limited and moderate success of predictive forecasting would hardly have evoked
the great interest of policy makers (national and international alike)
in contemporary methods and practices of scientific studies of the
future,  were it not for the fact that the social sciences (in particular) have developed during most recent years a wide array of
promising and partly already successfully realized approaches to the
kind of analytic forecasting which seems -- at least at present --
much more closely related and more relevant to the problems which
are faced by policy makers in their day-to-day activities.

Analytic forecasting -- as distinct from predictive forecasting -- is not directly concerned with what will happen tomorrow or
at any specified date.  Its main concern is to survey as systematically as possible, and as completely as can be done, what chances
for developments, and what options for action are open at present,
and then to follow up analytically to what alternative future outcomes these developments and actions, if realized, might lead.  As
opposed to predictive forecasting, analytic forecasting is often
coached in the form of disjunctive hypothetical inferences: "Given
A  or B or C; if A then E and F.  If B then G and H.  If C then
again G but K instead of H..." etc.  This approach to problem solving is in fact often instinctively followed by the reasoning or
the intuition of the policy-maker, at least in part  in many cases.
The policy-maker is normally trying to assess his options (such as
through, e.g. technology assessment procedures), and the (natural)
consequences of his own (or his adversaries') possible action (e.g.
technology forecasting and planning) in just such a way.  I will
return to these methods and techniques later on.  Their applicability within the context of analytic forecasting is directly related
to the degree to which analytic forecasting itself lends itself to
futures problems, given the correctness of the above argument.

As is known, the major problem of futures work is the element
of uncertainty, which is due to the time dimension, since decisions

and their consequences are being displaced over time.  The problem
is compounded by the element of complexity, which by itself often
characterizes the decision-making process.  Examples would be difficulties of exchange of knowledge, storage and retrieval of large
amounts of information in connection with the behaviour the planning
unit or mechanism itself.  Last, but not least important is the element of measurement and values.  The former plays of necessity a
central role within the context of long-term planning and forecasting, and may occur due to conflict of measurement standards and
techniques within individual sub-groups or organs in the planning
mechanism.  The latter may of course occur due to conflict of values
both between such organs, groups or individuals involved.

     Part of this is primarily technical in nature, and can be sol-
ved as such, at least as long as the technical component in the particular problem is recognized as such, isolated if necessary, and
attacked accordingly.  There are several methods and techniques
available for this, as I shall show.  That part of the problem which
is not technical, hence not subject to being solved,must of necessity
be attacked differently.  Here the question is one of devising methods of resolution which will, temporarily at any rate, settle the
problem.  The case then becomes one of designing, concurrently, approaches, methods of techniques to deal with- and accommodate- conflict with the aim of its resolution.

     Over and above this, a major problem in all futures work is,
in short, that its basis has never, so it seems, been properly articulated and defined.  The result has been somewhat of a proliferation of methodologies, the relationships among which seem far from
clear, as well as the tendency for one particular method to be cham-
pioned by an exponent, a consulting company or a group of experts.
Consequently, when an organization, whether an agency, organization,
or government department gets involved in futures work, the emphasis
has often tended to fall on one or more particular methodologies,
often determined by easy access and availability, at the expense of
a more comprehensive, hence possibly more conceptually 'mature' overview such as, for example analytic forecasting.

     It might be useful to make one more distinction at this stage,
that between well-structured and ill-structured problems.  Problems
of the past and present, as well as those which can be handled by
the methods or normal science fall into the former category, that of
well-structured problems.  Social science problems, on the other
hand, are often characterized by many variables, interactions and
qualitative and/or holistic aspects.  Futures problems share this
characteristic, with the added complication that the number of possibilities increase very substantially depending upon the extension
of the time-frame involved.  The latter two categories constitute
ill-structured problems, and must be attacked as such.  Whereas well-
structured problems are often almost completely defineable, being

either posed in this form as in theoretical treatments or through agreements by participants/observers, the case of ill-structured problems is very different indeed.  The most important observation to be made here is that the definition of the problem is often different from one individual researcher to the next, and that hence there may often be no commonly agreed set of assumptions.  Consequently, this is the area in which perception is often as important as, if in fact not more important than, reality itself.  I may be belaboring the obvious at this point, but it is very important to bear these distinctions in mind.  Pretending to define the future explicitly is to exclude its very central element and chief characteristic -- its basic unknowability.

It may perhaps be tempting to despair at this point.  However, for several reasons which I hope to demonstrate, this will not be necessary.  Suffice to say, at this stage, that well-structured and ill-structured problems must be defined as such and treated differently, and that governments, say, understand them fully, so that in the future the 'success' of political progress is not predicted in terms of short term cost-benefits, which often may have more or less disastrous consequences.


## METHODS AND TECHNIQUES

Faced with complex situations in which ill-structured problems will occur with increasing frequency, prime value will have to be put on the ability to define, measure and communicate, in short to somehow incorporate the learning process itself.

Although human, social, technical and political affairs do not by themselves lead to an experimental approach which will guarantee deeper insight and a better mastery of our social, technical and political environment, some of the social sciences have been able to produce a certain number of relatively sophisticated approaches for the analysis of complex systems.  Trend analysis, linear projection, dynamic programming, operational research, model-building, gaming, human- and computor simulation, scenario-writing, brainstorming, the evaluation of alternative outcomes (alternative futures), methods of "prospective," i.e., the assessment of impacts, are all quasi-experimental methods which may be gathered and organized under the rubric of analytic forecasting.  As such they may be able to produce options for actions and alternative future outcomes of developments and actions.  Let us look at some of them a bit more closely.

The Delphi technique is so well known within futures work that it is at times considered synonymous with futures research itself. This is a fallacy which should be corrected.  Delphi is a versatile and generally applicable technique of consensus forecasting, developed by Olaf Helmer, Norman Dalkey, and their associates at the Rand

Corporation.  The technique attempts to improve the utilization of
experts in analysis, evaluation, and forecasting by using a format
other than the committee meeting.  As is known, many of the drawbacks
of committee operation are due to the interaction of the personali-
ties and psychological characteristics of the participants themselves.  This would seem to imply that perhaps, in certain cases, a better situation for the utilization of the experts would be a panel
meeting without the face-to-face confrontations which might occur
in the committee, but with adequate communication and feedback between the individuals involved.  It should be stressed that mere elimination of face-to-face contact alone is of course insufficient.

     Stated simply, the Delphi technique is a carefully designed se-
ries of individual interrogations, usually best conducted by questionnaires, interspersed with infcrmation and opinion feed-back.  It
lends itself best to such questions as: When do you think this invention (x) will become reality? or, Estimate the year by which the
growth rate of country x will exceed that of country y.  There is no
manner in which the degree of 'correctness' may be controlled for.
Several 'rounds' follow in which group consensus or dissensus are
considered.  The respondents ere asked to examine new estimates and
the arguments offered and to revise (if they so choose) their answers.

     The Delphi technique has evolved extensively in recent years
and is generally applicable to most situations to which quantitative
values may be assigned, whether these are dates, weightings, or sca-
lings.  Consequently it is applicable to nearly all areas of inquiry.
Recent developments associated with the Delphi technique have included the computor-assisted Delphi, which enables one to participate
over an undetermined period of time.  This particular development,
largely the result of work by Murray Turoff, then of the U.S. Office
of Emergency Preparedness is described in the literature of futures
work.

     As with any generally applicable technique, Delphi is subject
to certain abuses, such as poor construction, inadequate testing,
hasty and inadequate analysis, and the lack of proper analysis of
the basis for divergent opinion.  More importantly, perhaps, is the
criticism that the technique may, and often does, lead to 'forced'
consensus which may seriously miss out important elements of consequence to the issue being questioned.  Despite this, the Delphi technique, if conducted properly, can be a useful guide for certain processes and a helpful aid in decision-making which involves consensus at different levels.  The technique has been applied in a few
instances to resource allocation where the critical element has
been the resource to be allocated.

     Cross-Impact Analysis is a more sophisticated and complex ver-
sion of the Delphi technique.  Whereas the latter has the advantage

that it tends to treat each component of the analysis as an independent variable, the Cross Impact Analysis evaluates components not
only independently, but in relation to each other.  Factor X may
have relatively low probability of occurrence.  However, the probability of factor X may be increased if factor Y is realized.  A further advantage with both techniques is that they are not limited to
experts, and hence can serve as devices for exploring the different
outlooks of experts and non-experts alike, or of widely heterogeneous groups.

   Trend extrapolation is a widely used futures technique applied
to different types of forecasts.  In a sense it is the 'easiest'
technique insofar as it presupposes quite often a very large number
of variables.  This is also its greatest short-coming.  This method
of forecasting has its legitimate place in, for example, technology
assessment when used as a guide to the analysis of present trends,
as well as a clue to normative judgements or "guesses."  Most economic forecasting has depended heavily on trend extrapolation.

   Mathematical methods have been utilized for planning problems
and forecasting for some time, and have attracted considerable attention.  It might perhaps be useful to emphasize that these methods
have primarily been applied to more limited (sectorial) aspects of
a larger problem.  Generally speaking, the supply of analytical models and quantitative tools of the above type is not inadequate.
The shortcomings connected with them consist more in the fact that
good mathematical tools have so far not been utilized on several
important problems where this could be carried out; that models applied are often ill fitted to particular problems (e.g. ill-structured problems); or that the model in question only covers certain
aspects of the problem to the exclusion of other, possibly more important aspects.  Both operational and illustrative analytical models may be utilized successfully, pending clear specification of
the main components of the problem, and the relationships between
these.

   Since the level of consensus within macro-economic theory is
relatively higher than that of other areas of social policy and planning, it might be instructive briefly to consider the state of the
art in this area of forecasting.  Several structural and definitional
issues of economic theory have been 'settled' (at times neglected)
for some decades.  Following the war, planning agencies in both market economies and centrally planned economies have applied considerable effort to construct econometric planning models.  The mathematical structures of such planning models used in Eastern and Western Europe are strikingly similar.  However, the precision of this
type of forecasting technique has not improved substantially, and
there have been diminishing returns to complexity.  While the flexibility of the models has decreased, the effort required to manipulate them has increased.  In periods when economic changes have been

the results of sudden political decisions, current models have often
proved quite inadequate.  The records of judgemental forecasters
has, in these conditions, often proved superior to more formal
techniques.  In general, it may now be stated that long-term economic modelling has been less successful than several short-term efforts, and that it is becoming increasingly clear that there is a
limit to the accuracy of economic theory or modelling technique,
however much it is improved.  Furthermore, the growing integration
of the world economy has resulted in factors exogenous to national
economies taking on increasing importance.  We know now that much
of the error in short- and medium-term economic forecasts is due to
inaccurate assumptions concerning foreign economies and foreign policies.  Some recent efforts have been made to improve precision by
including these in the forecasting model.  Even if it could be made,
a detailed study of, say, the combined economies of Europe would be
of little use if major externalities are poorly understood and predicted.  Now, even if this is recognized, in practice it does not
automatically lead to even the common availability of satisfactorily
detailed data.

   Physical and demographic feedbacks, which are not normally included explicitly in the above-mentioned economic models, must in
fact be taken into account.  In demographic forecasting, for example,
some of the errors arise because demographers are unable to take proper account of economic factors, and of other factors which may af-
fact fertility rate.  Both demographic and economic forecasting have
a relatively long history, and, so it seems, similar limits to prediction or analysis.

   Morphological analysis is a technique of exhaustive search for
effects which has been relatively little used, except by its inven-
tor, Fritz Zwicky, the aeronautical engineer who has used it is systematically searching for aircraft propulsion systems.  The method
consists in asking a set of questions exhaustive of technological
development.  This gives a full range of all possible answers to
each question.  A hypothetical case might be the location of a community clinic, the questions being asked being 'What kinds of people
would it serve?', Where should it be located? What types of services
should be offered? etc.  Against each of these questions, an attempt
is made to provide the maximum conceivable answers.  The next step
is to proceed with a morphological analysis i.e. systematically selecting all the permuted combinations of the answers to each question.  Clearly, the possibilities become extremely large, but one
quickly realizes that several combinations are incompatible or un-
realizable.  The result is that the total number of alternative systems explored in depth diminishes rapidly, although it may remain
quite large.

   Decision- and relevance trees are related methods which attempt
to develop all possible options and alternatives with regard to, say,

a given technology.  The branches of the tree are critical decisions
which exclude other major options.  Robert Ayres, Joseph P. Martino
and Erich Jantsch have discussed these methods and their implementation.

Systems analysis is one of the more powerful tools available
for forecasting and analysis.  According to Alfred Blumstein, systems analysis focuses successively on:

1. A particular system, a collection of people, devices and
   procedures intended to perform some function.

2. The function of the system, e.g. the jobs which it is supposed to perform.

3. Measures of effectiveness by which may be measured or calculated how well alternative system designs perform the
   function.

4. Alternative system designs to be compared.

5. A mathematical model with which can be calculated the measures of effectiveness associated with each alternative system design.

This type of broad-sweeping conceptual approach may serve as a
unifying principle in any analysis. Its emphasis on mathematical
modelling, quantitative measures etc. precludes the adequate treatment of many factors central to comprehensive analytic forecasting.
This method is more appropriately used for sub-categories, and not
as an over-arching conceptual framework of analysis.

More advanced systems theorists today advocate the use of holistic forecasting models, which link together social, economic, technological and physical processes either in a single model or via
social 'decisions' entered systematically during the forecasting
procedure.  The Mesarovic-Pestel Strategy for Survival model is an
example of the inability to satisfactorily build and link together
a model with 100 000 equations, where the inability comes not from
limitations of computer capacity, but from organizational, conceptual and theoretical constraints.

A shortcoming with the method above is that while it may help
to improve understanding, it may not bring about any great improvement in predictive accuracy.  However, for the purpose of analytic
forecasting this point is not of paramount importance.

As is to be expected, an increasingly important area of forecasting is anticipation of surprises and discontinuities brought
about by, for example, the incompatibility of technical and socio-
political factors, or by the different characteristic time scales
of change precluding easy resolution.  In the physical sciences as

well, transition phenomena are the most difficult to forecast, and
complex social systems are even more difficult to analyse in these
terms.  Even understanding of the stability of highly linked, "ideal"
mathematical networks is at an elementary stage.  However, some broad
indication as to the relationship between social structures and their
level of stability may eventually be gained from such studies (e.g.
catastrophe theory).

     Scenarios are mainly attempts to systematically develop complex
statements of future world states.  The technique differs from that
of the Delphi, insofar as it is much less well-defined.  In fact,
it is more an approach or even a style of working with the future
than a technique as such.  In principle, a scenario is a piece of
future history – a series of events postulated according to some
principle of action.  Successful scenario work makes special demands
on imagination (maybe even phantasy) by the participants; its shortcomings being derived from the same sources.  Scenarios can vary from
very brief statements of a few hundred words to very elaborate, complex, worked-out descriptions of situations with specific initial
positions, attitudes, data, and orientations for participants to the
scenario also specified.  In the latter case, the situation is often
referred to as gaming.  Particular kinds of "serious games" have
been described in some detail in the book by the same title by Clark
Abt.

     Scenarios and games have been played with the assistance of
computors, and are particularly applicable to the evaluation of
military hardware in simulated combat situations.  Some observors
have gone so far as to suggest that psycho-drama may be a useful
tool for certain assessment purposes.  While generally serious games
of the strategic sort are of somewhat limited value as research tools,
they often have substantial value as educational devices, and in sensitizing the interested player ot some of the complexities of situations with which he will be dealing.  An interesting example of
this is the Office of Technology Assessment game developed by Craig
Decker, which includes role playing running over several hours.  Gaming is often used as a supportive technique once scenarios have
been adopted in principle.  In the military context, the technique
of gaming is almost as old as military history itself.  Several decision makers may be involved in acting out a game,although at times
as few as two persons are allotted (e.g. diverter-inspector gaming
for safeguarding of nuclear materials).


                    CONCLUSIONS AND FUTURE DIRECTIONS

     There are several institutional factors which may determine
the context of issues examined by research groups working within
governments or within an international organization.  They also set
constraints on the choice of technique and the possibilities for

the wider checking of results.  Particular methods which might seem
most appropriate for the discussion of specific issues may be in
conflict with the nature of the policy process itself.  Since time-
liness and urgency of information are characteristics of government,
decision-makers faced with urgent problems may be unable to wait for
a comprehensive analysis to be concluded.  Although a more elaborate
method is theoretically justified, the time constraint alone may be
enough to preclude it for a more straightforward and simple forecasting method.  Furthermore, it is not always evident which of the
techniques available may be warranted.  There are often pressures
which exaggerate the value of elaborate methods, in the same way
vested interests may develop in introducing or excluding such methods
and techniques to be adopted.  Competition for influence within, as
well as - in certain cases - between departments, may determine the
fate of any one or more methods and techniques.  Also, often experiences do not live up to expectations or ideals.  Programme policy
budgetary systems (PPBS) and large long-term planning models have
frequently resulted in relocation of personnel within government departments, but little else.

     I have touched on bureaucratic and departmental rivalry.  Very
often personal fears, limited understanding, controversy and uncertainty surrounds situations in which forecasting is involved.  For
certain types of bureaucrats it is often convenient that controver-
sial issues should remain inadmissible.  It is quite likely that one
of the reasons of relative popularity of systems analysis as a method
is directly related to the fact that it seems to carry with it an
air of apolitical objectivity.  This, as much else, is often simply
an illusion.  The fact of the matter is that pressures such as these
exist, and should be taken into account.  As has been proven conclusively (by for example Cole, Clark and Archer) no nations and very
few corporations care to rely on the forecasts of others.  There is
also often pressure on outside agencies to make otherwise dismal
forecasts more optimistic (as in the case of the OECD at times).
The only remaining forecasting, as mentioned in the outset of this
paper, is that of meteorological forecasting, where there does seem
to exist a consensus view on possibilities.

     It may be that, therefore, good communication, with a research
team, on a peer group, and within the administration, is the best
we can hope for.  This is, by itself, no small matter and should
not be downgraded.  Simply to clarify and systematically extend the
network which already feeds information into the administration
through such devices as "workshops", experts meetings, brain-storming
and the like is more likely to improve the quality of advice offered,
than is achieved through the use of other methods.  It is of course
very important that information should find the right channel, and
that different channels are being kept open.  In my estimation, however, we should go further than this.

Strategies which leave options open, such as analytic forecasting, have the essential purpose of making better use of information about the short term while working within comprehensive yet still flexible long-term goals. The evidence indicates that in recent years there has been a trend towards this type of incremental planning despite the apparent increased interest in long term issues.

However, recognition of uncertainty implies that policies must be continually updated. Whereas overcommitment may be as much a barrier to change as it is a stimulant, more frank public discussion can lead to the awareness and acceptance of the trade-offs upon which policies are based. The clearest example of the latter can be seen in the relatively new fields of technological forecasting and, more appropriately for our purposes, technology assessment, an area which has taken on increasing importance.

The number of ways in which current methods of forecasting have failed to foster full public debate about the future (assuming this is always desirable, which I myself doubt), suggest that improvements can be made as regard the methods of presentation and modes of public involvement. From a methodological point of view it may be that analysts and forecasters alike should spend as much effort on developing methods which may be integrated into society rather than on internalising social processes within their own models.

Since forecasting and the political process are inseparable, it is open to debate whether the correct balance is being struct in attempts to link forecasting to a larger debate. Several persons, including Cole has argued this point and I find myself in agreement. Techniques and methods may often be used, deliberately or not, in such a way that they may mislead even a relatively sophisticated audience. In the sense that the methods are unrepresentative, they may add to the underestimation of the true range of option, hence in the long run lead to insufficiently flexible choices. This premise underlines the basis of my argument for analytic forecasting. The result may be more messy, more "open-ended", taking longer to develop, and carrying with it wide social implications. In the long run, however, it is safer. As Cole has argued, the "extensive" approach to forecasting employs a collective in the sense of _pluralist_ (my emphasis) social theory which has diverse origins throughout society. Its application involves both wider and more enlightened public discussion about the future as well as greater deliberate empirical searching for alternatives.

The above mentioned behavioural science methods, tools and techniques belong to the armory of _analytic_ forecasting, and may help us to explore actual and potential situations. Their limitations are evident, only relatively small, circumscribed and simplified abstracts or models of the social, economic, and technical reality are accessible -- as of now. However, the treatment which can be given

to problems is more exhaustive and more systematic within the given
limit than the unaided human mind is normally capable of.  Furthermore, the complexity and difficulty of the problems faced at this
stage of social and political evolution by national or international
decision-makers are such that some help and assistance from this
type of scientific methodology and research simply <u>must</u> be sought.

# THE DEVELOPMENT OF POLICY-SENSITIVE MODELS

Hans-Peter Weber

Bundesministerium für Verkehr

Bonn-Bad Godesberg, West Germany

This paper is concerned with the status of development of policy-sensitive models for traffic forecasting in Germany covering
national and international research projects. The development of
policy-sensitive models is urgently needed to quantify the effects
of various transport policies, which we have to consider as parts
of our general transport strategy.

There is a specific necessity not to concentrate on day-to-day
questions but to develop a strategy in the long run, which will give
us the instruments we have to use in the future for every aspect of
political planning. I should like to give some preliminary remarks
before dealing with the structure of our research programme in detail.

1.  The influence of policies on the development of traffic seems
    to be limited. I will come back to this statement later on.
    Much more important are obviously cycles and trends in the
    socio-economic development which influence transport as a lin-
    ked demand. The dependence of the transport pattern needs a
    permanent methodological improvement of our instruments to ex-
    plain these relations between socio-economic parameters and
    traffic generation, - distribution and the modal split. This
    is especially difficult in goods transport, where the scenarios
    need a much more refined analysis than for passenger transport.

    In the case of passenger transport we may limit ourselves to
    assumptions about population, employees, education, leisure
    time behaviour and so on in a regional diversification. The
    patterns are much more uniform and have less cycles than we
    find in goods traffic.

In the field of understanding or forecasting of goods traffic
it is necessary to have detailed information about the development of economic sectors. I will illustrate this with the following example: We must know not only whether the iron and steel
plants at Rhine and Ruhr in 2000 will exist and if and where
competing plants in the neighbourhood will be, we have to know
about the inputs - ore or pellets -. In addition we must know
which are the sources of supply and the areas of sale, which
modes are involved and which infrastructure will be existent.
There is urgent need to forecast in fields, in which the decision makers haven't made any decisions until now. On the other
side this can become the source of many errors, which will influence the profitability of many different projects of our
infrastructure.

2.    In connection with these lines we have to see the behaviour
orientated models. The usefulness of policy-sensitive models
depends on the ability to develop such models, which reflect
the real behaviour of individuals, so that we can get realistic
answers about the reactions of changing parameters. Here again
we have different problems in passengers and goods traffic in
respect of the data base. Especially concerning passenger
traffic the availability of data is still very limited. This
begins with the equality of data, the low response on surveys
and goes on to the ability of the people to give valid information about their own real behaviour. It seems to me to be
very useless to develop modal split models which are based on
time and/or cost advantages if the passenger does not have any
realistic knowledge about travel times and costs. We have already problems concerning the use of our own cars, not to
speak about potential alternatives. The assumption of rational
behaviour which normally is the basis of transport models needs
detailed checking. This involves very complicated consequences.
So our research brought up that the efficiency of publicity
measures on modal split is comparable to that of expensive investments in infrastructure (1).

In goods traffic the assumption of rational behaviour is more
useful though the knowledge about costs and potential alternatives is very restricted - not only in traffic on its own
account (2).

But additional problems exist. We know from our surveys that
decisions of shippers do not only depend on prices but also
to a great extent on quality factors, which are of different
importance for the different commodities. This means that in
respect of the diversity of commodities we need a high amount
of data. No wonder that we still are at the beginning of our
research.

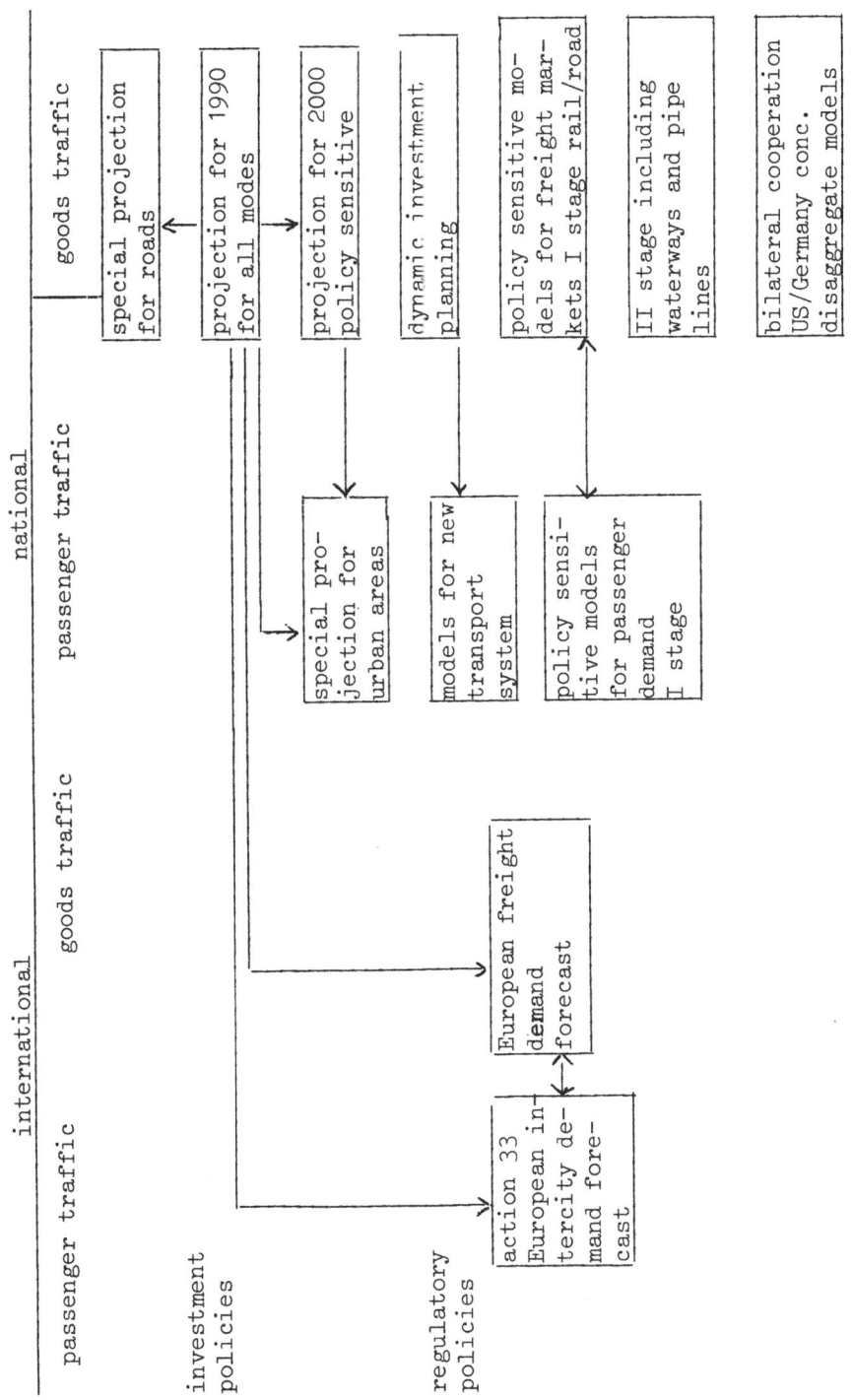

Figure 1

Now I will come back to the structure of the research programme in detail.

Figure 1 shows the interconnections between our different projects. The classification is self explanatory. However, it seems to be useful to make some remarks on the differentiation between investment and regulatory policies. This differentiation is not only a historical one but has its realistic reasons. The effects of these policies are extremely different in their regional impact. Investment policies - especially with concentration on special projects - have regional restricted impacts. They influence only some traffic flows and a subset of the regional infrastructure. On the other side the effects of regulatory policies are of a much more global nature. Changes in costs of car ownership through changes in taxes on cars and/or fuel have effects in the whole area. All car drivers are concerned. As one exception we may see parking policies, as far as we have a regional differentiated proceeding. We may call such sort of policy the link between investment and regulatory policies, because their instruments are taken from both of them. An absolutely neutral classification therefore seems to be impossible.

The nucleus of our forecasting research efforts is the development of an integrated projection for passenger and goods traffic (3).

The integration is necessary because the target was already - and will remain - at least to "estimate" the future use of the capacities provided in all parts of our infrastructure, which in most cases have to serve passenger and goods traffic. Our forecast had to tackle several hard complications, mainly because of lack of data from official records which we had to replace by special estimates. Difficulties arose also in forecasting socio-economic data in a very detailed form.

I will explain only some of our results according to ADP-outputs which are available. We have outputs for the years 1970 and 1990 showing traffic flows between 79 inland-regions and 18 foreign countries for passenger and goods traffic. These are differentiated for modes, 12 groups of commodities in freight traffic and 6 travel purposes in passenger traffic, and converted lateron into cars or train units between these regions. This projection is not yet based on a policy-sensitive model but is the result of modified status-quo forecase with one exception: the forecast for inland waterways transport does not depend on the network of today but on the network of 1990 which contains additional links.

Recent changes in the socio-economic data (e.g. a much more severe decrease in population than projected and the demand for an ongoing planning process) lead to the result to start already now and again work on additional projections.

We intend to improve by

1.  stronger formalisation of the (in many parts only qualitative)
    model and

2.  setting the horizon of forecasting now to the year 2000.  To
    better take into account the passenger traffic in urban areas
    we have dealt with some scenarios with different assumptions
    (4).

3.  We intend to substitute the status quo forecast for a policy-
    sensitive projection.  We have to include in our calculation
    the influence of the network of tomorrow and insert the neces-
    sary steps to reflect the effects of an extended or improved
    infrastructure on trip generation, - distribution, modal split
    and assignment.

4.  The discussed forecast procedure comes from the national level
    down to the details.  This has the advantage of remaining com-
    parable to other national economic goals - assumptions are not
    admitted if they are not matching realistic speed and volume
    in development of population and other socio-economic structures.
    Of course we have the problem of regional disaggregation and
    different developments.  Especially in the field of road traf-
    fic, though having already a rather dense network we are still
    fighting for many additional projects.  We need special and
    sometimes subregional projections for which we take also into
    account the results of automatic counting on the road.  We
    have in our general approach insisted on a strong connection
    between regional socio-economic data and global data including
    traffic behaviour to ensure compatibility - at least as much
    as possible with our data.

5.  Another problem is the mainly static nature of todays planning
    which ranks the proposed projects only according to benefits
    in a fixed time period without taking into account the optimal
    timing of investment.

By dynamic investment planning we try to rank projects according to their optimal time of investment under budgetary and other
constraints.  We have sponsored a pilot study to collect todays'
methods and theoretical solutions as well as data requirements and
the limits of todays' ADP-research possibilities and proposals for
further field studies. (5).

We found that optimal timing of investment by the method of
integer programming is favored.  Especially branch and bound algorithms give exact solutions for 200-500 binary variables and 100-
1000 constraints.

We will have test runs for several regions to get more accus-
tomed to practical use of this method.  This will be very useful

for the development of quick investment planning considering more possible solutions than before. Dynamic investment planning research was tried in a special case, the introduction of a new transportsysteme (NTS) in long distance traffic in the corridor Hamburg – Rhine – Ruhr – Frankfurt – Munich, where 50 percent of the population of Western Germany is concentrated.

The model used includes several submodels for the modal split and a passengers decision model. As it is a feedback algorithm the results are used as an input for the next iteration.

The structure of such a submodel is described in Figure 2.

Possible results of the introduction of a new high speed transportation system are shown in Figures 3 and 4 for several traveller categories and different O-D-relations. The model reacts to changes in the input assumptions such as price reactions of the competitors to the introduction of the new system, improvements of the quality of the competing systems and different consciousness of the travellers regarding costs. In the model a first step integration of investment and regulatory policies for a corridor was successfully used. Special importance is seen in the possibility to introduce feedbacks and delays which allow a better description of the development and setbacks of the system over time and the opportunity to estimate whether there will be a new equilibrium or a self exploding system.

Now we have a first version of a model which may be used to reach the following targets:

1. measure the impact of single projects on the development of traffic,

2. measure the feedbacks on the project resulting from the reactions of competing modes,

3. quantify interdependencies with other projects and ranking of projects including determination of optimal investment times and the impacts of changes in the assumptions on it.

In the field of international cooperation similar studies were carried out to forecast intercity passenger transport for Western Europe for the years 1985 and 2000 which was done by the ECMT, the OECD and the Commission of the European Community (called the action 33). The study was finished recently. In this study the impacts of several strategies were tested.

The main strategies studied:

– Status quo (SQ)
  Assumes that market demand must be met and the resulting problems remedied as far as possible by measures that do not interfere with

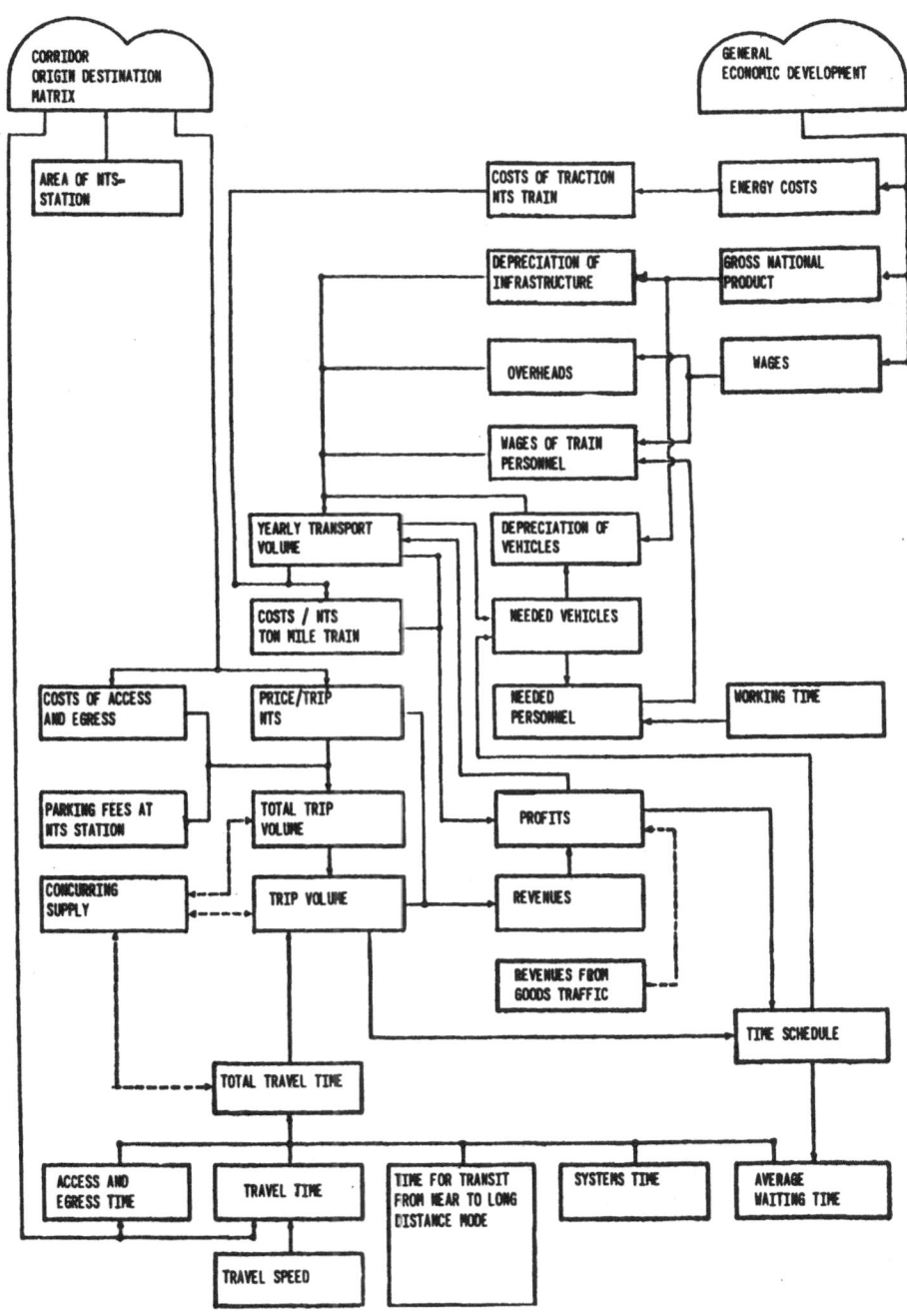

Figure 2.  Submodel NTS Passenger Transport.

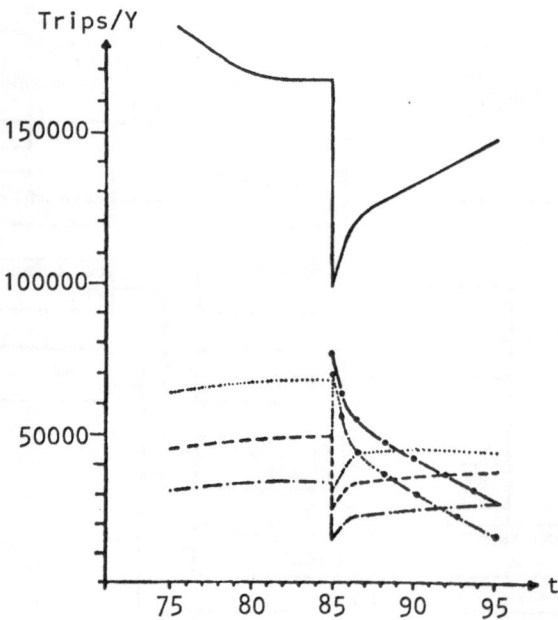

Figure 3. Relation: Dortmund – Frankfurt; Purpose: One day business
travel

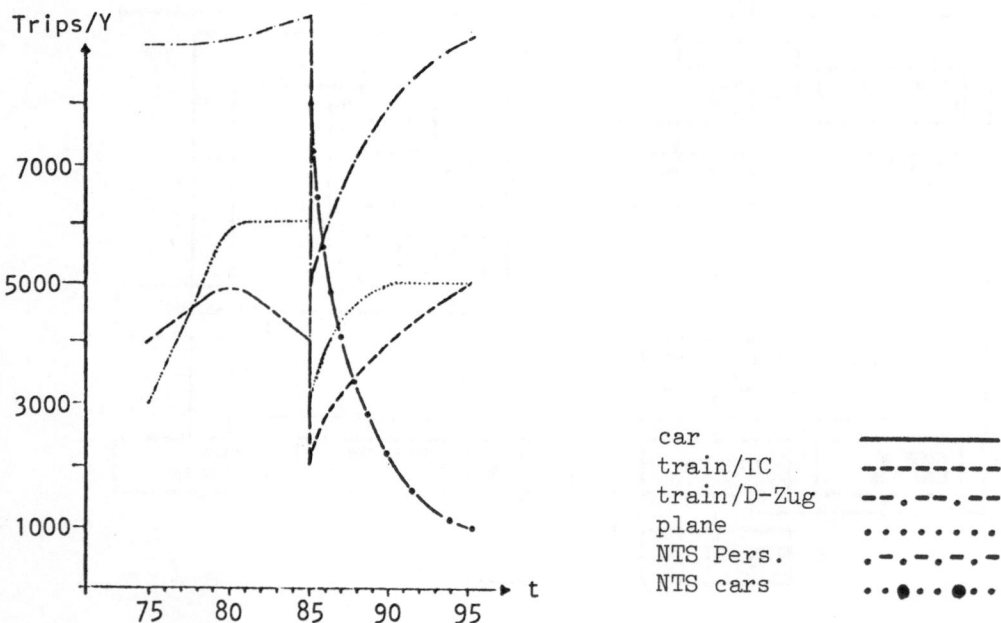

Figure 4. Relation: Hamburg – Munich; Purpose: Holiday Traveller/
Low income, no car.

- the market system.  In this strategy the growing demand for transport, by all modes, is matched by new roads, airports etc. in a
conventional way as far as the budget allows.  Two additional
variants of this strategy are a regulated status quo strategy
(particularly to reduce environmental effects) and a status quo
V/STOL strategy (to study the possibilities of a V/STOL system).

- Controlled mode strategy (CM)
Assumes that the main problems are caused by excessive demand for
car and air and that the basic solution is to find ways of attrac-
ting demand from car and air to rail.  It implies a smaller road
network than the status quo strategy and strict anti-nuisance
regulations are applied to road vehicles and aircraft.  This strategy has two variants: a) improved railway using conventional
technology with only a few new lines (CMA); b) a large network of
very fast services requiring new infrastructure and new technology (CMB).

- Controlled demand strategy (CD)
Assumes that excessive demand for car and air travel cannot in
practice be satisfied by other modes and must therefore be restrained directly.  A variant of this strategy studies a decentralised
air network as a means of reducing the pressure of demand on the
major hub airports.

- Planned demand strategy (PD)
Assumes that the problems are largely due to excessive concentration of demand in certain places at certain times.  The solution
is to disperse demand over time and space, trying in the process
to shorten trip lengths.  A variant of this strategy assumes low
economic growth together with low growth in investment budgets.

These strategies are partially a mixture of investment and regulatory policies.

The forecasting model used is in some way a conventional four
step algorithm.  The model has two parts representing supply and demand.  The supply model consists of a detailed, quantitative description of the transport system and the services it offers, including
prices.  The demand model consists of all the main factors determining decisions to travel, including the transport services and prices.  The demand model reacts on changes in the supply, to reach
consistency with the services offered, but the supply model does not
react on alteration of the demand.  The structure of the demand model is described in Figure 6.

The supply model consists of a complex network which includes
the three sub-networks: road, rail and air, on which six transport
modes (for long distance passengers) are distinguished, namely car,
bus, first class rail, second class rail, scheduled aircraft and

| Trip purpose / Model phases | Personal trips | Business trips |
|---|---|---|
| Trip generation ($t_i = \sum_j t_{ij}$) | Category analysis<br>Categories: household structure<br>household income<br>car ownership | Gravity model:<br>$t_{ij} = \alpha(GRP_i \cdot GRP_j)^\beta \cdot \overline{T}_{ij}{}^\gamma$ |
| Trip distribution ($t_{ij} = \sum_m t_{ij,m}$) | Zonal attractions ($A_j$): attraction weights<br>Travel impedance ($TI_{ij}$): function of minimum generalised travel time<br>$(G_{ij,min} = T_{ij} + C_{ij}/\lambda)$ | Measures of generation and attraction:<br>zonal income, GRP;<br>travel impedance –<br>modally weighted time, $\overline{T}_{ij}$ |
| Modal split ($t_{ij,m}$) | Category analysis: categories   unimodal<br>bimodal<br>multimodal<br>Modal split functions: diversion curves with:<br>$P(t_{ij,m}) = f\left(\dfrac{G_{ijm}}{G_{ijm}}\right)$ | |
| Trip assignment | Route assignment (for each $t_{ij,m}$)<br>Link and corridor volumes by network (i.e. rail and air) | |

Figure 5. Structure of the Demand Model.

chartered aircraft. The modelling of the network required a simplification of three data groups: first the quality of service offered
to users of the six recognized transport modes, second the time and
cost of access to the networks and third the utilization of network
capacity.

The total input to run the model, for any prognostic time period
consists of specified socio-economic data for the traffic zones
(109 in all) and specified network data, including the basic (short
distance) traffic on each link. The output consists of the number
of trips generated or attracted in each zone by trip purpose, zone
to zone movements by modes, traffic volumes on every link of the
network, travel distances, times and costs.

I should like to describe some results about the different impacts of strategies on the different stages. The influence on trip
generation (as shown in Figure 7) is very limited.

Somewhat more important is the impact on trip distribution
(see Figure 8).

Rather complex are the effects on modal split (Figure 9), but
some general remarks are possible.

The impact of all strategies (but limited in the case of the
planned demand strategy) tends to decrease the share of flights and
car trips. In absolute and relative terms the effects are most important in the controlled mode strategies for business trips and
sensitive to changes in strategies concerning trip generation and
modal split while holiday trips do not change very much in all strategies.

As a result we may say, that we need very intensive investment
to influence traffic in a remarkable way. It is obvious that we
need much more information about benefits and costs of investments
into transport networks before deciding about the advantages of one
or another of these strategies. In addition there are many changes
in traffic flows which influence each other in the aggregated form
of the model described. However, the results prove the thesis, that
the impact of policies on long distance transport are relatively
small compared with the influences of socio-economic trends.

A comparable project to forecast the freight transport in Western Europe currently is carried out by the Commission of the European Community with assistance of the E.C.M.T. As the project is
in its first stages we should not go into details; a rough description of the model is included in Figure 10. Some difficulties may
be expected during the preparation of scenarios which have to be much
more detailed than for passenger traffic.

| Strategy | | Business | Holidays | Short stay personal | Total |
|---|---|---|---|---|---|
| 1970 | SQ | 183.7 | 329.7 | 639.6 | 1 153.0 |
| 1985 | SQ | 363.9 | 444.4 | 987.8 | 1 796.1 |
| | CD | 360.6 | 444.4 | 979.0 | 1 784.0 |
| | CD(DA) | 377.1 | 444.4 | 979.0 | 1 800.5 |
| 2000 | SQ | 526.3 | 534.4 | 1 211.1 | 2 271.8 |
| | CM(A) | 536.4 | 534.4 | 1 200.2 | 2 271.0 |
| | CM(B) | 570.2 | 534.4 | 1 203.3 | 2 307.8 |
| | CD | 520.2 | 534.4 | 1 202.2 | 2 256.8 |
| | PD | 605.9 | 534.4 | 1 214.3 | 2 354.6 |

Figure 6.

| Strategy | | Business trips | Holiday trips | Short stay personal trips | All trips |
|---|---|---|---|---|---|
| 1970 | | 63 | 175 | 112 | 350 |
| 1985 | SQ | 130 | 236 | 178 | 544 |
| | CD | 140 | 241 | 178 | 559 |
| | CD(DA) | 161 | 242 | 179 | 582 |
| 2000 | SQ | 209 | 305 | 224 | 738 |
| | CM(A) | 219 | 312 | 234 | 765 |
| | CM(B) | 236 | 308 | 236 | 780 |
| | CD | 209 | 313 | 228 | 750 |
| | PD | 269 | 306 | 234 | 809 |

(1) Estimates based on approximate road distances.

Figure 7.

| | 1970 | 1985 SQ | 1985 CD | 1985 CD-DA | SQ | CM (A) | CM (B) | CD | PD |
|---|---|---|---|---|---|---|---|---|---|
| **Business:** | | | | | | | | | |
| plane | 25 | 51 | 51 | 66 | 97 | 97 | 102 | 82 | 129 |
| train | 44 | 91 | 101 | 104 | 137 | 164 | 185 | 158 | 166 |
| car | 79 | 141 | 138 | 136 | 190 | 172 | 175 | 181 | 194 |
| | 148 | 293 | 290 | 307 | 424 | 433 | 462 | 421 | 469 |
| **Holiday:** | | | | | | | | | |
| plane | 22 | 35 | 29 | 33 | 45 | 48 | 49 | 36 | 41 |
| train | 76 | 89 | 93 | 91 | 96 | 102 | 100 | 102 | 85 |
| bus | 17 | 20 | 23 | 23 | 21 | 21 | 21 | 28 | 28 |
| car | 187 | 254 | 253 | 251 | 310 | 302 | 303 | 307 | 310 |
| | 302 | 398 | 398 | 398 | 473 | 473 | 473 | 472 | 473 |
| **Short stay personal:** | | | | | | | | | |
| plane | 8 | 37 | 19 | 21 | 63 | 58 | 47 | 45 | 64 |
| train | 95 | 120 | 123 | 123 | 142 | 176 | 197 | 147 | 143 |
| bus | 26 | 30 | 45 | 45 | 34 | 34 | 34 | 53 | 46 |
| car | 345 | 506 | 500 | 498 | 587 | 550 | 546 | 574 | 576 |
| | 474 | 693 | 687 | 687 | 826 | 818 | 824 | 819 | 829 |
| **All trips:** | | | | | | | | | |
| plane | 55 | 133 | 99 | 120 | 206 | 203 | 198 | 163 | 234 |
| train | 215 | 300 | 317 | 318 | 375 | 442 | 482 | 407 | 394 |
| bus | 43 | 50 | 68 | 68 | 55 | 55 | 55 | 81 | 74 |
| car | 611 | 901 | 891 | 885 | 1 087 | 1 024 | 1 024 | 1 062 | 1 089 |
| | 924 | 1 384 | 1 375 | 1 391 | 1 723 | 1 724 | 1 759 | 1 713 | 1 791 |

Figure 8.  Trips by Mode, Purpose and Strategy, 1970–2000, Study Area (interzonal trips only) in millions.

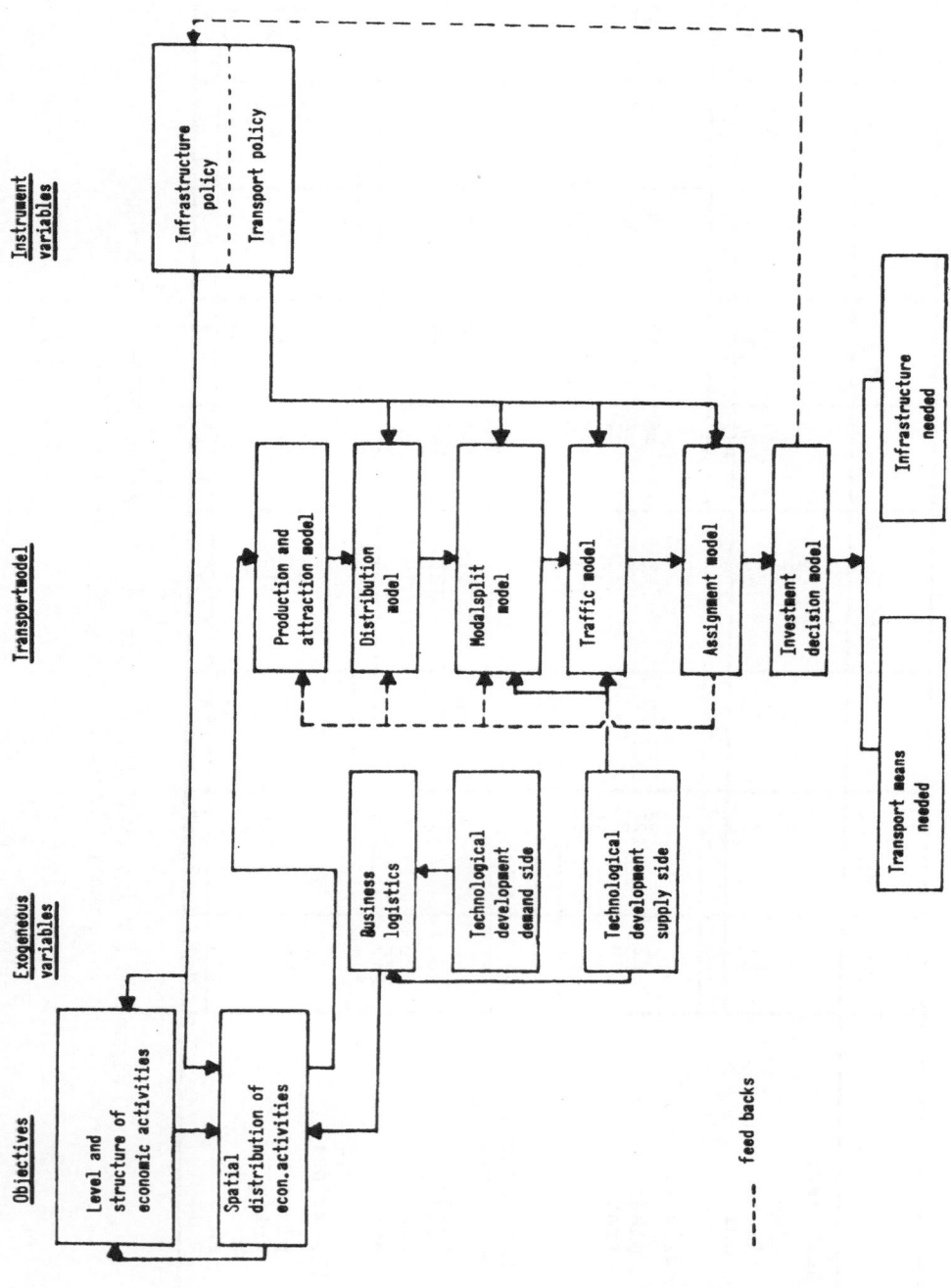

Figure 9.  Transportmodel as a tool for decision-making.

Concerning regulatory policies we tried to develop a policy-sensitive model fcr the inland freight transport market (2).

The model in its first stage is limited to the competition between rail and road including special consideration of combined traffic.  It can te described as a model, which has the target to quantify measures to change the modal split.  These measures influence at first the supply side.

The model consists of two submodels

- the parameter model
- the behaviour model.

The parameter model describes the impacts of the policies on the input parameters which decide the choice of mode.  The decision model describes the impacts cf changed parameters on modal split.

The model has the following character:

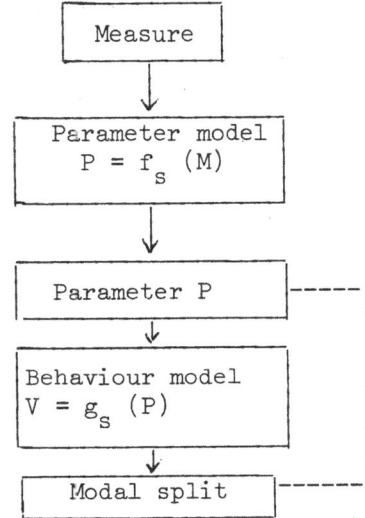

The parameter function $f_s(M)$ and the behaviour function $g_s(P)$ depend on several variables (s) which characterise each transport movement.  These variables (s) are the relevant decision attributes of the carriers and the transport systems and are parameters of supply which do not change with the measure.  For each group of variables (s) we need different functions.  The whole transport procedure has to become disaggregated to represent groups of homogeneous behaviour.

The basis for the disaggregation are
- the group of commodities,
- the unit of load,
- the distance of transport.

The regulatory policies influence directly or indirectly the cost parameters for long distance road transport. This limitation is the result of the assumption that only changes concerning shifts from road to rail are politically relevant and in addition till now we have no model to quantify effects of policies concerning capacity regulation.

The decision maker is the shipper who has to decide on the choice of mode of transport for the individual loading. The basis of the decision is the value of the parameters, i.e. the costs in road haulage. The behaviour model consists of the functions to describe personal reactions on changes of parameters.

These functions of behaviour describe

- the elasticities of demand for the different groups of commodities,
- the commodity independent factors influencing decisions (as volume of load, distance, structure of areas).

If no real alternative for the transport mode is given, the result of the function of behaviour is zero.

The behaviour functions were based on results of

- a research of the transport market,
- a statistical traffic analysis,
- a behavioural analysis depending on surveys of shippers
- and a statistical analysis of time series of the years 1950-1974.

The following policies were tested in the simulation model

- the introduction of a tax on the ton miles in long distance road transport,
- the rising of the tax on fuel,
- the decrease of the tariffs of railroads.

The impact of the introduction of a tax on ton miles on the modal split can be characterized as follows:

| Mode of Transport | Modal split (1973) amount (1000 t) share (%) | | shifted amount | | | new modal split share amount | |
|---|---|---|---|---|---|---|---|
| | | | to common carrier | to rail conv. | to rail cont. | | |
| | | | Tax 1 Pfg/tkm | | | | |
| R CONV | 157 456 | 50,4 | – | + 1309 | – | 158 765 | 50,8 |
| R CONT | 1 825 | 0,6 | – | – | + 150 | 1 975 | 0,6 |
| A | 53 920 | 17,3 | +586 | – 82 | – 4 | 54 419 | 17,4 |
| C | 99 032 | 31,7 | –586 | – 1227 | – 146 | 97 074 | 31,1 |
| Total | 312 233 | 100,0 | – | – | – | 312 233 | 100,0 |

R CONV = conventional rail transport
R CONT = container rail transport
A       = transport on own account
C       = common carrier in road transport

Somewhat smaller are the impacts of the rise of the tax on fuel as is shown in the following table:

| Mode of Transport | Modal split (1973) amount (1000 t) share (%) | | shifted amount | | | new modal split share amount | |
|---|---|---|---|---|---|---|---|
| | | | to common carrier | to rail conv. | to rail cont. | | |
| | | | Rise of fuel tax per 10 Pfg/l | | | | |
| R CONV | 157 456 | 50,4 | – | +944 | – | 158 400 | 50,7 |
| R CONT | 1 825 | 0,6 | – | – | + 83 | 1 908 | 0,6 |
| A | 53 920 | 17,3 | – 104 | – 88 | – 2 | 53 725 | 17,2 |
| C | 99 032 | 31,7 | + 104 | –856 | – 81 | 98 198 | 31,5 |
| Total | 312 233 | 100,0 | – | – | – | 312 233 | 100,0 |

These results again prove the thesis that regulatory policies have very small impact on the modal split.

They are somewhat smaller than the effects of business cycles considering that the German railroads lost every fifth ton between 1974 and 1975 owing to the general decrease in production.  The real

effects should be somewhat smaller as the model doesn't include the
reactions of competitors on these changes of parameters.

The model will, in an improved version, describe the whole goods
transport market in Germany including international transports.

In this context we intend to check the possibilities of construction of full disaggregated (variety of 50 400 classes of structure) but that is a disaggregation resulting from statistical records which do not go back to the individual transport case.  With
a special survey including shippers and the transport industry we
hope to obtain the data to calibrate a disaggregated model (9).

A comparable research was started in the field of passengers
transport.  Here we could rely on the work already done in the field
of disaggregated models (10).

REFERENCES

1.  Sozialforschung Bróg: Verkehrsmittelwahl Berufs- und Ausbildungs-
    verkehr München 1975.

2.  Messerschmitt-Bölkow-Blohm: Aufbau eines Instrumentariums zur
    Ermittlung und Bewertung des Einflusses ordnungspolitischer
    Maßnahmen, Stufe I Containerisierbare Güterströme, München 1976.

3.  DIW: Integrierte Langfristprognose für die Verkehrsnachfrage im
    Güter- und Personenverkehr in der Bundesrepublik Deutschland
    bis zum Jahre 1990, Berlin 1975/76.

4.  Battelle-Institut: Langfristige Entwicklungstendenzen für den
    Nahverkehr in Ballungsräumen (in progress).

5.  IABG: Entwicklung eines Verfahrens zur dynamischen Investitions-
    planung, Ottobrunn 1975.

6.  Trapp Systemtechnik: Die dynamische Modellierung als Methode
    der Verkehrsprognose dargestellt am Beispiel der Hochleistungs-
    schnellbahn (HSB), Wesel 1975.

7.  OECD: European Intercity Transport Study, Paris 1976.

8.  Nederlands Vervoerswetenschappelijk Instituut Vorläufige Unter-
    suchung zur Entwicklung eines Vorhersagemodells für den Güter-
    verkehr in der EG, Rijswijk 1974.

9.  The work done in this field is listed up on Paul O. Roberts
    Forecasting freight flows using a disaggregate freight demand
    model. - MIT 1976.

10. The literature on disaggregate models for passenger traffic is
    widespread, one of the last publications is by Martin G.Richards,
    N.Mars and P.Terpstra: An application of disaggregate techniques
    in the calibration of a trip distribution and modal split mo-
    del, Deventer 1976.

INTUITIVE PREDICTION OF GROWTH

William A. Wagenaar, H. Timmers

Institute for Perception TNO

Soesterberg, Netherlands

## INTRODUCTION

Many world-wide problems of today are related for a considerable part to growth. Economical growth and growth of populations induce shortages of energy, raw materials and food, and an increase of cost of living and pollution. These processes show a marked exponential character: it is going faster and faster. Any attempt to control these processes will depend on the cooperation of individual citizens; they should first appreciate how fast a growth process will be, before they can reasonably weigh the growth problem against a number of alternative issues, such as religion or personal comfort.

This paper describes some studies on perception of exponential growth by human subjects. The growth processes were presented directly, by means of graphs or in tables. Subjects were asked to predict future events on the basis of prior history. Would a normal person be able to extrapolate an exponentially growing process? Would he be willing to believe a most likely extrapolation? From the literature (Peterson & Beach, 1966; Tversky & Kahnemann, 1971, 1973; De Zeeuw & Wagenaar, 1974) it is clear that man has severe problems in dealing with quantitative data in an intuitive way. Normative rules as provided by mathematics or statistics, are usually too complex; rather Ss simplify the problem by using a repertoire of heuristic strategies which in some cases induce deviations from the 'optimal' behavior prescribed by normative theory. To what extent do such effects occur in the perception of exponential growth? And if they occur, to what extent can people be helped to make better predictions?

A SIMPLE ILLUSTRATION OF THE MISPERCEPTION PHENOMENON

Pollution in the upper air space was measured in five consecutive years.  The outcomes were:

| year | pollution index |
|------|-----------------|
| 1970 | 3 |
| 1971 | 7 |
| 1972 | 20 |
| 1973 | 55 |
| 1974 | 148 |

If nothing will stop this growth process, what do you expect the index to be in 1979?  The estimates of 30 $\underline{S}$s are presented in Table 1 (group 1).  You will notice that 20 out of 30 $\underline{S}$s estimate below 2500, which is only 10% of the best normative extrapolation. If you present the best extrapolation and ask when it will be reached, 20 out of 30 $\underline{S}$s expect that event not within the next ten years. Half of the $\underline{S}$s do not expect the event before the year 2000! (Table 1 group 3).

| GROUP 1 | | | GROUP 3 | | |
|---------|---------|------------|---------|---------|------------|
| estimated index | number of $\underline{S}$s | cumulative percentage | estimated year | number of $\underline{S}$s | cumulative percentage |
| 250–500 | 11 | 37 | 2500 | 5 | 17 |
| 501–1000 | 5 | 53 | 2000–2499 | 9 | 47 |
| 1001–2500 | 4 | 67 | 1990–1999 | 3 | 57 |
| 2501–10.000 | 5 | 83 | 1985–1989 | 3 | 67 |
| 10.001–25.000 | 2 | 90 | 1980–1984 | 8 | 93 |
| 25.000 | 3 | 100 | 1979 | 2 | 100 |

Table 1. Estimates of group 1 and 3 in the condition a = 1, b = 1.0, c = 0.

A more extensive study of this phenomenon should explain
1) which factors determine the degree of underestimation, and
2) where the enormous individual differences come from.

A more practical question is
3) how can underestimation be avoided?


A SYSTEMATIC STUDY ON NUMERICAL PRESENTATION OF EXPONENTIAL GROWTH

The study reported here was published more extensively by
Wagenaar & Sagaria (1975). The essentials are these.

The growth series presented in the previous example is described by
$$y = e^x \tag{1}$$
where y = the pollution index; x = the number of years (1 to 5).
Complication of this function may lead to three different sets of
stimuli.

Absolute size of the pollution index (the a-factor) is varied
by using
$$y = a\, e^x \quad a = 1,2,4 \ldots 128 \tag{2}$$

The growth tendency itself (the b-factor) is varied by using

$$y = e^{bx} \quad b = 1.0,\ 1.1 \ldots 1.7 \tag{3}$$

The initial level (the c-factor) is varied by using

$$y = e^x + c \quad c = 0,\ 100 \ldots 700 \tag{4}$$

The illustration examplifies the condition a = 1; b = 1.0; c = 0.

Three different modes of extrapolation can be attempted:
if nothing will stop the growth,
group 1: how large will the index be in 1979?
group 2: how large will the index be in 1975, 1976 .... 1979?
                                            (stepwise extrapolation)
group 3: in which year will the index surpass ....... ?

Following this design 4.620 responses were collected from 90
Subjects. The results led us to propose a mathematical model for
misperception of exponential growth which will be presented first,
in order to facilitate presentation of the results.

The model assumes that underestimation is a consequence of misperception of the b-factor. The subjects can partly compensate for
this misperception by adjustment of the absolute size of his responses a.

For a-series the predictions in group 1 would be described by

$$\bar{\bar{y}} = a.e^5 . (\alpha e^\beta)^5$$

or $\ln \bar{\bar{y}} = \ln a + 5\ln\alpha + 5(1+\beta)$                    (5)

which means that plots of $\ln \bar{\bar{y}}$ vs. $\ln a$ would yield linear functions with unit slope.

For b-series the predictions in group 1 would be described by

$$\bar{\bar{y}} = e^{5b} . (\alpha e^{\beta b})^5$$

or $\ln y = 5 \ln\alpha + 5b(1+\beta)$                    (6)

which means that plots of $\ln \bar{\bar{y}}$ vs. $b$ would yield linear functions with a slope equal to $5(1+\beta)$ and an intercept equal to $5 \ln\alpha$.

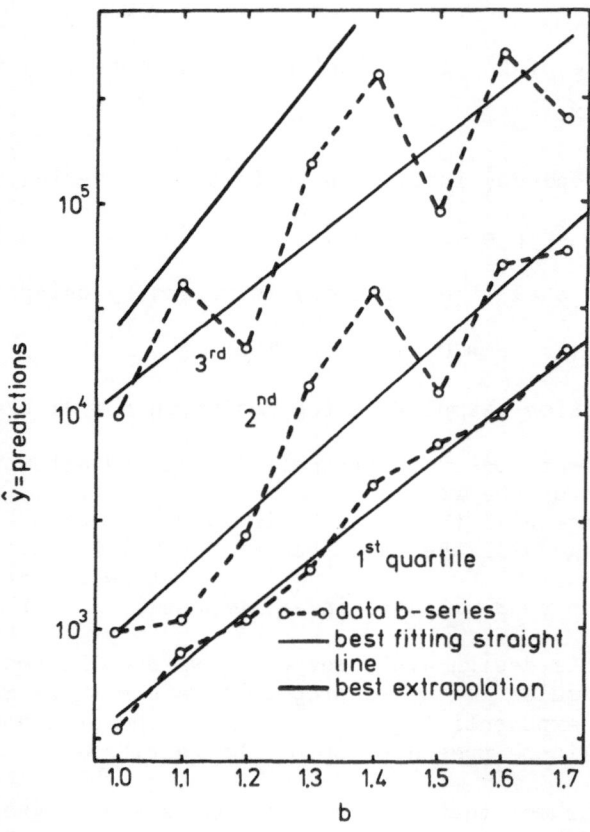

Fig. 1. A plot of predictions of group 1 (on a logarithmic scale) versus b (the growth rate).

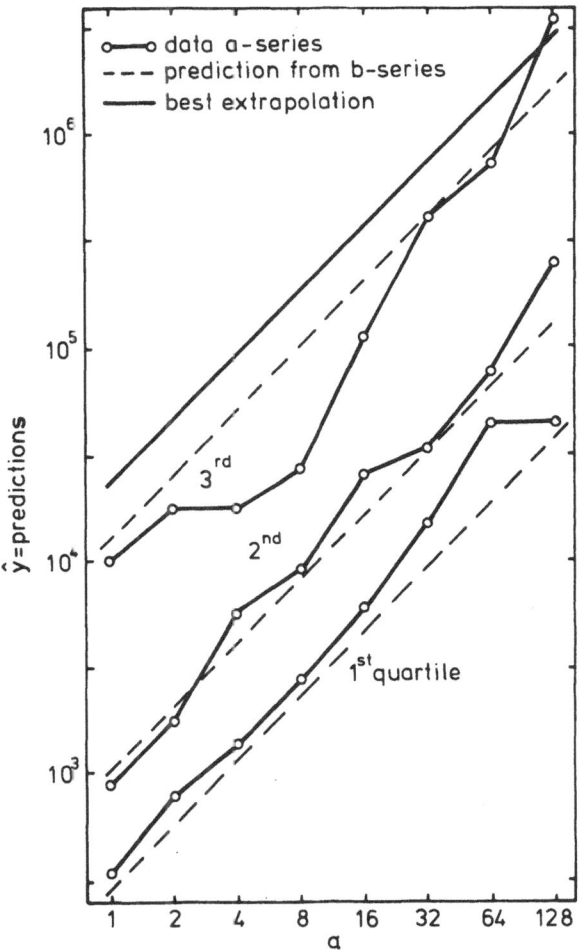

Fig. 2. A plot of predictions of group 1 (on a logarithmic scale) versus a (absolute size of the indices, also on a logarithmic scale). Dotted lines have unit slopes; intercepts are based on values of $\alpha$ and $\beta$ derived from Fig. 1.


In Fig. 1 the plots of lny vs. b are presented for the 25%, 50% and 75% points of the distributions of responses. The linear components explain 96, 82 and 69% of the variance (1st to 3rd quartile). The differences between subjects is caused by the values of $\alpha$ (intercept differences) and not by differences in $\beta$ (slope). The values of $\beta$ were approximately 0.20. The plots of lny vs. lna (Fig. 2) indeed have unit slopes as shown by the dotted lines. These lines have unit slope and intercepts based on the values of $\alpha$ and $\beta$ estimated from Fig. 1.  Thus the absolute size of the indices does not influence the effect of misperception.

Tentatively we may conclude:

1) Misperception of exponential growth presented through tabulated data is caused by the fact that people take into account only 20% of the exponent.
2) Individual differences are caused by different adjustments of the absolute size of the responses.

The results of the c-series and of group 2 and 3 will be only summarized.  The highlights are these.  In group 1 the c-series elicited strong effects of underestimation; in Fig. 3 this is illustrated by the drop of y when c goes from zero to 100.  Thus we conclude that exponential growth is more liable to induce misperception when superimposed on a constant, non-zero, level.

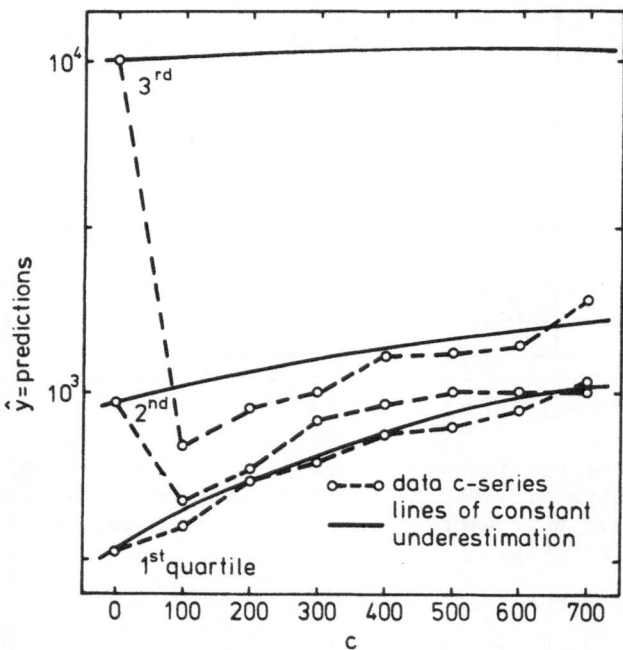

Fig. 3. A plot of predictions of group 1 (on a logarithmic scale) versus c (the additive constant). The drawn lines represent the response level to be expected when addition of c does not change underestimation.

The results of group 2 (step by step extrapolation) were equal
to the results of group 1. Additionally they reveal, as prescribed
by the model presented above, that underestimation is already present in the first step predicted, and that it remains the same for
all the following steps (Fig. 4). The implication is that subjects
correctly apply a constant multiplier to obtain the next extrapolation; only the multiplier is too low.

The results of group 3 ("when will a certain level be reached")
were in accordance with the results of the other groups; the responses were even more conservative than expected on the basis of
values of α and ß in the other groups.

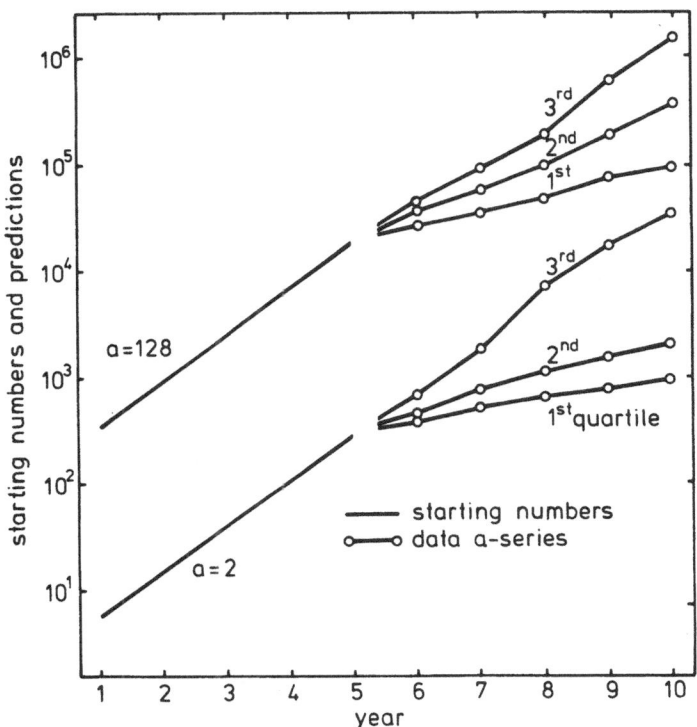

Fig. 4.   A plot of predictions of group 2 (on a logarithmic scale)
as a function of x (years).

GRAPHICAL PRESENTATION OF EXPONENTIAL GROWTH, AND THE FACTOR OF
EXPERIENCE

"A picture is worth a thousand words."  Keeping this in mind
it was attempted to improve perception of growth by the use of graphical representations.  The graphs, representing the b-series described in the previous section, had three length-to-width ratios (Fig.
5) each presented to a different group (group 4, 5 and 6).  Again
values of α and ß can be derived from plots of lny vs. b.  The

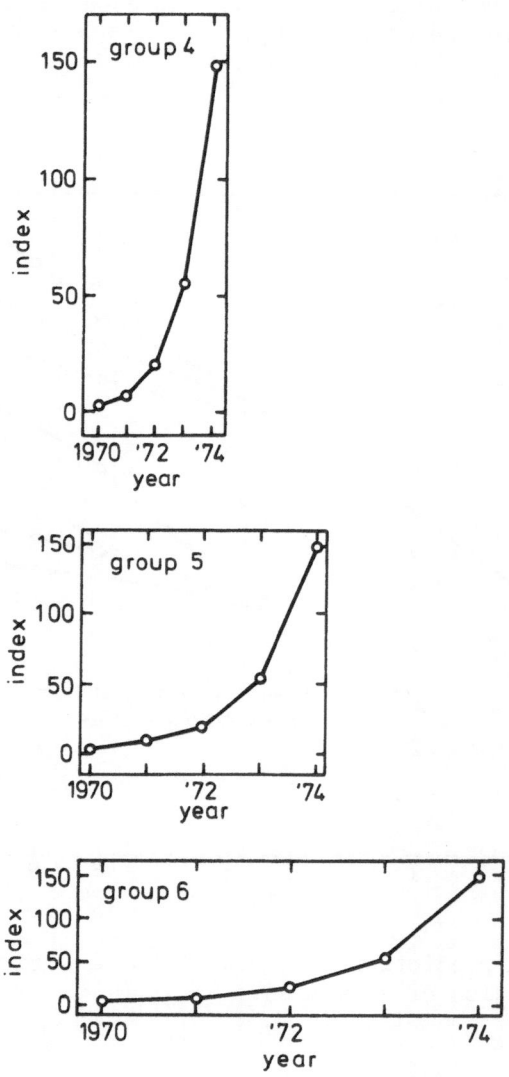

Fig. 5. Examples of graphical material presented to groups 4,5 and 6.

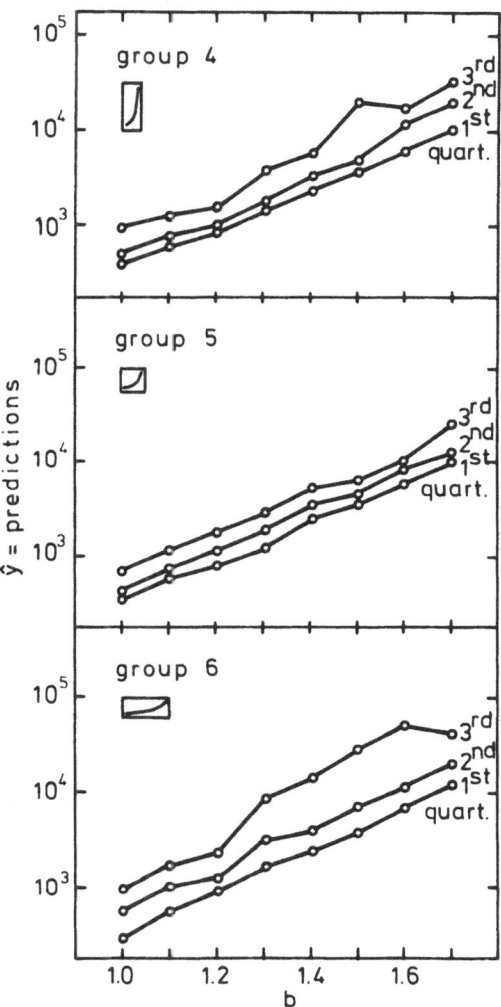

Fig. 6.  Plots of predictions of group 4, 5 and 6 versus b.

results, presented in Fig. 6, show that the plots are extremely linear.  The individual differences were again due to $\alpha$ and not to ß.
The values of ß, which were virtually similar in the three groups,
ranged from -0.07 to 0.28 with a median value of 0.04.  This means
that compared to numerical presentations, even worse underestimations
are elicited by graphs.  The low values of ß demonstrate that man is

not very sensitive to differences between the various growth rates,
as presented by graphs.  This sensitivity is not related to the
length-to width ratio of the graphs.

Considering these data, one factor that comes to mind immediately is naivety or amount of experience on the part of the subjects.
With numerical presentation we did establish already the fact that
estimations are not related to the amount of training mathematics.
The next question then becomes: does specific information on the nature of exponential growth reduce the underestimation?  Some answer
is obtained from a partial replication of the graph experiment
(square graphs only) with two additional groups.  Group 7 was a group
of students who received a  one-hour briefing on the nature of exponential growth and the underestimations displayed by previous
groups; group 8 was a group of members of the Joint Conservation
Committee of the Senate and House of Representatives of the Commonwealth of Pennsylvania U.S.A.  The results in Fig. 7, if compared to
Fig. 6, show that experience only affects the intercepts ($\alpha$) and

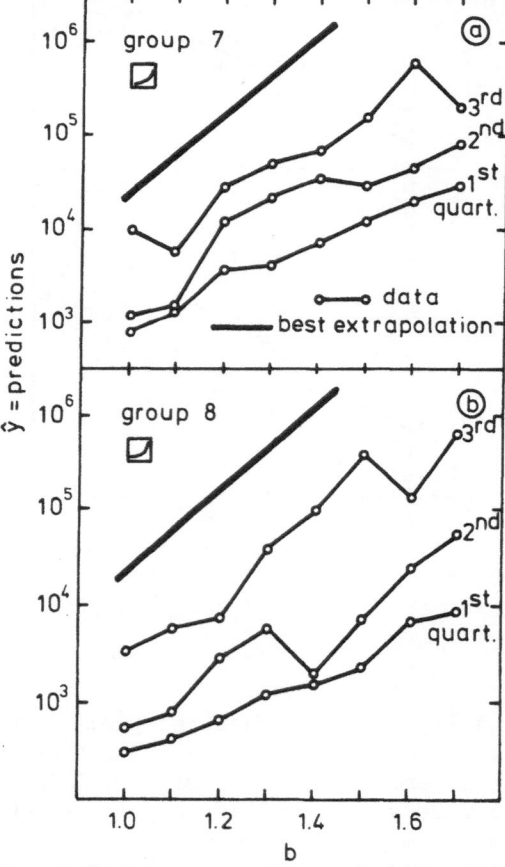

Fig. 7. Plots of predictions group of 7 (non-naive students) and
        group 8 (American senators) versus b.

not the slopes of the plots (ß).  Subjects can adjust the overall
level of the responses, but sensitivity to the different growth
rates appears to be a constant factor.


## THE TIME FACTOR: AN ILLUSTRATION

     Growth processes take time: an explosion develops in microse-
conds; the process of milk boiling over occurs within a few seconds;
colonies of influenza bacilli develop within some hours; price indices grow over years and populations accumulate over ages.  Presentation of processes by tables or graphs excludes this factor; however, people are often confronted with the present state of a process, while for the history they have to rely on memory.  As an illustration of the importance of this effect I will present an experiment on the prediction of prices by housewives.  The task was to predict the prices of bread, fish, milk, postage and a haircut.  Two
groups of 14 housewives were employed, that had been involved in
the economical process for at least 6 years.  The first group (group
9) got only present price levels while the other group (10) obtained
tables with prices in the years 1969 to 1975.

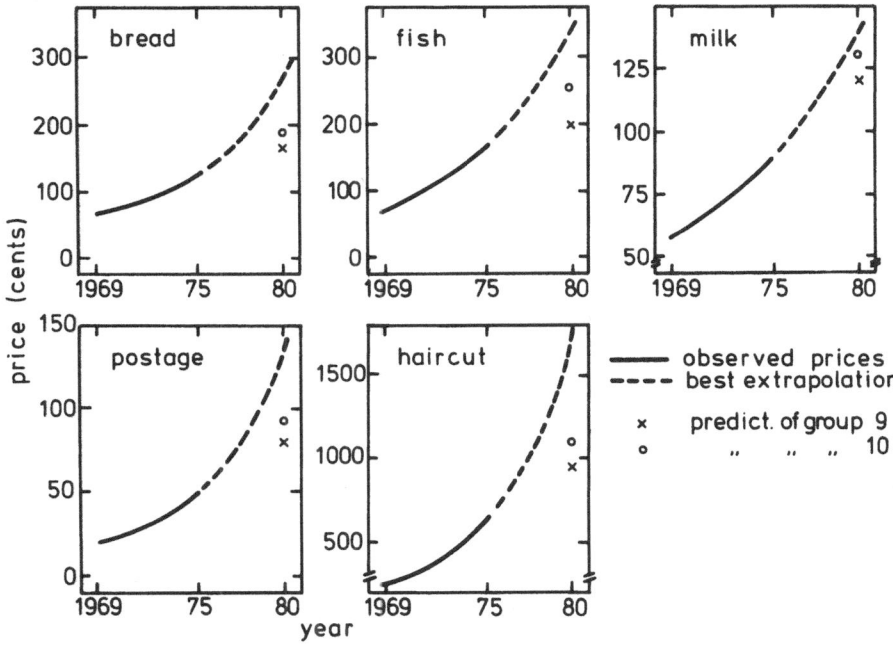

Fig. 8. Results of the price prediction experiment (groups 9 and 10).

Although the best extrapolations are somewhat arbitrary, the predictions (Fig. 8) seem to be conservative in all cases; the most interesting effect, however, is the difference between group 9 and 10 which indicates that insufficient storage of past information leads to lower predictions. Still the predictions of group 10 reveal that some of the past information has been stored. The value of b for milk and postage was respectively 0.13 and 0.23 (when x is counted in years). This difference is reflected by the subjective price indices of group 10 (subjective price index of 1980 is defined by

$$100 \times \frac{\text{predicted price 1980}}{\text{price 1975}}):$$

the median index for milk was 142, for postage 186. Thirteen out of 14 subjects displayed this difference, without being confronted with any growth data!

## DIRECT PRESENTATION OF GROWTH PROCESSES

Usually people observe growth processes themselves instead of numerical or graphical descriptions. We are all confronted with rising prices, fuller parking lots, dirtier waters, and so on. Does direct observation lead to the same sort of misperception?

The basic experiment that should provide some answers was inspired by the fable about the Chinese mandarin who as a youth planted some duckweed in a pond. At a great age he was quite satisfied to observe that one eighth of the pond was covered. He did not realize that in the three years to come one quarter, one half and finally all of the pond would be covered with weed. A similar process was simulated on the scope display of a PDP-8. The pond was a large square that could contain 256 small squares (duckweeds) in rows of 16. The number of duckweeds (n) increased as a function of time according to:

$$n = ae^{bt} - a$$

According to this formula $n = 0$ at $t_o(t=0)$; a was chosen such that $n = 256$ when $t = 10$.

The values of b were $b = 0.1$, 0.3 and 0.5; it should be kept in mind that these values have no absolute meaning: if we divide the time axis such that $n = 256$ at $t = 1.0$ we would obtain the same functions by choosing $b = 1$, 3 and 5.

The present value of t was chosen because this way the values of b can be compared with the values used before. The range of b includes the values used in the price prediction experiment.

The time unit (the time needed to arrive at $t = 1$) was chosen

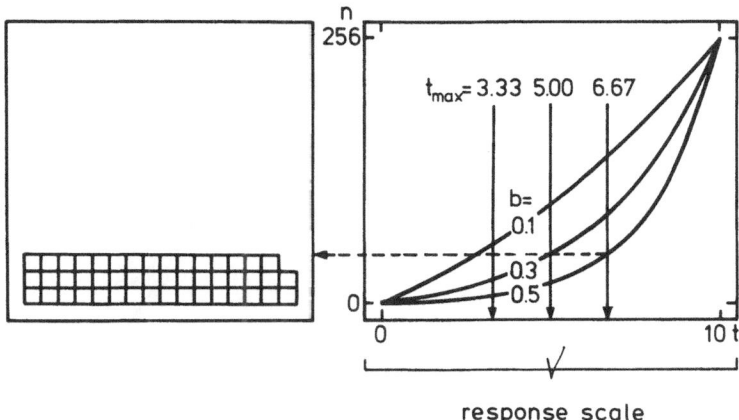

Fig. 9.   The pond-and-duckweed paradigm.
          The number of squares has been accumulated according to the
          function $n = 1.74e^{0.5t} - 1.74$, until t = 6.67 (n=47).   The
          question now is: how far are we between the beginning and
          the moment at which the pond will be completely filled? If
          you think we are about halfway (mark in the middle of the
          response scale) you are assuming a growth function with
          b = 0.3.   In that case you take into account

$$\frac{0.3}{0.5} = 80\% \text{ of the exponent.}$$

such that the pond would be completely filled (t = 10) after 1, 2,
4 or 8 minutes.  However, the completely filled pond was never shown
to the subjects, since the presentation was stopped at t = 3.33, t =
5 or t = 6.67.

When the presentation stopped the subject had to indicate on a
linear scale which proportion of the time $t_{10} - t_0$ had elapsed. That
is: if he estimated that the time needed to get the pond completely
filled was about equal to the time consumed so far, he would place
his mark halfway on the scale.  An example is presented in Fig. 9.
The responses are translated into ß-scores on the assumption that
the subjects will perceive the exponential character of the process;
only the value of the exponent may be misperceived.  In that case a
response can be translated into the exponent of the growth function
the subject thinks to see.  The quotient of this value and the veri-
dical value indicates what proportion ß of the exponent is taken into account.  The basic experiment employed 36 subjects (group 11).

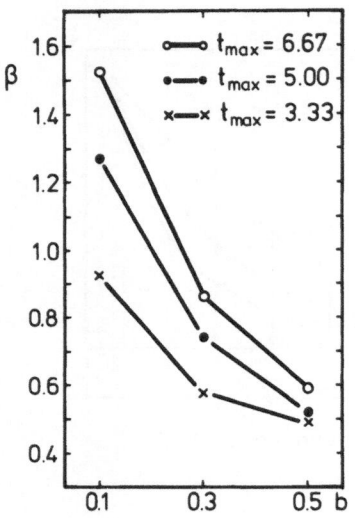 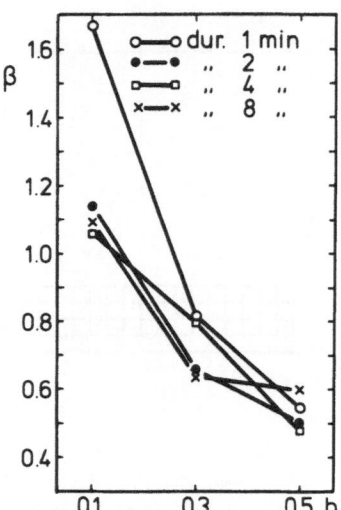

Fig. 10. ß-scores in the pond-and-    Fig. 11. ß-scores in the pond-
         duckweed experiment                  and-duckweed experiment
         (group 11) as a function            as a function of dura-
         of b and $t_{max}$.                   tion.

All subjects produced 36 estimates (3 values of b, 3 values of $t_{max}$, 4 durations).

     The results are presented in Fig. 10 and 11. The scores are around 1.0 when b = 0.1; they decrease rapidly when b gets bigger. The previous estimates of ß < 0.2 when b > 1 do not fit in too badly with these results. When a greater proportion of the process is shown, underestimation tends to decrease, but this effect is almost lost when b is large. The effect of total duration is nihil, with the exception of the condition b = 0.1, duration 1 min. (fast processes that grow almost linearly).

     Thus the conclusion is: inspection of the processes themselves instead of tables or graphs does neither increase nor decrease the misperception problem. The underestimation gets worse when either b is increased of $t_{max}$ decreased.

     In a subsequent experiment the effect of sampling rate was investigated. In the basic experiment updating occurred for each new duckweed to appear, that is: $(ae^{bt_{max}} -a)$ times.

Hence the number of data points varied from 7 to 142. In reality
processes that involve a reasonable amount of time will not be sam-
pled continuously, and the number of samples taken may bear some relation to the amount of underestimation. The experiment was almost
similar to the previous study. Again values of b were 0.1, 0.3 and
0.5; values of $t_{max}$ were 3.33, 5.0 and 6.67. Information about n
was presented 3, 5 or 7 times (we will call this variable d). Example: in the case b = 0.5, t = 5.00, d = 3 information was presented at t = 0.00, 2.50 and 5.00. The number of duckweeds shown at
these moments ($1.74e^{0.5t}-1.74$) was 0, 4, 19. The same process updated seven times would be presented by the amounts 0, 0, 2, 4, 7,
12, 19. In total 1620 responses were obtained from 30 subjects
(group 12); transformation of responses into scores followed the
same philosophy as before. The results are shown in Fig. 12: the
general picture is very much like the one shown before. Surprisingly, underestimation increased when the rate of updating was increased: the more information, the worse the performance! A possible
interpretation of this phenomenon is that subjects in estimating
growth rates in some way weigh successive differences.

The practical implication of this finding is interesting and
not incongruent with our daily experience: developments are more
easy to detect when you are confronted with information only from
time to time; persons continuously involved in a process often fail
to detect gradual changes that are easily discerned by people who
have been away for some time. Long and intensive experience with a
process does not qualify a person as a reliable predictor!

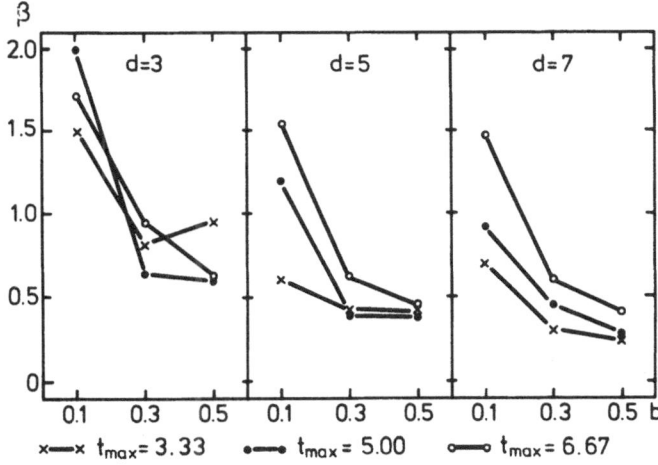

Fig. 12. ß-scores in the pond-and-duckweed experiment with limited
         updating rates (group 12).

INVERSE STATISTICS

The suggestion that subjects look at differences rather than at ratios of successive observations has an interesting practical implication, namely that it could be helpful to present inverse statistics. In the case of population growth you would not tabulate the number of people per square mile (increasing function) but the number of square miles per individual (decreasing function). The larger differences would then occur in the first few steps instead of later on. Other examples of this solution are not hard to find: crime rate can be presented by average time between two crimes; water pollution can be presented by the decrease of the abundance of fish; price increases can be presented by the amount you get for a dollar. Would extrapolations be less conservative when people are confronted with such data?

I will describe one experiment in which inverse numerical representation of exponential growth was used.

The general function describing the stimulus series was

$$y = ae^{bx} \qquad\qquad (8)$$

$x = 0$ to $4$; $b$ is negative; the 21 combinations of a and b that were used are

$$a \approx \quad 1.000; \quad b = -0.1, -0.2, \dots -0.7$$
$$a \approx \quad 25.000; \quad b = -0.3, -0.4, \dots -0.9$$
$$a \approx 625.000; \quad b = -0.5, -0.6, \dots -1.1$$

An example of the condition a    25.000 b = 0.9 is:

| year | index |
|------|-------|
| 1971 | 25.500 |
| 1972 | 10.368 |
| 1973 | 4.215 |
| 1974 | 1.714 |
| 1975 | 697 |

What level do you expect in 1980?

A group of 35 subjects (group 13) produced 735 responses. The estimates of $\alpha$ and $\beta$ are derived from the model presented before. In the present case we get:

$$\ln \bar{y} = \ln a + 5\ln\alpha + b(4+5\beta) \qquad\qquad (9)$$

The results are shown in Fig. 13. The most striking effect is

that the best quart of the subjects (first quartile) produce predictions according to the norm. The other subjects are still conservative, but this time the effect is due to low values of ß, whereas α
is almost equal for all subjects (see Table 2).

Table 2.
Results of the experiment with inverse
numerical presentations (group 13).

| a | quartile | α | β |
|---|---|---|---|
| 1.000 | 1st | 0.87 | 0.72 |
| | 2nd | 0.92 | 0.54 |
| | 3rd | 0.93 | 0.23 |
| 25.000 | 1st | 0.92 | 0.89 |
| | 2nd | 0.92 | 0.65 |
| | 3rd | 0.89 | 0.22 |
| 625.000 | 1st | 0.93 | 0.86 |
| | 2nd | 0.96 | 0.65 |
| | 3rd | 0.95 | 0.30 |

It is seen that only the subjects above the third quartile do
not profit from the inverse representation. (ß is still around 0.20).
Again the absolute size of the indices (a-factor) does not change
the misperception phenomenon.

The conclusion is evident: growth data are better understood

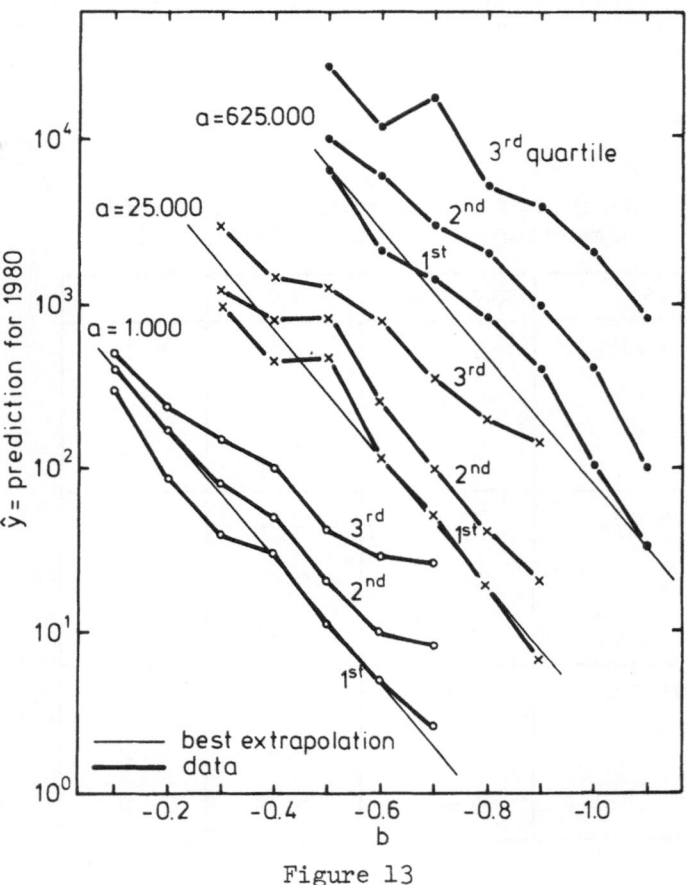

Figure 13

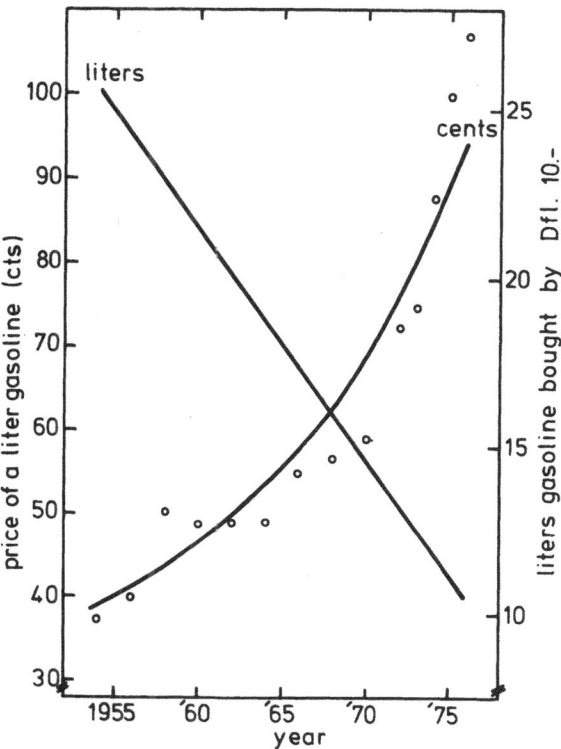

Fig. 14.    The price of gasoline in Holland (going up exponentially)
            and the amount of gasoline bought by Dfl 10,-- (going down
            almost linearly).

when the inverse process is presented.

An additional advantage is that the inverse of an exponential
process involving a positive constant (c-factor) is inverted S-
shaped, i.e. the first part cf the decreasing function is positively
accelerated, thus suggesting an ever accelerating decrease.  An example is shown in Fig. 14: the decrease of the amount of gasoline
bought by Dfl 10,-- is almost linear; whereas the price increase is
markedly exponential.

## SUMMARY AND GENERAL CONCLUSIONS

People tend to underestimate exponential growth considerably.
The degree of underestimation varies a little with the manner the
growth process is presented: directly, by tables or by graphs, the
latter being the worst!  The sensitivity to growth phenomena is not
enhanced by mathematical training, or experience with exponentially
growing processes.  When the processes themselves are observed instead of tabular or graphical representations a time factor is involved; this factor is not very important when the process develops
within some minutes, but a clear effect has been demonstrated when
it spreads over years.  There is a suggestion that, in judging growth
processes, people tend to concentrate on successive differences.
Two practical consequences follow from this suggestion:

1)   it helps when people do not observe processes continuously but
     only from time to time.
2)   it helps when inverse representations of growth are used instead
     of the usual increasing functions.

Both consequences were experimentally illustrated.


## REFERENCES

De Zeeuw, G. and Wagenaar, W.A.  Are subjective probabilities prob-
     abilities?  In: C. -A.S. Stael von Holstein (Ed.): The concept
     of probability in psychological experiemnts.  Reidel Publishing
     Company, Dordrecht, 1974.

Peterson, C.R. and Beach, L.R., Man as an intuitive statistician.
     Psychological Bulletin 1967, 68, 29-46.

Tversky, A. and Kahneman, D., Availability: A heuristic for judging
     frequency and probability.  Cognitive Psychology 1973, 5, 207-232.

Tversky, A. and Kahneman, D., The belief in the law of small numbers.
     Psychological Bulletin 1971, 76, 105-110.

Wagenaar, W.A. and Sagaria, S. Misperception of exponential growth.
     Perception and Psychophysics 1975, 18, 416-422.

# MODELS FOR PREDICTION UNDER CONDITIONS OF INTERACTION

P. C. Roberts

Department of the Environment

London, U.K.

## INTERACTION

History is replete with examples of legislation enacted to achieve specific changes but which resulted in small or negative change in the desired direction while generating unanticipated secondary effects. The attempt to introduce prohibition of alcohol consumption in the US was such a case. Analogous cases have begun to be noticed in the field of ecology. The following example is given by Holling and Goldberg[1].

> "The inland Dayak people of Borneo live in large single homes or long houses with up to 500 or more under one roof. This concentration of population allowed WHO to develop a thorough and orderly spraying of every long house, hut, and human habitation with DDT. The effect on health standards was dramatic with a remarkable improvement in the energy and vitality of the people - particularly those remote tribes who had not previously had access to medical aid. Nevertheless, there were interesting consequences that illuminate some of the properties of ecological systems.
>
> There is a small community of organisms that occupy the thatched huts of these villages - cats, cockroaches, and small lizards. The cockroaches picked up the DDT and were subsequently eaten by the lizards. In consuming the cockroaches, the lizards concentrated the DDT to a somewhat higher level than was present in the cockroaches. The cats ate

the lizards and, by eating them concentrated the
level of DDT still further - to the point that it
became lethal.  The cats died.  When the cats
disappeared from the villages, woodland rats in-
vaded, and it suddenly became apparent that the
cats had been performing a hidden function - controlling rat populations.  Now with the rat came
a new complex of organisms - fleas, lice and
parasites, and this community presented a new
public health hazard of sylvatic plague.  The
problem became serious enough that finally the
RAF was called to parachute living cats into
these isolated villages in order to control the
rats.

The story isn't finished at this point, however,
since the DDT also killed the parasites and pre-
dators of a small caterpillar that normally causes
minor damage to thatch roofs (Cheng 1963).  The
caterpillar populations, now uncontrolled, increased dramatically, causing the roofs of the huts
to collapse."

It is a truism that ecological systems have high internal resilience because those systems which we observe are the ones that
have survived through adaptation to the various traumas and shocks
in the course of their evolution.  There is natural selection among
potential eco-systems paralleling the natural selection of species.

At an elementary level this argument is applicable to the behaviour of human beings.  For example an acceptable answer to the
question of why human parents care for their children is to hypo-
thesise a society in which the parents abandon their offspring at
birth.  Clearly this society would be short lived or indeed could
never have arisen.  We observe only these societies where care for
the young is genetically inbuilt or culturally imperative.

At a higher level it is interesting and useful to enquire
whether those conventions and institutions which are characteristic
of human societies correspond to those properties of systems which
confer the attributes of resilience and survival.  The first substantial attempt to justify the truth of this proposition in the
field of economics was made by Adam Smith [2] in 'The Wealth of
Nations' two hundred years ago.  Smith followed through the feedback chains which, operating in a commercial society, produced a
total function working so well as to merit the description of a
'hidden hand'.  It is possible that the system model described by
Smith was adequate for explaining the general behaviour of economic
systems.  However, we can regard the perturbations in eco-systems
with more equanimity than those which occur in our economic systems

and hence the attempts by subsequent economists to devise models
which could provide explanations of inflation, recessions and income
distribution.

PREDICTION AND PARADOX

Prediction of the effect which will result from a given action
contains an inherent paradox. It is necessary to assume that nearly
all of the system continues unchanged - which implies that it lies
in a zone of stability so that the negative feedbacks are damping
out any perturbations; and yet for that part of the system where
interference is directed, the feedbacks will not negate the action
which is planned. A striking example of this paradox occurs when
some portion of the road network in a city is made unusable to
traffic. The common assumption is that the traffic which previously
used the banned links and nodes will redistribute itself over the
surrounding roads. In those experiments for which measurements of
traffic flows have been made before and after the interference, it
is found that the additional loads which ought to be observed on
the surrounding roads fail to appear. In order to understand this,
it must be appreciated that the measured flows of traffic in a
modern city are equilibrium flows. The volume of cars available is
more than enough to saturate the road system. Trips are made by car
in sufficient number to create that level of congestion at which the
next tranche of potential drivers is deterred by the delays which
will be experienced. Thus the subtraction (or addition) of road
capacity will require a change in the actual number of trips made
by car in order to re-establish a new equilibrium state. This state
appears to be determined largely by an approximately constant 'acceptable delay time' in the attitudes of motorists. It follow that
the traffic densities on roads tend to remain at unchanged levels.

A similar effect has been observed after the introduction of 8
carriage trains to replace 6 carriage trains serving the rush hour
commuter demand. The object of the change was to diminish congestion, but the result was to change the shape of the demand density/
time profile. The shape of the distribution altered to become nar-
rower with a higher peak. The same number of people travelled but
they now concentrated their start times into a shorter duration, so
that actual congestion levels in the trains were as high as before.
In most of the examples of action failing in its intended effect
(or the perversity of people - depending upon one's point of view)
the folly is insignificant. Laws can be repealed, the squandering
of resources can be put down to experience and ecological systems
usually re-establish themselves after the most dire insults. One
can reconcile one's self to this state of affairs - and I have noticed that intellectuals past the age of 40 indeed do so - by adopting the posture of an indulgent parent towards the mistakes of a
child during the learning phase. While the mistakes are trivial,

and more specifically where they are not lethal, this posture is not inappropriate. But no parent wishes her child to learn about electricity by running the risk of electrocution: and we should not regard the follies of technology and management with indulgence. Some of our current activities hold the danger of irrevocability; it has become important to avoid mistakes simply because we now have the power to make irretrievable errors on the global scale.

There is one simple recipe for avoidance of errors and this consists of generating change very slowly; only well tried methods are used. This course is currently advocated by numerous minorities - 'low impact technology', 'small is beautiful' etc. While these movements might possibly generate a grass roots reaction they are unlikely to impress those individuals responsible for the decisions which initiate change. Only quantitative methods based on evidence is acceptable coin.

## PREDICTIVE MODELS

The reason for poor prediction, whether of the effect of larger carriages in commuter trains or of blanket insecticide spraying lies in the use of poor models. We need models which incorporate feedback; models which are close enough to real system behaviour for the inspection of model behaviour to influence decisions. The Forrester and Meadows models are unconvincing because important feedback mechanisms are omitted. The influence of 'World Dynamics'[3] and 'Limits to Growth'[4] has been confined mainly to the lay public. Before there can be general acceptance of a world model, there must be the fulfillment of the criteria by which models whether in physics, ecology, or economics are judged. The objection to Worlds II and III and to the Mesarovic Pestel[5] model is that no demonstration is offered of these structures actually generating behaviour comparable with real world time series. One of the reasons for this failure is the desire on the part of the modellers to confirm the conclusions they had reached even before starting the modelling exercise.

If we suppose a model of the world system to be constructed which is not intended to prove that disaster is imminent, but primarily to represent as faithfully as possible the behaviour that is actually observed then this would offer a much higher chance of credence - and the possibility of 'interfering' actions which produce the result intended. As a basis for such a model we can do no better than quantify the argument advanced two hundred years ago by Adam Smith that the operation of the system depends on the role players pursuing enlightened self interest. A model in which this rule is applied rigorously has been constructed (SARUM 1976)[6] and it appears capable of explaining in quantitative terms much of macro-economic behaviour. In some ways the results of projection runs

into the future using this model are a comfort to those who feel
sceptical about world collapse in 2020 à la Forrester.  Taking the
constraints on natural resources set by planetary inventories,
there appears to be no good reason for thinking that a dramatic
overshoot is likely.


## DISCOUNTING

In order to formulate the decision algorithms in SARUM it is
necessary to define the rate at which future benefit is discounted.
Observation suggests that the real discount rates used in industrial
and commercial operations are of the order of 10% pa.  In the area
of social services such as health and education the discount rate
appears to be lower - about 2 or 3% pa.  The SARUM structure has
been used to follow the time path of a society faced with depletion
of a prime resource - but able to provide a substitute at a substantially higher price.  If the onset of depletion is rapid, pre-
valent discount rates are too high to avoid temporary scarcity and
consequent hardship.  There appear to be no clear cases of essential
resources for which this is likely to occur.  This result justifies
the conventional economist's position in respect of resources.


The conventional economic wisdom on pollution is to suppose a
balance between the marginal real or notional costs of the pollutant
and the marginal cost of abatement measures.  The concern of economists has not been with the rationality of this proposition, but
with the difficulty of ensuring equity between those responsible for
production of the pollutant and those who suffer the consequences.
Few would dissent from the marginal costs argument in the case of
noise as a pollutant.  Aside from the equity problem noted above, we
can aver that the noise to be tolerated is such that additional expenditure to lessen it by 1 dB is equal to the value placed on a
decrease of 1 dB (i.e. the amount the sufferer is prepared to pay
for such a drop in discomfort).

In the case of pollutant hazards that take some time to manifest (like thalidomide) the sufferers are compensated and legislation is enacted to avoid future occurrences.  Thalidomide took only
one human gestation period to manifest and provide an unambiguous
signal.  If a pollutant takes 10 years or 100 years to deliver its
signal what corrective measures exist?  At a 100 year interval, even
the knowledge of impending danger will cause no corrective action if
the decisions are taken consistent with a 10% pa discount rate because the present value of the cost to be endured a century ahead
is diminished by a factor of about 20,000.

## POLLUTION AND SURVIVAL

The problem of the slow acting poison has to be faced in
accounting for the survival of the human species over a period of
about one million years.  All sorts of practices like widespread
drug addiction or inbreeding could generate the source of extinction
- and yet the race has persisted.

The answer, if the parallelism between eco-systems and social
systems is accepted, is that with extensive heterogeneity present
there will be natural selection favouring those societies which
avoid harmful practices - even though these practices might take
centuries to manifest.  Two factors have altered significantly in
recent times.  Heterogeneity is being rapidly diminished and anthropologists have to be quick to make their records before the last
isolated communities vanish into the few big homogeneous cultures
that mark this century.  The second factor is the global character
of our modern pollutants.  Substances are discharged into the biosphere in such quantities that changes in global parameters become
measurable in a few years.

The conventional economic wisdom offers no answer to the problem of the delayed action pollutant.  The 'natural' answer through
heterogeneity and selection eliminating the offenders cannot operate
in modern conditions with global pollutants.  What is required now
is the widespread acceptance of propositions - concerning the systems
of the biosphere - which are not revealed by incrementalism nor do
they follow from extrapolative methods.  We shall not discover the
ultimate penalties of increasing the carbon dioxide of the atmosphere by monitoring the effects carefully and regulating emissions,
because the danger arises cumulatively and halting emissions is not
followed by rapid decline to the background level.  The carbon
dioxide emission/absorption model will become of steadily growing
interest in coming decades, and it will be one of several models
for situations which require decisions before the full doubt-
dispelling signals are generated.

Some of our models in the natural sciences have enabled laws
of great generality to be enunciated.  The best examples of these
are the first and second laws of thermodynamics.  It is unnecessary
to carry out trials with all possible versions of perpetual motion
machines because we know that the underlying structure of this universe precludes such a device.  It is unnecessary to do trials of
gambling against even chances with a strategy of stake doubling at
each loser, because we know that ultimate ruin is a certainty.

The SARUM structure approaches generality in the description
of industrial societies more closely than the models which preceded
it.  However, it offers no solutions to the more deep seated ecological problems.  The need now is for generalised models of pollutants.  If it were necessary to construct a new model for every new
substance that is being devised then the labour to trace pathways

and discover the secondary and tertiary effects is so great that the
race is lost - it becomes comparable with trying all the conceivable
perpetual motion machines.  Construction of generalised models
relating to eco-systems under threat is a formidable challenge (we
do not yet have any powerful theorems from general system research).
However, unless the challenge is met, the prospect is poor, because
the 'low impact' and 'zero growth' schools appeal to the emotions
and not to the intellect.

REFERENCES

1   Holling C.S. and Goldberg M.A. (1973) Managing the Environment
    pages 31-32, EPY-600/5-73-010   US Government Printing Office

2   Adam Smith  Wealth of Nations, Everyman, Dutton

3   Forrester J.W. (1971) World Dynamics,  Wright-Allen, London

4   Meadows D. (1972)  The Limits to Growth, M.I.T. Press,
    Cambridge, Mass.

5   Mesarovic M. and Pestel E. (1974)  Mankind at the Turning Point

    E.P.Dutton & Co.  New York 1974

6   Roberts P. et al., Paper to IIASA 4th symposium on Global
    Modelling,  September 1976.

# Session III

# METHODOLOGY: ASSESSMENT

# UNIVERSAL CHECKLISTS IN THE CONCEPT OF "IMPACT TREES"

U. Liebermeister

Industrieanlagen-Betriebsgesellschaft mbH

Ottobrunn, West Germany

On planning a Handbook for the Environmental Impact Analysis the theoretical foundations and the usual difficulties of the assessment procedure have to be examined and differentiated with particular care.[+]

**Procedure diagrams for the assessment**

For such assessments a considerable amount of procedure diagrams have been made available. What I would like to consider here is based on the procedure diagram as shown in Figure 1. This particular diagram emphasizes the "grammar" common to all assessments. Therefore, only minor modifications are necessary for adapting it to any analogous problem like technology assessment, cost-effect-analyses, assessments of goal systems or the assessment of socio-economic systems.

**Discovery of the types of effects**

The diagram enumerates various logical steps, such as problem analysis (steps 0 and 1), search for alternatives (steps 2 and 6) and decision making (steps 7 and 8). I shall not discuss any of them here. In this paper I want to concentrate on step 4, the analysis of the consequences of individual measures. Or to be more precise, I want to concentrate on the problem of how to discover what "types" of effects are to be expected.

---

[+]
This paper is based on research done for the Umweltbundesamt of the Federal Republic of Germany.

| 0   Rough orientation |
| 0.1 Type of problem and margin of action |
| 0.2 Environmental aspects of the problem |
| 0.3 Exclusion of trivial cases and parts |

| 1   Statement of objective |
| 1.1 Framework and nature of the problem |
| 1.2 Topical goals |
| 1.3 Basic constraints |

| 2   Description of proposed measures |
| 2.1 Approaches to the solution |
| 2.2 Alternatives to be checked |
| 2.3 Reasoning for the selection of these alternatives |

| 3   Check of relevance to the environment |
| 3.1 Conflicts between topical and environmental goals |
| 3.2 Preassessment of the measure |

| 4   Determination of environmental effects |
| 4.1 Analysis and prognosis of the state in question without |
| 4.2 Prognosis for the state with measure              (measure |
| 4.3 Comparison of the prognoses |

| 5   Ecological and sanitary valuation of impacts |
| 5.1 Compilation of aspects |
| 5.2 Compilation of criteria |
| 5.3 Evaluation under these criteria |

| 6   Search for and evaluation of alternatives |
| 6.1 Search for mitigation measures |
| 6.2 Search for alternatives |
| 6.3 Control steps |

| 7   Assessment of impacts |
| 7.1 Compilation of non-ecological and non-sanitary contributions |
| 7.2 Trade off between all contributions |

| 8   Decision on the measure |
|     (Recommendations; Selection procedure) |

Figure 1.   Procedure diagram as intended for the "Handbook on
            Environmental Impact Analysis"

As to any further assessment steps, especially the choice of models cr the approach of the
prognosis and last but not least the evaluation
and comparison of alternatives, we take it for
granted that all types of consequences are perceived well enough, regardless cf whether their qualitative or quantitative discussion should pose any
problems.

As assessment project will be a failure or of no
value at all if one single important response has been
overlooked.  It might even become a danger since in the
final decision it often serves as an unquestionable
authority.  It is the purpose of this paper to reduce
this very risk.

*Risk of overlooking important effects*

According to the procedure diagram, it is taken
for granted that in step 2 the measures[+)] to be examined
or the range of alternatives in question were given an
adequate definition which is as complete as possible.
Consequently, this means that any activity having to
do with the measure in question or any mitigation
measure is also subject to the assessment procedure
from the very beginning.

*Subject of the assessment: sensible package of measures*

In any case, the assessment procedure of such a
measure consists of the discovery, the analysis and
the evaluation of its effects.[++)]

*Definition of "Assessment"*

As it is generally known, the term "effect" has
to be defined as follows (regardless of the goals pursued by the measure.):

For any "state", which is intended to be influenced by a measure, two possible courses of development have to be confronted with each other.

*Definition of "Effect"*

---

[+)] The term "measure" is used synonymously for "action",
"plan", "proposal", "alternative", etc.

[++)] There is absolutely no assessment task which cannot
be fitted to this definition of the problem; it
might, however, be necessary to define a suitable
fictitious measure.

1)      The course of development which this state would
        take anyway, that means the course it would pre-
        sumably take under the influence of all the con-
        ditions already existing.

2)      The course of development the state will take
        presumably, if the planned measure is carried
        out in addition to the conditions mentioned above.

The "difference", i.e. the totality of the dis-               Difference
similarities between these two courses of development,       between
is an equivalent to the "unbiased and complete pattern       two courses
of effects" of the measure in question; it is this          of develop-
complete pattern of effects which we have to analyse.        ment
For obtaining a short list of items indicating the
major impacts we must simplify it by integration and
by some heavy cut-offs.

According to the type of measure the states to               Examples
be analyzed comprise such heterogeneous phenomena as

- the state of equilibrium of a section of our natural environment (e.g. when assessing the disposal
  of waste water or the construction of roads).

- the legal situation and the behavioral dispositions
  of the people (e.g. when assessing safety regulations in public transportation).

- the concepts about goals for mankind (e.g. when discussing "limits to growth" or some utopian states).

- the state of knowledge in science (e.g. when discussing additional support for cancer research).

It is well known that, due to their enormous num-            Assessment
ber and complexity, the interactions in nature and so-       risks
ciety are basically able to break through almost any
boundary of physical, scientific or systematic kind.
Thereby, they may allow a measure to have impacts in
most surprising areas.  Boundaries of systems which had
been taken for granted for a long time have had to be
revised in hundreds of cases, e.g. we have been using
all kinds of sprays for quite some time, but it has
only been recently that we have become aware of a very
indirect impact path affecting the ozone shield of the
earth and maybe leading to cancerogenic effects.  Or
we want to ensure our supply of energy and are surpri-
sed by side effects like citizens' movements at the
sites of projected power plants.

Every model concentrating only on the intentions

followed by a specific project, automatically excludes
large sections of the complete pattern of its effects.
This narrowing leads to an assessment risk which is
often hard to evaluate.

It would be safest but most demanding to consider each time the whole of the universe as the
"state" to be analyzed. Any confinement to a sub-
system (e.g. to the immediate surroundings of a production plant or a building site, to a small selection of detrimental factors or to the near future
and the like) implies that outside this reduced "region of integration" all the dissimilarities caused
by the measure may be considered as nonexistent or
as negligible. It is true, that in many cases we
know of fairly good reasons for such cut-offs. But
we also know that it was such risky cut-offs which
led to the extensive formalism of the assessment.

*Universe: safest approach*

*Cut-offs concerning the area of integration*

The result is a strong conflict of goals between the required correctness and the required
practicability. To be able to carry out thoroughly
the required comparison of the courses of the development with and without the measure, we should have
to understand our whole universe, at least all phenomena related to the "spaceship earth", and transform this knowledge into a supermodel.

*Conflict of goals: correct-ness fea-sibility*

Instead, all we can do is to attempt a fast scan
of the universe for possible problem areas, thereby
drastically reducing the original pattern of effects.

*Scan of the universe*

Experience shows that the elimination of supposed impacts by arguments is considerably easier for
us than the creative listing up of possible impacts.
This is true even if each of these types of impact
(e.g. noise, pesticides etc), looked at separately, is
well known to us. Therefore, it is much safer to
choose a strategy of elimination based on a suitably
generalized checklist system than any intuitive procedure based only on our individual experience and
knowledge.

*Strategy of elimi-nation in-stead of strategy of search*

An approach complementary to the above-
mentioned "pattern of effects" approach makes use
of the obvious relationships between the effects
hitherto regarded as isolated dissimilarities. As
in the first approach here again a measure or a
package of measures is supposed to assume the role
of the "cause".

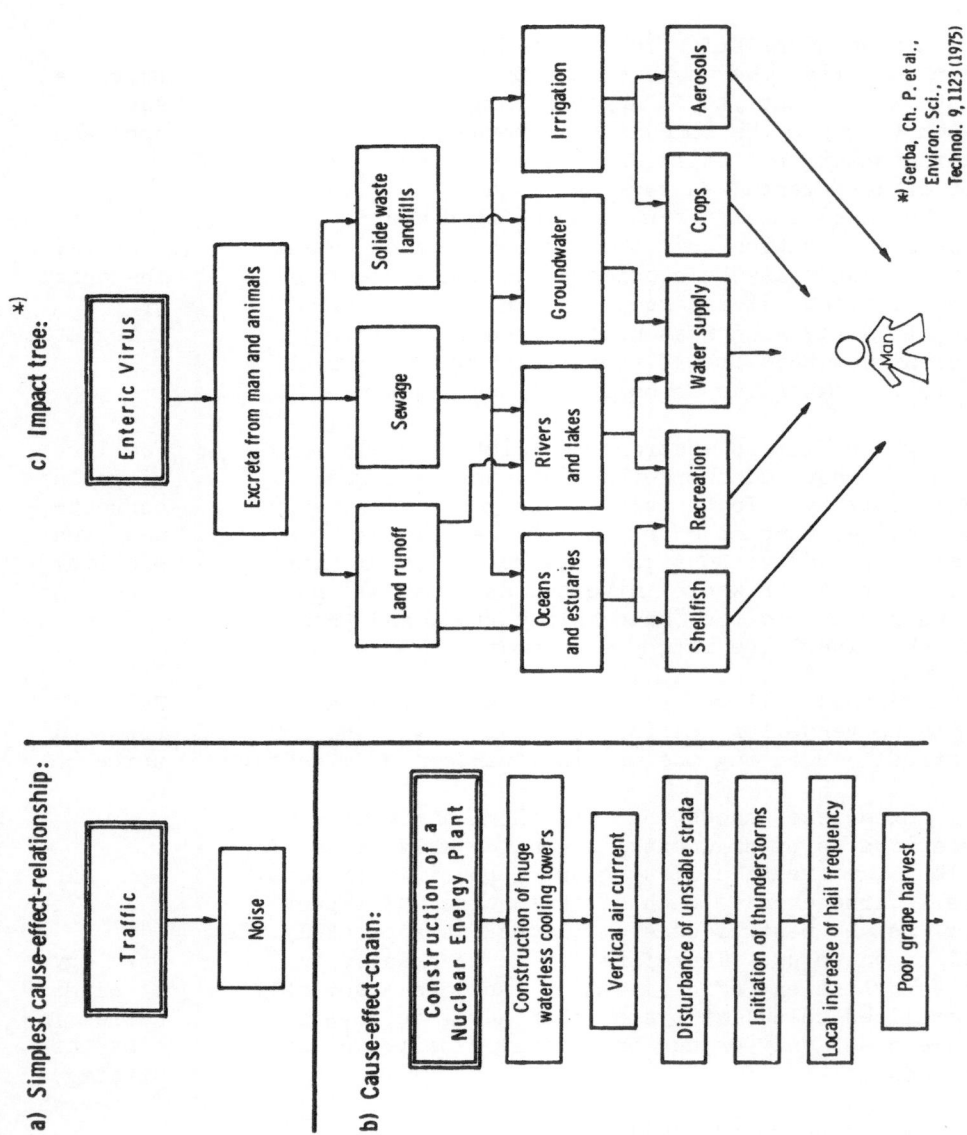

a) Simplest cause-effect-relationship:

Traffic → Noise

b) Cause-effect-chain:

Construction of a Nuclear Energy Plant → Construction of huge waterless cooling towers → Vertical air current → Disturbance of unstable strata → Initiation of thunderstorms → Local increase of hail frequency → Poor grape harvest

c) Impact tree: *)

Enteric Virus → Excreta from man and animals

Land runoff, Sewage, Solide waste landfills

Oceans and estuaries, Rivers and lakes, Groundwater, Irrigation

Shellfish, Recreation, Water supply, Crops, Aerosols → Man

*) Gerba, Ch. P. et al., Environ. Sci., Technol. 9, 1123 (1975)

Figure 2.  Three types of graphs for cause-effect-relationships

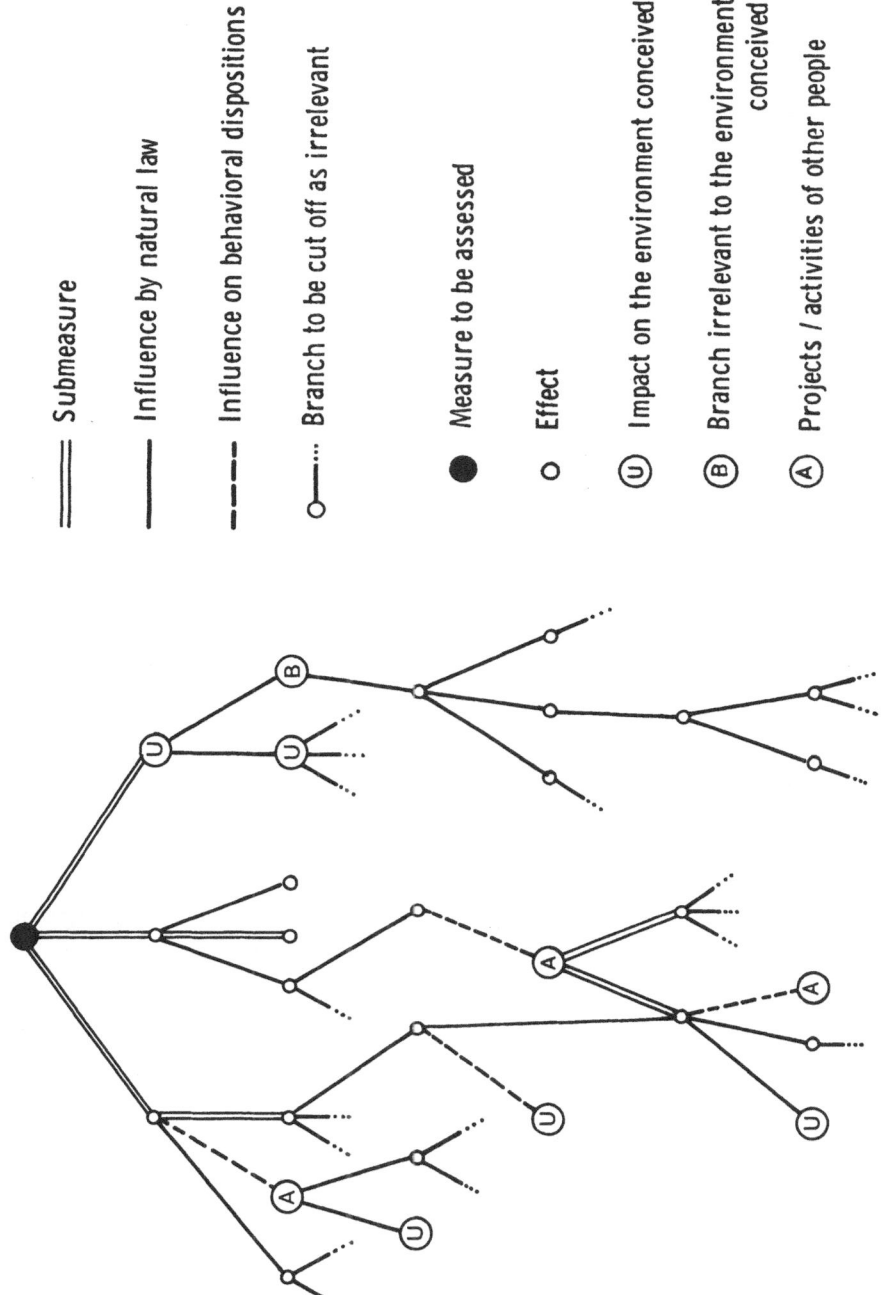

Submeasure

Influence by natural law

Influence on behavioral dispositions

Branch to be cut off as irrelevant

● Measure to be assessed

○ Effect

Ⓤ Impact on the environment conceived

Ⓑ Branch irrelevant to the environment conceived

Ⓐ Projects / activities of other people

Figure 3.   Basic Principle of Impact Trees

The simplest cause-effect-relationships (Figure 2a) do not meet the requirements described above because only one single effect is left.

The cause-effect-chains (Figure 2b) which contain many more details are not satisfactory either, since they are also orientated too much towards one single final effect or towards one single group of people concerned by the measure.

On the other hand, a complete graphical representation of the full pattern of effects without any cut-offs is not realizable.

A third possibility, lying somewhere between these extremes, is the concept of graphs with a tree-like structure (Figure 2c). They are frequently used in technical literature. Their purpose is to present and to link all those problem areas which remain after some kind of major cut-off procedure.

For these graphs I want to enumerate some basic qualities using the general principle of impact trees as shown in Figure 3. These qualities will be useful for the composition and structuring of the checklist system to be presented later on.

There are three basic types of cause-effect-relationships which differ profoundly as to the degree to which they can be influenced and predicted.

1) A measure can be split up into project elements or "submeasures" (e.g. a law contains submeasures dealing with its goals, with regulations of technical details, with the timing of its entry into force, with legal possibilities of making objections, or with regulations on jurisdiction). This type of relation is determined and controlled completely by the person or the group of persons in charge of the planning of this measure.

2) A measure, a submeasure or an intermediate effect (e.g. the disposal of waste water) operates on the environment according to the natural laws. In this case the course of effects is not determined by human expectations at all. This is true, unless additional supporting measures are applied (e.g. construction of a sewage treatment plant or artificial aeration of water bodies). On the other hand, these natural laws represent potential "services" of nature which we take for granted and often heavily exploit in best interest of our projects.

3) By means of the measure or a submeasure the behaviour of other people is influenced in reference to their own projects (Figure 3, label A ). The measure may enforce stimuli, it may interdict some action, it may put up standards etc. These relationships have their direct effect only on men not on nature. They may eventually result in a "change in a market situation" or in "boycotting" or in "the search for a loophole." Such measures often consist only of a piece of paper and are therefore negligible as far as their direct impact on the environment is concerned. Nevertheless, such "paper measures" can have a strong effect on the environment. This is the case if they condition the range of liberty of action of other people in a manner beneficial or detrimental to the environment. Accordingly, this third type of cause-effect-relationships may not be discussed without considering the effects of the stimulated activities.

*Influences on activities of other people*

A useful quality of impact trees is the fact that they can be split up freely into modular subtrees. By influencing one single branch, e.g. by a protective counter measure, the adherent subtree can be changed essentially without significantly changing the remaining tree. Similarly, a subsequent alteration of a measure (e.g. by passing an amending bill or by changing its regulations) changes only certain subtrees, whereas the remaining tree stays as it is. On the other hand, any impact tree can be regarded as a part of more comprehensive impact trees. For a measure can be seen as part of a higher plan, a superordinate project, part of a division of labour or the result of an ideology.

*Modularity of impact trees*

It is an important handicap that there is no unambiguous and transferable formulation of an impact tree: The same measure applied at different times, in different surroundings, in connection with different initial states can have effects which differ widely. Furthermore, any variation of the thresholds for importance necessarily means a major change in the complexity and in the structure of the impact tree. Supposing the threshold is set high enough, a tree which was originally rather complex, will be reduced to a few elements only. On the other hand, the more we lower the threshold and approach the unbiased and complete pattern of effects the more its graphical representation turns into a network.

*The relation between measure and impact-tree is not unambiguous*

In general it is only a minor portion of the
effects which lies in that specific section of the
universe which we declare to be the "environment"
interesting for the assessment at present (i.e.
that section for which thresholds of relevance and
criteria for evaluation had been defined before or
will be defined).  In Figure 3 I have marked these
cases with this sign: (U). Any of those "impacts"
may become part of the final impact statement, all
other effects are likely to be disregarded very soon.

The list
of impacts
depends on
the speci-
fied en-
vironment

In our Handbook for the Environmental Impact
Analysis, (U) obviously concerns the natural environment and the human health in terms of "classical environmental protection."  But any other sort of environment (e.g. an environment consisting only of
budgets) could be chosen.

If we restrict the meaning of the term environment very much, certain subtrees will not influence this
environment any more.  Thus, subtree (B) in Figure 3
does not contain any (U) ; it could present the chains
of effect connected with the date of entry into force
of a new taxation procedure.

Accordingly, there are lots of measures which contain only such subtrees and have no impact whatsoever
on such an environment.  Or the impacts are so unim-
portant that any improvement of these measures would
not mean any noticeable improvement of this environment.

Unimpor-
tant
measures

I have presented to you two approaches so far:
first the approach consisting of the attempt to lay
open the unbiased difference between courses of development with and without measure and, secondly, the
"impact tree approach".  The question is now how to
amalgamate these two approaches effectively to provide a practicable procedure.

Amalgama-
tion of the
two approa-
ches

To keep the risks as low as possible I would not
like to do without the strategy of the scanning of the
universe as part of approach number one.  The final
procedure must, therefore, be based on the union set
of a l l possible effects in one way or the other.

Union set
of effects/
phenomena

As we learned from the "impact tree approach",
this union set comprises a l l natural proceedings
in the universe (which in itself is nothing but the
union set of a l l imaginable environments).

Union set
of environ-
ments

Because we can split the measures, as mentioned above, and because these measures do assume important roles as far as the regulation of behavioral disposi- tions is concerned, the union set of a l l measures is part of it as well.

*Union set of measures*

At first sight any attempt to get complete hold of such union sets seems in vain. They are so compre- hensive and so heterogeneous, that we will not even suc- ceed to enumerate the various coordinate axes for the mathematical spaces which they belong to.

We can escape this dilemma by covering both mathematical spaces by suitable networks and thus subdividing them. Figure 4 captures the essence of this process. If we decrease the size of the meshes as shown in sector 1 or 2, our description, abstract enough in the beginning, will become more and more concrete.

*Disaggre- gation of union sets*

This network divides the space and its sectors into a number of patches (to be more precise: into a number of multidimensional fragments with very fuzzy boundaries).

By enumerating only these patches (or the con- struction principle involved) the intended network can be described in a reasonably comprehensive manner. For this purpose one chooses one representative term for each patch (as shown for sector 3) and combines these terms to establish checklists.

Naturally, there is no indisputable set of rules for the subdivision and the principle of selection of the representative elements.

What we want to achieve, however, is an intensive stimulation of adequate chains of associations. When the checklists are made use of, i.e. when a representa- tive term like "groundwater" or "traffic" is read, some associative scanning of the patches implied by these terms will take place in the brains of the user. This process is indicated in sector 4. Thus, items like "traffic sign," "bicycle" etc. will be covered, too, even if they are not explicitly mentioned in the checklist.

*Stimulation of mental associa- tions are required*

The more one disaggregates the items of the check- lists, the smaller every patch is, the less is demanded as to the associative ability of the user. On the other

*Associa- tive abili- ties of*

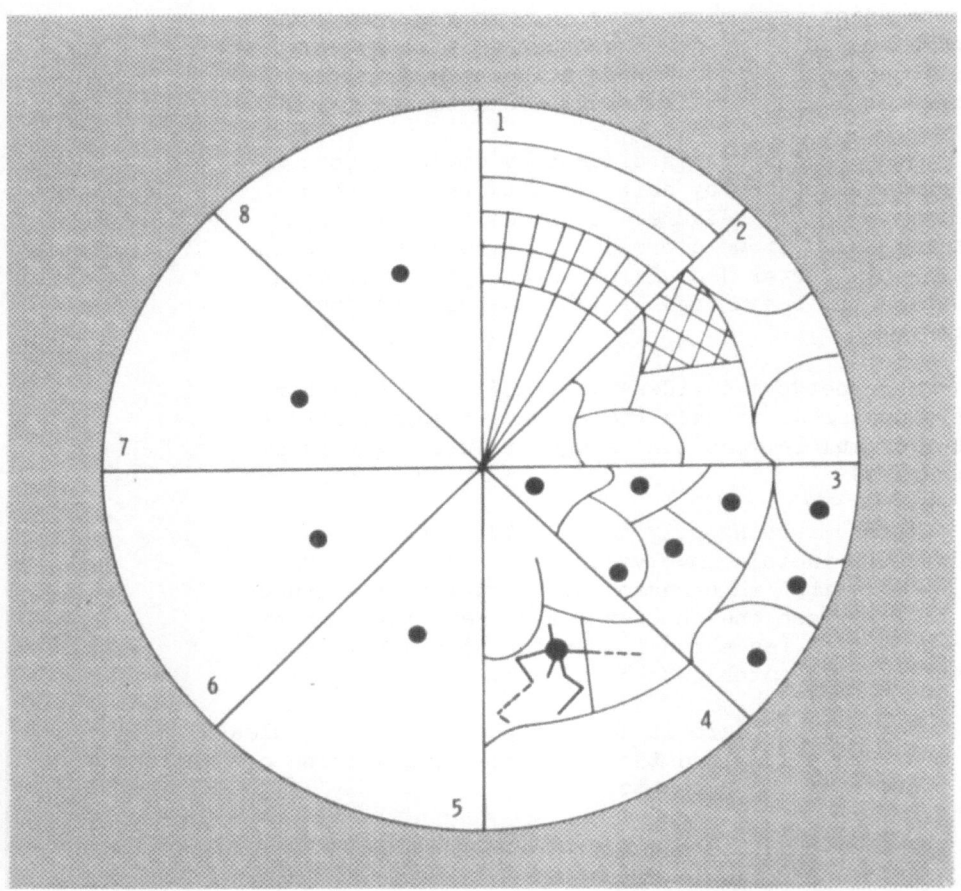

Figure 4.   Disaggregation and Representation of a Union Set
            - 2-dimensional reduction of the problem -

hand, the more items you introduce, the less practicable the system of checklists gets.

To handle this handicap, three levels of abstraction, i.e. three different sizes of meshes, were
deliberately chosen for this project.  The corresponding checklists or modules are given the names:

   I     Checklists for orientation
   II    Framework checklists
   III   Magnifying glass checklists.

For the prototypes developed so far, I have tried

- to choose the representative items as system-
  consciously as possible in order to provide
  them with an associative potentiality as versatile as possible,

- to choose the level of aggregation and the number of items so that the duration of one round
  of application of these checklists can be kept
  in the order of magnitude of an hour,

- to offer frequently useful suggestions for
  further disaggregation.

Now I would like to present some sections of our
prototype for the checklist system[+).

Let's begin with the
- orientation checklist for the environment
  (see Figure 5a).

It divides the universe into a reasonably small
set of sections which can be distinguished fairly easily.  This division is not so much a question of right
or wrong but whether it is easily adaptable to the
requirements of practical application.

Orientation
checklist
for the
environment:
modular
items

_____

[+) For the environmental checklists the material, i.e.
the suggestions  for the choice of patches and representative terms are mainly taken from scientific
systematisms and already existing checklists for
special purposes.  The checklists of measures are
mainly based on the analysis of annual handbooks,
diagrams of organisational structures, on daily
newspapers and so forth.  In the selection of this
material, special efforts were made to represent the
two union sets as well as possible.

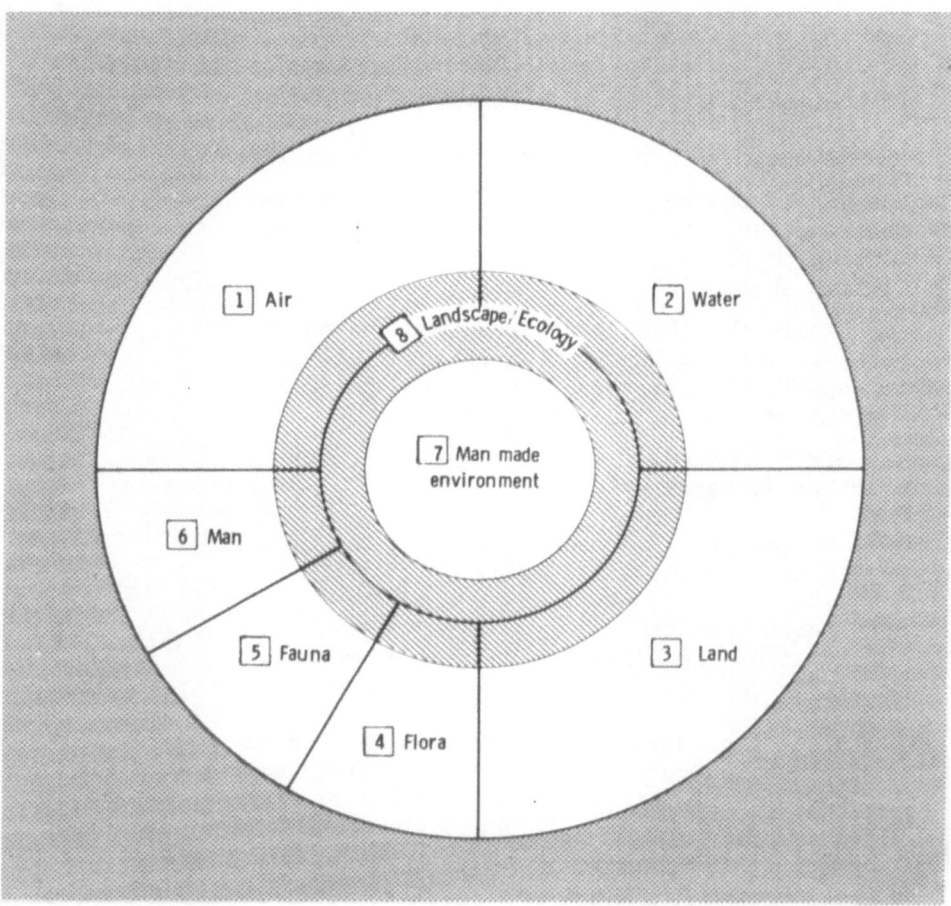

Fig. 5a.   Subdivision of the space of environments – Orientation
           checklist.

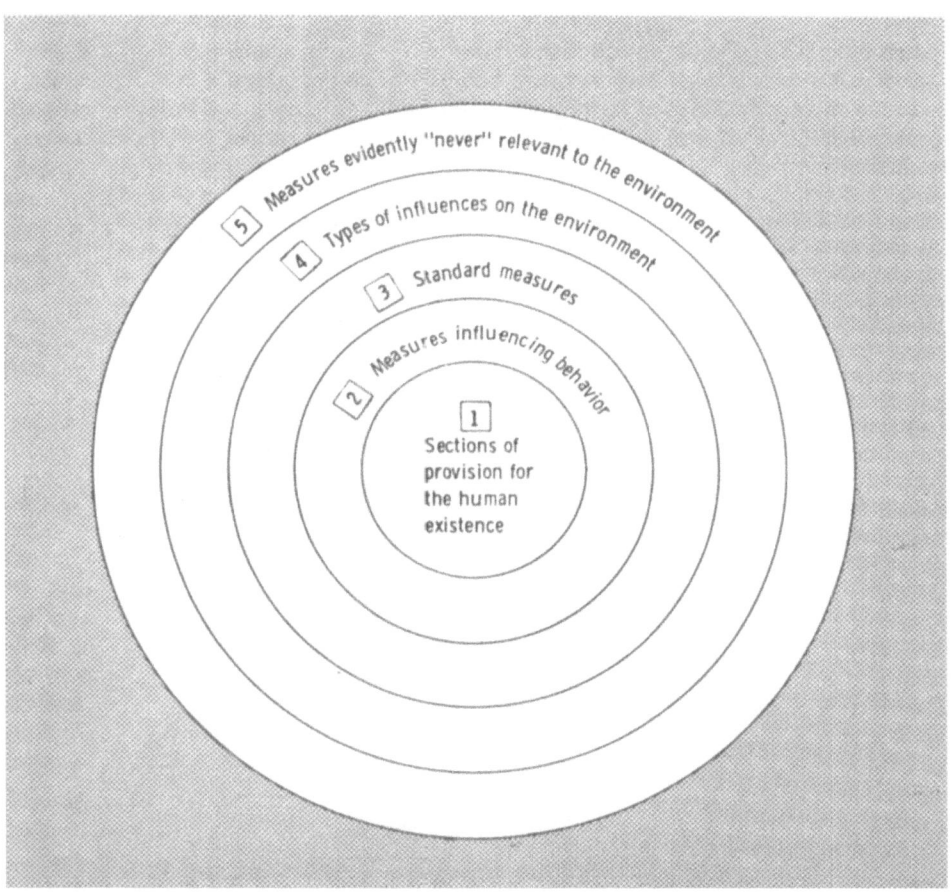

Fig. 5b.   Subdivision of the Space of Measures — Orientation
           checklist.

Thus we easily find the

- inanimate world (1 + 2 + 3)
- animate world (4 + 5 + 6 + 8)
- the human world (6 + 7 + a specific subset of 8)
- the technical world (a specific subset of 7)
- the socio-economic systems (another subset of 7)
- and so on.

I have disaggregated each of these sections into a particular checklist of environmental realms worthy of protection with a resolving power about hundred times stronger.  Two of the eight sections are presented in Figure 6.

*Framework checklist of the environment*

The items under the heading "Components and Expressions" are supposed to remind you of the kind of subsystems the sections of the universe can be divided into.

The items under the heading "Basic Processes", however, are supposed to remind you of the kinds of natural interaction within or among the subsystems.  It is mainly the ability of these interactions to form connections to other sections and subsystems of the universe which is responsible for the enormous complexity of the pattern of effects.

Thus, the eutrophication of water bodies is essentially an excessive increase of such a basic process as the "growth", and waste water treatment as a protective measure can be looked at as the obstruction of another basic process - the diffusion of pollutants in the medium water.

For dealing with more detailed questions, the third part of these checklists offers items of a different kind.  Instead of enumerating additional expressions or processes they just stimulate major aspects for further disaggregation[+)]!

*Aspects of disaggregation*

Some of these aspects have continuous expressions like the various time aspects or spatial aspects which are to be regarded more closely, e.g. when discussing time horizons of impacts or geographical subsystems.  Other aspects find their expressions in what we called "magnifying glass checklists," such as the periodic table of elements, toxic substance lists, lists of diseases, lists of products or of pollutants etc.

*Time and Space*

*Magnifying glass checklists*

2 WATER

Components/Expressions

- Flowing waters
- Stagnant waters
- Groundwater
- Oceans/seas
- Ice/snow areas
- Water involved in
  the weather process
- Physically/chemically/
  biologically bonded
  water
- Sewage/effluent

Basic Processes

- Water cycle (Air, Land)
- Absorption/transporta-
  tion/conversion/preci-
  pitation of gases/sub-
  stances
- Self-purification
- Aquatic flora/fauna
- Biological processes
  in organisms

Aspects for Disaggregation →

- Natural components
- Impurities                    +
- Subsystems (regional/
  topical)                      +
- Classes of quality
- Distribution of
  indicator values
- Geographical distri-
  butions and gradients
- States of aggregation        +
- Degrees of sensitivity
- Interfering variables
- Degrees of menace
- Stratifications
- Aspects of "time" and
  their 1 and 2 deriva-
  tion
- Roles in cause-effect-
  relationships
- Timescales of processes
- Basic process in water       +
- etc.

Corresponding checklists of
the handbook:

Indicators: Content of Impurities; BOD; pH, etc.  → —

Utilizations: Drinking, cooling, sailing etc.  → —

Loads: Waste water; waste heat etc.  → —

Goal Systems: — —  → —

Figure 6a.  Environmental framework checklist for the
section "Water"

| Components/Expressions | Aspects for Disaggregation | → |
|---|---|---|
| - Buildings<br>- Products<br>- Systems<br>- Structures<br>- Procedures<br>- Jobs | - Type of building, product, system, structure, procedure, job<br>- Stresses/Sensitivity<br>- Types of utilization<br>- Aspects of "time" and their 1 and 2 derivation<br>- Time scale of processes<br>- Roles in cause-effect-relationships<br>- Scientific disaggregations<br>- Group dynamical processes<br>- Planning steps<br>- Aesthetical viewpoints<br>- Types of interaction<br>- etc<br>- etc | |
| **Basic Processes**<br><br>- Production/Utilization of goods<br>- Commerce/Storage<br>- Traffic/Transportation<br>- Supply/Disposal<br>- Information/Communication<br>- Planning/Control<br>- Division of labour<br>- Property/Disposition<br>- Emission/Pollution<br>- Accidents/Catastrophes<br>- Use of areas and resources<br>- Population dynamics<br>- Education /Research<br>- Religion/Culture/Ideology<br>- Changes of attitude and behavior<br>- Politics/Participation<br>- Confinements of liberty and action<br>- Stimulation/Punishment<br>- Standardization/Categorization<br>- Group formation/Uniting interests<br>- Revolt/War<br>- Ideas/Suppositions/Hypotheses/Fears<br>- Life patterns | | |

(Left margin, vertical text: 7 MAN MADE ENVIRONMENT)

Corresponding checklists of the handbook:

| | | |
|---|---|---|
| Indicators: Social indicators etc | ⟶ | — |
| Utilization: Interests of individuals and groups. | ⟶ | — |
| Loads/Reserves: Disparities, Flexibilities, etc. | ⟶ | — |
| Goal Systems:  —  — | ⟶ | — |

Figure 6b. Environmental framework checklist for the section "Man Made Environment"

Such lists can obviously be made up any time
following your own knowledge and experience. But
usually it is easier to adopt them from suitable
manuals, from regulations and so forth. Some of
them may even be included in the handbook mentioned
above.

Only by the device of splitting off dozens of
such homogeneous lists and replacing them by the
corresponding "aspect" the length of the framework
checklists could be kept within reasonable limits
without giving up the claim of universality.

As far as the establishing of the second group
of checklists is concerned, i.e. the group which deals
with the space of measures, the material available is
considerably less systematized.

When conceiving the Handbook for the Environmental Im-          Checklists
pact Analysis we proceeded as follows.                          for the
In accordance with the basic qualities of impact trees         space of
the 5 following groups of measures form an orientation         measures
checklist (see Fig. 5b).

1   - Realms of provision for the human existence.             Provision
      Similarly to the item "provision for public              for the
      health", all other items of this group com-              human
      prise a fuzzy aggregation of most different              existence
      activities and measures, aiming at one com-
      mon goal.

2   - Measures influencing behavioral dispositions.            Influencing
      Such dispositions lead to consumer responses,            behavioral
      citizens' movements, boycotts and so forth.              dispositions

3   - Standard measures.
      This group contains very common items like               Standard
      "construction work," the "running of a produc-           measures
      tion plant." Because of the role they frequent-
      ly play with decisions these items deserve a
      simplified assessment procedure. Because of
      their relative conformity and the vast amount
      of experience and valuable rules available for
      them, they allow the development of specific
      modules.

---

+) The term "aspect" is used here as an equivalent to the term "coordinate axis", or "variable." It leaves complete freedom as to
   the "expressions" or "values" these variables (or groups of variables) may finally take.

4    - Types of influences on the environment.               Influences
       This group contains items like the emission           on the
       of noise, the drainage of groundwater and             environ-
       other items which describe measures and pro-          ment
       cesses which affect the environment rather
       directly.  It is these items on which exten-
       sive basic research has been done. They cor-
       respond to (U) in Figure 3.

5    - Measures which are very likely to be irrele-          Measures
       vant to the environment.                              of no en-
       This group contains items like the "appointment       vironmental
       of an ambassador", the "decision on the signi-        importance
       ficance of a certain computer bit" and the
       like.  These activities belong to the space of
       measures, too.  They are sure to have effects
       and side effects.  But in all probability, mea-
       sures of this type will have no negative conse-
       quences for the natural environment.

       Within this paper I shall only illustrate
the framework checklists for   1 ,  2   and  4 .

       The first framework checklist contains about          Framework
thirty realms of existence which appeal to the neces-        checklists
sities of life such as "food", "accommodation", "per-        for
sonal security" etc.  Moreover, it contains about 50         measures
items dealing with parts of the man-made environment
such as "provision for transportation", "for production plants" etc.

       Since we have applied the above-mentioned "mag-
nifying glass technique", a considerably greater resolving power is achieved.

       The framework checklist on "measures influenc-
ing behavioral dispositions" contain about 50 groups
of measures, the main quality of which it is to alter
the attitudes and activities of individuals or groups.
Among them we find such heterogeneous items as the
passing of bills, raising of prices, fines, advertise-
ments, education and provision of recipes.

       The framework checklist on "influences on the
environment" comprises submeasures and activities
which, quite independently of their way of execution,
are sure or very likely to have impacts worthy of special scrutiny.  The items selected for Figure 7 deal
with the man-made environment only, i.e. section 7 of
Figure 5a.  In the complete checklist all 8 sections
can be found.

Here again, by means of representative items and
the "magnifying glass technique" an adequate resolving
power is achieved.

It would not be surprising if somebody who peruses
such checklists without any purpose in mind would respond to them with the question "so what?", admittedly
they don't hold any scientifically new items.

If such checklists are, however read with a spe-       Measure
cial project or measure in mind, the reaction at each
item should be "Why not?." Any effect evoked by such                Why
a list which cannot be excluded on good grounds, rep-               not?
resents an issue worth closer examination, an issue
which might lead to an additional entry into the final    Checklist
impact statement[+).                                        item

Let me sum up. Three basic ideas were presented to you:

- the idea of the unbiased and complete pattern
  of effects regarded as the difference between
  the "without" and the "with" condition,

- the idea of lists of impacts, impact chains,
  impact trees or other graphs for cause-effect
  relationships as a means of representing those
  effects and interdependencies as they remain
  after cut-off procedures,

- the idea of checklists for the space of environments and the space of measures as a means of
  stimulating as many suitable associations as
  possible.

It is these three basic ideas which find their expression in the proposed system of checklists, especially
in the central framework checklists. And they lead to a
characteristic combination of features:

1) The checklist system is highly invariant. In their      Independ-
   abstract form, the framework checklists can be used     ence of
   in all countries, regardless whether they have to       problem
   deal with problems of the east, the west, the north
   or the south. They are devices for the assessment of
   international measures, of federal, state or local
   measures of private organizations or individuals.

2) The framework checklists are highly invariant, no       Independ-
   matter how dynamic the development in the scienti-      ence of
   fic or the technical field will be. For what the        scientific

| Measure/Activity | Aspects for Disaggregation |
|---|---|
| - Construction/Assembly/Finishing ⟶ | 3 - Land |
| - Area planning ⟶ | 3 - Land |
| - Resource planning | resources; uses |
| - Course of traffic | components; carriers; subprocesses |
| - Storage/transportation of solid/liquid/gaseous substances | types; substances; procedures |
| - Procurement/processing/commitment of resources | resources; procedures |
| - Biological/chemical/physical transformation of substances | procedures; substances |
| - Processing of substances/goods | procedures; substances; goods |
| - Treatment/disposal of gaseous/ liquid/solid wastes/rests/ discarded goods | procedures; wastes/goods |
| - Recycling of goods/resources | procedures; goods/resources |
| - Washing/cleaning of goods/tools | solvents; goods/tools |
| - Operation of furnaces | types; uses |
| - Operation of motors | types; uses |
| - Operation of industrial plants | types |
| - Utilization of durable/consumer goods | goods; utilizations; processes |
| - Installation of filters/stacks/ noisewalls/barriers/corridors | types; purposes |
| - Nonprevention of predictable risks | types of risk |
| - Explosions/leaks/operational failures | types |
| - Settlement | types; subactivities |
| - Places of employment/lodging/ meeting | types; subactivities |
| - Use of leisure/recreation/ sport | types |

⟶  1   Provisions for the human existence

⟶  3   Standard measures

Figure 7.  Framework checklist of submeasures and frequent activities worthy of special scrutiny
(Section 7, dealing with the "Man Made Environment")

items of these checklists present, only concerns
elements which frequently play important roles with
impact trees.  Whereas assertions as to specific effects between these items are started by way of associations.  Thus, new ideas on cause-effect mechanisms rather update the knowledge the user has to
contribute to the stimulation process than the very
items of the framework checklists.  Or they update
the magnifying-glass lists and other sources he will
draw upon.

*(margin note: and technical development)*

3)  It is not only the single measure which can be
examined by means of the checklist system, but
every programme, every development, every human
activity as well as any natural process (such
as the outbreak of a volcano, and the like).  Any
possible subspace within the space of measures
(e.g. measures which could be circumscribed by
the adjective "socio-economic") can be assessed.
That is why even those measures can be dealt with,
for which no immediate reason for an intensive
assessment has been felt so far.

*(margin note: Suitable for all kinds of assessments)*

4)  There are no unambiguous aspects for disaggre-
gating and making the checklists concrete.  This
is due to the enormous complexity and the fuzziness of the universe on the one hand and the space
of measures on the other hand.  For this reason
the orientation checklists, the framework checklists, the magnifying glass checklists and all the
other devices of the handbook are only considered
as stimulating modules.  Everybody can select
modules according to the special requirements of
his project and combine them to lists, tables,
texts just as he chooses to.

*(margin note: No tutelage of the user)*

Basically, no task, no standpoint, no method of
description is excluded.  Even the types of effects
in question, once they are registered as completely
and as replicably as possible, may and must be mani-
pulated freely.  For the subsequent use in models,
the terminology and the aggregation may be changed
at will.

_____

[+)] The numerous cut-offs involved are liable to cause uneasiness and
a feeling of uncertainty with the applicant.  Therefore, what is
planned for the handbook is to add a survey of dependable reasons
and reasons seemingly sensible, but which are often deceptive in the
end.

5) Even if any single measure c a n be assessed by
   means of the checklist system, it often is unneces-
   sarily complicated to do so.  The impact trees for
   one chosen type of similar measures are obviously
   not identical; but, nevertheless, they contain more
   or less the same central elements.  On the other hand,
   there are major differences if the type of measure is
   changed.  Pesticide use has different patterns of  effects depending on application or geography.  But these
   patterns of effects are basically different from the
   pattern of effects of e.g. new technologies of iron
   casting.  And the patterns of effects of construction
   work or car driving are basically different again.
   Therefore, it is planned to include specialized          Specialized
   modules for several types of measures. The types in      checklists
   question are enumerated in the framework checklist
   of standard measures mentioned in Fig. 5b.

   Each time a well-chosen mixture of orientation
   items, framework items and magnifying glass items
   has to be combined.  They may be freely reformu-
   lated as goals, indicators or the like. By means
   of specific technical knowledge, proper terminology and flexible aggregation, one can help to
   meet better the requirements and to take into account the abilities of association of the would-
   be user.

   A well-known checklist concerning the subset of
   measures dealing with water resource planning may
   serve as an example for such special checklists
   (Figure 8)[+].

   For illustration I have marked some of its items
   according to their relation with the checklist
   system presented here.

6) In environmental assessments one always ends up
   with descriptions of extreme complexity, as long
   as one wants to proceed correctly and at the same       Assessment:
   time concretely.  In our innate fear of complex-        Fighting
   ity we are confronted with almost unmanageable          complexity
   numbers of details and tend to react by oversim-
   plification.  Admittedly, I have not eliminated
   this weak point, but by means of structuring the
   problem and by laying aside the great number of
   homogeneous magnifying glass lists, it has become
   more manageable than it was before.

7) The application of the checklist system still requires considerable analytic and associative
   abilities.  Even with a precise measure in mind,
   the associative step from a checklist item to a

specific effect often tends to be comparatively
difficult.  Therefore, this system appeals mainly     Direct use
to experts on planning and to scientists, who         in top-down
have to examine questions of unusually high com-      strategies
plexity or questions of a completely new type.
Besides, they also serve the purpose of cross-
checking special checklists developed otherwise
(compare Figure 8).  They serve for validation of
models and system diagrams.

They also help to avoid or to settle time-consuming
arguments by supplying suitable parameters for in-
variance tests.  For the specialist in federal,      Special
state or local administration, however, the special  checklists
checklists mentioned above should be preferred.       for the
                                                      specialist

8) The framework checklists have a great deal in common with thesauri as used with documentations.  The
   magnifying glass checklists have a great deal in    Integration
   common with classifications as used for official    in science
   statistics or in systematisms in sciences. The check- due to key
   lists basically aim only at suggesting cause-effect  word quality
   relationships.  But due to their keyword quality,
   they also invite the selection of reference books,
   documentations, data bases or research projects where
   details required for a thorough analysis of the cause-
   effect relationships in question may be found.  At
   the same time, cross references to further planning  Integra-
   devices such as goals systems, generalized models    tion in
   or indicator systems are easily integrated, too.     planning
   This is indicated in the lower section of the        systems
   prototype framework checklists in Figure 6.

By this enumeration of qualities I have tried to show
that the characteristic features of the checklist system
are a combination of

   - comprehensiveness for the environment as well as
     for the measures

   - stimulation and association

   - stability and flexibility

   - abstraction and applicability

   - integration into the realm of planning as well as
     into the various topics of science.

With all these features in mind, it does not seem to
be too pretentious to call this checklist approach a universal tool for the assessment.

---

[+)]Dee, Norbert, et al, "Environmental Evaluation System for
    Water Resource Planning," Battelle-Columbus, 1972.

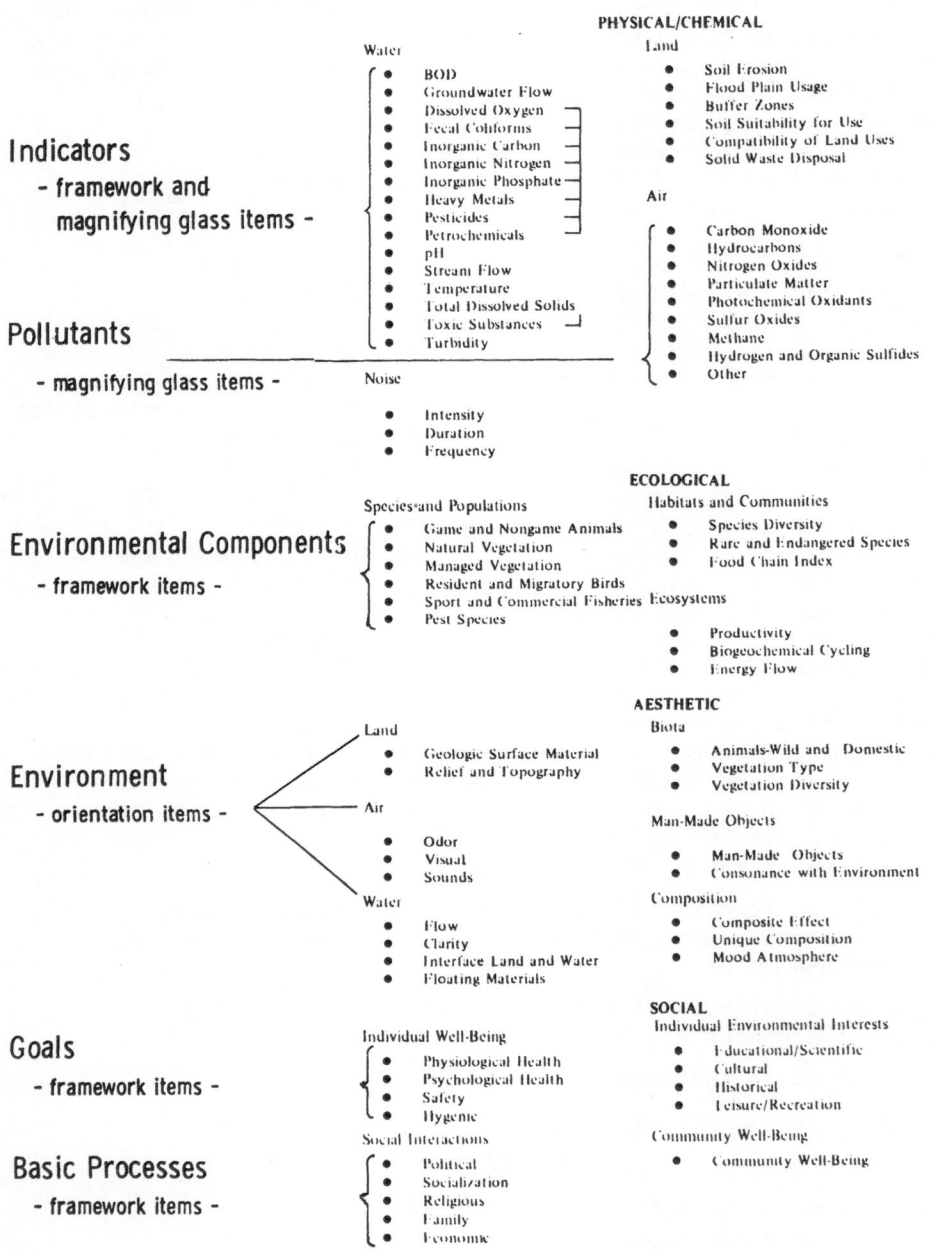

PHYSICAL/CHEMICAL

Water

Indicators
- framework and
  magnifying glass items -

- BOD
- Groundwater Flow
- Dissolved Oxygen
- Fecal Coliforms
- Inorganic Carbon
- Inorganic Nitrogen
- Inorganic Phosphate
- Heavy Metals
- Pesticides
- Petrochemicals
- pH
- Stream Flow
- Temperature
- Total Dissolved Solids
- Toxic Substances
- Turbidity

Land
- Soil Erosion
- Flood Plain Usage
- Buffer Zones
- Soil Suitability for Use
- Compatibility of Land Uses
- Solid Waste Disposal

Air
- Carbon Monoxide
- Hydrocarbons
- Nitrogen Oxides
- Particulate Matter
- Photochemical Oxidants
- Sulfur Oxides
- Methane
- Hydrogen and Organic Sulfides
- Other

Pollutants
- magnifying glass items -

Noise
- Intensity
- Duration
- Frequency

ECOLOGICAL

Species and Populations

Environmental Components
- framework items -

- Game and Nongame Animals
- Natural Vegetation
- Managed Vegetation
- Resident and Migratory Birds
- Sport and Commercial Fisheries
- Pest Species

Habitats and Communities
- Species Diversity
- Rare and Endangered Species
- Food Chain Index

Ecosystems
- Productivity
- Biogeochemical Cycling
- Energy Flow

AESTHETIC

Land
- Geologic Surface Material
- Relief and Topography

Environment
- orientation items -

Air
- Odor
- Visual
- Sounds

Water
- Flow
- Clarity
- Interface Land and Water
- Floating Materials

Biota
- Animals-Wild and Domestic
- Vegetation Type
- Vegetation Diversity

Man-Made Objects
- Man-Made Objects
- Consonance with Environment

Composition
- Composite Effect
- Unique Composition
- Mood Atmosphere

SOCIAL

Individual Well-Being

Goals
- framework items -

- Physiological Health
- Psychological Health
- Safety
- Hygenic

Social Interactions

Basic Processes
- framework items -

- Political
- Socialization
- Religious
- Family
- Economic

Individual Environmental Interests
- Educational/Scientific
- Cultural
- Historical
- Leisure/Recreation

Community Well-Being
- Community Well-Being

Figure 8.   Specialized Checklist
            (Dee, N. et al, Battelle-Columbus, 1972)

PROBLEMS WITH PEOPLE,DECISION, CONFLICT, LANGUAGE AND MEASUREMENT

K. C. Bowen, J. I. Harris

Ministry of Defence

United Kingdom

## 1. INTRODUCTION

Every now and then, a researcher should stop and consider where he is going. He will find that his various, self-imposed, tasks have spread out into a number of superficially different areas, and that the focus of his main concern has changed. He will feel that his past and current interests are coherent, but he will not be able immediately to define that coherence. He will perceive that they are interconnected, because they have stemmed from one another, but he will not be able to identify their relative importances to the purpose that now motivates him.

The authors' present motivation can be defined, in the context of this conference, as a desire to define the proper role of analysis in aiding those faced with the complicated problems of social systems and their environments, and to say why certain aspects of research and its application should be given more attention than is at present the case. The nature of the proper role will be derived subjectively from the authors' work, from their reading, and from their observation of the work of the analyst and decision-maker. It can only be a tentative statement.

We believe that, to make progress towards the ideal, a lot of "different" subject areas, and the techniques within each, need to be looked at as a whole, instead of being separately developed and then pieced together pragmatically for each new problem. We therefore discuss why certain research has been carried out, how it has led to diverging studies, and how the research might be rationalised into a system of interrelated items.

This rationalisation leads to some general statements about operational research methodology, and the lack of techniques to carry out some of the analysis desired.  Finally, we reformulate our own research programme, bearing in mind that, on its own, it cannot move far or fast towards the ideal, but that it must eventually be seen as part of a wider research area in which the efforts of many research groups are integrated.


## 2. A BRIEF HISTORY OF A RESEARCH STUDY OF CONFLICT

In 1967, a proposal was made to carry out research into the precursors of war.  The aim was to develop realistic scenarios and to provide a coherent rationale for the initial conditions of battle models.  By the time research effort was available, there had been a shift of emphasis in analysis tasks, from a major concern with national commitments East of Suez to a concern with joint commitments in a European/NATO context.  It was decided to restate the research programme as follows:

> to find ways of determining the facilities (political/military) that might help to provide flexibility of decision, at all stages of a developing situation, for moving in a desirable direction, not necessarily to avoid combat but certainly to minimise the risk of long-term or large-scale warfare.

In 1970, such a programme was started.  Almost immediately it was found that, before we could think clearly about the problems of "escalating conflict", we had to have a clearer idea of what conflict was, what crisis was (and when it might or might not be perceived), and what were the more important factors in this that could be identified, related and studied.

In 1972, this part of the work was presented at a Social Sciences seminar at Edinburgh University.[1]  It did not deal primarily with Defence problems.  The arguments and analyses that led to the conceptual models had ranged over a wide field, and ideas developed in one area had been tested in others.  The work led us to a new, diagrammatic systems-structuring process, and forced us to delve into psychology, political science, sociology, linguistics and many other subjects.  Two words describe what we saw as the key, essential, factors - "perceptions" and "language".  Ironically, in extending this work into practical applications, we found that, despite our every effort to use a sort of meta-language, those for whom we wrote wrongly perceived our intentions and misunderstood our language.  We had created a conflict within our study of conflict, although we had used our own work to try to avoid it!  We were also aware, at this stage, that measurement would be limited, for the forseeable future, to set-theoretic statements and classification processes.  We recognized that, even so, many of the important sub-sets would be fuzzy.

Over the same period, work was done to understand better what the playing of games could offer. This was stimulated by associated work on command and control[2] - studies which were carried out because we were, and still are, far from understanding the association between data received and decisions taken. We realized that, eventually, we might want to play games using some of the conflict situations studied, but this was not, at the time, a motivation. There was, however, an important link between the three studies. Diagrammatic description of the command and control process led to similar attempts to describe conflict. What turned up was different, but it was successfully used to describe the structure of a game and so assisted in the development of a classification of games[3]. In turn, this has proved to be of value both in the furthering of command and control work, and in the more recent attempts to state what a conflict game must be and how it should be developed and controlled.[4]

Reverting to 1971, there had been two other areas seen as of potential importance to conflict studies - decision theory and game theory. In a small research group, very little breadth of study is possible. We therefore sought to form exploratory contacts between our own and other research groups.

The application of theories of decision had been a topic often discussed by the NATO APOR (now SPOSS) around 1970. In 1971, the first author obtained the agreement of the Panel to the sponsoring of a Conference. Professor D.J.White, of Manchester University, was the chosen Scientific Director, but circumstances made it a joint endeavour. The tasks of planning and partaking in this Conference, and the subsequent editing of the Proceedings[5], provided stimuli in many directions. The conflicting views - on Bayesian statistics and Decision Analysis; on Decision Methodology[6], particularly the choice of models (for which there is negligible theory); and on whether an implicit need for fuzzy concepts led people to reject theories of decision as inapplicable to many real problems - all seemed to have relevance to any study of decisions in conflict situations.

Game Theory, applied to conflict, had been used in a most painstaking research study by Ackoff, Emshoff, Howard and others at the University of Pennsylvania in the second half of the 1960's[7,8,9]. We decided to involve ourselves, with the cooperation of the University of Sussex, in similar work, replicating parts of it with different cultural groups of subjects, and trying to extend the behavioural theories developed in the earlier work. So far, the Sussex work[10] has been kept separate from our other work on conflict, except for the inevitable cross-fertilisation of ideas, although it had always been intended to link it, if possible, to the main research programme, which is itself a speculative and changing endeavour.

Since 1973, Ackoff and Emery's work on purposeful systems[11] has influenced our thinking. We were already aware of the essential behavioural science content of what we were attempting. We were also unhappily aware of the lack of structure (as it would be understood in the systems science sense) in much of the social science literature. This new book offered a light at the end of one of our many tunnels, and it has been thoroughly studied[12,13]. Its treatment of information and information measurement, for example, seems to be of great importance to the study of conflict. We discuss this later.

In 1974/75, a major change took place in the emphasis of the main study. It had been intended to play games set in pre-combat contexts, and to use similar games to develop scenarios[4]. However, we did not have people available to do this. It then dawned on us that we had a scenario within our own organisation. We had already seen conflict as a pervasive process (not always malevolent), and it abounded both within any OR department, and between that department and those whom it served. What did our research say about the communication aspects of the OR process? And could we cast any light on the decision processes in a complex organisation, processes which, in turn, might confuse or even make redundant a lot of OR advice? Were we paying too much heed to detailed measurement, and were we measuring the right things in the right way?

Some papers have been written as provisional thoughts on these questions[14,15,16]. They give views on possible reasons for some of the difficulties of extending methodology to the process of implementation of decisions (see also White[6]), and they start to examine the ways in which the perceptions both of the analysts and of the various echelons of the decision-making system might create conflict. The conflict notation[1], as extended for discussing the logic of games[4], has been important to the attempts so far made to extend the conflict work into the field of OR Methodology. The basic problem always seems to be how to get a lot of information into a small space (preferably diagrammatic, since any mathematical formulation would make much essential communication impossible). Such compression is valuable both because, in doing it, one necessarily leaves out what is irrelevant, for reasons of space, and because it allows the essential structure of systems to be examined in a richer and fuller context.

Finally, the efforts of the second author have switched (late 1975) to a full-time study of the literature of fuzzy sets and their applications. We have hovered on the edge of this subject for many years, and now we are convinced that there is so much that is fuzzy in all the processes that we have studied - conflict, decision, language, measurement, information and subjective probability (there are some who say that all probabilities are subjective!) - that we can no longer be frightened off by the inherent logical difficulties of the subject.

It is now time to attempt to give a rationalisation of this wide, but not random, coverage of broadly associated themes, which can be seen as coherent in the context of this Conference. In so doing, we shall see our initial research concept as one special case in a broader enquiry into how analysis can help with large scale socio-economic problems.

## 3. RATIONALISATION IN THE CONTEXT OF SOCIAL SYSTEMS AND THEIR ENVIRONMENTS

### 3.1   Problems with People

People enter into the problems of social systems in many ways, but it is convenient to classify them into three groups:

a.   those who control the system, changes in which may resolve the problem;

b.   those who are not part of the system being controlled but who are affected by it and who will react to changes; and

c.   those who ideally are outside the system but who still interact with it in the process of studying it and proposing changes to it.

The first group includes the decision-makers at all levels, down to those operators of the system who have minimal decision-making authority.  At the top level of responsibility, there exists a great deal of experience which is not explicit, including knowledge about the interactions that continually take place between the system to be changed and other systems, which may suffer what they consider to be undesirable changes as a result of such interactions.  Competition for scarce resources, ideas of status and independence, and concepts of priority are some of the issues that will arise.  Personal considerations are also of great importance; since no one can regard his own position as other than important, each individual must see his own retaining of authority and confidence as an essential element of any decision.  The concept of "the greatest good for the greatest number" is one which will be interpreted in a personal context by nearly everyone, and it will be looked at in a time scale that will vary with the individual standpoint.

The second group includes those whom the system intends to serve, or to control indirectly by influencing their behaviour.  The reactions of these people are obviously of great importance to the success of changes to the system.  If they do not perceive that service

has been given, or if their behavioural response is very different
from that predicted, the changes can lead to an aggravation of an
existing problem, or to new problems which are even more difficult
to resolve.

Finally, there is the group that contains the analysts.  It also
contains pressure groups, who either - in a limited analytical way -
perceive the problem and seek its resolution, or react - adversely
or otherwise - to proposed ways forward.  It is perhaps wise to regard the analysts also as a very particular pressure group, since
they too tend to seek ways towards 'better' futures from a particular standpoint.  Their difference is that they claim to be "unpre-
judiced"; but this is never wholly possible, and they too will have
introduced personal preferences in dealing with the secondary decisions[6] that the analysis/modelling process demands.

Ideally, the influences of all these people on the problem and
its resolution should be measured, and included in any analysis.
Since this cannot yet be done in any satisfactory and explicit manner, it is necessary to ask what this implies about the relevance
of current ways in which decision-making organisations operate, and
particularly the ways in which data and analysis are used.

It is suggested implicitly in the above that, because there are
always these three groups of people involved, all problems involve
the essential elements of social systems and their environments.
Clearly, in dealing with smaller systems, the difficulties will be
less.  However, many of the principles that emerge from detailed
study of smaller systems may be capable of generalisation.  Further,
if the resolution of problems with smaller systems shows up any
shortcomings in traditional procedures, it is not unreasonable to
expect that large-scale social system analysis may need a different
approach.

3.2. The Relevance of the Research Already Carried out

Essentially, our research has been a systems study of conflict,
in a very general sense.  There is conflict inherent in the resolution of any problem, and in the making of any decision.  Conflict
is an essential ingredient of any progress or change. It need not
be other than benevolent, but it is very difficult, in anything that
matters (when personal feelings will be strong), to keep it from becoming malevolent.  Although we set out to determine the facilities,
both physical and conceptual, which could be useful for retaining
flexibility in political/military situations, we found that we were
dealing with certain general features:

    a.  physical facilities that would enlarge options for future
        actions;

      b.   data acquisition and interpretation facilities; and

      c.   an understanding of behavioural attitudes, and of the limitations resulting from prior judgements ("prejudice").

In particular, internal rationalisation of the actions of systems external to the controllable system assumes some "knowledge" of the attitudes of the people who control those systems, and any prejudice here tends to be unaffected by data acquisition, since interpretation is coloured by the views already held.

     These features have relevance to almost all problems. Unless the conflicts that generate a problem, or make resolutions difficult, are themselves understood, actions may be taken to provide a "better" physical system which still performs "badly" from the point of view of the people who affect it, or whom it affects. In particular, these same general features have relevance to the analysis process itself, and to the way in which analysis systems relate to the decision-making systems they serve. Decision theory[5] and decision methodology[6] do not at present offer what is required; the latter, however, seems to provide a basis for linking analysis more firmly to the total decision process, including the inherent conflicts.[17]

### 3.3. The Roles of Analysis

     Elsewhere[14], operational research has been defined as "the process of giving aid to decision-makers through measurement." "Measurement" is intended to be taken very broadly, from absolute measurement to pure description, passing through preference ordering and classification. It should not be thought that analysts need to concentrate on numbers that can be handled by computers. Structured descriptions of complicated concepts and systems may be as far as measurement can reasonably go. In the same way, aid to decision-makers is not to be interpreted as providing analysis which defines the appropriate decisions : not only is this often impossible, it is also generally inadvisable. In well-understood problems, however, ongoing models used as management aids may quite properly become prescriptive, when they are adopted by decision-makers as reflecting an acceptable decision process.

     There is clearly a very great range of "levels" of analysis. In agreement with White[18], we would hypothesise that an explicit, objective process, $\theta$, through which we move from a set of possible courses of action, $\{C_i\}$, and associated relevant data, $K$, to selection of a particular course of action, $C_s$, is a more "desirable" process than one which is implicit and subjective. However, there is no a priori reason why it should, in any particular case, give better results; much depends on the relationship of $K$ to $\{C_i\}$ and on the relevance of $\theta$ to any desired outcomes. If decision metho-

dology can lead us to decision-theoretic choice processes which we
perceive as good, we should use them, but we must always be careful
that they are valid.

Consequently, from the definition of OR above and its extended
versions[14], "operational research" can be taken to include advice
of all sorts, whether given by operational researchers, statisticians, economists, or any other advisers, provided that the level
of objectivity and explicit modelling is chosen for sound reasons.
A "scientific" approach is not reasonable if its logic depends on
doubtful assumptions, neglect of unmeasurable parameters, and excessive distortion in modelling real-world processes.

In most problems concerning large-scale social systems, behavioural factors play a major and increasing part; accordingly, a
great number of standard analytical/numerical techniques are now
less relevant and less attractive.  If, however, we were able to
measure such behavioural factors, even at lower levels of measurement, the same or similar techniques could again become as applicable as they are for the study of well understood "mechanical"
systems (i.e. systems in which human involvement is minimal or can
be statistically represented from detailed observation over a long
period of time).

It is in this context that the development and use of games
and game-theoretic analysis seems to be very important; although
the difficulties must not be underrated[3,7,8,9,10,19,20], there is
no other way than to develop such experimentation on people's behaviour as has been done, over the years, on the performance and
variability of physical systems.  It will take time, perhaps a very
long time, to achieve similar results and to establish confidence
in theories in this new area.

Meanwhile, we do have choices about the directions in which
developments of methodology should go, and about what, as analysts,
we should model and describe in order to aid decision-makers.  The
proposals [14,15,16], to extend the conceptual models of conflict are
of a very general nature; broadly, they offer a way towards the
examination of

a.  the nature and limitations of decision-making systems;

b.  a methodology of problem formulation;

c.  more explicit rules for the choice of models (related to
    the type of advice that seems possible, relevant and
    acceptable); and

d.  an improved process, or language, for communication be-
    tween analysts and decision-makers.

In all that is said above, no criticism is intended of the existing range of techniques or their usefulness.  What is being doubted is whether they are quite as useful as many analysts seem to believe. Since we do not have adequate understanding of a, b and c above, and since in all three cases we need the support of behavioural science to develop our understanding, it seems sensible to have such doubts.

The final section of this paper will summarise briefly our own future research programme and the associated research with which we are in close touch.  At the time of writing, we are still in the planning stage but our intentions are firm.

## 4.   FUTURE RESEARCH

### 4.1.    The Place of Current OR Techniques

It is suggested[15] that what is available in statistical, mathematical and computer modelling techniques can, in general, be considered as sufficient for the moment.  We can already model, with conventionally measurable variables, at a level which provides us with artificial worlds as complicated and as difficult to understand as the real world that we seek to study.

There are however two areas, mentioned already, where we have techniques but need more understanding as to how to use them in the study of behaviour and decision in the real world.  These areas are gaming and game theory.  There is a third area which seems likely to be crucial, and that is the ability to put operational measures on information, i.e. on that set of data which effects the probability of choice in specific choice situations[11,21].  These three topics are dealt with separately below.

A fourth area that concerns us is the general one of being able, eventually, to put measures to concepts that are at present only describable in words or in diagrammatic form.  Such measures are needed so that statements can be processed and implications drawn, in a rigorous fashion, reverting eventually to natural language. This is the essential role for the application of fuzzy sets, which, albeit not in the UK, is being furthered by work in other NATO countries.  This is also discussed in more detail now.

### 4.2. Why a New Type of Game is Necessary

The type of game that has often been played to study complicated systems has been a learning rather than a research game[19].  The players have been placed in a simulated "real" (and therefore com-

plicated) situation.  While these players certainly gain experience,
the problems for the experimenter - of understanding which data have
been used and of controlling the variables adequately, so that analysis of the results is possible (and statistically reasonable) -
have been unresolved.  For this reason, it has seemed sensible to
determine the simplest possible useful type of game, and to understand both the difficulties inherent in this and how, when these
difficulties have been seen to be surmountable, to move logically
to more complicated games.

A detailed study has been made of what could be involved in
creating and playing a conflict game; two versions of this work are
now available[22],[23].  The diagrammatic notation developed by Smith[1]
has been extended, and its descriptive and logical power demonstrated.

It is concluded that it is advisable, and perhaps essential, in
such experimentation, using people in a game setting, to start with
a very simple, single-decision-point game, with tightly controlled
data input.  Only a few such games have been developed and in all
cases they seem to have been fairly successful, although full consideration of logical control has not been attempted.  Such games
have been reported elsewhere (see references in reference 3 and a
brief example quoted in reference 20) and another, the Wessex Ward
Management Game, is described in a recent publication compiled by
D. Hicks and issued by the Wessex Regional Health Authority[24].

There are certain issues in game-playing which are easily overlooked.

a.  The purpose of a game must be clearly stated in order for
    its success to be measured.

b.  The different roles that people play in games must be under-
    stood if proper control is to be realised.

c.  The scenario and rules of a game must be designed to fit
    its purpose and to encourage realism for the player.

d.  The variables in a game must be identified and measured if
    any valid conclusions are to be drawn.

e.  Some theoretical basis is necessary for the planning and
    control of games, and to provide a language for communi-
    cating about them.

Besides taking these issues into account, it is believed that
the simple games we propose to develop may be useful for drawing up
"scenarios", using "expert" players to provide the full information
context to be used in later games.  Until we are certain that we can
play these simple games efficiently in settings with fairly compli-

cated sets of data, we do not propose or recommend the playing of
games involving group decision-making, more than one level of decision-making, or several stages of decision.

Finally. we believe that, apart from some early games to develop the actual process of game-playing, all games should be testing
some stated hypotheses, preferably ones derived from the real-world
behaviour of decision-makers. This is discussed further in the context of game-theoretic studies.


4.3.   Game-Theoretic Experiments and Theories of Behaviour

The research being carried out at the University of Sussex
started as a speculative venture aimed at extending the work of
Ackoff and his associates at the Wharton School, University of
Pennsylvania[7,8,9], and originally sponsored by the US Arms Control
and Disarmament Agency. The purpose was to develop theories of behaviour in 2 x 2 games and then to see how these behaviours could
be related to real-world situations, via a sequence of models of
"artificial realities"[25]. The results of the Sussex work are fully
documented by Dando and Bee[10,26,27,28]  Only a very brief summary
will be given here.

The direction of the work has diverged from the Wharton School
Study. This is not because of doubts about the basic logic of the
Ackoff and Sasieni proposals[25], but because the validation problems
involved in generalising system functions to other environments of
different complexity are not adequately resolved and because progress
would necessarily be slow. Dando has therefore adopted a methodology proposed by Milburn[29], who argued that it might be possible, in
attempting to assist decision-makers in conflict situations, to derive hypotheses about decision-making behaviour from records of real-
world conflicts and test them in controlled crisis-like conditions.
Well-tested hypotheses should then be used to produce decision rules
which do not tell the decision-maker what he should actually do, but
rather what possibilities he should check. The hypotheses would indicate behaviour characteristics of potential relevance and the de-
cision-rules (non-prescriptive) would provide explicit means of
dealing with them. The provision of a monitoring process to enable
a decision-maker to check his behaviour was also suggested by earlier
conflict studies[1].

Dando and Bee have shown that in "Chicken" games, conflict-like
situations with similarities to real-world conflicts only arise when
players have radically different aims. Players experiencing crisis-
like decision-making in such games typically exhibit behaviour which
would exacerbate real-world conflicts. To have explicit knowledge
of such behavioural tendencies seems to be important.

Another importance, stressed in a more general context by
Emshoff[30], is that there is a great need for explanations of phenomena wherever possible, and that empirically based descriptive studies are not enough if we wish to generalise, since the logic of
the process we would then be using is elusive.  In the mixed-motive
games played, the advances in understanding that have been achieved
have been in large measure due to the theory-based research strategy
followed.

Finally, it is accepted that conflict/crisis studies do not
cover the whole of the difficulties of effective planning, and that
study of all aspects of decision-making behaviour in the face of
complexity and uncertainty will be required.

### 4.4. The Information Content of Data

In the study of information systems for decision-makers, emphasis has been placed on making large quantities of data available
and easily retrievable.  What data a decision-maker actually uses;
when he perceives it; and whether or not he knows anything explicitly about the process he uses to transform data to information and
information to decision, all seem to be questions which are largely
unstudied.  It is doubtful whether a decision-maker necessarily
knows, a priori, what part of the available data he should examine;
it is certainly true that his decision process is rarely made explicit, and it is quite likely that for him to do so is difficult
or impossible.

It is useful to define information, in a choice situation, as
a function of the probabilities of choice from among an exhaustive
and independent set of actions.  Ackoff and Emery[11] develop such
definitions with great care, introducing concepts of "net" and "gross"
information to allow for shifts of opinion as incoming data appear
to support or contradict earlier data.  They define "determinate"
and "indeterminate" choice-situations (respectively, with one of
the choice probabilities equal to unity, or with all probabilities
equal to one another) and take as their unit of information, the
amount required to change an indeterminate two-choice situation to
a determinate one.  Thus they arrive at a maximum number of $(m-1)$
units to resolve an m-choice situation.  They comment that Shannon's
measure of minimum information, $\log_2 m$, does not seem to be usable
in the general behavioural context.

It has recently been found[21] that, by introducing the concept
of "prior information" inherent in the structure of the data, and
using Shannon's uncertainty measure, a step can be made towards
merging the Shannon and Ackoff-Emery concepts of information.  The
difficulty is in identifying what the structure of a choice-situation
is in any given case.  For the unstructured situations, the Shannon-

type measures developed have some advantages over the Ackoff-Emery
measures, which, being based on moduli of differences of probability,
tend to have some undesirable characteristics.

A more important difficulty is that of establishing probabilities of choice.  It is hoped that games of the type described in
section 4.2 may help to identify the way in which different types
of data build up information on which a decision can be based, but
whether it will be possible to get useful measures of probability
of choice remains to be seen.

Differences in behaviour between different players, in similar
situations in a game, may lead to insights into other difficulties.
What prior information does a decision-maker "hold" due to his experience, his optimism and even his dogmatism?  What misinterpreta-
tions or rejections of data take place, and under what circumstances?
These behaviours are important, not only when considering what
management information systems are needed for decision-making, but
also when considering what type of analysis it is worth while presenting (see c. in section 3.3).


4.5.   The Concept of Fuzziness and its Potential Value

The theory of fuzzy sets[31] is relevant to the study of problems involving people, and the language in which they express themselves.  It is a way of looking at subsets of some defined set which
are not themselves precisely defined.

Elements of the defined sets are regarded as having only a
degree of membership in the "fuzzy" subsets, and are accordingly
assigned a grade of membership in each of these subsets, somewhere
between the values 0 and 1.  Although this involves attaching an
actual number to each element, this may imply little more than an
ordering of the various degrees of membership (and even this may
sometimes behave rather oddly); in particular, fuzziness is not to
be identified with any concept of probability.  It is treated, as
far as possible, as an equivalent of ordinary set theory for vaguely
defined subsets.  Its axioms define, in fuzzy terms, the basic operations on fuzzy sets, such as union and intersection, and fuzzy
relations such as containment between subsets.

The reason why something similar to fuzzy set theory seems to
be needed to handle "problems with people" is that, although these
problems should apparently be amenable to scientific treatment, they,
and the constraints affecting their possible solution, are liable
to be expressed in natural language, which is intrinsically vague.
This need not be the resolvable vagueness that would result from a
sloppy use of language, which could be made precise.  On the contrary, it may be an essential part of the meaning of the words, re-

flected in a slight variation, from occasion to occasion, in the
objects to which the words - say "little", or "hard", or "good" -
are applied.  Nor need it be a flaw in language, as some philoso-
phers might have it, but a potential advantage;  it allows normal
communication to proceed, directing attention to approximately the
right kind of degree of properties, without descending to an unnecessary and even unnatural level of precision.  It is not a question of probability: it is not that the "correctness" of applying
some term in a given situation is knowable in principle, but at present uncertain, but rather that that "correctness" is to some extent
a matter of human choice, allowed by the meaning of the word. Treating the meaning of words in terms of aggregate probability distributions for their use by various people, in various situations, is
at least superficially possible, but would probably lead to a mis-
interpretation of "meaning".

When language is essentially vague, and not just carelessly
used, forcing problems expressed in vague terms into over-precise
mathematical formulations may well amount to distortion. Not only
will the language in which the problem is expressed be altered, but
since human language, in its precision or lack of it, probably tends
to reflect human thought, it may not be possible to make an accurate
match between the problem as originally conceived and a more precise
formulation.  In addition to the difficulty in translating the problem itself into precise terms, any solution or recommendations resulting from analysis of the problem will have to be communicated
to somebody who is able to use them, and this will involve an appropriate translation back into ordinary, and therefore imprecise,
language.

Some means of manipulating vague or "fuzzy" human concepts
with something approaching mathematical consistency, but without
destroying their vagueness, should be of use in handling any problem arising from human thought.  Fuzzy sets seem appropriate for
this purpose in that they allow a certain amount of vagueness over
membership of sets, and are not based upon probabilities.  From our
reading up to now, it seems that there may be two main pitfalls in
applying them.  First, even if we can interpret language in terms
of fuzzy sets, and manipulate these back into ordinary intelligible
language.  Second, it may be all too easy to define convenient fuzzy
properties or entities, without discovering that they imply extraordinary conditions in the space where they are defined.

There is no lack of problem areas where language is normally
and naturally vague - for example, local government statements of
ideals, goals and objectives[33]; subjective descriptions used for
identification purposes or the planning of research programmes[32];
and the relationships and definitions of subsystems in any organisational or conceptual structure.  Any of these would provide natural openings for the use of fuzzy sets in the future.  It should be

noted that certain uses of language, e.g. diplomatic negotiation, may have rather different characteristics: their vagueness may be intentional, to enable flexibility of interpretation and action to be maintained.

We believe that the concept of fuzziness is relevant to many of the practical problems that concern this Conference. It might seem attractive to remove fuzziness completely and to seek precision of measurement or probabilistic statement, but doing this without there being an underlying precision of meaning is dangerous (e.g. experiments for establishing axioms of choice or "subjective" probabilities[5]). We therefore seek to learn whether acceptable and comprehensive methods of using fuzzy set concepts can be found[32].

### 4.6.    The Shape of A Research Programme

The plan for future research activity on behaviour and games, for the three years beginning with the academic year 1976-77, will be as follows:-

a.  continuation at the University of Sussex, under Science Research Council funding, of studies of individual behaviour in mixed-motive games to provide further theories relevant to the monitoring of behaviour in real-world conflict/decision situations;

b.  the development of games (Royal Holloway College, University of London/DOAE) to develop theories on how decision behaviour is affected by the patterns of data provided in realistic conflict situations, and to study some aspects of crisis development; and

c.  a study of the effects of the structure of an organisation on its internal behaviour and conflicts (University of Sussex/DOAE).

The third, and new, systems-theoretic study, is central to the whole programme.  Its end purpose will be to describe, at various system decision levels, the kind of analysis relevant to decision-making and implementation, and to develop a theory-based analysis of the various types of interaction that will seriously reduce the "goodness" of analysis and decision as perceived from the several sub-system points of view.  Clearly, in large-scale social system studies, the effects of "public opinion" could be considered either as external (environmental) interactions with the organisational structure, or as internal interactions of a more broadly defined organisation.  Thus our intention is to seek extensions of the existing diagrammatic notation into a language to aid development of theories of interaction and conflict which will be of general applicability.

The findings of the gaming study will be important to this new study in that they should give information about the types of data (and therefore also of data sources) which can have important effects on decision processes, and about the types of data that may be neither sought nor used. In reverse, the systems-theoretic study can provide concepts which will help to identify relevant information about the system's internal problems and conflicts. If we can achieve a facility for playing more elaborate games, involving more than one decision-maker (at the same or at different levels), the link between the two studies will become even more direct.

The studies of individual behaviour can be of value in the initial choice of conflict situations for games, viz. situations for which theories of behaviour tested in game-theoretic studies are relevant. They can also, it is hoped, cast useful light on how and why individuals' perceptions of the systems (organisations) in which they operate will vary. Conversely, theories of such perceptions may eventually be testable in the game-theoretic context, with suitable extensions of experimental game-formats.

None of the above should lead the reader to believe that mountains are suddenly to be moved. What is indicated is the need for a three-pronged assault covering individual behaviour, interactions between information and decision situations, and organisational effects. In all cases, decision-makers are not necessarily single individuals, although exactly how to reproduce the complexities of group decision-making, other than by constraints on, or additional interactions with, an individual player, are as yet uncertain.

The relationship between the various main and other subsidiary, studies is shown in figure 1. The inherent conflicts between adjoint sets are assumed at present to be constructively benevolent; they will not necessarily remain so, without a suitable common direction!

· The other main element of our research programme will be the study of the application of fuzzy sub-sets (section 4.5). The programme also includes, necessarily, a watch on the relevant literature and the carrying out of such work as is needed to integrate ideas which seem to be usable or potentially usable into the existing or a suitably amended framework of research.

Figure 1 also indicates the need to keep a close liaison with independent research that is covering some of the vast field of enquiry of which we hope our own proposals are a basic, even though a small, component. A link exists at present with the University of Bath, where work is in progress on studies of organisational behaviour and of OR Methodology[33,34,35]. Within DOAE, a study is using

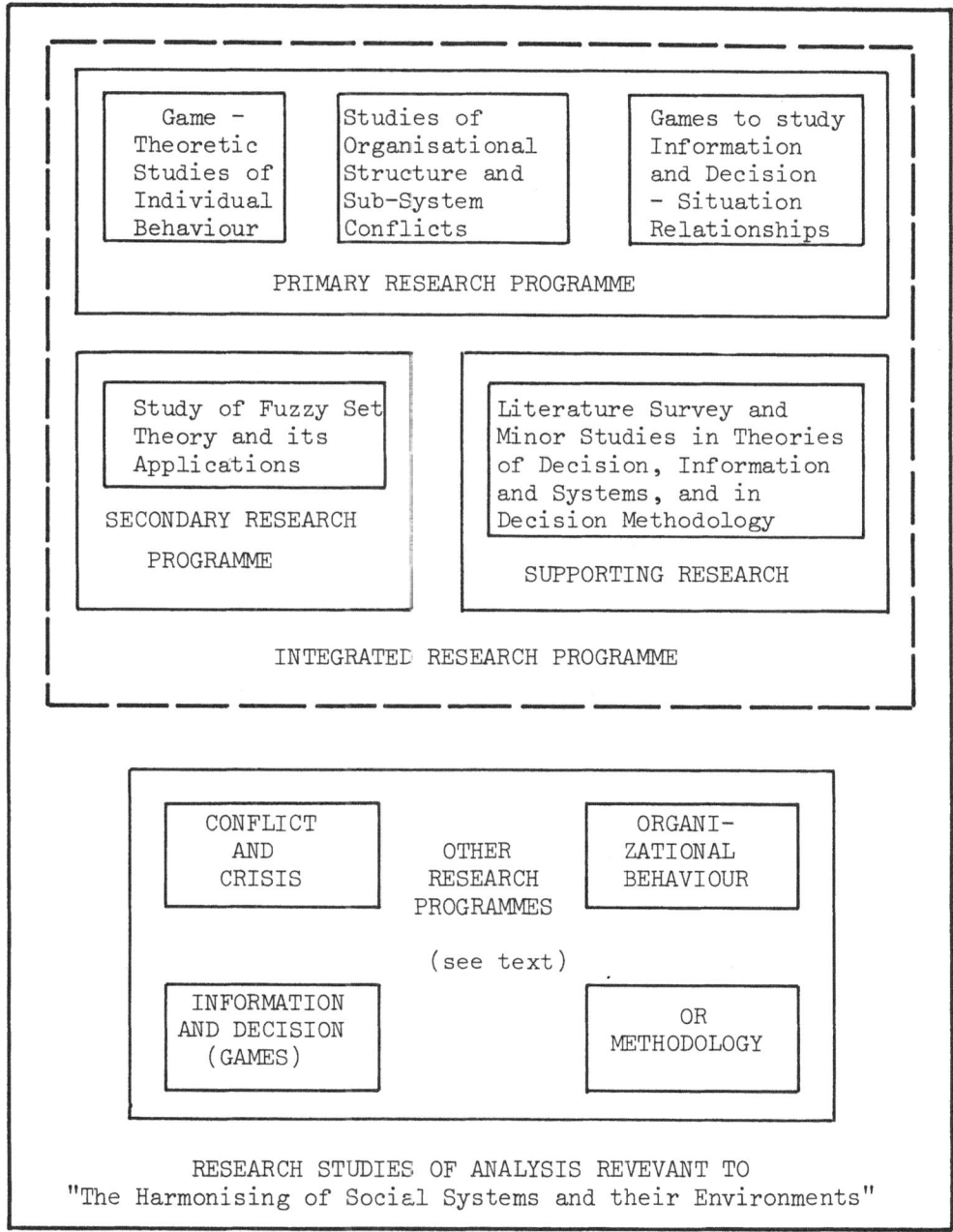

Figure 1. The diagram uses the notation developed for conflict
          studies[1]. It contains only the main sets and subsets which
          are relevant to the theme of this paper.

gaming techniques similar to those we have proposed, to examine information and decision in an Army command and control context; we
are closely interested in this and in any other similar work that
may arise.  Additionally, a first interchange of ideas has taken
place with Mr. D.A.Conway (Hatfield Polytechnic Centre for Operational Studies),who is carrying out a research programme within a
large organisation; his study aims to examine, in the real world,
some of the issues of the analysis/decision process.  Finally, there
have been discussions with Dr. A.N.Oppenheim, Department of Social
Psychology, London School of Economics and Political Science, on
studies of conflict and crisis; he is very much concerned in his
current research with real world studies and group decision-making.
In earlier work[36], Dr. Oppenheim made a detailed study of industrial
conflict in a factory, and his observations on the possibilities of
the transfer of understanding from this field to larger socio-
economic systems are of considerable interest (see section 3.1).


## 5.   CONCLUDING REMARKS

Conclusions in the usual sense are inappropriate.  What has
been said implies our belief that conflict, decision and information
are not subject areas to be studied separately, and that they need
an integrated behavioural science/operational research approach. We
also suggest that there are major language implications.  Inadequacies of language, as well as being a basic cause of conflict, are
a limitation placed on our abilities to structure complicated concepts for analysis, and a barrier to the use of logical/mathematical
processes of measurement.

We believe that in all these fields there are valid, basic concepts which can be used in examining any large scale social system
and its environment.  We also believe that, with care, what is learnt
at one level of complexity can be translated to other levels, not
necessarily in the same context.  Neither of these beliefs is capable
of 'proof' of any sort at this stage of research, but progress towards theory must be based on such intuitive feelings; if they are
not justified, to some extent at least, theories of conflict and
behaviour, and of decision methodology, are not going to be theories
in the true sense[37].

We have exposed the pattern of our thinking and why we are
doing and have done the research reported, in the hope that we shall
stimulate others to follow similar paths or to tell us, with some
insight into our philosophy, why they think we are right or wrong.
Our views on the limitations of current OR procedures[14] are echoed
in a recent report by Stockfisch of RAND[38]: "The serious problem ...
is often not one of methodology.  Rather it is one of people, who
are constrained to behave in peculiar ways due to organisational
settings and incentives."

We should finally observe that the economic aspects of social systems are included only implicitly in this paper. Cost, however, is only one of many factors in any assessment. It is also one among many whose utilities have to be considered as affecting decision behaviour. Economics is also a social science; like operational research and behavioural science, it is necessarily limited in the understanding that can be reached by standard methods of observation and applied mathematics, and it too is loaded with fuzzy concepts, such as 'confidence' and 'stability', that make theory difficult to establish. It is time that we openly admitted the limitations of our various arts, and studied what best to do in the explicit light of these limitations.

It is proper to mention that in 1962, Gregson[39] read a most important paper on the problems of communication for operational researchers and those they serve. It was an excellent message, but its impact was slight because few operational researchers knew of it, because it was published in an occupational psychology journal. Any barriers between disciplines, which in this way hinder communication, must be removed. Facts, rather than beliefs, must also be sought; Noor[40] has made a start with a detailed statistical survey of relationships (defined through questionnaires and interviews) between accountants and EDP and OR specialists.

There is everything to be gained and nothing to be lost by seeing science and the humanities as "two aspects of one culture"[41]. The art of social science is to make the best use of both of these aspects in studying problem situations. We are, presumably, as Ackoff is, concerned with redesigning the future. We should in particular remember that, as he has argued elsewhere[42], if we can move towards participative planning, which requires insight into information, decision, conflict, and all the issues discussed in this paper, we can avoid the impossible task of trying to quantify the quality of life.

## ACKNOWLEDGEMENTS

In planning a research programme of the kind discussed here, it is inevitable that the authorship of ideas is very widely spread. The importance of what has been borrowed from colleagues whose work is listed in the references is acknowledged.

## REFERENCES

1.    Bowen, K.C. and Smith, D.G., Development of Models for Application to Conflict Problems, in Collins, L. (ed.), Social

Issues in the Seventies, No. 4., The Use of Models in the Social Sciences, Tavistock Publications, London, 1976.

2.    Strong, J.P., Application of the Concept of Pattern Recognition to Provide a Framework for Research into Command and Control at the Tactical Level, DOAE Memorandum, M7040, May, 1970.

3.    Bowen, K.C., The Structure and Classification of Operational Research Games, DOAE Memorandum, M7117, June, 1971.

4.    Harris, J.I., Some Concepts of a Conflict Game, DOAE Memorandum 7420, December, 1974.

5.    White, D.J., and Bowen, K.C. (eds.), The Role and Effectiveness of Theories of Decision in Practice, Proceedings of Conference held in Luxembourg in August 1973 under the aegis of the NATO Science Committee, Hodder and Stoughton, London, 1976.

6.    White, D.J., Decision Methodology,  John Wiley & Sons, London, 1975.

7.    Ackoff, R.L. et al., Towards a Quantitative Theory of the Dynamics of Conflict, Management Science Center, University of Pennsylvania, Philadelphia, February, 1968.

8.    Emshoff, J.R. and Ackoff, R.L., Explanatory Models of Interactive Choice Behaviour, J.Conflict Resolution, 1970, XIV, 1.

9.    Howard, N., Some Developments in the Theory and Application of Metagames, General Systems, 1970, XV.

10.   Dando, M.R. and Bee, A.J., Operational Research and Human Behaviour: Towards the Monitoring of Decision Making in Conflicts through the Study of Experimental Games, OR Department, University of Sussex, 1976 (for publication).

11.   Ackoff, R.L. and Emery, F.E., On Purposeful Systems: An Interdisciplinary Analysis of Individual and Social Behaviour as a System of Purposeful Events, Tavistock Publications, London, 1972.

12.   Bowen, K.C. Purposeful Systems and Operational Research, Opl Res.Q., 1975, 26, 1, ii, pp. 125-131.

13.   Bowen, K.C., Some Notes on Purposeful Systems, DOAE Research Working Paper CM 10, February, 1975.

14.   Bowen, K.C., Helping with Decisions on Complicated Systems, 2nd. Inter-American Conference on Systems and Informatics (II CIASI), Mexico City, Mexico, November, 1974.

15.  Bowen, K.C., Where Techniques are Lacking, Commonwealth Defence
     Science Organisation Symposium on Operational Research, New
     Delhi, India, December, 1975.

16.  Harris, J.I. and Bowen, K.C., Notes on the Analyst and Decision-
     Maker: Interaction and Perceptions.  Unpublished notes. Dated
     19 May,1975.

17.  Bowen, K.C., Operational Research Methodology, DOAE Research
     Working Paper CM 11, January, 1976.

18.  White, D.J., Decision Theory, Aldine Pub. Co., Chicago, 1969.

19.  Bowen, K.C., Some Comments on the Scope of Gaming, DOAE Research
     Working Paper CC5, July, 1972.

20.  Bowen, K.C., Models and Decision Processes, in ref. 5.

21.  Stratton, A. and Bowen, K.C., The Information Content of Data,
     DOAE Working Paper D/WP/7502 (Revised), November, 1975. (to
     be published).

22.  Harris, J.I., Some Concepts of a Conflict Game, DOAE Memorandum
     7420, December, 1974.

23.  Harris, J.I., Games for Research into Conflict, a supporting
     paper, NATO Conference on Environmental Assessment of Socio-
     Economic Systems, Istanbul, Turkey, October, 1976.

24.  Hicks, D., The Management of 120-Bed Clinical Nursing Units:
     An Account of Research Carried out in the Five Years 1970-74,
     Wessex Regional Health Authority, Winchester, 1976.

25.  Ackoff, R.L., and Sasieni, M.W., Fundamentals of Operational
     Research, Wiley, New York, 1968.

26.  Bee, A.J., Testing Emshoff's Choice and Policy Matching Theories
     of Behaviour in Simple Mixed-Motive Games with Subjects from a
     Different Culture, BSc thesis, University of Sussex (OR) 1974.

27.  Dando, M.R. and Bee, A.J., An Investigation of a Methodology
     for the Study of Decision-Making Behaviour in Conflicts and
     Crises, Research Report 75-4, University of Sussex (OR), May,
     1975.

28.  Bee, A.J., The Development of a Methodology for the Study of
     Decision-Making in Conflicts, M.Phil thesis, University of
     Sussex (OR), 1976.

29.  Milburn, T.W., The Management of Crises, in Hermann, C.F. (ed.),

<u>International Crises, Insights from Behavioural Research</u>, The
Free Press, New York, 1973.

30.  Emshoff, J.R., Behaviour Theory for OR Applications, <u>Opl Res.Q.</u>,
     1975, 26, 4, i, pp. 675-692.

31.  Bellman, R.E., and Zadeh, L.A., Decision Making in a Fuzzy En-
     vironment, <u>Man. Sci.</u>, December, 1970, 17, 4, pp. B.141-B.164.

32.  Harris, J.I., A Preliminary Enquiry into the Relevance of
     Fuzzy Set Theory to Operational Research, August, 1976. (This
     paper is a later version of the material originally made avail-
     able to Conference delegates as Annex A to the paper printed
     here.).

33.  Eden, C.L. and Birley, T., The Modelling of Decision-Making in
     Organizations: A Local Government Example, Centre for the Study
     of Organizational Change and Development, University of Bath,
     1975.

34.  Eden, C.L. and Harris, D.J., <u>Management Decision and Decision
     Analysis</u>, The MacMillan Press, Ltd., London, 1975.

35.  Eden, C.L., An Exploration of the Role of the Analyst, Centre
     for the Study of Organizational Change and Development, Univer-
     sity of Bath, 1975.

36.  Oppenheim, A.M. and Bayley, J.C.R., Productivity and Conflict,
     Proceedings of the International Peace Research Association,
     Third General Conference, Van Gorcum and Company N.V., Assen,
     The Netherlands, 1970.

37.  White, D.J., The Nature of Decision Theory, in Reference 5
     (page 6, section 4).

38.  Stockfisch, J.A., Models, Data, and War: A Critique of the Study
     of Conventional Forces, R-1526-PR, Rand, Santa Monica, Califor-
     nia, March, 1975.

39.  Gregson, R.A.M., Operational Research, Psychological Informa-
     tion and Managerial Decision, <u>Occupational Psychology</u>, 1962,
     36, 3, pp. 109-123.

40.  Noor, A.E.E., <u>Accountants-Electronic Data Processing and Manage-
     ment Science/Operational Research Specialists Relationships: an
     Empirical Investigation with Special Reference to British
     Industry</u>, PhD Thesis, University of Sheffield, July, 1975.

41.  Ackoff, R.L., <u>Redesigning the Future: A Systems Approach to
     Societal Problems</u>, Wiley, New York, 1974. (Chapter 2, discus-

sion of Interactivism).

42.  Ackoff, R.L., Does Quality of Life have to be Quantified? (In-
     ternational Committee on the Unity of the Sciences, London,
     1974.), Opl Res.Q., 1976, 27, 2, i, pp. 289-303.


Requests for papers not formally published should be made through
Mr. K.C.Bowen, DOAE, Parvis Road, West Byfleet, Surrey, KT14 6LY,
England.

LONG-TERM POLICY ASSESSMENT OF ENERGY/ENVIRONMENT SYSTEMS:

A CONCEPTUAL AND METHODOLOGICAL FRAMEWORK

W. K. Foell

International Institute of Applied Systems Analysis
(IIASA), Austria
University of Wisconsin, Madison, Wisconsin, U.S.A.

## INTRODUCTION

Late in 1974, a research study on Management of Regional Energy/
Environment Systems was initiated by the Ecology/Environment project
at the International Institute for Applied Systems Analysis (IIASA).
The study was structured to take advantage of IIASA's international
and multidisciplinary character.  In addition, during 1975 and early
1976, it served as a rich source of case studies for what has been
the dominant aim of IIASA's Ecology project since its inception --
the development of a coherent science of ecological management that
could be applied to similar problems throughout the world.  The research was founded upon the following four key presumptions:

- Energy-use limitations could result from unacceptable costs
  and consequences, not from physical resource exhaustion;

- Strong relationships exist between energy systems and economic development.  Energy and its environmental corollaries
  will exert increasing influence on technological, economic,
  and environmental dicision-making bodies throughout the
  world;

- Many significant social and environmental consequences of
  energy systems arise from embedding the system in a specific
  region or human environment;

- There is a need to study alternative human patterns and life
  styles in connection with energy/environment systems.

The study, designed to integrate energy and environmental management considerations from a system perspective, has four primary

objectives:

1) To describe and analyze existing patterns of regional energy
   use and supply, and to gain an insight into their relationships to socio-economic patterns;

2) To analyze and compare alternative methodologies for regional energy and environmental forecasting, planning, and
   policy design;

3) To develop new concepts and methodologies for energy/envir-
   onment system management and policy design;

4) To use these methodologies to examine alternative energy
   policies and strategies for test regions, to explore their
   environmental implications from various perspectives using
   sets of indicators related to environmental impacts, energy-
   use efficiencies, etc., and to investigate whether these
   strategies represent a viable choice for the society in
   which they are being considered.

In this report on the way this study has developed over the
past eighteen months, emphasis is placed upon the conceptual and
methodological framework within which it has been conducted.  A companion paper at this conference describes some early results (1).

THE RESEARCH FORMAT

The Comparative Case Study

One of IIASA's strengths is its access to research institutions
and scientists throughout the world and its mandate to interact with
them in applied and policy-oriented research.  To take advantage of
this capability and as a vehicle to sharpen the research, the Energy/
Environment study was organized on a comparative basis, three distinct geographical regions being chosen as the first case studies:
the German Democratic Republic (GDR), the Rhone-Alpes region in
southern France, and the state of Wisconsin, in the U.S.A.  The regions were chosen in part because of their greatly differing char-
acteristics--socio-economic and political structures, technological
base, geographic and ecological properties, and institutional approaches to environmental and energy planning management -- and partly
because of the presence in each region of an institution with a policy-oriented research program, examining energy/environment systems
from a broad resource-management perspective.

## A Research Network

A small core team of IIASA scientists, cutting across several research projects, conducted the in-house research in collaboration with research institutions in the three regions under study, namely:

- The Energy Systems and Policy Research Group of the Institute for Environmental Studies and the College of Engineering, University of Wisconsin-Madison, U.S.A.

- The Institut fuer Energetik, Leipzig, GDR

- The Institut Economique et Juridique de l'Energie (Centre National de la Recherche Scientifique - CNRS), Grenoble, France.

Each of these institutions plays an active role in its country or region in conducting policy-oriented energy research and in advising decision and policy makers.

The interaction between IIASA and the collaborating institutions is shown in Figure 1. As indicated, there was a flow of models, data and personnel. The vigor of this flow reflected positively upon IIASA's potential coordinating role in the international scientific community. Planning for a follow-up phase was begun in 1976 with preparations for participation by additional countries.

## Research Components

The research activities can be broken down into five components:

- Description of the energy/environment systems of each region: past and current energy use, energy supply modes and flows, environmental quality indices (air, land, water, etc.), economic activity, demography, human settlement patterns, and so on;

- Description and comparison of institutional and organizational structures within which energy and environment planning, management, and policy design are conducted;

- Comparison of energy/environment modeling tools used in each region, according to methodology, domains of policy and planning applications, relation to the decision-making structure, transferability to other regions, etc.;

- Development of alternative futures (scenarios) for each region as a tool to examine alternate energy and environmental policies and strategies;

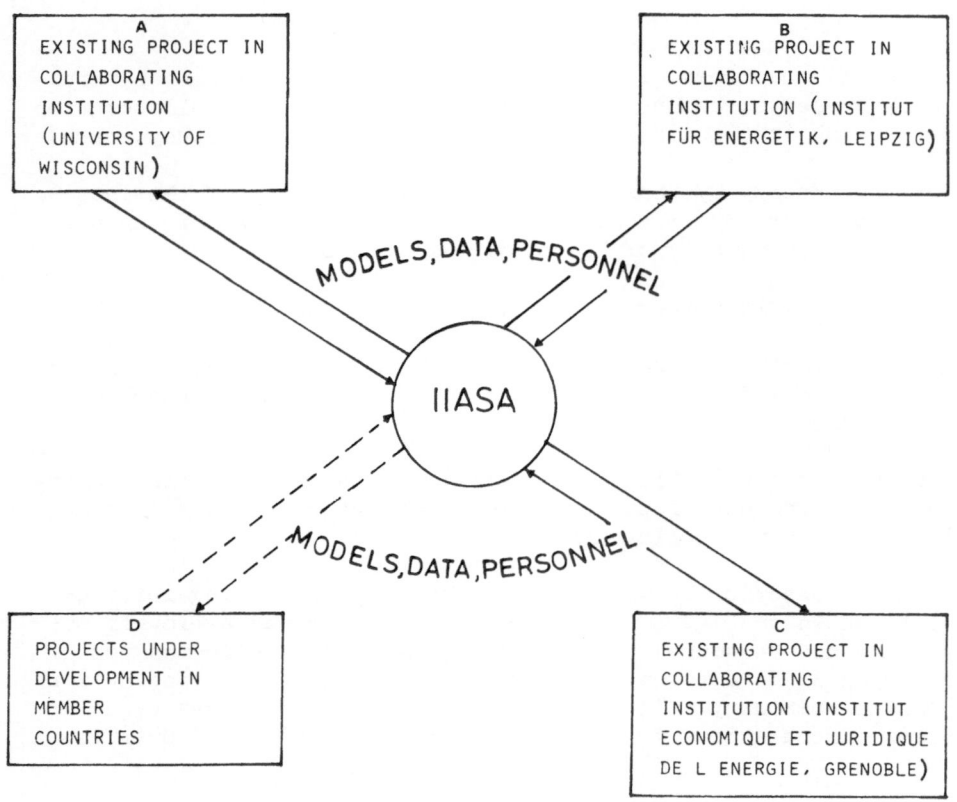

Figure 1. Interinstitutional Relations within the Energy/Environment
System Study.

  -   Development of methods and concepts for communicating and
      evaluating energy/environment strategies and options.


DESCRIPTION OF REGIONAL ENERGY/ENVIRONMENT SYSTEMS

      A detailed comparative descriptive analysis was developed for
the three regions.  It focused on relating differences in energy
use, supply, and environmental conditions to socio-economic activity
and geographic properties.  There are dozens of ways to aggregate
and display the characteristics of the energy/environment system of
a region: from an economic perspective, on an energy-flow basis,
with material-economic flows (input-output), and so forth.  For the
purposes of this study, the system structure shown schematically in
Figure 2 was used.  The major components are socio-economic activities, energy demand, energy conversion and supply, primary energy,
and environment.

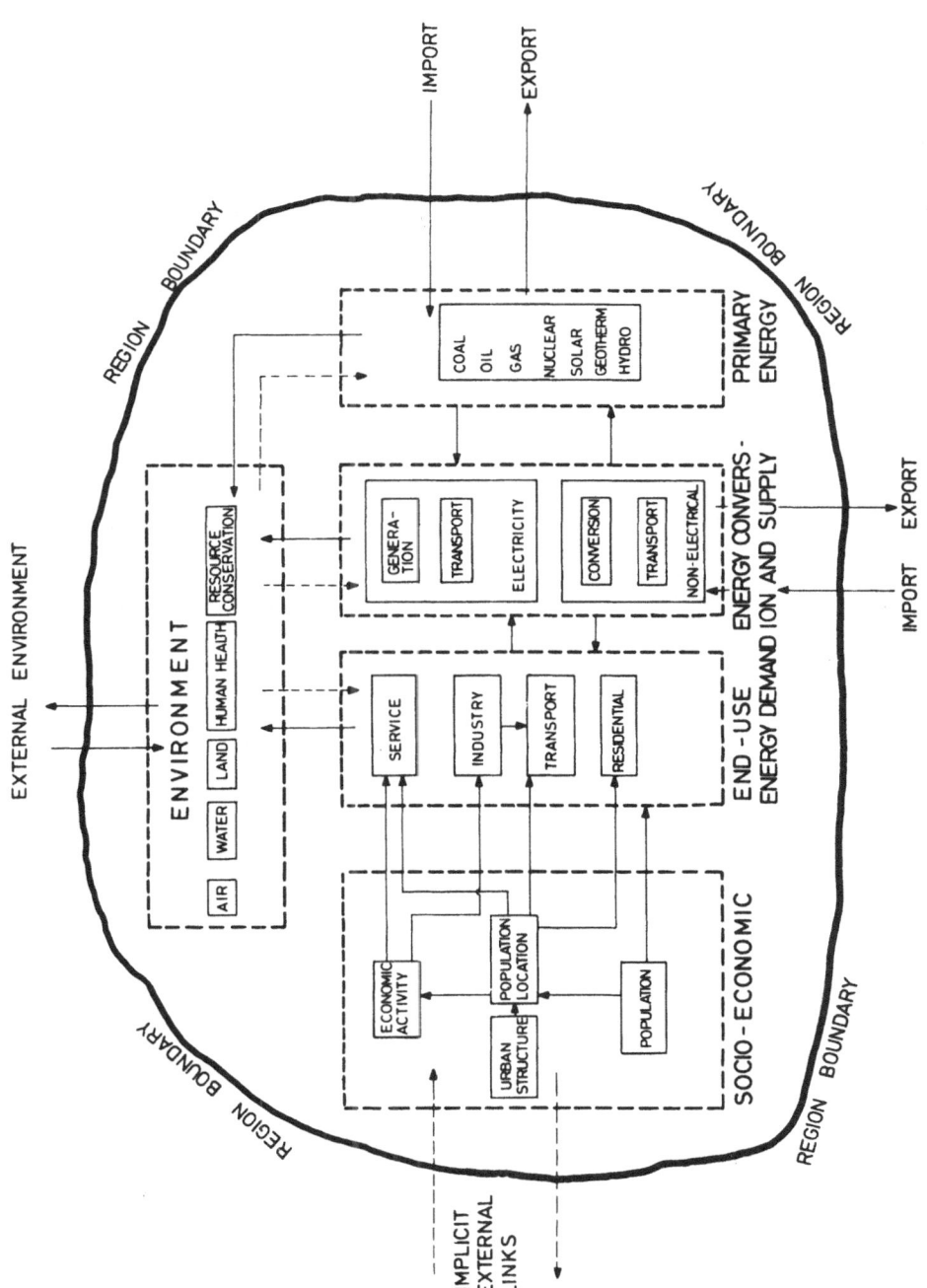

Figure 2.  Structure of Energy/Environment System.

The hierarchical structure within each component is complex; details are given in a forthcoming work (2).

As indicated earlier, one component of the research program was to describe the institutional and organizational structures associated with planning and policy analysis in the energy and environmental areas in each region. Although this was one of the smaller parts of the research effort, it was a significant one. As work progressed, it became apparent that there was a strong relationship between the institutional and decision structures of a region and the formal models and planning tools that were used. This was demonstrated quite vividly by the contrasts in the structure of the three regions. These structures, and the models and planning tools, are described in several papers by regional energy experts and policy makers (3).

## APPRAISAL OF ENERGY/ENVIRONMENT MODELS

A major objective of the project was to appraise and compare the energy and environmental models of the three regions studied. This appraisal would be valuable to each region in assessing its potential use of models from other regions, and would reveal how the models are tied to the policy objectives and characteristics of the region.

In order to emphasize the transferability of the models, the appraisal was divided into two parts. Each collaborating institution described its own system of energy/environment models; and each appraised the models of the two other groups from the perspective of its own system and methodological requirements for planning and policy analysis. For example, the Wisconsin group identified the types of information it desires and examined whether the French models treat these areas adequately.

### The Models

Although each region uses a large variety of models, only the major planning models were appraised (2).

The GDR models appear to be aimed at long-term planning activities, with emphasis on the economy/energy (as opposed to the energy/environment) relationship (2). They combine demand projections, technological development estimations, and investment planning in a system that allows for analysis of alternative growth strategies. Although energy-related environmental modeling activities are apparently going on in various institutions and planning organizations in the GDR, these models have not been integrated into the central energy-planning models. The highly integrated GDR energy model ap-

pears to be quite advanced in its capability to examine and model
the significant interrelations of the various economic sectors.  An
economic objective function, the minimization of social expenditures,
forms the basis of the optimization procedure used.

In Wisconsin, the multiplicity of decision-making units means
that it is impossible to construct a single model with a unique objective function, or even with a common constraint set, since not
all the various agents in the system are constrained by the same array of factors (2).  The need is for a comprehensive well-integrated
system model, but one that explicitly recognizes the fragmentation
of decision making.  For Wisconsin, one must talk in terms of a set
of energy/environment models and the means by which they can be integrated.  The modeling activity comprises efforts in both the public and private sectors, only some of them coordinated.  One exception is the work of the Energy Systems and Policy Research Group at
the University of Wisconsin, which has resulted in the development
of a computerized dynamic simulation model of the entire Wisconsin
energy/environment system.

Although there is considerable centralization and a very important public sector in energy planning in France, the private sector plays a significant role, and hence the energy modeling activities are somewhat more directly akin to those in Wisconsin.  However,
as has been mentioned earlier, the economic and energy activities of
the Rhone-Alpes region do not comprise an autonomous economic system
since the institutional and economic structure of France is very
centralized.  Therefore no energy modeling exists exclusively for
the Rhone-Alpes region.  Consequently, the model evaluation dealt
with models for the nation rather than the region.  Particular attention was given to the linear programming model developed at the
Grenoble Energy Institute.  It provides for an optimization of the
total energy system, subject to constraints on availability of particular primary energy fuels.  It also provides for the inclusion
of environmental constraints although not at a level of complexity
which make them amenable to regional analysis.


SCENARIO BUILDING

The writing of alternative futures, often referred to as scenario building, was chosen as a methodological device in this research
because of its value in the study and evaluation of the interaction
of complex and uncertain factors (4).  Broadly described, scenario
building is a detailed examination of the possible futures and the
consequences of alternative assumptions about them.  This set of
futures may provide a better view of what is to be avoided or facilitated, the types of decisions that are important, and the points
in time after which various decision branches will have been passed.

The techniques used assumed that the region under study could
be described as a system comprised of socio-economic, technological
and environmental components coupled to each other with various degrees of strength.  The system description used for our work is
shown schematically in Figure 2.  The scenario-building process was
one of imposing given policies on the systems within the framework
of the initial conditions and constraints characteristic of the
region, and then evaluating the resulting development and evolution
of the region.


                        The Policy Issues

     The policy issues were chosen on the basis of two criteria.
They had to be of special interest to at least one region and of
general interest to the other two; and they had to have sufficient
focus and data that they could be approached in at least a semi-
quantitative manner with methods available to the IIASA team.  They
also had to be relevant to mid- and long-term planning and policy
analysis (5 - 50 years).

     The procedure for choosing policy issues satisfying the above
criteria was an iterative one, beginning with discussions with the
collaborating institutes.  After several issues were identified,
they were explored by the IIASA team to see whether they could be
approached within the time frame of the project and with the expertise available at IIASA.  After general decisions were made on these
policy issues and on the types of scenarios that would help illuminate important policy questions, some months were spent gathering
data and developing relationships with which to describe the alternative futures.  Several of the major issues are listed below.

Urban Settlements:

  - How are energy use and environmental impact related to urban
    density, urban size, types of housing and energy supply tech-
    nology and type?  In all three regions the answers to these
    questions are useful for policy analysis related to land use,
    building standards, district heating strategies, and the like.

Transportation Systems:

  - What are the energy and environmental implications of con-
    tinuing present trends and policies for inter- and intra-
    city passenger transportation?  How are these modified by
    policies favoring alternative transportation modes, including
    mass transit systems?
  - What will be the energy and environmental implications of
    higher efficiency automobiles?

Energy Supply:

- What are the consequences and implications of satisfying future demand through alternative energy supply options and
  strategies?
- What is the feasibility of the introduction or expanded use
  of alternative heating technologies, including district heating, combined thermal-electric plants, and waste-heat use
  systems?

Environmental Protection and Resource Conservation:

- Are there environmental limits associated with various patterns of energy demand and supply?
- What are the environmental effects of various pollution control policies associated with alternative energy system strategies?
- What are the major environmental tradeoffs associated with
  alternative fuels for the production of electricity? How
  will a policy encouraging expansion of district heating influence air quality?

In order to specify a "policy set" within which a scenario was
built, it was necessary to develop a notation for expressing a
policy in terms of a number of characteristics. In a functional
form, the framework for a given scenario is described in the following terms:

- Population,
- Economic Growth and Structure,
- Human (Urban) Settlement Location and Form
- Technologies of Energy Use,
- Transport Systems for People and Goods,
- Primary Energy Conversion and Supply Technology (including
  electricity generation),
- Environmental Control and Protection.

This framework then is used to provide the exogenous functions,
boundary conditions, and constraints for the models used to build
the scenarios.

The policy issues listed above were addressed by two specific
paths. First, three alternative policy sets were developed, and
each applied to each of the three regions. In selecting a limited
number of scenarios for study, we tried to choose rationales that
were meaningful in all three regions, combined the majority of the

policy issues described, and could conveniently be compared.  Second, sensitivity studies were developed to evaluate the effects of variations in one policy variable while holding the others constant.


## Models and Methodology

The main quantitative tool used for scenario building is a large-scale simulation model, originally developed at the University of Wisconsin (5) and extended at IIASA to treat regional Wisconsin. For each region, the model was driven by alternative socio-economic patterns provided by the collaborating institute in that region. For example the Institut fuer Energetik in Leipzig provided extensive economic input based upon their preparation of the GDR long-term energy plan.

The WISconsin Regional Energy Model (WISE) is a computerized simulation model designed to describe the technological-economic-environmental interactions in a regional energy system (5).  It is built of a hierarchy of submodels.  Its simulation structure provides considerable flexibility in both the modeling process and the application; it makes possible the modification of selected components without the necessity to rework the entire model, and the focusing of attention on specific system areas as well as on the entire system.  Although there are numerous ways to describe the overall structure of the WISE model, one of the more revealing is by component subsystems (Figure 3).  Within this scheme, the model can be conveniently subdivided into five major components:

1) Socio-Economic Activity.
   This component includes population and demographic characteristics, economic activity, and the geographic distribution within the region.  Most of the inputs to these components are exogenous and were derived from other macroeconomic regional or national models or studies.

2) Sources of Primary Energy.
   This component contains information on fuel sources within and external to the region, according to origin, composition, energy content, price, availability, etc.

3) End-Use Energy Demand.
   Demand is modelled according to various sets of energy use categories; the specific choice of these categories depends upon the application of the model.  The sets include class of use according to economic sector, physical process, fuel type, location within the region, and time period when use takes place.

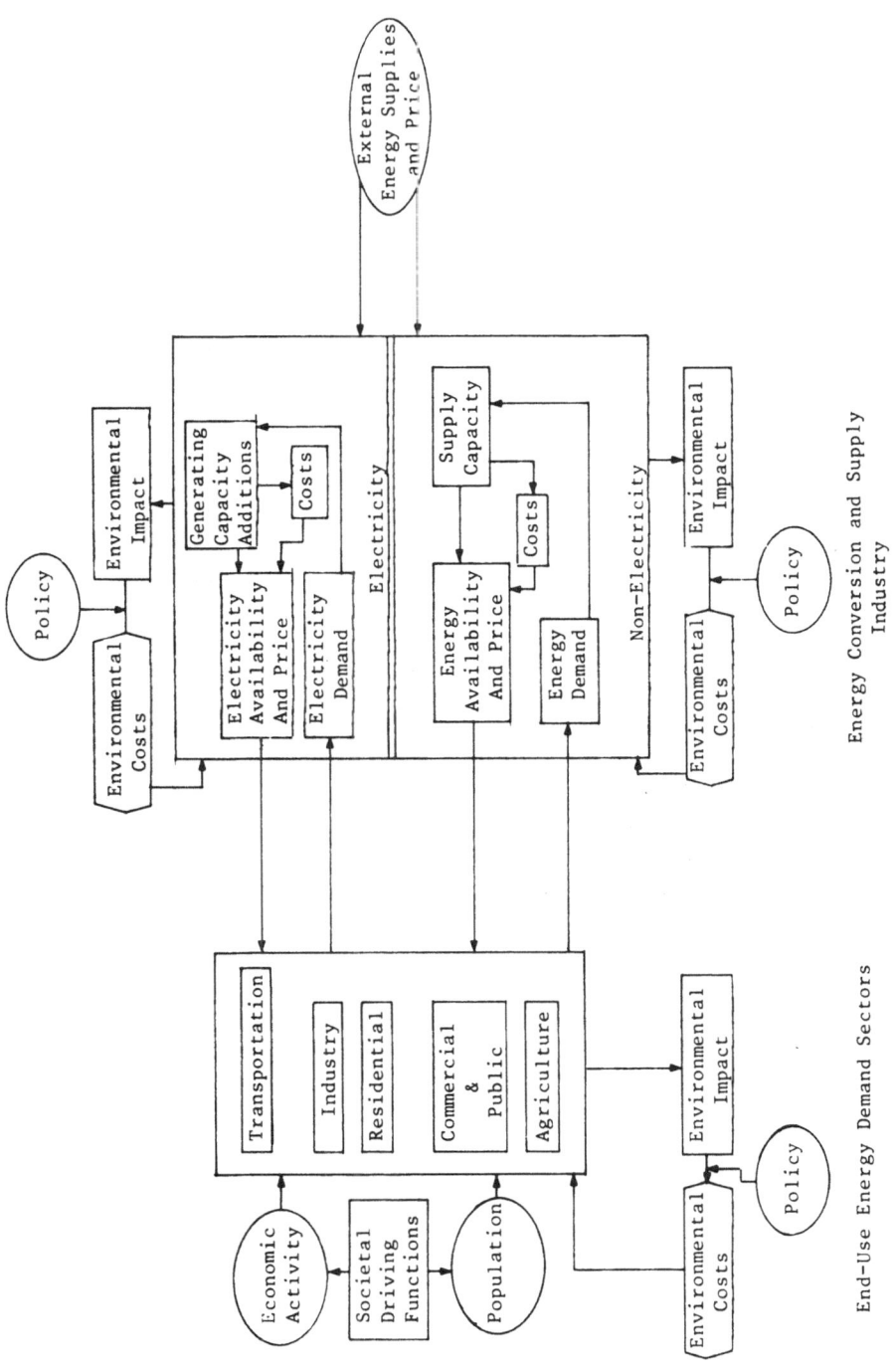

Figure 3.  The WISconsin Regional Energy Model (WISE).

4) Energy Conversion and Supply.
   This component treats the conversion of primary fuels into
   intermediate or final forms of energy for end use. The two
   main groups of submodels treat _electrical_ and _non-electrical_
   energy.

5) Environmental Impact.
   This component simulates on a year-to-year basis the "quantified" environmental impacts resulting from a region's use
   and the associated energy supply and conversion system (6).
   As an example of the information contained in the impact
   models, Figure 4 illustrates the impact pathways which the
   model associates with electricity use in a region. The calculation of quantified impacts from electricity use in an
   energy system in a particular year is based upon impact factors that relate impacts to a unit of electricity generation
   for a reference plant in the specified year. The impacts
   are classified into the general categories of land, air,
   water, human health and safety, and fuel resource use.

The flow of information in the model begins with the exogenous
specification of population, human settlement patterns and economic
activity. These variables provide a basis for calculating end-use
energy demand. A second group of models calculates characteristics
of supply systems necessary to meet that demand, including supply
capacities, primary energy, etc. The environmental impact models
use population and human settlement data, as well as outputs of the
energy demand and supply models, to calculate environmental impacts
(indicators), including human health and safety. A growing literature exists on the structure and applications of the WISE model,
(e.g. 7,8) and on the IIASA extensions. The use of the model in
scenario building is described in more detail in Reference (4).
When a submodel or set of submodels was not applicable to a particular region, other alternatives were used. Since in the GDR a plan
for energy use through 1990 exists, some end-use demand scenarios
were obtained from the Leipzig Institut fuer Energetik rather than
calculated with the WISE model. In addition, because the Rhone-
Alpes region is not a distinct political unit, some types of data
were difficult to obtain; here the models had to be simplified accordingly. In addition, a number of additional models were developed at IIASA during the course of this research, e.g., energy/en-
vironment preference models (9, 10) and air pollution analysis methodology (11,12).

## The Scenarios

The three scenarios can be briefly characterized as follows:
S1, the base case, represents a continuation of current socio-
economic trends and policies (the "Plan" in the GDR). S2 results
from policies encouraging a high-energy future, based on presumed
low or moderate energy costs and placing little or no emphasis on

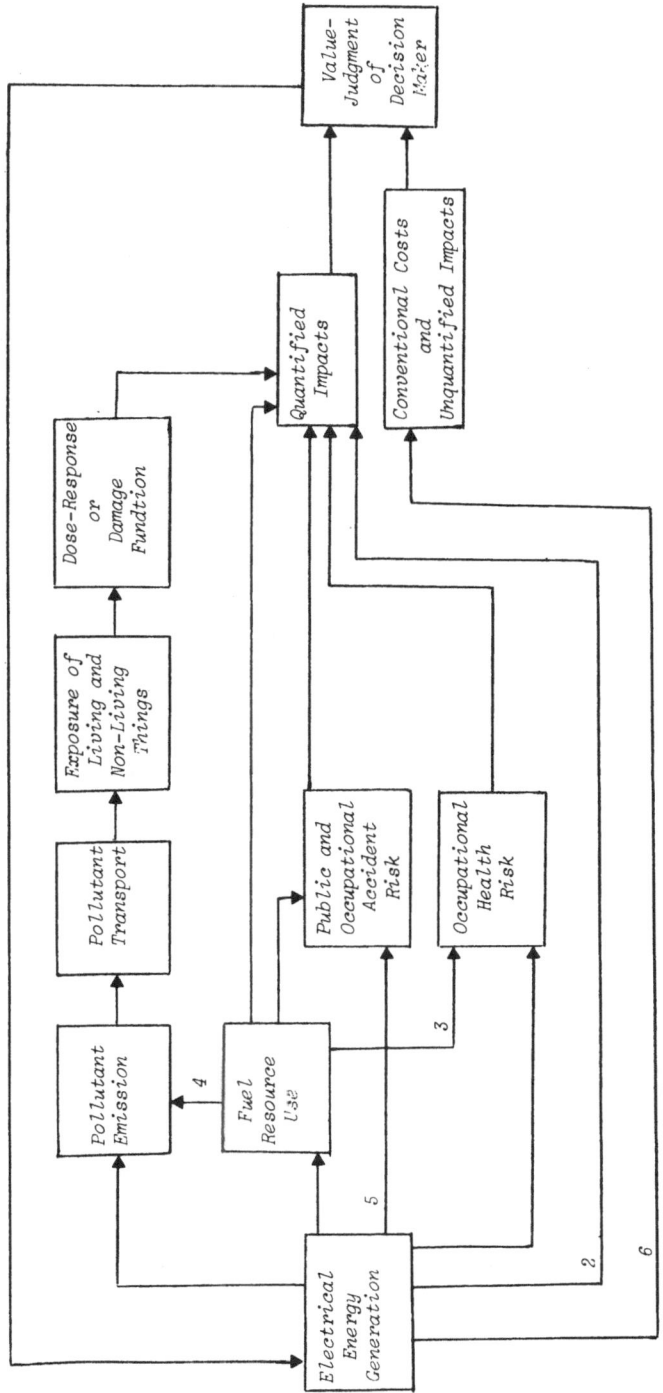

Figure 4. Electrical Energy Impact Pathways.

improving efficiencies of energy use.  Low environmental controls
are also assumed.  S3 is a low-energy future, resulting from policies encouraging energy-saving technologies of transport, heating
and industry, and promoting increased environmental quality through
conservation and stricter pollution controls.

Other scenarios could have been chosen for the initial study,
perhaps for equally good reasons.  However, these three could be
applied consistently across the three regions and focus attention
on many important issues.  They were built at IIASA for Wisconsin,
Rhone-Alpes, and a composite region ("Bezirk"X") typical of the
heavily industrialized southeastern GDR.  They were discussed in
November 1975 by energy and environmental experts and decision
makers from the three regions at the IIASA Workshop on Management
of Regional Energy/Environment Systems.  The final step of the scenario-writing process is still under way: examination of the internal
consistency of the dynamic evolution of the energy/environment system.  The scenarios are discussed in more detail in a companion paper
presented at this conference (1).


EVALUATION OF OPTIONS AND STRATEGIES:   IMPLEMENTATION OF RESULTS

It has been pointed out that scenario writing in no way represents a forecasting or prediction procedure.  The scenarios are meant
to stimulate discussion and to provide a better basis for evaluating
alternative futures.  The success of their use in design or management depends on feedback between the scenario builders and the managers and designers of the energy/environment systems.  Feedback in
scenario writing is similar to the mechanism by which man's knowledge grows.  In that sense, the cycling is a process that rarely
stops for long; new knowledge evolves continuously.  Time also affects
feedback, to the extent that hypothetical future events as laid out
in the scenarios either do or do not occur.

From the methodological description in this paper, it is obvious that no formal method has been applied for including uncertainty in the procedure.  Rather, the uncertainties must be judged subjectively by scrutinizing the scenarios and the sensitivity studies.
Clearly there is ample opportunity to exclude major components and
events that can completely change the evolution of the energy/
environment system.  This is a well-known hazard of scenario writing.

The scenario writing process is <u>descriptive</u>. To explicitly
transform the scenario output into <u>prescriptive</u> forms, additional
steps are obviously required.  One of these is the embedding of the
scenarios into an institutional and decision framework where <u>prefer-
ences</u> and <u>values</u> must be applied to the results.  This is a complex
task that would differ considerably across the three regions studied
because of their very different social and institutional structures.

Decision Analysis - An Evaluation and Communication Tool

It has been a major task simply to describe these systems and their possible evolution. If one then adds the difficulty of embedding the descriptive and prescriptive into an institutional structure for implementation, the management problem is truly formidable. Some of its important characteristics are:

1) Interdependencies among economic, technological and ecological characteristics of a region;

2) Difficulties in identifying costs and benefits and associating them with specific societal groups;

3) Uncertainties and changes over time;

4) Difficulties in communicating complex material;

5) Multiple decision makers.

Each region studied provides a wealth of examples of the complexity of the management problem. Decision analysis has been applied in this study as one approach to the evaluation and communication of alternative policy designs. The method used was based upon multiattribute utility theory (13). In this approach, a so-called preference-model is introduced into the evaluation process. The relationship between the energy/environment impact model and the preference model is illustrated in Figure 5. The outputs of the impact model are impact levels of the attributes, i.e. the altered system states. Examples are the sets of environmental impacts associated with the various regional scenarios. To the extent possible, impact models are meant to be objective and to exclude value-judgment content. The construction of the preference model for a decision maker requires the assessment of a utility function for each attribute.

Assessment requires personal interaction with the decision maker, since his utility function is a formalization of his subjective preferences for the attributes (impacts). One of the advantages of this evaluation framework is that recognized but unquantified impacts can be identified and included in the analysis by determining an appropriate proxy variable that can be measured. The overall preference model, based on the measured utility function for a particular individual, allows the calculation of the individual's expected utility associated with the combined impacts of a given policy (scenario). The expected utility calculated for an alternative is a measure of the relative desirability of that alternative for the assessed individual.

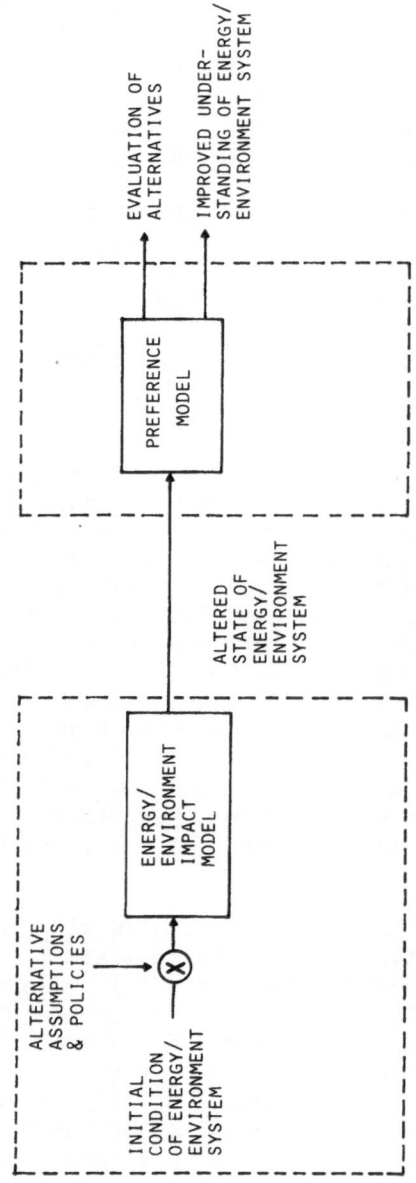

Figure 5. Relationship between Impact Model and Preference Model

The first application of the above method to regional energy/ environment systems was based upon a set of policies related to the choice of electricity generation systems for Wisconsin. The Electricity Impact Model (6) was used to generate 11 attributes of a set of scenarios based upon alternative policies. Utility functions were determined for two individuals from Wisconsin and used to evaluate the set of scenarios (14). In a follow-up study (10), preliminary utility assessments were completed for five individuals from Rhone-Alpes, the GDR, and Wisconsin, over a set of four attributes selected from the above set of 11. The group included decision makers and energy/environment specialists. The utility functions for the five individuals were used to evaluate their preferences for several hypothetical energy supply and environmental policies. More detailed results of this application are described in the companion paper at this conference (1).

## Implementation and Transfer of Research Results

Although each research component described in the preceding sections has the potential to contribute to improved management of regional energy/environment systems, none of them should stand alone. It is essential that each be used as a complement to the others and more importantly, that they be linked in a coherent research format that promotes <u>frequent interaction with the institutional and decision clients for which it is intended</u>.

The need for interaction with the client is vital and was thus given primary emphasis in our research. From inception of the program, information was solicited from the appropriate users; and at the conclusion of the first phase, they were asked to evaluate the scenario-building results. Frequent workshops encouraged this interaction. This process, shown schematically in Figure 6, was perhaps the key element in integrating the several components of the research program in providing a communication interface between the modelers in the three regions.

Figure 6

The research program reached a milestone in late 1975, when a workshop held at IIASA brought together 25 scientific experts, policy makers, and members of the public from the three regions. Figure 6 is also representative of that workshop. Apart from the socio-technical interaction of specialists and policy makers from the GDR, France, and the USA, the workshop provided an opportunity to introduce the comparative scenarios and alternative models into planning and policy design procedures in these countries.

Papers presented by the collaborating institutions included appraisals of the others' modeling procedures and comparisons of some of the energy and environmental planning practices in the region, e.g. pricing, environmental standards, and building practices. These papers are being prepared for publication (2).

A research transfer process tacked onto the tail end of a research program has almost no chance of success. It is essential that the transfer process be given high priority at the very beginning of a study whose ultimate objective is improved policy design and that this priority be preserved through the entire process. The objective of this transfer is not specific policy recommendations, but rather the transfer of concepts, models and methodologies, evaluation procedures, and a range of policy analyses. Our efforts to do this, perhaps only partially successful, have been terrible demanding of time and energy and, occasionally, even frustrating. At times, they may seem to distract us from the substantive research activities which have traditionally been the domain of specialists in each of our fields. But without exception, there is agreement among the research team that even a partial success in embedding the research outputs into the actual policy-design processes would be more than adequate justification of our efforts.

REFERENCES

(1)   Foell, W.K., et al. Assessment of Alternative Energy/Environment Futures: A Comparative Case Study of Wisconsin (USA), the German Democratic Republic, and Rhone-Alpes (France), Presented at NATO Conference on "Environmental Assessment of Socio-Economic Systems", Istanbul, Turkey, October 4-8, 1976.

(2)   Proceedings of a Workshop on Integrated Management of Regional Energy/Environment Systems held November 10 - 14, 1975. International Institute for Applied Systems Analysis, Laxenburg, Austria, forthcoming.

(3)   Born, S., C. Cicchetti, R. Cudahy, and J.Pappas (Wisconsin); P. Hedrich, K. Lindner, and D. Ufer (GDR); J.-M. Martin

and D.Finon(France), Energy/Environment Models and their Relationship to Planning in Wisconsin (USA), the German Democratic Republic, and Rhone-Alpes (France), RM-76-21, International Institute for Applied Systems Analysis, Laxenburg, Austria, 1976.

(4)  Foell, W.K., Scenario Writing: One Component of a Systems Approach to Energy/Environment Management, RM-76-20, International Institute for Applied Systems Analysis, Laxenburg, Austria, 1976.

(5)  Foell, W.K., J.W.Mitchell, and J.L.Pappas, The Wisconsin Regional Energy Model: A Systems Approach to Regional Energy Analysis, University of Wisconsin-Madison, Institute for Environmental Studies, Report 56, Sept. 1975.

(6)  Buehring, W.A. and W.K.Foell, Environmental Impact of Electrical Generation: A System-Wide Approach, RM-76-13, International Institute for Applied Systems Analysis, Laxenburg, Austria, 1976.

(7)  Mitchell, J.W. and D.A.Jacobsen, Implications of Commercial Building Codes for Energy Conservation, University of Wisconsin-Madison, Institute for Environmental Studies, Report 42, December 1974.

(8)  Hanson, M.E. and J.W.Mitchell, A Model of Transportation Energy Use in Wisconsin: Demographic Considerations and Alternative Scenarios, University of Wisconsin-Madison, Institute for Environmental Studies, Report 57, December 1975.

(9)  Keeney, R., Energy Policy and Value Tradeoffs, RM-75-65, International Institute for Applied Systems Analysis, Laxenburg, Austria, 1975.

(10) Buehring, W.A., W.K.Foell, and R.L.Keenes, Energy and Environmental Management: Application of Decision Analysis, RR-76-14, International Institute for Applied Systems Analysis, Laxenburg, Austria, 1976.

(11) Dennis, R.L., Regional Air Pollution Impact: A Dispersion Methodology Developed and Applied to Energy Systems, RM-76-22, International Institute for Applied Systems Analysis, Laxenburg, Austria, 1976.

(12) Buehring, W.A. and R.L. Dennis, Evaluation of Health Effects from Sulfur Dioxide Emission for a Reference Coal-Fired Power Plant, RM-76-23, International Institute for Applied Systems Analysis, Laxenburg, Austria 1976.

(13) Keeney, R.L. and H. Raiffa, <u>Decision Analysis with Multiple Conflicting Objectives: Preferences and Value Trade-offs</u>, Wiley, New York, (in press).

(14) Buehring, W.A., <u>A Model of Environmental Impact from Electrical Generation</u>, unpublished Ph.D. dissertation, Dept. of Nuclear Engineering, University of Wisconsin-Madison, 1975.

ECONOMIC AND ENVIRONMENTAL ASSESSMENT OF A WATER QUALITY MANAGEMENT

SYSTEM (River Neckar)

J. Klaus

Volkswirtschaftliches Institut Universität Erlangen-
Nürnberg
Nürnberg, West Germany

## 1. TECHNICAL PROJECT, ECONOMICS AND ENVIRONMENT

### 1.1 Project Characteristics and the Evaluation Problem

By order of the German Federal Minister for Research and Technology a team from scientific and administrative institutions is
working on the "Prognostic Model of the River Neckar", technically
coordinated by the Dornier System GmbH. This team's task is to find
methods by means of which an optimal water quality management can
be achieved. Factors relevant to water quality are to be regulated
in such a way that an optimal state and usage of this common property resource is guaranteed.

There is a wide scale of measures which aim at water quality
improvements. They range from a relatively simple intensification
of waste treatment on the one hand or constraints for water users
on the other, right up to whole groups of simultaneous measures,
the optimal mix of which can only be found by complex evaluation
systems.

Within such a wide range the research work reported in the present paper concentrates on regional intensification of waste-water
treatment with fixed location of the (already existing) cleaning
plants. The resulting strategy of water quality management depends
on the physical, technical, economic and social characteristics of
the whole river basin. This report has to deal with the question
of the role economists have to play in the context of such a decision.

Only recently has it been acknowledged that water quality management measures can only be evaluated by comprehensive investigations using a complex mathematical model of the physical system that is to be controlled.

By means of a simulation model the direct and indirect consequences of possible measures concerning specific states of the system can be traced.  In order to find the optimal performance of the system – defined either by a minimum of costs or a maximum of benefit items relative to certain constraints – an optimization model is an additional requirement.

In the case of the River Neckar, a simple data-based simulation model has proved to be suitable as a basis to the optimization model. For choosing the optimal measures, an elaborate technique of linear programming is being used.  By means of this method the optimal combination of quality measures at the different sections of the river can be found by maximizing a linear objective function, the variables of which are subject to a number of constraints in the technical, environmental and other fields.  The economist has to develop an evaluation method which makes it possible to find the optimal socio-economic state of the system in question.

Here, special problems will arise as to the decision on:

- the total set of evaluation criteria

- the construction of evaluation techniques

- the empirical identification of the advantages and
  disadvantages

of the project in question.

From a socio-economic point of view neither the minimization of economic costs nor the restriction to an economic net benefit can be considered as satisfactory, and that is why the development of methodical foundations to a more complex evaluation method is necessary.  By means of this method it will be possible to connect water management optimization models (for example LP-models for the minimization of costs) with techniques for the maximization of both economic _and_ non-economic benefit.  Within such a framework the economic benefits and the environmental advantages of alternative measures for the improvement of water quality must be subject to simultaneous optimization calculations.

Thus it is necessary to develop

- methodical foundations for the integration of both cost-benefit
  calculation and environmental assessment into one optimization
  model,

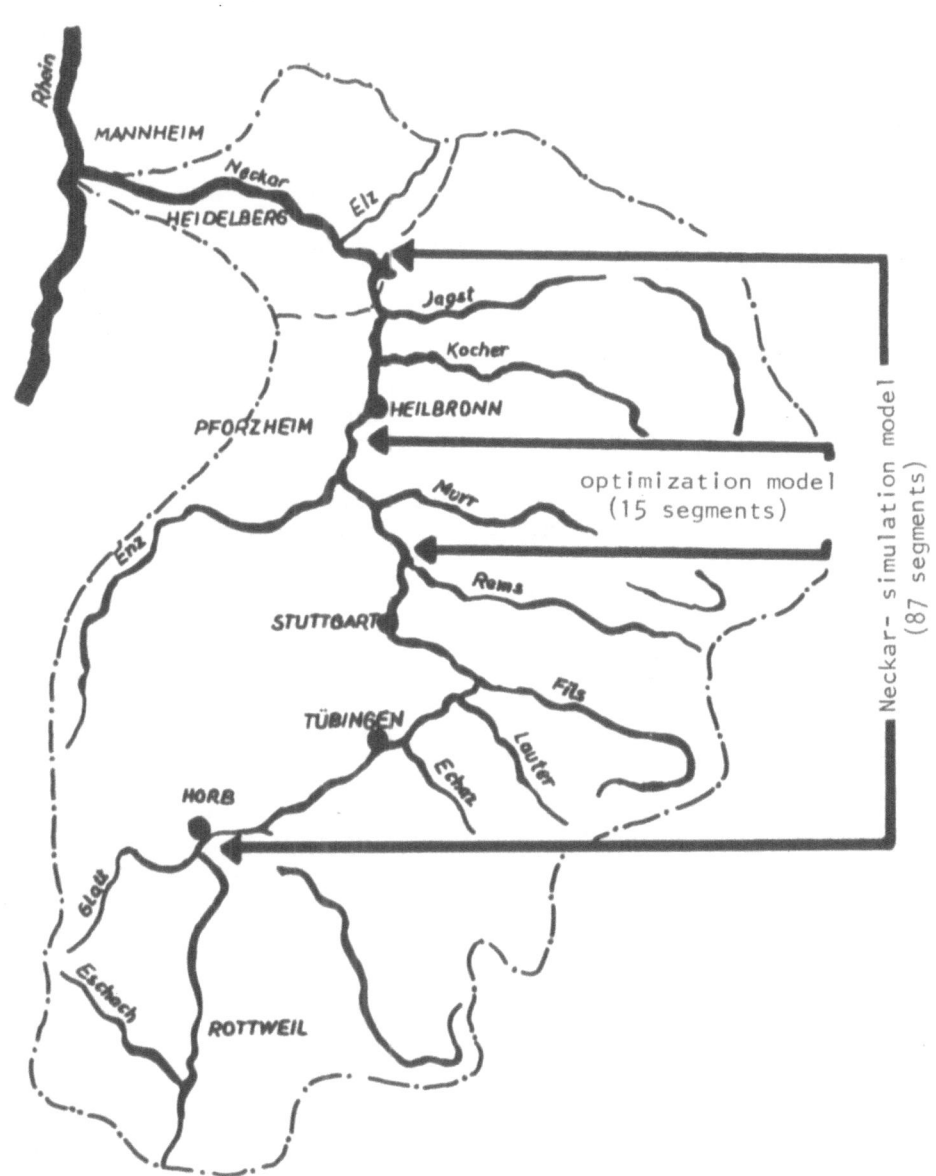

Fig. 1  The Neckar River Basin

techniques for the comprehensive identification, quantification
and evaluation of the benefit elements both on the economic and
the environmental side (a distinction which is theoretically problematic but in practice inevitable as a result of the intangible
character of certain benefits and costs).


## 1.2 Proposal of a new method

For the project in question a two-stage optimization model
shall be proposed: during the first stage a ranking and a selection
of water management projects is to be found according to their efficiency in terms of cost-benefit-analysis, and during the second
stage shifts in the ranking according to environmental effects are
to be identified and necessary changes in the decision analysed.  A
variation in the objective function, that is to say of the preferences (weights) between economy and environment, makes it possible to
test the sensitivity of the optimal solution with regard to the relative importance of both objective fields.  Thus the decision-
maker obtains a bunch of possible decisions,with which he will have
to deal in the end.  One of the most important tasks for modern decision theory is to show clearly the conflict that often arises between economy and ecology – and to solve this conflict by means of
an adequate optimization method.

If an analytical apparatus of that kind is actually to offer
some help for the decision-making authorities, sufficient possibilities for the identification and quantification of at least the most
important elements must exist.  Yet there are considerable difficulties.  As literature only provides minimal support in this respect,
a method has to be developed by means of which the benefit elements
in economy and ecology can be identified, whether they are measurable in monetary units or not.  Several partial problems have to be
solved, the first of which is the setting-up of a catalogue of project consequences, which are to be measured and evaluated in further
investigations.

For the analysis of economic benefit elements, which can be expressed in money terms, a detailed catalogue has to be set up containing savings in factors of production, productivity and also
consumer advantages which may result from improvement measures. Important benefit elements may, for example, accrue to users

- in the drinking-water sector

- in the processing industry

- in the tertiary sector (leisure).

The problem is, now, to find the monetary indicators expressing
the savings in economic resources, productivity and consumer advan-

tages. It is according to the rules of a free enterprise economy that these effects are evaluated by the market. Yet there are often difficulties in obtaining the proper market values, especially if substitutes have to be found, which are necessary if there is no market or if the market cannot be relied on because of social, sometimes even ideological reasons. Whereas it is possible to measure some benefits directly as savings in cost, other benefits (for example the effects concerning outdoor recreation) can only be obtained by means of a detailed study of the relevant items, their measurement in physical units and their assessment in value terms.

Still greater problems arise on the side of the so-called non-economic effects. Because of the intangible character of many environmental effects, (as far as they can be expressed in monetary units, they can already be included in the economic sector), it is rather difficult from a technical as well as from a methodical point of view to measure these advantages (advantages for the environment). In order to find a comprehensive expression for the quality development in a relatively large river basin, the concept of a quality index is to be used. By means of this concept it should be possible to express in one figure both the different meanings of quality changes at each river section and the evaluation of each quality standard itself. Such figures, which might be called weighted quality points, are to be compared with the economic benefit variables in order to find an optimal solution. Both this comparison (based on the weights of the variables within the objective function) and the construction of an index measuring the environmental effects of water quality management, are closely connected with subjective or political preferences respectively. Thus, from a methodical point of view, it is of importance to know how these preferences can be obtained.

As a whole the evaluation concept within the water management model can roughly be summarized as follows:

- Assessment of the economic benefits (net benefits) according to a catalogue of concrete possible usages in consequence of an improved water quality. This catalogue may be obtained by using previous information gained by analysing published quantitative data and by direct inquiries to local water users.

- Assessment of the environmental benefits by means of an especially developed evaluation method. This method will have to make use of the specific knowledge of the disciplines concerned in order to formulate judgements and to work out an evaluative relation between environment and economy.

- Integration of the two fields - economy and ecology - into the water management model. This means the appropriate adaption of the resulting variables to the model structure, and the development of an optimization rule suitable for the calculation itself.

The method as a whole has to be constructed in such a way that the new technique can be transferred to other areas, too. In principle, so far as the method is at present constructed, it will answer this requirement. If this is confirmed in further investigations, a first step for an enlarged decision method in the field of water resource management will have been taken.

## 1.3 A Concrete Goal Structure for Water Quality Projects

The question arises as to how to obtain the concrete objective function which can serve as a basis to the optimization calculation. By means of the graphic representation of a so-called "relevance tree" a demonstration can be won of how the goals are reduced to operational items.

Any benefit assessment of water quality measures has to start from the following questions: which of the individual and social needs can be satisfied by an improvement in quality and to what extent will this be possible? According to a general catalogue of social goals, a catalogue of water quality effects, that are relevant because of their social importance can be obtained. As far as the single effects meet the various goals the total benefit of quality management measures will vary correspondingly.

Apart from environmental goals concrete goal identifications can be developed for four efficiency oriented economic fields (see figure). The fields of "transport" and "production of energy", however, show only an indirect connection with projects concerning water quality management. For the evaluation of quality measures it is of special importance to look at the range of alternatives in water utilisation: "water for production needs including water works" and "water for immediate usage". The most important step is to obtain information concerning the goal items which belong to these fields of utilisation. These items must be operationally formulated, i.e. empirically measurable and mathematically manageable.

The item "irrigation water" is of no relevance in this context, as it has become apparent already at the beginning of the empirical investigations that the positive and negative effects of the quality management of river water concerning agriculture neutralize each other to a large extent, and thus they do not influence the evaluation of the projects any further.

For the economic evaluation of quality management projects according to the objective of an improvement in economic efficiency, it is necessary to identify the following kinds of benefits together with their goal contributions:

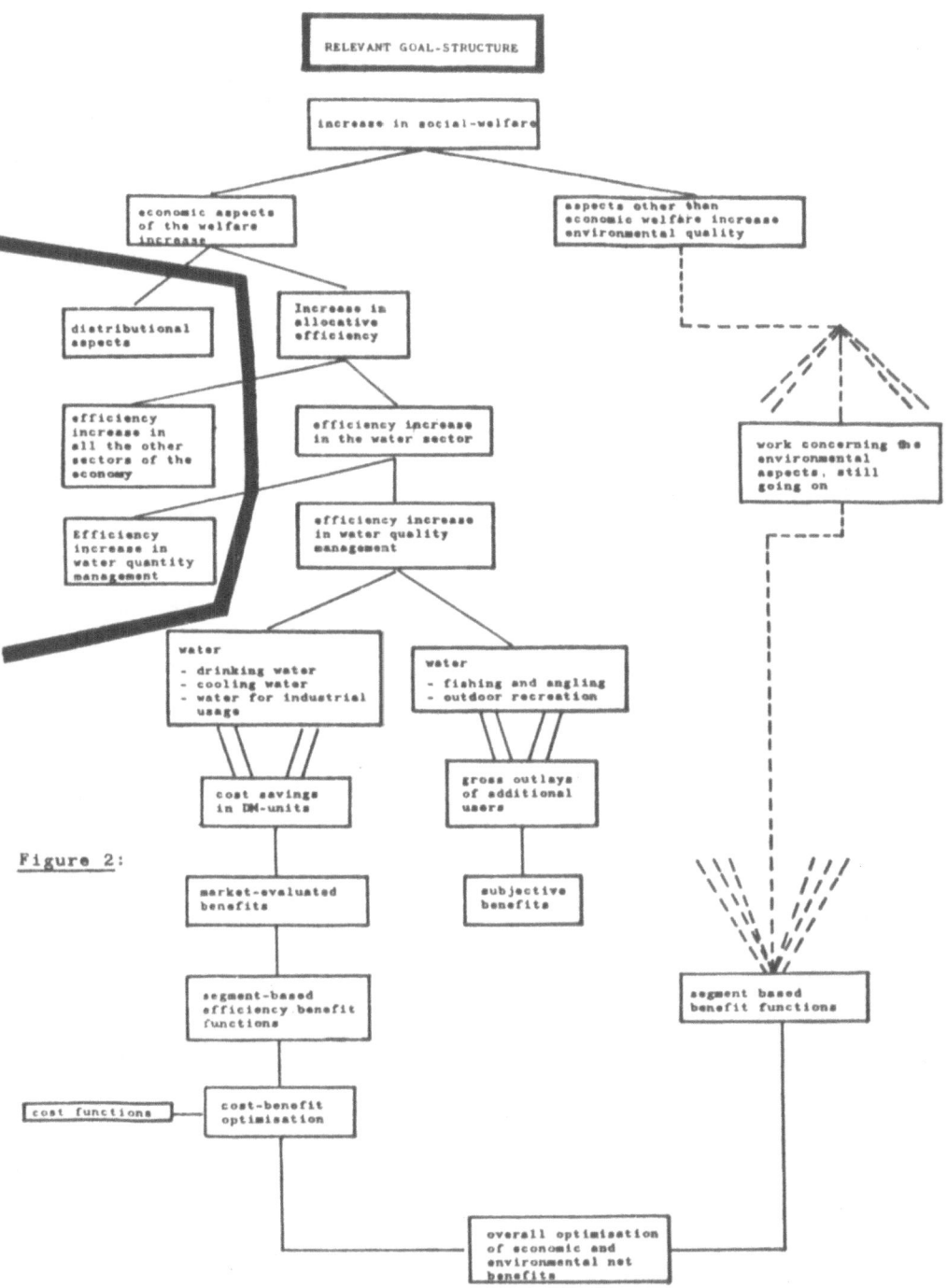

Figure 2:

- water for production needs including water works
  - benefit from drinking-water
  - benefit accrueing to power stations (cooling water)
  - benefit from water used in industry

- water for immediate usage
  - benefits from fishing and angling
  - benefits from boating and yachting    )
  - benefits from swimming and bathing     )   outdoor recreation

    The overall economic benefits manifest themselves in possibilities for increases in production, in consumption advantages and in savings of economic resources, which result from the measures taken.

    As a consequence of the improved water quality:

- water of better quality is available as a production factor
- savings in other production factors become possible
- more productive methods become technically available
- water of the new quality can be used as a consumer good or as a complementary good together with other consumer goods.

That is to say the benefits result in savings in costs which would otherwise emerge if water of the new quality were not available, e.g.:

- costs for the removal of damage resulting from pollution
- costs for cleaning the water before usage
- costs for switching to other resources
- renunciation costs for the usage of water.

2. ECONOMIC BENEFIT FUNCTIONS[*)]

2.1 Methods of Measurement

    The purposes of the Prognostic model of the River Necker, here used as a basis for a case study, demand a systematic approach which connects the benefits of quality management measures with different water quality conditions. It is necessary to set up a formal economic benefit function for each of the fifteen sections of the river basin, in order to take into account regional peculiarities. These sections have been subdivided according to the location of waste influents, especially by waste treatment plants.

    Now the DM-values related to different kinds of utilization of the river water have to be identified. Thus an expression of the increase in economic overall utility as a consequence of improved water quality in terms of BOD values (Biochemical Oxygen Demand) for each of the 15 sections cf the model basin (efficiency-oriented first step of the evaluation concept) can be found.

The benefits are assumed to spread over thirty years. They are estimated for this period and discounted at a rate of 4 percent per annum, inflationary effects being disregarded. The benefits in the production sector are calculated from the input side. The usage of river water for the supply of drinking water, for the cooling of power stations or for industrial purposes reduces costs, leading to gains in resources and possibly to a rise in the gross domestic product. As the saved cleaning costs are imposed in monetary terms - being evaluated by the market - this part of the utility function needs no further discussion.

It is more difficult, however, to measure the benefits in the consumption sector. Here benefits are concerned which will emerge from the expanded activities (or new ones) of fishing, boating and bathing under conditions of improved water quality. These benefits have to be quantified and evaluated in DM-values, by means of a suitable method in order to become smalgamated with the benefits from the production sector.

From a methodical point of view it would be the best to measure the willingness to pay as an expression of individual preferences. But as the preferences and the willingness to pay with relation to public goods - in this case water or the quality of water respectively - are not easily discernible, alternative methods must be found. The construction of hypothetical demand curves is an elegant method, but requires much empirical work. By means of the so-called "gross expenditure method" results can be obtained in a relatively easy way without a significant loss of accuracy and generality.

This method is based on the following considerations: if men, as consumers, are assumed to behave rationally according to the economic principle, the monetary outlay for obtaining a consumer good can never be larger than the resulting subjective utility. By identifying all expenses in DM-values for the different kinds of benefits of fishing, boating and swimming, a lower limit of the actual benefit of these activities can be estimated. The exact amount could be calculated by adding the consumers' surplus to the gross outlays. As it is extremely difficult to identify the consumers' surplus, it will only be taken partly into account in this study. The method

---

*)
Empirical findings presented in a research report by J.Klaus and W.Vauth delivered to Dornier System GmbH within the Prognostic Model of the River Neckar, by order of the German Federal Minister for Research and Technology.

of gross expenditures provides the means for the calculation and
evaluation of the benefits of the consumption sector in DM-values.
Thus a large part of those effects that were formerly considered as
intangible can be integrated into the economic utility function.
This means that the applicability of the cost-benefit-criterion in
the field of water quality management has been enlarged.

In the following, the specific identification and evaluation
problems are roughly demonstrated by showing some typical problems
and ways to solve these problems.

## 2.2 Drinking-Water Benefits*)

Benefits arising from producing drinking water more economically
in the case of step-wise improvements of water quality will play a
decisive role within the model. The calculation of these benefits
may become difficult if the production process in the water works is
only feasible after a certain degree of water quality management has
been installed. Cost curves of general validity for the conversion
of river water into drinking water are not available. In the case
of the River Neckar it has been possible to use the experience of a
water work in Stuttgart-Berg as a suitable model for a water-works.
Although this water-works does not produce drinking-water at present,
it was possible to estimate hypothetical cost savings according to
alternative states in the quality of the River Neckar. Thus a reference scheme for the calculation of benefits, arising from the conversion of river water into drinking-water, became available.

By means of this method it was possible to calculate discounted
total benefits for a large-scale water-work, which would supply water for the region of Ludwigsburg-Stuttgart (North). For an improvement in quality of the River Neckar (medium degree) the total benefit
amounted to 40 to 120 million DM. In another case the benefits for
a smaller water-works, which supplies water only for the region of
Heilbronn, amounted to 30 million DM. The amount of cost savings
proved to be clearly dependent on the degree to which improvement
measures were taken. Thus a veritable benefit function, i.e., a
schedule of cost-saving as a consequence of different degrees of
water quality improvement could be derived.

## 2.3  Fishing and Outdoor Recreation

There are benefits resulting from improved fishing conditions
which amount to millions of DM if the level of the river water quality is raised. These benefits, however, are not so great as one

*) Benefits from a higher quality of water used in power stations
   and industrial plants have also been calculated and included in
   the analysis. The description is omitted here.

might have expected according to other comparable investigations, (for example at the River Trent). The quality development of the River Neckar causes a change in the structure of the fish population in favour of useful fish, which anglers are only interested in. As a result, the number of anglers, who fish or who can be admitted to a certain river section, will increase. In order to calculate the annual benefit from increased fishing possibilities, the number of additional anglers per annum is multiplied by the amount of the estimated gross outlays of every angler (corrections concerning consumers' surpluses are necessary).

One of the most interesting and most difficult problems is the estimation of outdoor recreation benefits. It has already been shown in the Saint-John River Study that benefits of a large amount can be expected in this field, though they are difficult to estimate. Instead of setting up a model showing hypothetical effects of this sort, a more practicable method was developed and worked out. This method is based on experiences gained at another river, the sections of which show different quality conditions. It is not easy to find a river which is suitable for a transference of empirical values, as several conditions must be fulfilled:

- the hydrological characteristics of the river must, to a large extent, correspond with those of the River Neckar

- the referential river must show all the quality conditions that are suitable for the various recreation activities

- the referential river must be similar to the model river concerning landscape, settlement and infrastructure.

As a result, a certain section of the River Main between Bamberg and Aschaffenburg was found as a referential river suitable for the recreation benefit evaluation. As there are no direct market prices as evaluation factors for these leisure activities, the method of gross expenditures was used once more. Thus DM 25 per hour for hiring a motor-boat were calculated, DM 3 per hour for canoeing and hiring a rowing-boat respectively. As a result the benefit functions of the river sections, resulting from these considerations, could be derived. They all showed increasing benefits along the scale of improving the quality of water.

## 2.4 Empirical Findings [*])

Apart from some special problems (which still exist) it can be said in summary that this method helps to identify benefit functions for the management of water resources in a river basin consisting of several sections. The instruments already mentioned have been developed in order to achieve as much practical relevance as possible,

and in order to apply it to any other river system without great
difficulty and without too great a sum of money, time and staff.
Although the empirical findings obtained by this method will be affected by errors of different kinds and by uncertainties, we can
nevertheless hope that the process of decision with regard to the
economic aspects of the management of water resources can thus be
greatly improved. This will become clear after it has been shown
how the integration of these economic functions together with the
environmental functions into a multi-level decision model will work.


### 3. TECHNIQUES FOR IDENTIFICATION, MEASURING AND SIMULTANEOUS EVALUATION OF ENVIRONMENTAL BENEFITS


#### 3.1 Empirical Issues

For the next step, in addition to the calculation of economic
benefits, the environmental effects of water resource management
have to be estimated. In order to integrate these environmental effects a technique has to be found, by means of which it is possible
to identify, measure and evaluate the changes in the relevant environmental factors and to amalgamate these effects into one single
number, which can then be compared with the economic benefit value
measured in DM-units. As the river basin consists of sections differing in structure and quality changes, a weighting scheme has to
be set up for the construction of a weighted index. This task is
to be fulfilled by teams of experts from different fields, e.g. ecologists, regional planners, economists. They will have to

- measure the amount of the environmental effects along various relevant scales

- estimate the weights of the different river sections and of the
  different scales

- estimate the marginal rate of substitution between economic efficiency and environmental effects.

Several techniques which have been developed in the fields of
decision theory, may be adapted for these purposes. For projects
of this sort a method for finding a group judgement must be constructed. In detail there are the following steps for setting up an evaluation technique:

a)   Which formal and operational indicators can be found to express
     the environmental effects resulting from a change in the water
     quality? Which variables can be used according to the present
     state of knowledge and information in the collaborating disci-
     plines?

---

*)  As examples see Figures I-IV in Annex A.

b)    Is it possible to identify so-called goal achievement functions
      for each river section (corresponding to the technique used
      for the benefit analysis?  Is it necessary to use the so-called
      "damage units" of the "Sachverständigenrat" or are there other
      suitable variables, such as the direct use of the average amount
      of pollution caused by an inhabitant, the BOD (biochemical
      oxygen demand) or DO (dissolved oxygen)? Which of them can be
      regarded as a suitable substitute for constructing environmen-
      tal benefit functions?

c)    How is it possible to consider the relevance of each river sec-
      tion and the differences with regard to possible dangers for
      men, animals, plants and landscape by using a kind of environ-
      mental index? it was not before 1974 that the "Sachverstädigen-
      rat" for environmental problems stressed the necessity of such
      a procedure in their report.

d)    Which persons and institutions are to be consulted for their
      knowledge and judgement so that they can participate in the
      construction of environmental benefit functions?

      What is looked for is a technique which is as practicable as
the one by means of which the economic benefit functions were obtained, and which can be transferred to other river sections, too.
First attempts have already been made, and it seems that further research can be started soon.

### 3.2  Institutional Approaches

      The following technique might be adequate for obtaining a meas-
urement of the total environmental effects.  This expression is to
be compared with the efficiency measure in an indifference curve analysis in order to identify the optimal project alternative.

- Every single quality state, which has already described the water
  quality for estimating the economic benefits and costs, is evaluated with regard to its environmental effect by a team of experts,
  for example by scoring on a point scale.

- The weights of every single river section, too, are identified by
  a team of experts, considering regional planning, ecological and
  social aspects.

- By means of these two steps a measure can be calculated for the
  total environmental effects of each river management project, which
  can be confronted with the economic efficiency of these projects.

      One main question arises as to which people shall belong to the
teams that have to take the evaluation steps.  Other problems are
how to fix the measurement scales and how to take the evaluations.

Apart from those who are directly affected by these projects, there ought to be among the participants

- experts (scientists, administrative officials and people employed in private industry) and

- politicians (federal, regional and communal).

The team of experts for evaluating the environmental effects of the changes in the defined quality degrees should at least consist of:

- hydrologists
- ecologists (biologists for example)
- physicians
- economists
- regional planners.

The team of experts and politicians, which are to identify the weights of the river sections in order to calculate the weighted aggregate environmental effects, should at least consist of:

- regional planners
- ecologists
- water economists
- politicians of the region concerned.

The group evaluation can either be obtained by anonymous or direct communication.


### 3.3 Integration of Partial Evaluations for the Derivation of a Total Optimum

The last step, which is the mutual weighting of economic and environmental benefits, proved to be especially difficult. A value ratio is to be set up in order to identify the optimal measures by means of the optimization technique. Strictly speaking, such a value relation can only be set up as a political judgement. Certain substitutes can be used, however, by which a decision within the optimization model can be made possible, if the political evaluation process is too complex. A solution of this kind is, if possible, to be developed as a final step.

One of the basic ideas refers to the principle of opportunity costs. According to this principle the DM-value that has to be given up because of the realisation of a certain amount of environmental improvement is to be identified. A possible measure, as an example, would be the average cleaning costs for the reduction of BOD by one unit in the river region in question. Yet this would mean a strong and very specific approach. Whether this is justifiable or whether other substitution rates are to be chosen is a matter which has to be considered very carefully.

An alternative approach can make use of the statements of the "Sachverständigenrat" for environmental problems and of the decision of the Federal Government concerning the amount of a water pollution tax. Here the price that has to be paid for the pollution of a certain quantity of water and which shall be an incentive to refrain from any pollution, would be applied to the setting up of a kind of social value relation. Advantages and disadvantages of the different methods must be investigated in detail.

4.   Final Results: The Complex Evaluation Method

With these methodical and practical considerations the evaluation method, proposed in this study, is complete. According to the relevance of the objectives, economic efficiency on the one hand and environmental quality on the other, the maximization of a social preference function consisting of these two components (with their subcomponents) can be achieved.

Within the analysis, which is in the first stage set up according to economic aspects, we have one cost and one benefit function for every river section as elements of the optimization model. The maximal cost-benefit difference, which corresponds to the respective functions <u>at each single</u> river section, is of no interest, as both the water quality for the whole river basin and the total cost-benefit difference only result from the combination of different cleaning standards and the respective techniques along the whole river. This maximal total difference between benefits and costs has to be calculated by a computer. So the computer has actually to search the optimal set of cleaning measures along the river, according to given technical and regional conditions. If, the costs and benefits are additionally discounted for future periods, the described technique will show the integration of cost-benefit analysis into the optimization model. Although the optimal solution is prepared in steps by using cost-benefit functions for every river section, it is performed simultaneously together with the whole water quality model.

For the second stage of this technique, by means of which the ranking of the alternative projects and their selection is to be obtained by simultaneously considering economic and environmental objectives, several possible ways are available. These are

- a ranking based on economic efficiency subject to an additional consideration of the environmental benefits

- a separate maximization according to both goals, and calculating a weighted sum afterwards (goal programming)

- simultaneous optimization within the framework of one objective function, which yields one single overall benefit measure.

Although it is possible to show the trade-offs between economic efficiency and environmental effects using the first way, it is entirely left to the decision maker to amalgamate both objectives and to choose the optimal solution according to an internal and non-transparent weighting procedure.  Within the framework of a computer-based optimization model this procedure cannot be used without additional qualifications.  The second way can lead to sub-optimality, as the separate maximization might yield solutions which are inefficient under a simultaneous approach.

Only the third procedure meets the rationality conditions, which are assumed to be valid in this study.  The different parts of this procedure are

- the setting up of a ranking of alternative projects with regard to economic and environment aspects, weighted in a specific way

- selection of the optimal project, i.e. of an optimal combination of improvement methods at each river section

- sensitivity analyses of variations in the weights with regard to efficiency and environment in the objective function.

Different subjective judgements of the decision-makers will lead to different selections of optimal measures.  It is of course of special interest to show eventual characteristic shifts of the decisions.

If this research project can be finished successfully, a multiple stage evaluation method for economic and environmental effects, integrated into a water quality management system, will become available for the first time.  It is expected that this method can also be applied to other kinds of environmental projects.

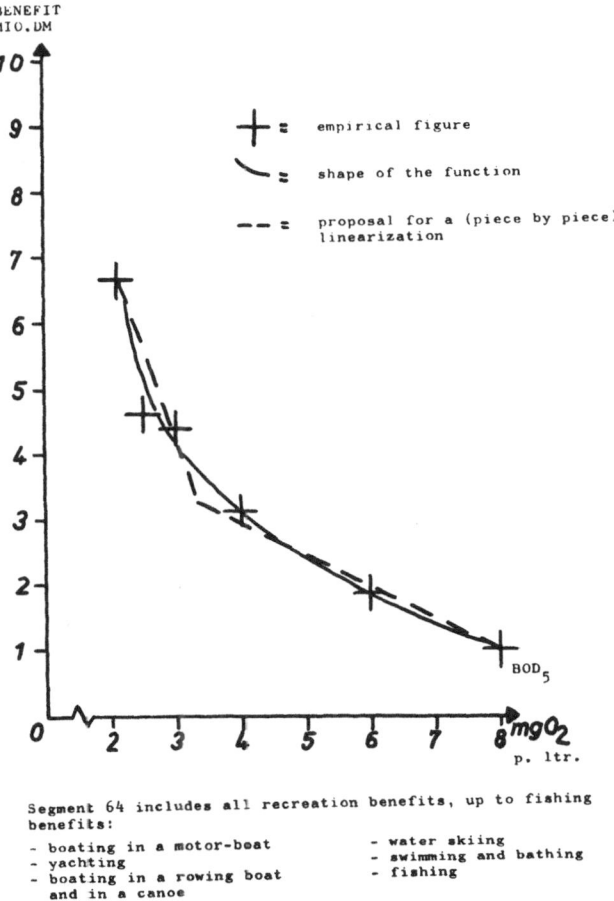

Figure I. Benefit function of segment 64.

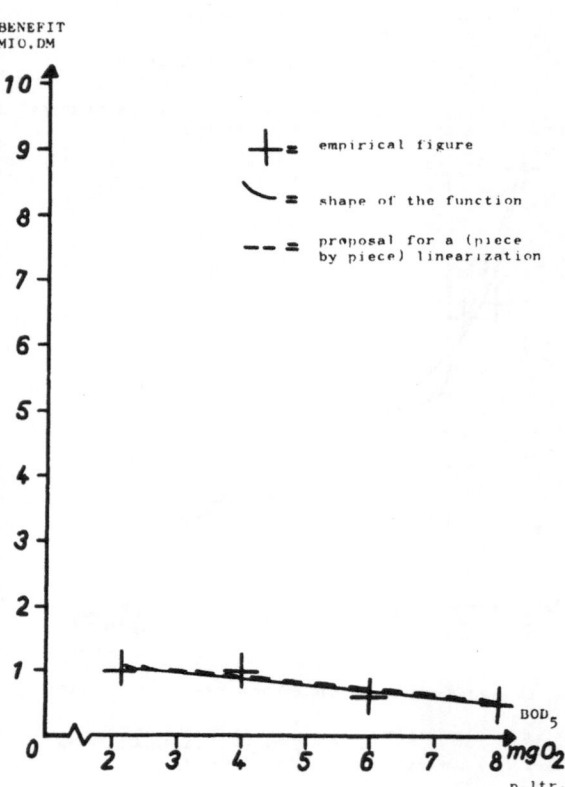

Figure II. Benefit function of segment 54.

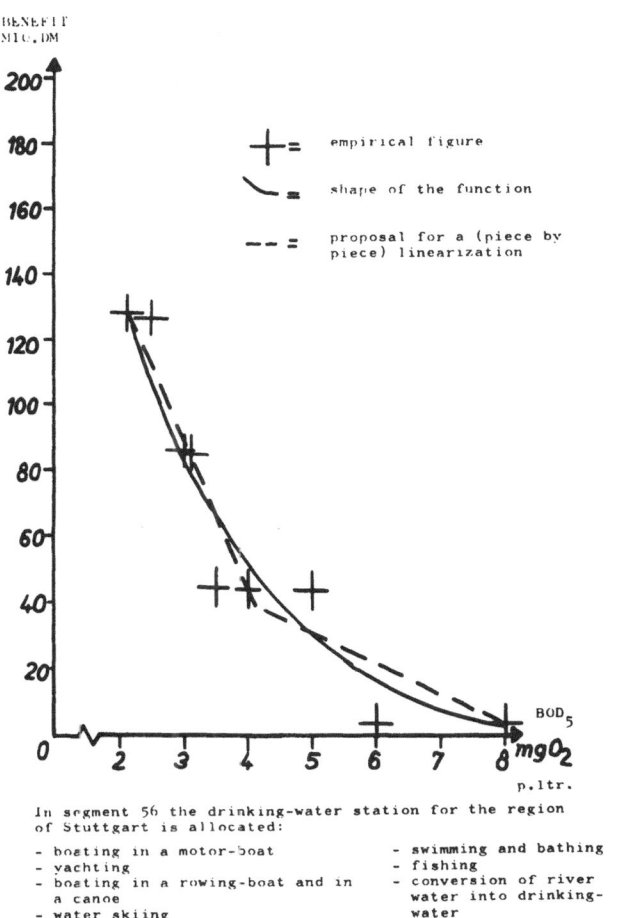

Figure III. Benefit function of segment 56.

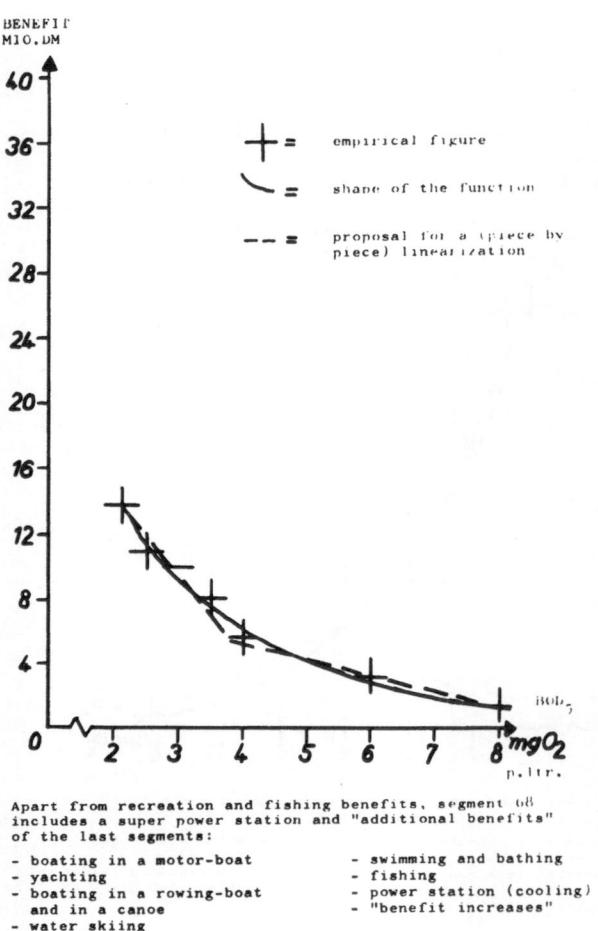

Apart from recreation and fishing benefits, segment 68
includes a super power station and "additional benefits"
of the last segments:

- boating in a motor-boat          - swimming and bathing
- yachting                         - fishing
- boating in a rowing-boat         - power station (cooling)
  and in a canoe                   - "benefit increases"
- water skiing

Figure IV.  Benefit function of segment 68.

SELECTED REFERENCES

Burton, T.L., Noad, P.A., Recreation Research Methods. A Review of
   Recent Studies, Edinburgh 1968.

Der Rat von Sachverständigen für Umweltfragen. Umweltgutachten 1974
   Stuttgart, 1974.

Dornier System GmbH (Hrsg.), Prognostisches Modell Neckar, Berichte
   1, 9 and 13, Friedrichshafen 1976 und vorher.

Gandenberger, W., Einfluß der Verschmutzung von Oberflächenwasser
   auf die Aufbereitungskosten der Wasserwerke, in: Gas- und
   Wasserfach (GWF) 1956, H. 12.

Klaus, J., Methodik der Nutzenbewertung, Bericht 11, Prognostisches
   Modell Neckar, Hrsg. DORNIER SYSTEM GmbH, Friedrichshafen 1975
   sowie Ökonomische Nutzenfunktionen (zusammen mit W.VAUTH), Bericht 12 der gleichen Studie.

Klaus, J., Lerch, B., Fehm, K., Freizeitnutzen und wirtschaftsförd-
   ernder Wert von Naherholungsprojekten, Berlin 1975.

Kneese, A.V., Bower, B.T., Managing Water Quality: Economics, Tech-
   nology, Institutions, in dt. Übers. Die Wassergütewirtschaft.
   Wirtschaftstheoretische Grundlagen, Technologien, Institutionen,
   München - Wien 1972.

Kneese, A.V., Smith, S.C., (Eds.), Water Research, Baltimore 1966.

Peskin, H.M.,Seskin, E.P. Cost-Benefit Analysis and Water Pollution
   Policy, Washington D.C., 1975.

Recktenwald, H.C., Die Nutzen-Kosten-Analyse. Entscheidungshilfe
   der Politischen Ökonomie, Tübingen 1971.

Reichenbach-Klinke, H.H., Die Kartierung eines Fließgewässers nach
   Art und Menge der Fische, Wasser- und Abwasser-Forschung 1,
   (1968).

Renner, E., Der Main und der Main-Donau-Kanal:Wasserstraßen für
   Freizeit und Erholung, in: Zeitschrift für Binnenschiffahrt
   und Wasserstraßen 98 (1971), H. 9.

Schmittke, R.F., Die Bedeutung der Nutzen-Kosten-Analyse im Ent-
   wicklungs- und Steuerungsprozeß wasserwirtschaftlicher Systeme,
   in: Wasserkalender 9. Jg., 1975.

Strassert, G., Turowski, G., Nutzwertanalyse: Ein Verfahren zur
    Beurteilung regionalpolitischer Projekte. Informationen des
    Instituts für Raumordnung, 21. Jg. (1971), Nr. 2.

Department of Environment, Government of Canada, Water Quality Man-
    agement Methodology and its application to the Saint John River,
    Niagara Falls (Ontario), 1971.

Water Resources Board (Ed.) The Trent Research Programme, London –
    Reading 1972/73.

Zangemeister, C., Nutzwertanalyse in der Systemtechnik.  Eine Methode
    zur multidimensionalen Bewertung und Auswahl von Projektaltern-
    ativen, München 1970.

Zwintzscher, K., Social Costs in der öffentlichen Wasserversorgung.
    Ein Beitrag zum Problem der Messung und Bewertung der social
    costs, Dortmund 1973.

# BEHAVIOR SETTINGS : THE BUILDING BLOCKS OF SOCIAL SYSTEMS

Mehmet Gürkaynak, W. Ayhan LeCompte

Technical University Ankara, University Beytepe, Ankara

Turkey

## INTRODUCTORY REMARKS

The focus of this paper is on those units of the ecological environment called "behavior settings" and on their role in assessment and evaluation. Prior to the definition of this unit and the discussion of its use in generating data on human environments some general remarks would seem to be in order regarding the relevance of this perspective to those interested in the modelling-systems orientation.

Clearly, this paper contrasts sharply with many of the others in the present collection. To begin with, it is presented by two psychologists interested in applying an ecological approach to the study of human behavior. Thus, its level of analysis extends from the person upward to the micro-ecological system of the small community, hospital, church or school, rather than the macro-ecology of the city, nation or world. Further, as a preliminary foray into the immediate socio-physical environment of the individual, it is more concerned with systematic description and quantification than with prediction and/or long-term forecasting of trends. Nevertheless, we believe that it has something to offer those whose interests are tuned to the large system and the long-term forecast. It is through work such as that presented here that the large system can be intelligibly linked to the level of individual experience and behavior. Conversely, the development of knowledge about the small system feeds upward to set parameters and constraints on the operation of the larger environment in which it is embedded.

## THE BEHAVIOR SETTING UNIT

As defined by Roger Barker (1968) and as used in previous research, the behavior setting unit consists of a reliably discriminated part of a human environment possessing both a pattern of extraindividual (e.g., social) activity and a surrounding and supporting physical and temporal location. Consider a typical conference
hall during a scientific meeting. Papers are presented in a predetermined arrangement, that is, the behavior setting "Conference Hall"
exists within a space-time locus. The extraindividual nature of
the activity pattern is apparent from considering the degree of
"substitutability" of the population in the setting. Speakers can
be changed and listeners come and go without a major interruption
in the ongoing activity. Physical structure enters in behavior
settings in two ways: first, every behavior setting is surrounded
by a physical and temporal milieu. The behavior setting "Conference
Hall" is enclosed within walls, ceiling and floor and these boundaries are embedded within the structure of a larger physical institution. Secondly, every setting has objects that are arranged to enhance the activities that occur. The chairs of the listeners in the
setting "Conference Hall" are arranged to permit the members of that
setting direct visual access to the speakers.

Although behavior settings such as the one described above can
be created for the special purposes of an investigator, the vast
majority of them exist independently of the world of the researcher
as objective parts of the ecological environment. They are part of
the structure of the real world and one of the problems of the person who deals with behavior settings for research purposes is to design ways of generating data without producing feedback from his
methods that operate to change the level of functioning of the setting.

Because behavior settings are jointly defined by both behavioral and physical properties, these units of action-and-situation
are complex and inherently multidisciplinary in scope. They have
clearly defined boundaries that separate them from each other, but
behavior settings are often in nested sequences within some part of
the ecological environment. For example, the treatment environment
of a rehabilitation hospital to be described later consisted of 122
behavior settings, of which 46 or 38 per cent were never inhabited
by patients although they were extremely important to the functioning of the hospital. One of these "back-stage" behavior settings
was the physician's weekly review of his cases with his rehabilitation team, at which decisions of great significance to patients
were made.

It is this attribute of behavior settings, namely, their tendency to be organized in significant ways vis-a-vis each other, that
prompted the subtitle of this paper. In the assessment of socio-

economic systems, the behavior setting concept is perhaps of greatest importance and relevance because it identifies a strategy that
allows a researcher or evaluator to assess an entire environment
with some confidence that he has not overlooked an important part.
If, in the investigation of the treatment environment of the hospital mentioned above, the researcher would have restricted himself
to studying settings that were actually inhabited by patients, he
would have been in the position of describing only the visible part
of the iceberg, while the "backstage" settings would have been passed over in silence. Human environments are often constructed from
such combinations of behavior settings, whose interdependent relation to each other suggests that the "building block" notion may
aptly describe an important part of the intersetting relationships.


                        THE ECOLOGICAL APPROACH


      Typically, an ecological investigation of a human environment
begins with an attempt to list all of the "parts" of the milieu that
exist within a certain defined degree of independence from each
other. Although this level of independence is subject to a precise
objective definition, it apparently also corresponds in most cases
to the phenomenal perception of an inhabitant that one is crossing
a boundary or entering a new place. An example might be the experience of a conferee when he leaves the previously described setting
"Conference Hall" and enters the setting "Hallways". There seems to
be a point of agreement between the normal perception of the inhabitants and the objectively described environment - a coincidence of
inside and outside perspectives. In any case of doubt, however, the
procedure itself leaves no doubt as to what is to be preferred by
the investigator. It is the ecological, preperceptual, objective,
public world that this approach is designed to elicit as opposed to
the psychological, post-perceptual, subjective, private world of the
individual. In this regard, Behavioral Ecology stands with respect
to the discipline of Psychology in the same way that the science of
Optics stands with respect to the Physiology of Vision.

      After completing the initial step of unitizing the environment
into behavior settings, the work of the ecological investigator usually consists of a systematic description of the behavior settings
that have been discriminated in terms of a number of measures or
indexes. Some of the measures to be mentioned in the present paper
include Occupancy Time (the person-hours-per-year that people spend
in behavior settings), Territorial Range (the number of settings
within a defined environment in which members of a given population
subgroup are found), and Level of Manning (the ratio of the available positions within a behavior setting to the number of applicants
to enter the setting).

It is to be hoped that the previous few paragraphs carry the
flavor, if not the substance of the ecological approach.  Space limitations necessitate bringing this preliminary discussion to an end
and turning to the issue of the relevance of the ecological approach
for assessment.  More detailed descriptions of the procedural rou-
tines for unitizing an environment into behavior setting units and
for generating data from the settings that are found are available
elsewhere (Barker and Wright, 1955; Barker, 1968; LeCompte, 1974).

## APPLICATIONS OF AN ECOLOGICAL APPROACH TO ASSESSMENT

There appear to be at least three levels on which an ecological
approach to assessment and evaluation can be discussed.  For the
sake of clarity, these will be presented in order of the inclusiveness of the attempt of the assessor to gather data on an environment.
At the simplest and most basic level, there is no attempt to assess
the environment at all, but rather a need to hold environmental stimuli constant and to allow stimuli from some other source such as
individual differences to vary.  This approach can be labeled "single
situation assessment".  At the second level of environmental assessment, a single characteristic of behavior settings is varied across
a range of values.  If the comparisons are valid, a general principle
may emerge regarding the effects of the variable on some class of
dependent variables.  This approach may be regarded as single variable assessment across a range of situations.  Finally, at the most
general level of inclusiveness, are the surveys of entire environments, institutions or systems.  They can be designed for the purpose of systematic description, comparison of two or more states of
the same system or comparisons between different types of environments.  For the sake of simplicity, these various methodological
strategies will be treated as inclusiveness levels I, II and III in
the following discussion although it is recognized that they differ
on other grounds than merely the inclusiveness dimension.

## LEVEL I: SINGLE SITUATION ASSESSMENT

Perhaps one of the most depressing developments in the recent
literature on psychometric assessment has been the belated recognition that validity correlations for many individual difference
variables, while possibly statistically significant, account for an
extremely small percentage of variance in the criterion variable.
Mischel (1968), after reviewing a wide variety of personality tests,
concludes that the average validity correlation is about .30 in personality test research, accounting for about 10 per cent of the total
variance in the dependent variable.  Similarly, Wicker (1969a) has
reviewed over 30 studies of the assessment and prediction of overt
behavior from attitude scales as administered in advance and found
a median correlation of about .30, predicting the same small percen-

tage of variance as in the case of the personality tests.  In contrast, measures of academic achievement typically correlate in the .7 to .8 range with later criterion variables such as grades, accounting for more than 50 per cent of the dependent variable variance.

Reflecting on the success of the cognitive measures in this instance and the failure of the personality and attitude measures leads to the speculation that at least part of the difference between them lies in the differences between the behavior settings in which the various measures are taken.  Consider the typical assessment situation for achievement testing.  These tests are administrated in quiet rooms to respondents who are seated and attending to the stimulus material with few distracting elements.  Turning to the typical validity criterion for achievement testing, it can be seen that most studies have used grades in some form - either cumulative grade point averages or grades in specific courses.  How are such grades usually determined?  Typically, from scores on exams administered in quiet rooms to students who are seated and attending to the stimulus material with few distracting elements.  Tests of personality and attitudes are also typically administered in settings similar to those described for achievement testing, but the criterial variables are usually dramatically different.  The latter may be based, for example, on rates of hospitalization.  In other words, the validity correlations for such psychometric instruments seem to be high in proportion to the degree that the behavior setting in which the respondent is assessed is similar to the behavior setting in which data are collected for the criterion variable.

Thus, the critical variable may well be an ecological one; with high homology (i.e., point by point structural similarity between assessment and criterion situations), the validity correlations seem quite predictive of behavior.  With low homology (high homology) on the other hand, validity correlations seem disappointly low.  Unfortunately, there seem to be little systematic attention paid to the effects of either focal or background stimulation in the testing situation itself and almost none to the concept of homology.  What little evidence that is available, however, suggests that background stimulation in behavior settings can have a subtle but profound effect on an inhabitant.  What seems to be needed is research in which the degree of situational homology is systematically varied and the consequent size of the various validity coefficients is studied as a dependent variable.  Clearly, for example, the phenomenon if not the concept of homology must be implicitly recognized in the field of situational assessment.

LEVEL II: SINGLE VARIABLE ASSESSMENT ACROSS SITUATIONS

In this strategy, a single dimension is isolated, operationally

defined, and its effects are studied across a number of settings
that are similar in other respects but differ in the degree to which
the independent variable is present.  It has been applied primarily
to factors related to population size in behavior settings, which
has been a critical marker variable in ecological theory, but the
same research strategy can also be used to investigate other variables as well.  In one recent study, for example, the effects of
the location of the treatment settings on attitudes towards mental
illness were investigated using this approach (Stembridge and
LeCompte, 1972).

    The population size variable has been conceptualized as the
ratio between the number of positions within the setting to the number of applicants available to enter and participate in the activities of the behavior setting.  When the number of applicants drop
below the number of positions the setting is said to be "undermanned";
a greater number of applicants than positions within the setting
hypothesized to create a condition of "overmanning" and as the ratio
between the two approaches unity, the setting is said to be "optim-
ally manned".  Due to the smaller variance of positions within the
setting than of applicants to enter the setting, variations in degree of manning can be easily observed in many cases, merely by
watching the same setting over time.  Studies of industrial organizations have consistently found that population size of the work-
place is positively related to absenteeism, but these data have not
previously been raised above the level of an empirical generalization.
The analysis according to manning levels, however, provides a theoretical interpretation to such data.  Presumably, as the size of the
working group increases, the opportunities for gratification of the
average worker decreases and he has consequently less motivation to
enter into and participate in the activities of the setting. Similar analyses have been done of voluntary organizations, such as
Rotary Clubs, Churches and extracurricular activities in American
high schools.  In each case, percentage of attendance as a dependent
variable has been found to be negatively correlated with size of
membership, as expected from the theoretical analysis of manning
levels (Barker, 1960; LeCompte and Barker, 1960).  More recently,
psychological consequences of long-term exposure of adolescents to
chronically over and under-manned environments have been investigated.  Not only have they been found to participate more, but greater participation itself also seems to lead to a more cognitively
complex person (Wicker, 1969b).

    Apart from the sexist terminology involved in this discussion
of "manning", there seems to be some awkwardness to the concepts as
well.  In particular, the links between the distal theoretical variable of behavior setting positions and available applicants and the
proximal operations of correlating membership size with percentage
of attendance seem quite loose.  Perhaps a reformulation of these

concepts along the line of information theory may be regarded as one type of input - output system and the population of members that are available for attendance as one type of input variable. The proportion of those members that are "used" to fill positions at one occurrence of the setting can be constituted as an index of the amount of transmission that was reached on that occasion. As the setting increases in size (input increases), the corresponding number of responsible positions from which members can get gratification shows a slower rate of increase. This relative stability of transmission seems to be a characteristic of many different types of settings.

Consider, for example, Rotary Clubs as one type of organization. Every club has just one position each for president, vice-president, secretary and treasurer, although its membership may vary from less than 25 to more than 500. To be sure, the larger club will have more committees and other signs of organizational complexity, but not in equal proportion to its increase in size. Many able members will be left functionless, reduced to the status of onlookers, with corresponding reductions in their own gratifications. This picture contrasts sharply with that of the small club, where attendance of every member is needed for survival and one person may even have to serve in more than one position.

When the population of the setting reaches a point at which the number of members balances the number of positions, but an increase in members is no longer matched by a corresponding increase in the number of opportunities and obligations provided by the setting. The setting can be defined as running at capacity. An increase in membership will result in over-population and the setting will initiate procedures for restricting entrance. In this regard, an overcrowded setting can be seen as engaging in strategies similar to an individual with "sensory overload". The setting will initiate a programme of screening its inputs, change its norms to emphasize participation less and selection more, spend less time with any one input, etc..

## LEVEL III: THE ENVIRONMENTAL SURVEY

The systematic description of behavior settings that have been unitized within an environment of interest to an evaluator or researcher may involve any of a wide variety of items. For example, the following data collection categories have been used in surveys of small communities, schools and hospitals:

1) Population - Who goes to the setting and why? Who performs in or has control over the setting?

2) Size - How big is the setting in a physical sense? How many people spend time there?

3)  Objects – What types of and how many behavior objects are
    used in the setting?  What are the ecological possibilities
    for stimulation?

4)  Actions – What kinds of things happen? How rich, varied or
    repetitive are the things that people do?

A survey year is taken in many studies as an appropriate interval for data collection, since shorter periods may involve fluctuation in behavior setting functioning due to personnel turnover,
seasonal variation or other short term factors.  Whatever specific
items and time period is chosen, every behavior setting is indivi-
dually surveyed.  Informants may provide valuable information; personnel reports, notices of meetings and internally circulated memos
all may serve to generate data about the settings but none seem able
to replace time spent in direct observation.  The generation of data
on 100 per cent of the behavior settings that have been discriminated is both a weakness and a strength of this method.  Although it
avoids problems of sampling, it requires more man hours for data
collection than other methods.  These data collection procedures result in a systematic, precise and quantitative data base that can be
used to construct ecological indexes for comparative purposes.  Such
indexes represent considerable progress over earlier research in
complex organizations.

Consider, for example, the research on "total" institutions,
that is, those environments that are inhabited by a subject population on a 24 hour-a-day basis.  Such studies have typically produced
reports that are rich in qualitative description and correspondingly
poor in quantitative analysis, with the result that different types
of environments cannot be compared.  The methods of the behavior
setting survey, however, provide a systematic description of an
environment that are directly comparable.  In a recent investigation
of a rehabilitation hospital (LeCompte, 1972), it was suspected that
much of the observed interpersonal conflict among personnel derived
from high population density.  The problem then became the difficulty of deriving a measure that would permit a quantitative comparison between the entire treatment environment and a normal community.
This was accomplished with the Occupancy Time Index (OT), defined
as the total number of hours spent in a behavior setting by all of
its inhabitants during a survey year.  For example, the hospital
setting "Treatment Area in Physical Therapy" included a total of 89
positions for all types of inhabitants during the 1968-69 survey
year.  Summing the time spent by individuals on an average day and
multiplying it by the number of days the setting was open for business during the survey year produced a total OT for this setting of
26,475 hours.

Occupancy Time can also be a useful measure for assessing
change in the same social system over time.  The hospital study men-

tioned above, for example, was replicated during the 1971-1972 survey year and OT measures were compared to determine what changes,
if any, had occurred during the three year period between the two
surveys.  Some of the data for these comparisons have been reproduced in Table 1.

Line 1 in Table 1 indicates that nearly a 20 percent increase
in population density has occurred during the period between the
two surveys.  Lines 2 and 3 divide this general amount into time
spent in back - and front - stage settings.  As discussed earlier,
the former term refers to those settings that are part of the treatment environment, but are never inhabited by patients.  Comparison
of Occupancy Time data presented in Table 1 indicates that the
behavior settings not inhabited by patients have lost Occupancy in
the same proportion that the settings inhabited by patients have
gained in Occupancy.

TABLE 1

Ecological Comparisons of the Treatment Environment
of a Rehabilitation Hospital from two
Different Behavior Setting Surveys

| Item | 1968-9 Survey | 1971-2 Survey | % Change |
|------|--------------|--------------|----------|
| 1. Total Occupancy (person hr./year) | 1,221,604 | 1,457,929 | + 19.3 |
| 2. Front-Stage OT | 1,011,296 | 1,304,815 | + 29.0 |
| 3. Back-Stage OT | 210,308 | 153,112 | - 27.2 |
| 4. Front Members | 2248 | 3730 | + 66.0 |
| 5. Back Members | 619 | 333 | - 46.0 |
| 6. Front Performers | 535 | 805 | + 50.4 |
| 7. Back Performers | 338 | 400 | + 18.3 |
| 8. Total Settings | 122 | 120 | - 1.6 |
| 9. Front Settings | 76 | 79 | + 4.0 |
| 10. Back Settings | 46 | 41 | - 10.9 |

Lines 4 through 7 of Table 1 trace this decrease in Occupancy
to the fewer members in behavior settings not inhabited by patients
in the second survey.  Inhabitants of behavior settings are classified as performers if they enact a vital role in the setting or have

some control over the setting.  Thus, leaders and active functiona-
ries are classified as performers.  Members of behavior settings
include customers, bystanders, and others who may be present on a
regular basis but have little individual power or control over the
behavior setting.  Lines 4 and 5 of Table 1 show a 66 per cent increase in members of front-stage settings, but a decrease of nearly
50 percent in back-stage member positions.  Lines six and seven show
increases in performer positions in both the front and back-stage.

   Taken as a whole, the data in Table 1 tell a very interesting
story.  While the hospital as a whole has become a more crowded
place, this increase in population density has occurred differenti-
ally for two important sectors of the treatment environment. Front-
stage settings have grown in both member and performer positions,
while back-stage settings have actually lost nearly 50 percent of
their member positions.  While the hospital has become 20 per cent
more crowded (line 1), the back-stage settings have shrunk to approximate the "smoky back room" of fame in political circles.  Finally,
lines 8 through 10 of Table 1 indicate that the extent of the treatment environment has remained remarkably stable during this period.

   An interesting example of internal comparisons is provided also
by the hospital data in the form of an ecological measure of staff
"presence".

   Territorial Range is defined as the number of behavior settings
in the treatment environment in which the members of a particular
population subgroup hold either member or performer positions.  The
larger the Territorial Range of a group, the more broadly its members are distributed around the hospital.  Conversely, the smaller
the Territorial Range, the more specialized and narrow is the role
of the group in the treatment process.  Table 2 presents data for
ten groups in the hospital; a number of smaller groups have been
omitted from these calculations because their size was too small to
permit valid comparisons.

   Inspection of the data in Table 2 indicates an increase in
Territorial Range occurred for every group between the two surveys.
However, the increase is an additive one, since the rank - order
correlation between the two columns in Table 2 is .95.  The relative positions of the ten groups has remained almost perfectly stable,
despite an average increase in Percentage Territorial Range of 11
percent.  Physicians, Clerical Workers and Nurses continue to domi-
nate the treatment environment, while Vocational Counselors and
Custodians remain specialized and hard to find.  Interestingly, the
variation among groups cannot be accounted for by knowing the size
of the group.  Rank Order correlations between Percentage Territorial Range and the population size of each group in Table 2 is only
.60 for the first survey and .03 for the second survey.

TABLE 2

Percentage Territorial Ranges for Ten Population
Subgroups from two Behavior Setting Surveys

| Group | 1968-9 Survey | 1971-2 Survey |
|---|---|---|
| Aides and Orderlies | 32.0% | 45.0% |
| Secretaries and Administrators | 48.4 | 54.2 |
| Nurses | 42.6 | 55.0 |
| Custodians | 17.2 | 26.7 |
| Occupational Therapists | 28.7 | 40.0 |
| Volunteers | 22.1 | 36.7 |
| Physical Therapists | 25.4 | 45.8 |
| Physicians | 55.7 | 57.5 |
| Social Workers | 39.3 | 46.7 |
| Vocational Counselors | 24.6 | 38.3 |
| Mean Percent Territorial Range | 33.6% | 44.5% |

CONCLUDING COMMENTS

Of the many types of measures that have been used in ecological
research with human environments, only two have been discussed in
detail.  The concepts of population density and territoriality have
captured much recent attention and the measures of Occupancy Time
and of Percentage Territorial Range were introduced with respect to
that topicality.  Interested readers seeking further information
are advised to consult Barker and Schoggen (1973) or Barker and
Wright (1955) for examples of other ecological indexes.

In addition to the behavior setting concept, a number of other
threads seem to tie the foregoing examples of research together.
These may be labelled as "strategies" in research that characterize
an ecological approach.  Many of these have been implicitly involved
in the previous discussion, but two particular threads deserve specific reference.

Foremost among these strategies is the precise attention to the
degree of internal differentiation of a system.  In a single situation research this implies a detailed examination of the background
stimulation possibilities in the setting, as potential confounding
factors.  In work within environmental systems, the internal differentiation strategy means that much work is done on unitizing the
system into comparable parts prior to any attempt to gather systematic descriptive data.  In the behavior setting survey, for example,

much work is done on structural and dynamic tests of activity synomorphs in order to produce precisely defined, equivalent behavior settings with a known amount of interdependency among them.

A second strategy in ecological research is the preference for generating data systematically from an entire environment over either 1) sampling units according to some prearranged scheme or 2) selecting specific units to survey on a theoretical basis. The advantage of systematic data collection lies in the confidence with which descriptive statements can be made and the determinate nature of comparisons, both within a specific environment and between systems. The disadvantage, however, is that larger and more complex environments become almost impossible to survey with such methods. Perhaps, with the cumulative knowledge generated from studies of smaller systems it may be possible to develop a taxonomy of types of behaviour settings that will permit a sampling plan of the larger system. Some efforts are currently underway along this line (Bechtel, 1970).

Both of these general strategies in data collection may limit the willingness of an assessor or evaluator concerned with complex systems to adopt an ecological approach. However, the rewards in terms of information gained would seem worth the cost of increased effort or time. The scientific understanding of those contexts that function to set parameters for human behavior and experience may be best gained through the use of an ecological approach.

## REFERENCES

Barker, R.G. Ecology and motivation. In M.R.Jones (Ed.), Nebraska symposium on motivation. Lincoln, Neb.: University of Nebraska Press, 1960. pp. 1-49.

Barker, R.G. Ecological psychology. Stanford, Calif.: Stanford University Press, 1968.

Barker, R.G. and Schoggen, P. Qualities of community life. London: Jossey - Bass, 1973.

Barker, R.G. and Wright, H.F. Midwest and its children. New York: Harper and Row, 1955.

Bechtel, R.B. A behavioral comparison of urban and small town environments. In J.Archea and C.Eastman (Eds.) EDRA-2: Proceedings of the Second Annual Environmental Design Research Association Conference. Pittsburgh, Pa.: Carnegie-Mellon University, 1970. pp. 347-353.

LeCompte, W.A. The taxanomy of a treatment environment. Archives of Physical Medicine and Rehabilitation, 1972, 53, 109-114.

LeCompte, W.A.  Behavior settings as data-generating units for the
        environmental planner and architect.  In J.Lang, C. Burnette,
        W. Moleski and D.Vachon (Eds.), Designing for Human Behavior.
        Stroudsburg, Pa.: Dowden, Hutchinson and Ross, 1974. pp 183-
        193.

LeCompte, W.A. and Barker, R.G. The ecological framework of coopera-
        tive behavior.  Paper presented at the American Psychologi-
        cal Association Annual Convention, Chicago, 1960.

Mischel, W. Personality and assessment. New York: Wiley, 1968.

Stembridge, D.A. and LeCompte, W.A. Attitude changes after brief
        encounters with mental patients in residential and institu-
        tional environments.  In W.J.Mitchell (Ed.) Environmental
        design: theory and practice.  Los Angeles, Calif.: University
        of California Press, 1972. pp. 1271-76.

Wicker, A.W. Attitudes versus actions: the relationship of verbal
        and overt behavioral responses to attitude objects.  The
        Journal of Social Issues, 1969, 25, 41-78. (a).

Wicker, A.W. Cognitive complexity, school size and participation in
        school behavior settings. Journal of Educational Psychology,
        1969, 60, 200-203. (b).

Session IV

# CASE STUDIES: ASSESSMENT OF

# TECHNOLOGICAL SYSTEMS

# GOVERNMENT REGULATION AND RAIL SAFETY *

H.P.Johri, J.D.Milne, R.E.Wright

Government of Canada

Ottawa, Ontario

## INTRODUCTION

In recent years there has been a growing concern regarding the environment and the quality of life. An important aspect of this concern relates to accidents within transportation systems. It is generally agreed that transportation accidents cannot be completely eliminated, therefore, one must focus instead on reducing the number of accidents and their harmful effects upon individuals and the environment. Another aspect of this problem is the extent of control which an individual may exert in protecting himself from a transportation accident. In this era of mass transit, an individual has only limited control in determining the level of risk in the transportation offering. Therefore, transportation safety, as a major public concern, must be examined within the context of the individual, the environment, and the society.

---

*

The authors would like to express their appreciation to Mr. G.H. Cooper, Executive Director, Railway Transport Committee, for several discussions and his comments on an early draft of this paper. The paper is based upon the authors' involvement in the Rail Safety Study, carried out on behalf of the Canadian Transport Commission. The assistance of the many consultants who worked as part of the project team is also gratefully acknowledged. Finally the authors would like to express their gratitude to the C.T.C. who made this research possible, and in particular, to Commissioner John A.D. Magee who provided guidance throughout the conduct of the study. The views expressed in the paper are the authors' own and do not in any way represent the views of the Commission or the Bureau.

In this paper, we examine the problems and issues related to
safety on the Canadian Railway System.  In Canada, the railway system is under regulation of the government and, therefore, a major
issue relates to the role of the government in the overall context
of safety.  Within this context, other issues emerge:  What exactly
is safety and how does it differ from the notion of risk?  What is
the socially optimal level of risk and what is an acceptable level
of risk?  What policy alternatives are open to society to induce
movement in the direction of the acceptable level?

We have used the Canadian Railway System as our point of reference, however, the arguments developed in this paper will apply
equally to other industries or modes of transportation and to other
countries, provided the interrelationships between the groups involved are not different.


## HISTORICAL PERSPECTIVE

In order to examine the subject of safety in the proper perspective, it is first necessary to trace very briefly the history of
government regulation of the Canadian railways.

In Canada, the railways have been the principal constituent of
the transportation system.  Rail transportation has played a role
which was not limited to the movement of people and goods but was,
in a significant way, responsible for the building of a nation.

When the railways were first introduced in the mid-nineteenth
century, their greater speed and independence of route quickly obtained for them a virtual monopoly in transportation over most of
the country.  The growth of the railways and the development of
Canada became highly interdependent.  This situation led to public
demands for controls to guard against the possible abuse of the
railways' growing power.  As a result, the Federal Government in
1888 assumed for the first time regulatory power over the railways.
In 1903, with the passing of the Railway Act, Parliament accepted
the principle that regulation of railway transportation should be
placed in the hands of a non-partisan, commission type organization
and created the Board of Railway Commissioners for this purpose.

The Railway Act also embodied another principle, that it was in
the national interest that railways subject to federal jurisdiction
should be constructed and operated safely.  To this end (among
others), the new Commission was vested with wide ranging powers of
regulation and control.  The intent, to provide safety, is best illustrated by Section 227 of the Act which says in part:

> "generally providing for the protection of property, and
> the protection, safety, accommodation and comfort of the

public, and of the employees of the company, in the running
and operating of trains ...."

In recent years, both the economy and the transportation system in Canada have changed considerably. The railways now operate
within a highly competitive environment. The significant development of other modes of transport was recognized by Parliament in
1967 with the passing of the National Transportation Act. This Act
created a new single agency, the Canadian Transport Commission*, to
regulate and control the various modes of transport. Within the
Commission, the Railway Transport Committee was established and charged with the responsibility to ensure that the railways observe the
provisions of the Railway Act and the Orders, Rules and Regulations
of the Commission.


SAFETY IN THE TRANSPORTATION CONTEXT

The demand for transportation services can be hypothesized to
be a function of two vectors: an activity vector comprising social,
economic and other variables of the user; and, a level of service
vector comprising variables such as comfort, safety, convenience,
cost, reliability, speed and frequency. Safety, is therefore, considered to be one component of this level of service vector. When
viewed in this light, one might unwittingly assume that:[5]

"Safety to a large extent, is a purchaseable commodity".

However, in the context of the individual user of a transportation
service, safety**cannot be bought and sold over the ticket counter.
A user can only trade indirectly for safety by influencing the demand for a particular service offering. The perception of the changes in demand may cause those responsible for the management of the
transportation service to alter the inputs into the transportation
offering, thereby producing a different level of risk and a different
combination of the other level of service variables. In other words,
the management of the transportation service can change the level

---

* The Board of Railway Commissioners was succeeded by the Board of
  Transport Commissioners in 1938 which in turn was succeeded by
  the Canadian Transport Commission in 1967.
** Safety is defined in the next section, however, at this point one
  must distinguish between safety and assurance. A user can buy
  assurance (e.g. a life insurance policy) in the market place but
  this does not directly mean that the user receives full protection
  from the risk inherent in the operation of the system. Assurance
  can only compensate the user in the advent that the bad consequences of the risk do in fact occur.

of risk in the system by spending more or less capital, but such
actions are not at the disposition of the user of the system.

In purchasing a transportation service, the user in a sense is
purchasing a "joint good" which includes safety and the other level
of service variables.  Safety is often indistinguishable from the
other level of service variables insofar as influencing the demand
for transportation services is concerned.  Thus, the influence an
individual user exerts over the level of risk in a transportation
system (not owned and operated by himself) is minimal and quite indirect.  Schelling,[11] in another context, has expressed similar
views:

> "Safety is indeed different from most consumer goods
> and its purchase different from most commodities."

The intervention of society in protecting an individual user of the
transportation service from risk of an accident is predicated upon
arguments of this nature.

Traditionally,* society intervenes either when the individual
participation in an activity is not on a voluntary basis, or when
the magnitude of the risk associated with a particular activity is
very high.**  Societal activities, according to Starr,[13] can be
categorized into those in which the individual participates on a
voluntary basis, and those in which the participation is involuntary,
and in some sense imposed by the society in which the individual
lives.  One extreme example of an involuntary activity is war, where
the society demands participation of an individual.  In the case of
voluntary activities, the individual uses his own value system to
evaluate alternatives.  Involuntary activities differ from voluntary
activities in that the criteria and options of the former are determined not by the individual affected, but by a controlling group.
As Starr notes:[13]

> "Such control may be in the hands of a government agency,
> political entity, a leadership group, an assembly of auth-
> orities or 'opinion-makers' or a combination of such bodies."

Public transportation systems fall somewhere between the two categories described above and, therefore, in most advanced societies
there exists some form of a control group.

---

* There are other economic reasons for the injection of a control
  group by society, these are discussed in a subsequent section.

**An extreme example of societal intervention is in the case of an
  individual who attempts suicide.

We explore the role of the control group in a later section, but first must clarify the definition of safety.

## DEFINING SAFETY

So far we have used the terms safety and risk without presenting a precise definition for them.  As a logical point of departure we will discuss a number of definitions of safety.

The Oxford English Dictionary defines safety as:

> "The state of being safe; exemption from hurt or injury; freedom from danger.  Used in active sense:  The action of saving."

Both of these meanings are reflected in the definitions of safety in the literature.

The most common approach has been to define safety simply as a state or condition; for example, according to Ling Suen:[14]

> "(Transportation) safety can be defined as an optimal state where there are no accidents, incidents and surprises in the (transportation) process."

A similar definition is given by the Office of Manned Space Flight, National Aeronautics and Space Administration,[10] as follows:

> "Freedom from chance of injury, or loss to personnel, equipment or property."

Other researchers have included in the definition the economic cost of preventing accidents as well as the social and political implications related to safety by qualifying the above definitions in the following manner:[7]

> "Safety (is) an economically, socially and politically acceptable level of the number and severity of traffic accidents."

Lister and Raisbeck[6] have further modified the above definition to conclude that:

> "Safety is an acceptable level of risk of system failure."

Babcock[1], in a paper for the Stanford Research Institute, has provided a more philosophical interpretation of safety:

> "Thus, safety as an objective quantity tends to be

replaced by a subjective comprehension of safety, which
I have chosen to call assurance, the state of being certain."

The second approach has been to use the word safety in the ac-[2]
tive sense, as the action of saving. For example, Bird and Germain
view safety as the control of any sort of damages, and approach safety research by an analysis of the "near misses" in industrial operations. The National Safety Council,[9] on the other hand, views
safety as the reduction of human and material loss. Ralph Nader,[8]
similarly approaches safety as the prevention of exposure of human
beings and their environment to the possibility of harm or injury.
The Intendes Group[4] defines safety as follows:

> "Safety is the function of anticipating undesirable
> consequences of behaviour and capability to alter
> those consequences."

We have found all these definitions lacking from the viewpoint
of our intent to review the role of the control group and as such
we have chosen to define safety as follows:

> Safety is the state which results from the management of
> the risk* of damage (to life, property or the environment)
> inherent in the operation of a system involving men and
> machines.

This interpretation of safety is broader** than the definitions
of safety reviewed above and is directed to examining the institutional or group relationships which both determine and affect the
level of risk in the system. Thus, we are concerned with the social
effects of harnassing technology and technological change. This definition views safety as a state resulting from a management process
within which the operation of interrelated constituent parts of a
system is controlled with the objective of achieving an acceptable
level of risk. The level of risk which results may differ from the
acceptable level in the actual operation of the system.

In the next section, we describe this management process*** with
particular reference to the Canadian Railway System. The process

---

* Risk is the liability to, or the chance of, injury, damage, loss,
pain, harm or peril. It refers to a situation in which future
outcomes are imperfectly known.

** It is interesting to note that the body of 23 members appointed in
October 1659 by the Parliamentary Army to conduct the Government
of England during an interregnum following the practical disposition of Richard Cromwell was designated as a "Committee of Safety."

*** Hereafter referred to as the process of safety.

of safety is characterized by the institutionalization of the role of various groups involved in the operation of the Canadian Railway System.

## THE SAFETY PROCESS - A CONCEPTUAL MODEL

To understand the complex interrelationships of the various groups involved in the safety process, we have developed a conceptual model of the safety process within the context of the Canadian Railway System. The model can be generalized to other industries or modes of transportation provided the interrelationships between the groups are not different.

The safety process consists of three distinct but interrelated components: the physical system; the economic system; and, the value system (see Figure 1). The three systems are controlled by railway management and the government. The interposition of the control group (the government) is done in recognition of the fact that the transportation activity is not totally voluntary. Wynholds,[16] commenting on the situation in the United States, observed:

"Part 601 B of the Federal Aviation Act of 1958 in calling for the highest degree of safety in airline operations, fails to define whether industry or the FAA is responsible for achieving this. The operating environment is, of course, provided by the government. Industry is responsible for producing aircraft, techniques and management to operate safely in this environment. And the ultimate responsibility for safety is borne by the individual working as part of the system."

The same situation, the failure to define whether industry or the government is responsible for safety, could be alleged to prevail in Canada, however, the Railway Act attempts to define the responsibility of the control group* in this respect.

At the heart of this problem is the lack of clarity in defining the role of the individual, the industry and the control group vis-a-vis safety. For this reason, we will explain our conceptual model in some detail. Figure 2 shows this model in diagrammatic form. In the diagram, the three systems of the safety process that were identified earlier are further broken down and their interrelationships established. We will now explain the three interacting systems in detail.

---

* The control group, as defined in the Railway Act, refers to the Canadian Transport Commission.

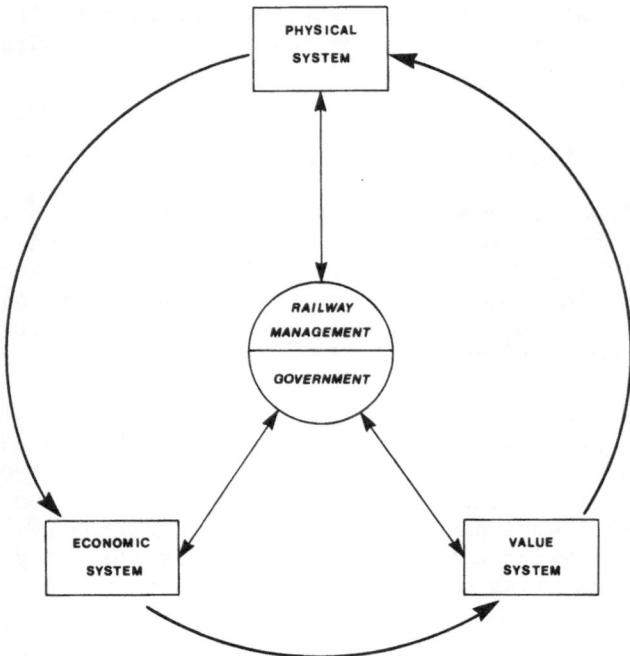

Figure 1.   The Safety Process Component Relationships.


The Physical System

     The physical system consists of the railways and their physical
environment.  There are five basic sub-systems within the physical
system; rolling stock, infrastructure, control, employees and meter-
ology.  As shown in Figure 2, these five sub-systems are highly interdependent.  The rolling stock represents a variety of equipment
that may use the railway right-of-way.  This sub-system is characterized as large[*] (over 200,000 elements of rolling stock) heterogeneous, dynamic and spatially distributed.  The rolling stock has
a long service life which makes the task of incorporating changes in
design difficult.  The infrastructure includes the rail, ties and
the roadbed.  In Canada, there are 44,000 miles of mainline tracks.
Several hundred of miles of track are located at high altitude in
mountainous terrain and must face the risk of rockfalls, avalanches,
and snowslides.  The railways employ a variety of control equipment
including such things as centralized traffic control, and automatic

---

[*]  All figures pertain to the Canadian Railway System.

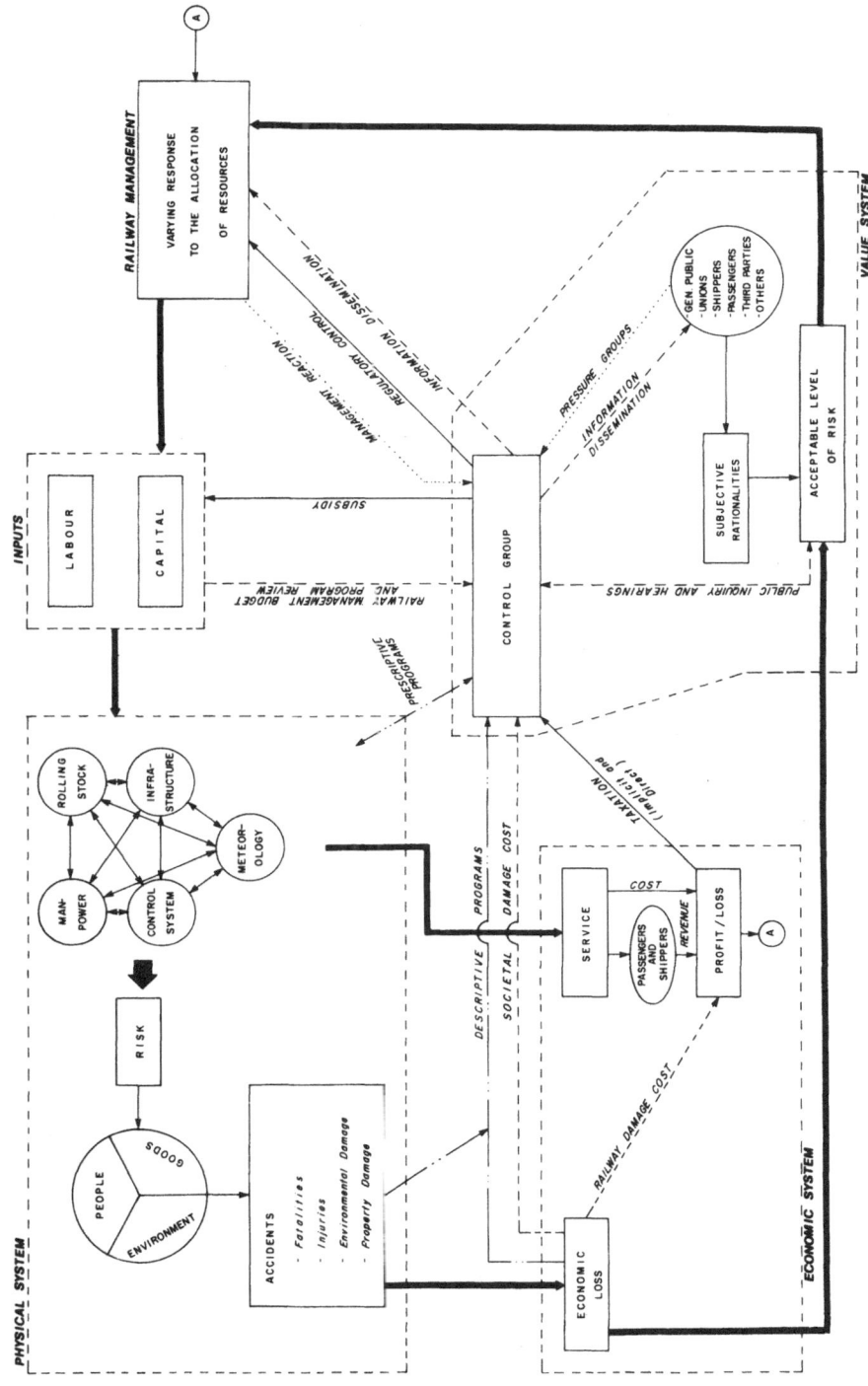

Figure 2.  Conceptual Model of the Safety Process.

block signalling. The railways are a highly labour intensive industry. They employ over 120,000 people. The railway system in
Canada faces one of the worst climatic conditions anywhere with
annual temperature differentials of up to 80°C (145°F).

The inputs to the physical system are labour and capital. The
railway management controls these resources directly. The control
group can influence the allocation of these resources in two ways:
directly, by increasing or decreasing a subsidy or tax; or, indirectly, through feedback from the value system.

The output of the physical system is the provision of transportation services. At the same time, this service gives rise to risk
to people, goods and the environment with which the system must
interact. The risk is manifested by accidents (and incidents) resulting in property and environmental damage and fatalities and injuries to people.


The Economic System

The economic system is characterized by an exchange of money
for the services provided by the physical system. Thus, the transportation of goods and people gives rise to revenue, cost and profits. Any taxation is proportional, in some way, to these variables.

For our purpose, it is important to identify the economic loss
caused by accidents. This loss is suffered both by society and the
railways. The cost of safety not only includes the damage loss but
also the cost of prevention.* The two components of this economic
loss are shown in Figure 2, as the societal damage cost accruing to
the control group and the railway damage cost accruing to the railways. The economic loss to the railways results from compensation
paid to their employees, passengers, shippers and third parties as
well as such things as the loss of their own property and the loss

-------------------------------------

\* It is important to differentiate between the social costs that
  accrue to society and the damage costs which are borne directly by
  the railways. There are two major ways in which the social costs
  of rail safety may differ from the costs borne by the railway company. First, the social costs of accidents may be greater than
  those considered by the railway company. The liability of the
  railways is limited in certain circumstances, for example, in
  cases of loss of life. In such cases, the liability of the railway can be substantially less than the value of lost productivity
  suffered by society as a result of a death. Second, the cost of
  accident prevention may be less with societal action than without.

of revenue caused by decreased system capacity as a result of equipment which is out of service.


### The Value System

One of the key components of the safety process is the value
system.  The value system represents the process whereby the acceptable level of risk is determined and includes the formulation of
policies and plans to induce the railways to operate at or below
this level of risk.  For this reason, the control group is considered
to be a subcomponent of the value system as shown in Figure 2.

It is important to recognize that the acceptable level of risk[*]
may vary from group to group and change with time.  At any given
point in time, however, it should represent a balance by the control
group of the expectations with respect to the level of risk of the
various groups in the value system.

The process of determining the acceptable level of risk by a
group is complex.  The process should recognize the factors related
to the current state of technology, economic factors and socio-
political factors which have been termed as subjective rationalities
in Figure 2.  The complexity of the process is such that we have not
observed any explicit attempts by the control group to establish and
communicate a level of risk, however, indirect inferences about the
level could be made by observing the response of the control group.
We will review some of the techniques that may be used to determine
the acceptable level of risk later in this paper.


### ROLE OF THE CONTROL GROUP

The conceptual model of the safety process developed above is
helpful in providing a better understanding of the role of the control group.

In most advanced industrial societies the existence of a control group is evident, but the role of this group is not always
clearly defined.  This is especially true in Canada and the United
States as is evident from the earlier quotation from Wynholds[**] in
which it is stated that the United States Federal Aviation Act of
1958 fails to define whether the airline industry or the government
is responsible for achieving the highest degree of safety in the
industry.

---

[*] We define this term later in the paper.

[**] See page 229.

This confusion between the role of the industry and the control group with respect to safety is evident in Figure 1 where both the railway management and the government appear to be responsible for managing the three systems of the safety process.  However, using the model developed in Figure 2, this apparent conflict may be resolved.

The function of the control group is hypothesized to be one of managing the value system.  In order to manage the value system, the control group must:

  (i)   determine the level of risk acceptable to the society;
 (ii)   measure the level of risk in the physical system; and,
(iii)   induce railway management to operate the physical system
        at or below this acceptable level of risk.

While conceptually it is possible to define the role of the control group in this manner, in practice, serious difficulties may arise in carrying out this role because of the intimate relationship of the value system, physical system, and economic system.  These aspects are more fully discussed below.

In order to manage the value system, the control group must monitor the level of risk in the physical system and, without itself operating the physical system, must influence this level to accord with that which it finds to be acceptable.  This is usually done by various safety programs of the control group.  We have categorized these programs as being either descriptive or prescriptive.

Descriptive programs are designed primarily to monitor the behaviour of the physical and economic systems.  An example of this type of program would be statistical data reporting which establishes the criteria by which accidents and incidents shall be reported by the railways to the control group.

Prescriptive programs are used to alter the behaviour of the physical system.  These programs usually take the form of such things as regulatory order, standards, rules and their enforcement.  Rules and regulations are usually devised on the basis of the results of monitoring the physical system through descriptive programs and various forms of inspection or accident investigation programs.  An example of this type of program would be the promulgation and enforcement of equipment standards.

There appears to be two fundamental reasons for conflict between the control group and those responsible for the management of the physical and economic systems:

(i)   Economic
(ii)  Nuisance*

The conflict arises if the level of risk as determined by the control group is below the level of risk as determined by the railway management.  To achieve a lower level of risk, railway management must make readjustments in the allocation of resources (the inputs in Figure 2).  If there is no compensating subsidy, the lower level of risk can only be achieved by either reducing the level of profits or by increasing the revenue through price adjustment. Both alternatives require careful consideration of many factors in the economic system.

The safety programs of the control group can be a nuisance to the management of the transportation system in a variety of ways. The nuisance results from inconvenience to the railway staff and their consequent resentment toward the control group.  These inconveniences include such things as delays in dispatch of trains warranted by repairs of equipment and extra maintenance; involvement of staff in public hearings, accident investigations and various related activities.  Programs requiring the reporting of accidents and incidents are also a form of nuisance to the management of the transportation system because of the time that employees devote to these activities.

In summary, the intertwining of the three systems within the safety process makes it very difficult for the control group to manage the value system in isolation.  The full extent of this problem may be better understood by examining the role of the control group in determining an acceptable level of risk and inducing the railways to operate at or below this level.  These aspects are discussed more fully in the following two sections.


DETERMINATION OF LEVELS OF RISK

In this section we define and discuss both the acceptable level of risk and the socially optimal level of risk.  In addition, a number of approaches that have been used in attempting to determine these levels are introduced and described.

The acceptable level of risk can be defined as:

> that threshold level of risk, as determined in the value
> system, above which the control group would intervene to
> reduce it.

---

* By nuisance we mean simply a source of annoyance or inconvenience.

As noted previously, the determination of this level is complex and
requires balancing the expectations of the various groups in the
value system.  And determination of this level must be based upon
the social, economic and quality of life factors which characterize
a particular society.

The socially optimal level of risk on the other hand is based
solely upon economic factors.  In deriving the optimal level, both
social accident prevention costs and social accident damage costs
are considered.  These two cost components include the costs to
the railway companies, the costs to the control group, and other
costs which do not accrue to either party.  It is defined as:

> that level of risk at which the sum of the social accident
> prevention and social accident damage costs is minimized.

A number of approaches have been advanced to analyze the acci-
dent/incident data of the railways and therefrom determine a level
of risk.  These approaches may be grouped as statistical, economic
and socio-economic.  The statistical approaches (trend analysis and
comparative analysis) are not directed towards defining the acceptable level of risk per se, but rather they are descriptive and provide only a measurement of present and past levels of risk.  The economic approaches are normative approaches and define a socially optimal level by considering the social accident prevention costs and
the social accident damage costs at various levels of risk.*  The
merit of the economic approach is in providing a clearer understanding of the policy options available to the control group.  The socio-
economic approach is based on drawing inferences from observations
of the collective behaviour of society or a group.  The method used
by Starr, although socio-economic in form, is not yet fully developed,
does not deal with all the socio-economic variables and is therefore
treated here as a comparative analysis.  Other than Starr's work we
are not aware of any research using the socio-economic approach.
These methods are briefly discussed below.

Trend Analysis.  Given a time-series of accident and incident data, one can ascribe a trend to the underlying process using
statistical methods.  Such an analysis is very useful in analyzing
and understanding the behaviour of the system.  This understanding
is achieved through building empirical correlations between the cau-
sative factors and the physical manifestations of an incident or
accident.  As has been noted previously, however, trend analysis
does not define an acceptable level of risk per se.

---

* The accident and incident data represent disaggregate statistics
  which in some way are a measure of the risk in the system. The correspondence between the risk level and accident and incident data
  cannot be easily derived.

Comparative Analysis. Statistical methods can also be used to
provide a comparative analysis of systems having something in common.
It is customary, for example, to compare the accident statistics of
one country with another or, within a country, to compare one industry system to another. In transportation, it is also not uncommon
to compare one mode with another. Such comparisons, however, do not
provide an indication of the acceptable level of risk as such.

Starr has used the comparative analysis approach, at an aggregate level, to determine the social acceptance of risk. Starr der-
ives a quantitative correlation between the benefits and risk associated with both voluntary and involuntary activities. At a very
aggregate level he compares such activities as general aviation, the
Vietnam war, skiing, hunting and smoking and concludes that:[13]

> "The disease death rate appears to play a yardstick role
> in determining the acceptability of risk on a voluntary
> basis."

This approach is aimed at determining directly an acceptable level
of risk, however, as Starr notes it is still in an exploratory stage.

Economic Approach. The economic approach is aimed at defining
an optimum level of risk. The total cost for any level of risk is
the sum of the cost of accidents occurring at that level of risk
and the accident prevention cost incurred in achieving that level of
risk. The general problem is illustrated in Figure 3. In this figure, the T function is the total cost of maintaining various levels
of risk; it is the sum of D, the total damage cost from accidents,
and P, the total accident prevention cost. The T curve is minimized
at that risk level S at which the marginal savings from damages pre-
vented equal the marginal cost of safety programs.

In the economic approach, it is assumed that the same party who
pays for accident prevention also suffers the total economic loss
from accidents. There are two major ways in which the social costs
of rail safety may differ from the costs borne by the rail company.
Second, the social cost of accident prevention may be less with societal action than without.

This gives rise to a key question: is the level of risk that is
most economical for the rail company identical with the economically
optimal risk level for society as a whole? Our research[*] indicates
that the economically optimal level of risk for the rail company will
be higher than the socially optimal level of risk if the proportion

---

[*] Not fully reported here.

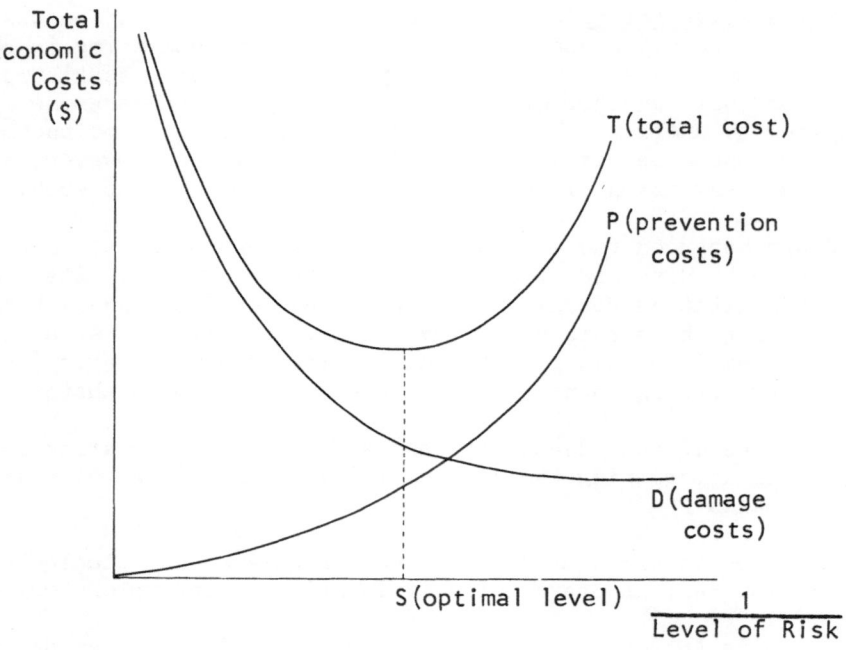

Figure 3.  Levels of Risk and Related Costs.


of the social accident prevention cost borne by the rail companies
is greater than the proportion of the social damage cost which accrues to society.

     In conclusion, our research, using the earlier attempts of
Tye[15], Canale[3] and Sinclair et al[12], indicates that in the
case of the railways it may be useful to consider the problem of determining the optimum level of risk from two viewpoints:  the railway companies; and, society.  Under certain conditions, the economically optimal level of risk for the rail company may be higher
than the socially optimal level of risk.

     It is also useful to remember that the acceptable level of risk
may or may not be coincidental to the socially optimal level of
risks.  In general there must exist some correspondence between the
two levels, since societal costs are based on societal preference.

     Except for the method used by Starr, none of the approaches de-
scribed above are directed at determining the acceptable level of
risk per se.  If the control group is to fulfill its role in determining the level of risk which is acceptable to society, then the

strengths and weaknesses of the above methods must be understood.

The word optimal is commonly used to connote, in some sense, that which is most desirable. One might therefore argue, based upon economic logic, that society is best served when the physical system is operated at the socially optimal level of risk.

At this point of our research the relationship between the acceptable level of risk and the socially optimal level of risk is not clear.

## POLICY ALTERNATIVES

The third role of the control group has been hypothesized as inducing the railway management to operates the physical system at or below the acceptable level of risk.

In the context of the railways, it has been found useful to consider the problem of determining an optimum level* of risk from two viewpoints: that of the railway companies and that of society. This is necessary because many of the accident damage costs which accrue to society do not accrue to the railway companies. As a corollary to this, it can be shown that under certain conditions the economically optimal level cf risk for the rail companies may be higher than the socially optimal level of risk.

There are three basic ways that the control group can induce the railways to operate at a lower level of risk than is economical for them.

Policy Alternative I - Subsidy: The control group can lower the marginal accident prevention cost of the railway, thus, inducing the railways to operate at a lower level of risk than that which they would otherwise select. This alternative requires subsidization of the railway company safety programs.

Policy Alternative II - Taxation: Alternatively, the railway companys' marginal accident cost curve could be raised by means of a tax so that the railway companys' economically optimal level of risk coincides with that of society. A wide variety of tax schemes may be employed towards this end, but all will have the effect of lowering the railway companys' desired risk level as long as the marginal cost of accident damages to the railway company is increased. For practical reasons, it may be more realistic to think of the tax

_____

* We are assuming a correspondence between the acceptable level of risk and optimal level of risk.

as a tax on railway company "safety violations" when violations are
proportional to the damages rather than a tax on accident damages.

   Policy Alternative III - Implicit Taxation:  Implicit taxation
is the result of the promulgation and enforcement of standards (for
the physical system) which when implemented result in a level of
risk lower than that which is preferred by the railway companies.
The control group, on its own, would establish standards which, when
enforced, would result in the railways operating at the acceptable
level of risk.  This is the alternative generally followed by the
control groups in Canada and the United States.

   Other Policy Options.  It is quite possible that some combination of the accident prevention subsidization and the accident damage taxation methods could be utilized concurrently so that the subsidy and taxation were equal.  Then there would be no, or minimum,
transfer payments from society to the railway companies, yet the
railway companies would still be induced to operate at the socially
optimal level of risk.

   The fiscal policy alternatives (subsidy and taxation) can be
applied incrementally so that the railway management is provided
with a positive inducement to move towards a lower level of risk.
The regulatory policy alternative (implicit taxation) requires that
a comprehensive set of standards covering all facets of railway operation be prepared and constantly adjusted to reflect changes in
technology.  While conceptually it is possible to prepare standards
to reflect a given level of risk, practically this is difficult and
a normal tendency is for the control group to prepare standards that
are based upon some incremental improvement over those which existed
previously.  Thus, the level of risk resulting from this implicit
taxation is often not predetermined.  In addition, the regulatory
measures necessitate enforcement, a task which is difficult on a
system as large as that of the railways.


                            CONCLUSIONS


   Safety, as we view it, is the state resulting from the management of risk.  In this paper, we have developed a conceptual framework to explain the role of the control group in relation to the
railway industry.  The roles of these two groups are highly interdependent and as a result this gives rise to several conflicts.  The
nature of these conflicts is not related to whom has ultimate responsibility for managing risk in the railway system, instead, it ultimately reduces to the question of whom should bear the additional
economic burden in achieving a level of risk lower than that which
is economically optimal to the railway management.  In the final analysis, a combination of policy and program alternatives should be
adopted by the control group so that the additional economic burden

is shared fairly and equitably. The exact nature of this sharing must, however, be determined by each society according to its own value system.

In the context of transportation safety there are some compelling reasons to interpose a control group. These reasons are not exactly similar to that of an economic regulator, who is supposed to balance "the wastes of unbridled competition with the evils of monopoly".[16] Instead, the reasons for interposing a control group are related to the notion that safety and the transportation offering cannot be bought and sold separately, and that rail transportation is partly an involuntary activity where the individual cannot use his own value system to evaluate alternatives, except in a very limited sense. Traditionally society has charged a control group to intervene when individual participation in an activity is not on a voluntary basis, when the magnitude of the risk associated with a particular activity is very high or when both of these conditions apply. Our research[*] has indicated that there is another and perhaps more compelling reason for the interposing of a control group. Based on an economic approach, it has been found that the economically optimal level of risk for the rail company will be higher than the socially optimal level of risk if the proportion of the social accident prevention cost which must be borne by the rail company is greater than the proportion of the social damage cost which accrues to society.

No one method which explicitly determines the acceptable level of risk currently exists, although there have been a number of approaches advanced to provide an input to the determination of this acceptable level. In addition, the correspondence between the acceptable level of risk, and the socially optimal level of risk is not clear. The acceptable level of risk may be less than, equal to or greater than the socially optimal level.

The relationship between the acceptable level of risk and the socially optimal level of risk can only be understood once the nature of the two terms has been precisely defined. In general, it can be said that in a democratic society there will be a tendency for the control group to align itself with the expectations of the various groups in the value system and thus choose an acceptable level rather than socially optimal level of risk. If the acceptable level of risk is higher than the socially optimal level, then decisions governing safety will be sub-optimal.

---

[*] This facet of our research is not fully reported in this paper.

BIBLIOGRAPHY

1.   Babcock, Dean F., <u>Airway Capacity: Some Reflections on Safety</u>
     <u>and Other Criteria</u>, Standard Research Institute, Menlo Park,
     California, 1970.

2.   Bird, Frank E. Jr., and Germain, George L., <u>Damage Control</u>,
     American Management Association, New York, 1966.

3.   Canale, S., <u>Hazard Risk Measurement and Optimization</u>, Presented
     at Government – Industry System Safety Conference, NASA,
     Washington, D.C. May 1968.

4.   Intendes Group Limited: <u>Pilot Research Study of Transportation</u>
     <u>Safety for the Transportation Development Agency</u>, Final
     Report March 1972.

5.   Lederer, Jerome, <u>Air Safety – A Study of Ethics, Economics and</u>
     <u>Attitudes</u>,  Flight Forum, Winter – Spring 1964.

6.   Lister, Simon F. and Raisbeck, Gordon, <u>An Approach to the Es-</u>
     <u>tablishment of Practical Air Traffic Control Safety Goals</u>,
     A.D. Little, Inc., Cambridge, Mass., May 1971.

7.   A.D.Little Inc., <u>Final Report on Cost/Effectiveness in Traffic</u>
     <u>Safety</u>, Cambridge, Mass., 1967.

8.   Nader, Ralph, <u>Unsafe at any Speed</u>, Pocket Books, New York, 1966.

9.   National Safety Council, <u>Status Report 1970</u>, Chicago, Ill.

10.  Office of Manned Space Flight, <u>System Safety Requirements for</u>
     <u>Manned Space Flight</u>, National Aeronautics and Space Admin-
     istration, Washington, D.C., January 1969.

11.  Schelling, T.C., <u>The Life You Save May Be Your Own</u>, Problems
     in Public Expenditure Analysis, The Brookings Institute,
     Washington, D.C., September 1966.

12.  Sinclair, T.C., Marstrand P., Newick P; <u>Human Life and Safety</u>
     <u>in Relation to Technical Change</u>, Science Policy Research
     Unit, University of Sussex, April 1972.

13.  Starr, Chauncey, <u>An Overview of Public Safety Problems</u>, ASSE
     Journal, July 1969.

14.  Suen, Ling, and Simon Bergen-Henengouwen, <u>Research Proposal for</u>
     <u>Transportation Safety</u>; Transport Development Agency, Decem-
     ber, 1970.

15. Tye, Walter, <u>Unresolved Civil Airworthiness Problems</u>, Institute
    of the Aerospace Sciences, Inc., 1959.

16. Wynholds, H.W., <u>Some Economic Aspects of Aviation Safety</u>. Index
    Serial No. 1110. Lockheed Missiles and Space Co. Inc.,
    Sunnyvale, Calif.

# TOWARDS A SOCIAL PSYCHOLOGY OF THE TRAFFIC ENVIRONMENT

C.K. Knapper, A.J. Cropley

University of Regina

Canada

Environment is often assumed to be static, but is more use-fully regarded as dynamic. Furthermore, people are often conceptualized as passive entities on whom environmental forces work, and it is frequently forgotten that they themselves form part of the environment. A good example of this can be found in studies of the traffic environment.

It has been common for social scientists to identify three types of factors which may cause traffic accidents. These are (1) the road and associated surroundings, (2) the vehicle, and (3) the driver of the vehicle. Vast sums of money are spent in efforts to reduce accidents resulting from defects in the first two of these, for example by improving the design of highways and traffic systems or by improving safety standards in vehicles. However, there are grounds for believing that road and vehicular factors are much less important in the driving situation than the third factor -- the driver himself.

One of the first investigations suggesting this was that by Ross (1940) who found that mechanical defects in the automobile and the road accounted for no more than ten per cent of all the motor vehicle accidents he studied. He concluded that the remaining 90 per cent of accidents had to be explained in terms of driver characteristics. Clayton and Mackay (1972) have recently made a similar point. On the basis of their analysis of police records in Britain, they reported that the environment alone or the vehicle alone accounted for a total of only 8.6 per cent of traffic accidents, as opposed to a total of 44.7 per cent attributed to the user (a further 31 per cent of accidents was accounted for by interaction between the driver and the environment, with the remaining accidents attributed to other combinations of factors).

In the USA, Treat and Joscelyn (1974) carried out a study of
215 automobile accidents in which the cause was investigated in detail by a multidisciplinary team of professionals.  They reported
that human factors were the most frequent accident causes, with
environmental factors ranking second and vehicular factors third.
Indeed, human beings were identified as a contributing "definite
cause" of the accidents investigated in 82.3 per cent of cases, and
as "definite or probable cause" in 96.7 per cent.  Furthermore, in
57 per cent of cases, human factors were the only cause of the accident which the investigating team could identify as being probable
or definite.  By contrast, environmental factors were definite
causes in only 16.4 per cent of accidents, the vehicle in a mere
4.2 per cent and, in the case of both these categories, the majority
of accidents involved contributions of other factors (primarily
human) too.

Even in the minority of cases where the physical environment
is primarily to blame, it is worth bearing in mind Malik's (1968)
point that there are many situations involving accidents where very
little can be done to eliminate environmental hazards.  Therefore,
she concludes:

> If accidental injuries are to be prevented in these environments.
> the task lies with the analysis of the situation which leads up
> to the accident, and the modification of attitudes, values and
> behaviour patterns which result in the accident ... These might
> include variables characterizing the interaction of the indi-
> viduals involved in the accident situation. (p. 112).

A number of writers have recently called for an increased emphasis on social and psychological factors in the transportation
and traffic domain (e.g., Krampe, 1975; Leibbrand, 1975; Michon, 1975).
Unfortunately, however, the efforts by traditional psychologists in
the field of driver behaviour have been rather limited in scope.
They have tended to regard the automobile driver as the operator of
a machine just like any other, and have concentrated on his skilled
performance in this role, often measuring variables which are of
limited interest, such as reaction time (see McFarland, Tune & Welford, 1964), or even trivial -- such as the investigation of driving
skill by Greenshields (1963) in which the dependent variables was
the amount that the steering wheel was turned over a set course.

An extension of the point of view which regards the driver as
a skilled machine operator is to see him as an information processing system (see Baker, 1963; Forbes, 1972; Perchonok, 1972; SWOV,
1971).  But as Michon (1975) has pointed out, this approach often
overlooks the fact that man must process information about the social
as well as the physical environment.  Indeed, although there has been

a recent resurgence in what is called "environmental psychology",
all too often the results of investigations in this field have provided very little in the way of insights into the driving situation,
precisely because they ignore what might be called the social environment -- the other people in the driving situation, their needs,
motivations, feelings and attitudes.

DRIVER CHARACTERISTICS: EXPERIENCE, DEMOGRAPHIC AND PERSONALITY
VARIABLES

A great number of investigators have attempted to relate
specific characteristics of groups of drivers to driver behaviour,
primarily evaluated in terms of involvement in accidents. One set
of studies has looked at experience of driving, involving such factors as mileage driven, success in advanced driver training courses,
enjoyment of driving, and even type of car driven (see for example
Hoinville, Berthonol & Mackie, 1972; Quenault & Parker, 1973). Of
particular interest in this regard has been the effect of driver
education and training programmes on subsequent performance. Harrington (1971), in reviewing a large number of studies on such programmes in the USA, concluded that those who obtained driver education or training have better accident conviction records than those
who did not, but that the attributes of those who opt for training
programmes are such that the differences in subsequent driving performance may be merely a reflection of the "superior personal characteristics" of this group.

Among the general demographic traits which have been investigated are the sex of the driver (e.g., Harrington, 1971; McGuire,
1969), intelligence (e.g., Egan, 1967) and age. The latter variable
has been of considerable interest, especially as it relates to very
old drivers and beginning drivers (for examples of studies of the
elderly driver see Gianturco, Ramu & Erwin, 1974; McFarland et al.,
1964; and Planck, 1972; for studies of the newly licensed teenage
driver see Carlson, 1973; Harrington, 1971; and Klein, 1972).

Not surprisingly, there has also been considerable interest in
the personality of good and bad drivers. Linares-Maza (1971) has
identified two basic types of study. The first involves those which
make use of paper and pencil tests to measure a battery of personality variables, and then relate these to some index of driving such
as accident rate or successful performance on a driving test (examples of this type of study are found in Benton, Mills, Hartman &
Crow, 1961; McGuire, 1969; Parry, 1968; and Undeutsch in Wagner &
Wagner, 1968). Space precludes a full and detailed description of
the results of these studies, and it is difficult to generalize
about their findings since most investigations made use of different
sets of personality variables. However, a number of researchers
(e.g., Conger et al, 1957) have summarized the evidence as indicating

that -- at least as far as involvement in accidents is concerned --
poor driving is characterized by generally unsatisfactory adjustment
to the social environment and a lack of conformity with institutionalized social patterns.  This would seem to imply that driving behaviour is an extension of day-to-day social behaviour.

The second approach to personality and driving described by
Linares-Maza has used more subjective, often psychoanalytic, techniques, and has tended to concentrate more on the personality structure and development of the chronically bad driver (see, for example,
Alonso-Fernandez, 1966; Hamilton, 1967).  Unfortunately it is very
difficult to establish any validity for the techniques used in the
studies on which Linares-Maza bases his analysis.

In fact the great majority of studies of driver characteristics
have not merely investigated individual variables, but have adopted
a multivariate approach in which some dependent variable relating
to driving performance has been correlated with every conceivable
factor that could be measured in the captive driving population under study.  Somewhat disappointingly, this "blockbuster" approach
has failed to yield any very definite conclusions about personality
syndromes underlying good or bad driving.  For example, Harrington
(1971), reviewing many dozens of studies in the area, found that
what he called "biographical variables" were poor predictors of accidents among young drivers, and Hoyos and von Klebelsberg (1965)
concluded that the best predictors of good and bad driving behaviour
tend to be relatively superficial characteristics, such as years of
driving experience, history of past court appearances, and so on.

It is thus perhaps a little ironical that one of the earliest
investigators in the field (Ross, 1940) was perspicacious enough to
conclude that the predominant cause of traffic accidents (and the
major influence on driver behaviour) is the attitude of the average
motorist.  Some 30 years later Malik (1968), who herself employed a
multivariate approach, echoed this opinion when she wrote that accidents are best viewed as social events to be explained primarily in
terms of social and cultural (rather than personality) variables.


                         ATTITUDES AND OPINIONS

Investigations of driving which focus upon the attitudinal systems of drivers can be divided into two main types.  The first of
these concentrates on cultural values, and the second on individual
attitudes.

The idea that driving behaviour is a special manifestation of
a general cultural norm is one that has fascinated a number of researchers, especially in the USA, where the automobile occupies a
special place in the history and culture of the nation.  Ross (1960-

61) has conceptualized the violation of traffic laws as a "folk crime" and relates such behaviour to risk taking in general, which he claims has always had a positive value connotation in American society. Thus risk taking in cars (often taking the form of breaking traffic laws, and sometimes leading to accidents) can be thought of as something which is encouraged by the prevailing mores of the general society. While the validity of this notion is difficult to establish (and indeed it might be asked why other countries with quite different cultural heritages -- such as West Germany -- have higher accident rates than the USA), the notion at least points to the importance of keeping in mind more subtle and hidden social norms which may dictate driver behaviour, as opposed to the overt norms represented in traffic laws and driver manuals.

Studies of individual attitude systems, as opposed to those imposed by the culture at large, are relatively more common in the field of driving. For example, Brown and Copeman (1969) investigated drivers' attitudes to the seriousness of road traffic offences in relation to the severity of penalties for committing them, Asai and Inayosh (1973) studied attitudes of high school students towards traffic safety, using a factor analytic approach, and Turrell (1957) showed that drivers who had had accidents had different scoring patterns on the Allport-Vernon-Lindzey scale of values than accident-free drivers.

It should be borne in mind, however, that an important consideration in studying attitudes is to determine if and how they are related to actual behaviour. In calling for a move away from the study of purely environmental factors and a focus on the individual driver, Treat and Joscelyn (1974) argue that:

> It is desirable to know in considerable detail what it is that drivers either do, don't do, or have trouble doing, which leads to their accidents. Similarly in developing state driver manuals and driver license tests, both written and behind the wheel, it is desirable to know which kinds of bad practises or errors are most likely to result in an accident and ideally, for a particular applicant or test subject, which of these problems or errors are most likely to be problems for him or her. (p. 8).

In practice the easiest way to do this is to talk with drivers, rather than merely observe what they do. The reasons for this are twofold. Firstly, there are many practical difficulties involved in carrying out reliable and valid observations of driving behaviour. And, even more important, for any thorough understanding of driving it is essential to go beyond what the driver actually does, and attempt to probe the underlying reasons for his actions. For example, it is well known that very few North Americans wear car seat belts when driving. But it is a matter of considerably greater interest (and one which is also more difficult to establish) to know <u>why</u> drivers fail to buckle up (see Knapper, Cropley & Moore, 1973;1976).


## THE IMPORTANCE OF SOCIAL INTERACTION

It is a striking fact about the few existing attitudinal studies of driving that they almost all tend to look at the driver as an isolated individual, rather than someone who is part of an ongoing system of social interactions. Malik (1968) is one writer who has strongly argued for a conceptualization of accidents as social events. She points out that, even when individuals are physically separated (as in the case of a single driver in the anonymity of his car), "a mutually shared set of expectations operates with respect to the normative behaviour required in any particular situation" (p. 124). Later she goes on to say that an individual's position in the total social system largely determines the attitudes which he brings with him to the situation of a potential accident, and that this social context may be a crucial factor in determining whether or not an accident actually takes place.

A similar point is made by Stewart (1958) who looked forward to the time when traffic research would become much more comprehensive, and would take into account the total personality of the driver as well as the social interactions involved in traffic. Stewart also expressed the hope that research into driver attitudes would turn towards the use of indirect and projective techniques, but this wish has gone largely unnoticed and unrealized.


## SOCIAL PSYCHOLOGY AND DRIVING BEHAVIOUR

Blumenthal (1970) combines a plea for a renewed emphasis on social factors with a protest against the naive assumption that the driver is always capable of making rational decisions -- a point of view often implied by the systems theory approach to driving.

Wilde (1971) has also suggested that social psychological principles should be an integral part of any major conceptual framework in the field of road safety research. Wilde is convinced that any adequate conceptualization of the driving process must pay due attention to aspects of the social environment, such as group density, personal space, communication and competition between road users regarding intended changes in speed and course of travel, both legal and informal norms of contact, and the effects of disparities between the two upon the likelihood of accidents. He notes that driving involves a high degree of "pacing", implying that what any individual driver does on the road is largely determined by the behaviour of others around him.

The evidence, then, is very suggestive, if not completely conclusive, that social factors play a very important part in determining moment-to-moment responses by automobile drivers, and that driving behaviour can usefully be conceptualized in social psychological terms. In an even wider context, Michon (1975) has argued for an approach to the whole transportation domain which would stress social psychological factors. He has also called for a dynamic analysis of the behavioural and social variables involved in driving and transportation where, he says, our efforts to date remain "very rudimentary at best". The purpose of the project reported here was to attempt to respond to this call, albeit in a modest way.


METHOD

The present authors have developed a psychological approach which, it is hoped, will permit better understanding of the psychological basis of behaviour in the traffic environment. The basic theoretical position implicit in the method is that people's actual behaviour is consistent with underlying attitude systems, regardless of the values and opinions that they express publicly. Argyris (1975) has recently distinguished between "espoused theories of action" and "theories-in-use". It is apparent that there are discrepancies between explanations of behaviour inferred from the actual observation of real behaviour, and explanations derived from the attitudes, values and opinions that people verbalize. Furthermore, most people do not seem to be aware of these discrepancies in their own behaviour, although they do notice them in the behaviours of other people. The fundamental point of view here is that behaviour can only be understood when the intangible and subjective psychological systems mediating it are identified. Attempts to change external behaviour without an examination of the underlying motivation are likely to be unsuccessful, and may result in public resentment and resistance to change.

The method has been described in greater detail elsewhere (see Knapper & Cropley, 1976). It involves a fundamental analysis of public opinions aimed at eliciting information about basic feelings and attitudes. The procedure consists of a sequential, funnel-like set of steps. In the first step, interviews are conducted with experts in the area in order to get some idea of the "official" opinions. Subsequently, members of the public are interviewed in a group setting, using a variety of techniques derived from such diverse sources as psychoanalytic theory, T-Group procedures, as well as objective psychology. On the basis of these interviews, schedules are constructed and administered in a series of semi-structured interviews with a small random sample of members of the public. Finally, a second instrument -- in the form of a questionnaire -- is constructed from the data assembled to this point, and administered to a large, random sample of the population. The content of this final questionnaire is thus derived from the actual views of the public, and not from the investigators' preconceptions. The purpose of the strategy is to allow people to specify, in their own words, the relevant psychological dimensions of the area under investigation. An important element in this approach involves using the methods of subjective psychology to yield data that are subsequently subjected to more objective, quantitative treatment.

For the construction of the questionnaire, what is required is a relatively objective instrument which, while not being totally self-administering, will be within the capacity of a wide range of respondents to complete for themselves with only minimum supervision from an interviewer. A major feature of measures developed in this way is that items preserve the actual wording of responses given in earlier stages and thus the contents of such instruments are couched in the ordinary English of the man in the street. These statements are typically presented in a format which can easily be understood by respondents, yield comparable data from subject to subject, require only a simple response, yet yield data amenable to expression in numerical form, suitable for subsequent statistical treatment.

This was achieved in the present study in a number of ways. One technique involved a check-list of potentially dangerous driving behaviours or situations derived from statements made in previous stages (e.g., "excessive speed", "children in the car"); the rating task involved simply checking those items which respondents felt were of concern to them, based upon their own real life experience as a driver or car passenger. A second task involved the presentation of slightly longer statements about driving behaviour (e.g., "people talk too much when they drive") and letting respondents indicate the extent of their agreement with each statement by use of a simple seven-point scale. A third technique comprised a variation on this task, in which respondents were asked to imagine themselves in various situations relating to a collision or near collision and then express their agreement with a series of statements concerning the situation, again done on a seven-point scale (e.g., "if I were involved in an accident...I would be embarrassed regardless of who else was involved"). A final technique made use of a series of seven-point semantic differential scales (e.g., "observant --- unobservant") on which respondents were asked to describe their image of the dangerous driver.

## RESULTS

An extended outline of the results of the study, including details of the method of analyzing data derived from the various scales, is to be found in Knapper and Cropley (1976). For the purposes of the present report, the major results are summarized below.

They are presented in three sections, corresponding to the three
broad areas of interest in the study: sources of hazard on the road,
reactions to a near accident or collision, and the image of the dangerous or bad driver.


### Social Psychological Factors in Perceived Road Hazards

1.   Respondents identified other people as one of the major sources
     of road hazards: such people could be other drivers, passengers,
     pedestrians, or even bystanders.

2.   Some aspects of people's driving behaviour which were seen as
     dangerous involved traits that are more or less directly obser-
     vable and directly relevant to risky driving (such as careless-
     ness or impatience).  Few of these were "fixed" qualities
     (e.g., sex) which could not be "corrected" by a driver who
     wished to do so.

3.   Respondents also referred to traits that are not readily obser-
     vable, and that are only indirectly related to observable be-
     haviour.  In particular, they made inferences about covert
     traits of personality and motivation in other drivers (e.g.,
     "Young drivers often speed up just to keep in front of you").

4.   Many of the traits identified as characterizing dangerous dri-
     ving behaviour referred to qualities of an interpersonal nature
     (e.g., discourtesy), as opposed to traits which do not imply
     the presence of another person (e.g., carelessness).


### Social Psychological Factors in Reactions to a Near Accident

1.   Respondents reported that, when a near accident occurred, their
     reactions would be dominated by their feelings and emotions
     (e.g., relief, anger).

2.   These feelings would be affected by the behaviour of the other
     driver and by the presence of other people.  Other people would
     have a particularly strong effect when the near accident was
     the fault of the other driver and when the "guilty" driver was
     apologetic.

3.   A major dimension along which reported reactions were organized
     involved the effects of variables such as the age, sex and ap-
     pearance of the other driver.


### Image of the Bad Driver

1.   Bad drivers were described mainly in terms of inferred personal

characteristics, rather than in terms of driving behaviours.

2.    One group of such inferred characteristics involved traits re-
      ferring directly to behaviours behind the wheel (e.g., "hesit-
      ant", "inexperienced").

3.    However, a second cluster of inferred characteristics involved
      traits which did not refer directly to driving, but to general
      personal properties (e.g., "arrogant", "selfish").


DISCUSSION


Social Psychology and Driving

     In general the results supported the notion that people are in-
deed affected in their behaviour behind the wheel by factors over
and above the merely objective variables like road conditions, state
of repair of their vehicle, and so on.  In particular, they see certain properties and actions of other drivers as major elements in
the traffic environment, and as major sources of hazard.

     This general finding and the line of attack out of which it
arises represent a relatively novel approach to the analysis of the
driving situation.  To date, approaches to road safety and to understanding problems in driving have concentrated on making improvements
to the traffic environment through, for example, the building of better highways, or improved design and placement of traffic signs. Recently there has also been increased emphasis on the design of vehicles themselves, for instance in the work of Ralph Nader and his
associates.  However, investigations in driving and transportation
have been dominated by planners and engineers, and have tended to
ignore what one respondent in the present study called, with a certain blunt eloquence, "the nut behind the wheel".  Even psychologists have tended to approach research in the area of traffic safety
from a human factors or human engineering standpoint, concentrating
primarily on problems of perception and learning.  Thus the driving
task has been conceptualized as one type of skilled performance,
and the car regarded as just a piece of machinery to be operated like
any other.

     Perhaps under the influence of the common folklore about the
dangerous driver there has been considerable interest in personality
variables associated with bad driving, and this has led to a number
of formulations which have already been discussed.  However, these
kinds of psychological findings appear to have had little effect on
such practical questions as remedial driver training, where programmes continue to concentrate mainly on improving technical and

mechanical skills, rather than emphasizing the social or intraper-
sonal aspects of driving. By contrast, the line of argument arising
out of the present research states that driving is just as much a
social as a mechanical skill. Furthermore, the present findings
suggest that this is well known to many people from their everyday
driving experience. Despite this fact, the social psychology of
traffic behaviour has received scant attention from behavioural or
social scientists, and little heed has been paid to the social implications of promoting safety, improving driver performance, training beginning drivers, detecting drivers requiring special remedial
treatment, and similar questions.

The findings of the present study have demonstrated some aspects of the social psychological aspects of driving, and have shown
that it is indeed a social situation in which interpersonal reactions
are of great importance. Moreover, these reactions appear to follow
some general principles which are well established in the field of
social psychology. Some of the practical implications of this position have already been mentioned (e.g., for driver training). However, there are also profound implications for the theoretical understanding of driving: for an approach which stresses interpersonal
factors allows researchers to bring to bear on their investigations
the insights provided by social psychologists in a number of different fields involving social interaction. For example, the present
survey revealed that one important determinant of driver behaviour
involves the emotional or affective state resulting from interactions
with other drivers. These reactions are not merely based on factors
such as whether or not the other driver was observing the speed
limit, but are also derived from inferences about the other driver's
personality and motivational structures.

Even more important, many of these kinds of inferences cannot
be made directly from observation of behaviour, but only deduced
from it. In fact, there were basically three sets of characteristics of other drivers that influenced respondents' perceptions. The
first set of factors involved what might be called "fixed" characteristics such as age, sex, and physical appearance. The second
involved perceived or inferred characteristics which can be more or
less directly observed from observation of concrete aspects of other
drivers' behaviour (e.g., carelessness or impatience). It was the
third kind of factor which was of greatest interest for the present
investigation. This involved perceived qualities of other drivers
which could not be observed, but only inferred or deduced (e.g.,
arrogance). All of these types of inferences, and the last group
in particular, are subjective in nature, and probably depend very
much upon the perceiver's prejudices, prior experiences, personality,
and similar elusive factors.

It was also apparent that respondents regarded the driving
situation as an extension of everyday social life. Of particular

importance in this regard were notions like courtesy or good manners.
Thus respondents saw relationships with other drivers not only as
partly interpersonal interactions, but also as bound by the same
rules as any social intercourse among people.  They saw the driving
situation as an extension of the ordinary social situation, and as
being regulated by the same kinds of variables and rules.

To give just one example, a concept from social psychology
which has interesting implications for the driving situation is that
of "social norms" (see Sherif, 1936).  In a formal sense, norms in
the driving situation might be thought to involve official laws or
codified "rules of the road", such as are found in the British "Highway Code" or the various Canadian "driver manuals".  Indeed human
factors and systems theory approaches to the study of driving have
tended to focus on normative behaviour defined in this narrow way,
studying driving skill in terms of a catalogue of "good" and "bad"
behaviours or manoeuvres (see Treat & Joscelyn, 1974, for an example).
This has led in turn to an assumption that defective driving must
result from either a failure on the part of the driver to conform
to these norms because of weak signal detention skills, poor decision making ability, and so on, or a failure on the part of authorities to communicate the norms effectively by means of such devices
as traffic signs or road systems.  It is especially interesting that
high school driver education programmes in North America have tended to concentrate almost exclusively on students' ability to learn
and recognize these formal norm systems, by stressing such skills
as knowledge of traffic law, recognition of basic road signs, etc.
and often in a classroom situation far removed from what would be
experienced when actually handling a car.  As Kurt Edvardsson, of
the Swedish Road Safety Office, has pointed out in a recent personal
communication, this approach is too optimistic;  the present writers
would also argue that it is oversimplistic.

This is because traffic behaviour cannot be understood merely
in terms of the formal set of norms described above (even if it
could be shown that people actually obeyed the relevant rules when
driving, in addition to being able to recite them in a classroom).[*]
If accidents could be explained purely in terms of rule breaking,
then the frequency of collisions would be vastly greater than it is
at present, and indeed it is quite impossible to police more than a
tiny proportion of the infractions of rules of the road detailed in
most driving manuals.  The fact is that there is another system of
norms operating in the driving situation -- a situation of informal
norms arising out of the interactions among drivers and, it is argued,
very similar to the general norm system which operates in all social
intercourse.  Of course there is an independence between the formal
and informal norm systems in the driving situation.  Not only do

[*] A somewhat similar point has been made by Wilde (1971; 1973).

drivers act in terms of both systems (for example by braking too
suddenly to let an old lady cross the road), but some particularly
interesting situations arise when a driver, acting in terms of
largely informal norms, encounters an official representative of the
more formal norm system, such as a police officer. This clash between the two norm systems is exemplified each time an impaired
driver is arrested or an indignant motorist is tecketed for exceeding the speed limit by three miles per hour. The existence of these
two types of normative behaviour has gone largely unrecognized in
the study of the traffic environment and driver behaviour, and the
area seems a fertile one for future research.

NB:  The study described in this report was supported by funds pro-
     vided by Transport Canada. Although the authors are grateful
     for this financial support, the views expressed here are those
     of the writers alone, and should not be construed as represen-
     ting the views or policy of the Government of Canada or its
     officials.

## REFERENCES

Alonso-Fernandez, F.  Configuraciones psiquicas peligrosas del con-
     ductor tremulo. Revista de Psicologia General y Aplicada, 1966,
     21, 947-960.

Argyris, C. Dangers in applying results from experimental social
     psychology. American Psychologist,1975, 30, 469-485.

Asai, M., & Inayosh, H. Some determinants of safety-mindedness: The
     attitude of high school students toward traffic safety. Mihon
     University, Department of Psychology, College of Humanities
     and Sciences, 3-chome Sakurajosui, Setagaya-ku, Tokyo, 1973,
     25-40.

Baker, J.S. Traffic accident investigators manual for police. Evan-
     ston, Ill.: Northwestern University, Traffic Institute, 1963.

Benton, J.L., Mills, L., Hartman, K.J., & Crow, J.T.  Auto driver
     fitness: An evaluation cf useful criteria. Journal of the
     American Medical Association, 1961, 176, 419-423.

Blumenthal, M. Traffic safety and the structure of a social problem.
     In Highway and traffic safety. Chapel Hill, N.C.: University
     of North Carolina, Highway Research Center, 1970.

Brown, I.D., & Copeman, A.K. Drivers' attitudes to the seriousness
     of road traffic offences considered in relation to the design

of sanctions. Cambridge, England: Medical Research Council, Applied Psychology Research Unit, 1969.

Carlson, W.L. Age, exposure and alcohol involvement in night crashes. Journal of Safety Research, 1973, 5, 247-259.

Clayton, A.B., & Mackay, G.M. Aetiology of traffic accidents. Health Bulletin, 1972, 31, 277-280.

Conger, J.J., Gaskill, H.S., Glad, D.D., Rainey, R.V., Sawrey, W.L., & Turrell, E.S. Personal and interpersonal factors in motor vehicle accidents. American Journal of Psychiatry, 1957, 113, 1069-1074.

Egan, R. Should the educable mentally retarded receiver driver education? Exceptional Children, 1967, 33, 323.

Forbes, T.W. Human factors in highway safety research. New York: Wiley, 1972.

Gianturco, D.T., Ramu, D., & Erwin, C.W. The elderly driver and ex-driver. In E. Palmore (Ed.), Normal ageing, Vol. 11. Durham, N.C.: Duke University Press, 1974.

Greenshields, B.D. Driver behavior and related problems. Ann Arbor, Mich.: University of Michigan, Transportation Institute, Highway Research Record, H-25, 1963.

Hamilton, J.W. The rear-end collision: A specific form of acting-out. Journal of the Hillside Hospital, 1967, 16,187-204.

Harrington, D.M. The young driver follow-up study: An evaluation of the role of human factors in the first four years of driving. Sacramento, Calif.: California Department of Motor Vehicles, Research and Statistics Section, Research Report HPR PR-1(7) B0113, 1971.

Heussenstamm, F.K. Bumper stickers and the cops. Trans-action. February, 1975, 32-33.

Hoinville, G., Berthonol, R., & Mackie, A.M. A study of accident rates amongst motorists who passed or failed an advanced driving test. Crowthorne, England: Ministry of Transport, Road Research Laboratory, Report LR H-99, 1972.

Hoyos, C.G., & von Klebelsberg, D. (with others) Psychologie des Strassenverkehrs. Bern: Huber, 1965.

Klein, D. Adolescent driving as deviant behavior. Journal of Safety Research, 1972, 4, 98-105.

Knapper, C.K., & Cropley, A.J.  Social and interpersonal factors in driving. Regina: University of Regina, 1976.

Knapper, C.K., Cropley, A.J., & Moore, R.J.  A quasi-clinical strategy for safety research: A case-study of attitudes to seat belts in the city of Regina, Canada. Regina: University of Saskatchewan, November, 1973.

Knapper, C.K., Cropley, A.J., & Moore, R.J.  A clinical/quantitative analysis of public opinions about seat belts. International Review of Applied Psychology, 1976, (in press).

Krampe, G. A cluster analysis of towns in Germany in terms of personal rapid transit systems. Paper presented at the NATO conference on transportation and urban life, Zurich, September 1975.

Leibbrand, K. Recent changes in transportation and urban planning. Paper presented at the NATO conference on transportation and urban life, Zurich, September 1975.

Linares-Maza, A. Psicologia clinica, psiquiatria y conduccion de automoviles.  Revista de Psicologia General y Aplicada, 1971, 26, 30-55.

McFarland, R.A., Tune, G.S., & Welford, A.T.  On the driving of automobiles by older people.  Journal of Gerontology, 1964, 19, 190-197.

McGuire, F.L. The understanding and prediction of accident-producing behavior. Chapel Hill, N.C.: University of North Carolina, Highway Research Center, 1969.

Malik, L. Social factors in accidents. (Doctoral dissertation, The American University, Washington, D.C.) Ann Arbor, Mich.: University Microfilms, 1968.  No. 68-13, 062.

Michon, J.A.  The mutual impacts of transportation and human behaviour.  Paper presented at the NATO conference on transportation and urban life, Zurich, September 1975.

Perchonok, K. Accident cause analysis. Final report, Cornell Aeronautical Laboratory, prepared under contract No. DOT-HS-053-1-109, NHTSA, DOT, July 1973.

Planck, T.W. The ageing driver in today's traffic: A critical review. Chapel Hill, N.C.: University of North Carolina, Highway Research Center, 1972.

Quenault, S.W., & Parker, P.M. Driver Behaviour: Newly qualified drivers. Crowthorne, England: Ministry of Transport, Road Research Laboratory, Report LR 567, 1973.

Ross, H.L. Traffic accidents: A product of social-psychological conditions. Social Forces, 1940, 18, 569-576.

Ross, H.L. Traffic law violation: A folk crime. Social Problems, 1960-61, 8, 231-241.

Sherif, M. The psychology of social norms. New York: Harper, 1936.

Stewart, R. G. Can psychologists measure driving attitudes? Educational and Psychological Measurement, 1958, 18, 63-73.

SWOV, Netherlands Institute for Road Safety Research. Psychological aspects of driver behaviour. Vols. 1 and 11. Voorburg, Netherlands: Institute for Road Safety Research SWOV, 1971.

Treat, J.R., & Joscelyn, K.B. An investigation of the causes of accidents. Paper presented at the meeting of the American Psychological Association, New Orleans, August 1974.

Turrell, E.S. Emotions: Personality's multiple facets. Traffic Safety, 1957.

Wagner, K., & Wagner, H.J. (Eds.) Handbuch der Verkehrsmedizin. Berlin: Springer, 1968.

Wilde, G.J. Road safety campaigns: Design and evaluation. Paris: Organization for Economic Co-operation and Development, 1971.

Wilde, G.J. Social psychological factors and use of mass publicity. Canadian Psychologist, 1973, 14, 1-7.

# CONTROL METHODOLOGY OF THE U.K. ROAD TRAFFIC SYSTEM

M.R.C. McDowell, Dale F. Cooper

University of London

England

## 1. INTRODUCTION

We have been engaged in one aspect of operational research on the U.K. Road Traffic system (hereafter "the system") for over a decade. It rapidly became apparent to us that our efforts, and those of similar groups, were directed towards understanding and possibly sub-optimizing only our own part of the system, without any real model of the links of this part with the whole. Of course, sub-optimization is not an unusual feature of O.R. on large scale systems with multiple or indeed fuzzy objectives. However, in our case the problem was not our blindness, but the refusal of certain elements in the control mechanism to allow the investigation necessary for constructing a wider model to proceed. This was our original motivation for looking at the whole system.

In § 2 we describe the system and we discuss some of its component sub-systems in more detail. In § 3 we examine the control elements, their ranges of action, and their interactions. Finally, in § 4 we state our conclusions.

## 2. THE SYSTEM

The system naturally involves vehicles, roads and people, but any model of it must identify the interactions between these and their control mechanisms.

Some aspects of the very complicated interactions involved are illustrated in Fig. 1. People appear in the system at two levels, firstly as road users, either as drivers, passengers or pedestrians,

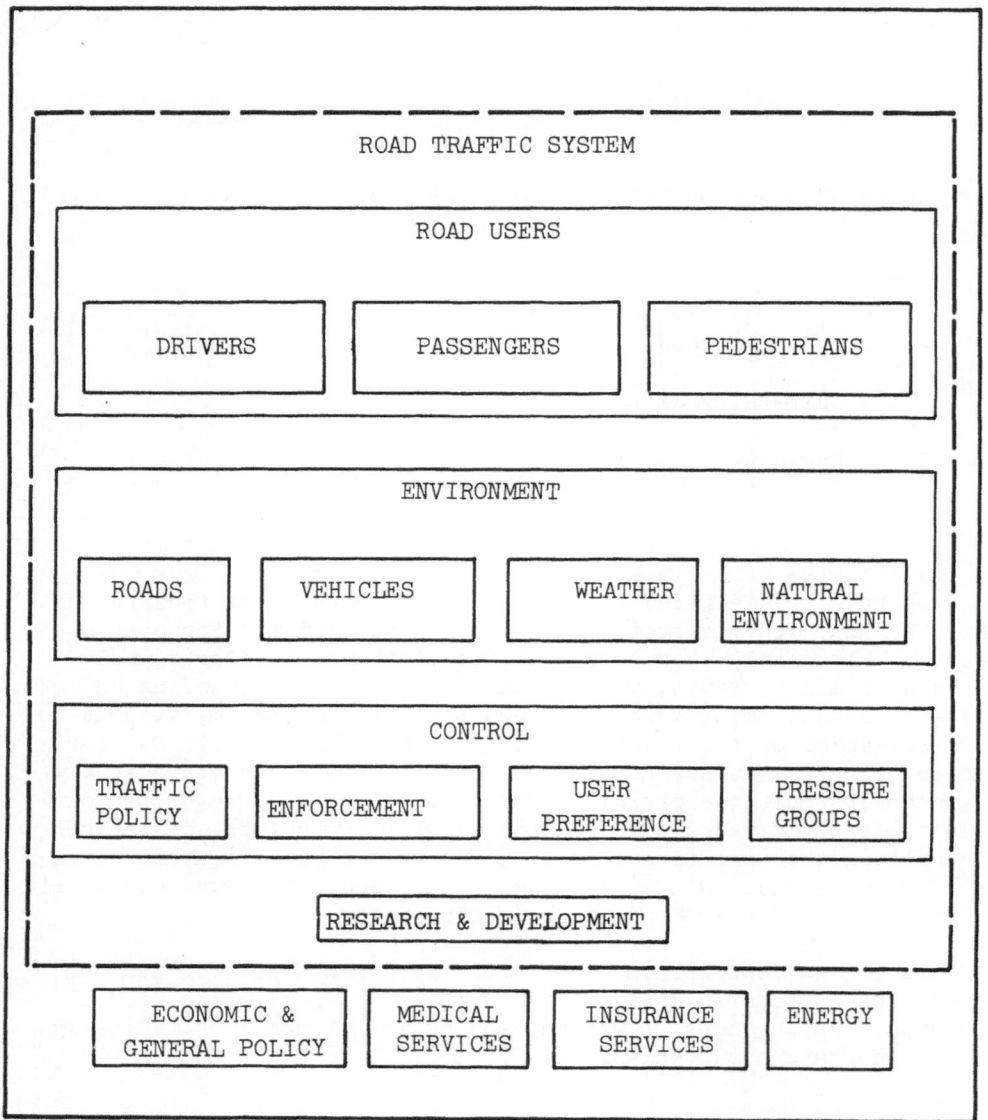

Figure 1.   The Road Traffic System.


and secondly as elements in the control circuits, which includes their role as designers of the system.

These people act in, or operate on the physical aspects of the

system, which we refer to as the internal environment.  This includes
the roads now existing, the weather, and the engineering and economic aspects of vehicle supply and use.  These are in turn modified
in time by research and development, and by economic and political
changes in the larger world in which our system is embedded.

The control elements of the system are both <u>ad hoc</u> and complex.
They include national transport policy [1], legal constraints laid
down by the U.K. Parliament and the E.E.C. and the mechanisms for
their enforcement, and of equal importance, user control through
exercise of choice.

### 2.1  Transport Policy

U.K. National transport policy is now under review, and this
review includes a reassessment of the proportion of the economic

| Vehicle | Value of working time [a] | | Time savings [b] | |
|---|---|---|---|---|
| | | | Working | Non-working |
| Car | Driver<br>Passenger | 331<br>287 | 26% | 23% |
| Bus | Driver<br>Conductor<br>Passenger | 166<br>158<br>168 | 3% | 6% |
| HGV | Occupant | 155 | 11% | |
| LGV | Occupant | 139 | 11% | |

Notes:     a.  Pence/hour, 1975 prices.
           b.  Benefit as percentage of total savings.

Source:   Transport Policy [1], Vol. 2, p. 99. Also Vol. 1, p.71.

Table 1.  Benefits of a Road Scheme in Time Saved.

|                  | Urban | Rural | Motorway | All   |
|------------------|-------|-------|----------|-------|
| Fatal injury     | 41900 | 47600 | 52000    | 44000 |
| Serious injury   | 2750  | 3800  | 3810     | 3060  |
| Slight injury    | 450   | 810   | 870      | 520   |
| Injury (average) | 1810  | 4240  | 4120     | 2360  |
| Non-injury       | 220   | 270   | 310      | 230   |

Source:  Transport Policy [1], Vol. 2, p. 100.

Table 2.  Cost per road accident (£'s, January 1976 price levels).


resources available for transport as a whole which should be devoted
to the road traffic system, primarily as compared with the railways
(air, coastal shipping, pipelines and inland waterways being of minor importance, or restricted to bulk goods).  Hence the expression
of policy involves decisions on the allocation of resources to provision of highways, maintenance of minor roads, and access to these
and to city centres.  The main control mechanism on level of use is
fiscal, expressed through Road Fund Licence levels, parking fees,
fuel tax, and VAT and other taxes on vehicles.

    The provision of additional or improved roads is assessed using
a Cost-Benefit Analysis (C.B.A.) in which the major benefits are
assumed to be time savings to individuals (80%) and savings due to
reductions in costs from accidents (20%).  Time savings are summarized in Table 1, and the accident costs currently in use are shown in
Table 2;  they are necessarily approximate.  The benefits are discounted over thirty years and the criterion for a positive decision
is a non-negative NPV.  This method of analysis is not applicable to
urban schemes, where trip distribution models are used (see, for example, Leblanc[2]) nor to motorway planning, where national economic effects must be evaluated.  The U.K. Government has said that it
will be seeking an independent assessment of the value of these procedures.


## 2.2  Legal Constraints and their Enforcement

    As in all countries, the laws governing the system, the super-

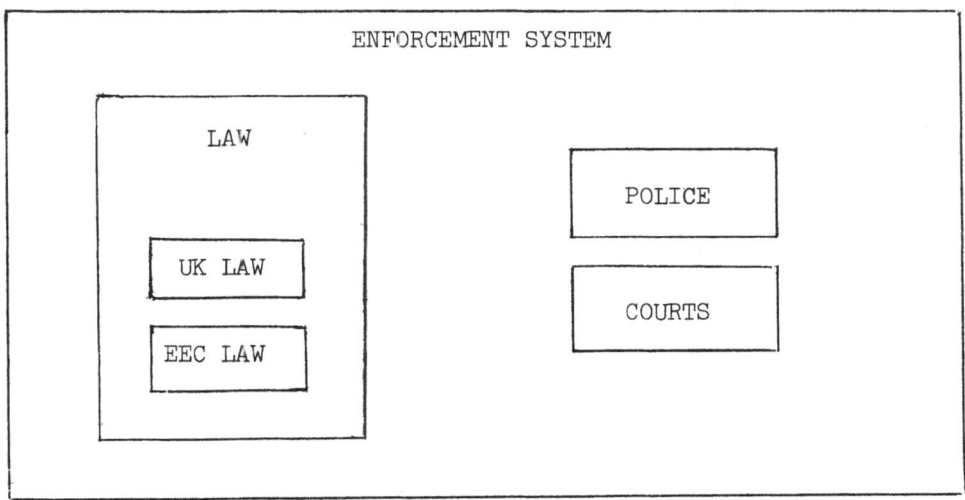

Fig. 2.   The Enforcement System


visory mechanism (the police) and the enforcement mechanism (the courts) must be considered as a single sub-system (Fig. 2).

The relevant laws, at least in the U.K., can be divided into two categories: laws on construction and use of vehicles and laws on driver behaviour, though at a deeper level, with which we shall not be concerned in this section, planning law sets constraints on road construction and improvement schemes. (Planning law does effect the system, since in many cases the planning process allows a clear expression of public preference, which can sometimes have a determining effect on new road schemes).

We shall be most concerned here with aspects of Road Traffic legislation.  The various construction and use regulations, which the police are required to enforce, cover such factors as vehicle roadworthiness, loading and operating of goods vehicles, documentation (licensing, insurance, taxation, etc.) and more recently, but to a lesser extent, environmental factors (noise, smoke and fumes, etc.).  The user of the system is normally more affected by laws relating to driver behaviour.  These contain provision for setting and enforcing speed limits, provision of and penalties for ignoring road signs, and regulations on the use of drugs (including alcohol).  In addition drivers must obey the directions of a police officer.

While these regulations have on the whole an unambiguous interpretation, offences of, for example, careless and dangerous driving

are less clearly defined; a common distinction is to prosecute for
dangerous driving only where serious or fatal injury has been caused,
or other strongly anti-social behaviour is involved.


### 2.2a  The Role of the Police

As far as the road traffic system is concerned, police duties
involve not only the detection of offences but preventive, persuasive
and punitive tasks.  The OECD group on Road Research list these as
follows[3]:

2.2a.1  Preventive duties.  The principal preventive duties are
directing and supervising traffic and investigating accidents.
Police are expected to:

(a)  Supply information to aid road users.
(b)  Indicate to drivers and pedestrians how to behave, especi-
     ally in congested or hazardous conditions.
(c)  Take emergency action to direct traffic.
(d)  Warn other traffic after an accident has occurred.
(e)  Determine how and when accidents happen.
(f)  Record and report all relevant information and draw appro-
     priate conclusions with a view to enhancing road safety.

2.2a.2  Persuasive duties.  Persuasive police duties may include:

(a)  Issuing written warnings when the offence is relatively
     minor, such as driving at a speed marginally above a speed
     limit.
(b)  Issuing verbal warnings when appropriate.
(c)  Visibly patrolling on foot, in cars and on motor cycles;
     this is believed to influence favourably road user behav-
     iour and to act as a deterrent to potential traffic law
     offenders.

2.2a.3  Punitive duties.  In attempting to enforce traffic laws,
punitive police duties may include:

(a)  Arresting a road user when there is a high probability that
     he is endangering others.
(b)  Immobilising vehicles and drivers.
(c)  Initiating action which may lead to a court appearance.
(d)  Applying on-the-spot fines and/or fixed penalties when mi-
     nor offences, such as parking offences, are committed.
(e)  Assisting the courts.

The use of "on-the-spot" fines, though common in other European
countries (and in the Southern United States) is not a part of the
U.K. system at present, presumably because of the additional

administrative load it would impose on the police and the opportunities it presents for corruption.

In the U.K. most Police Forces operate within one Local Authority area, except in London and the Thames Valley, and the emphasis as to which laws and regulations shall be most strictly enforced varies with the Chief Constable.

Whatever the area of emphasis the police do not, in principle, make judgements on whether an offence has been committed (that is a matter for the courts): they are supposed to confine themselves to a decision, often with legal advice, on the existence of a prima facie case. In practice the level of penalty is set so low that it is uneconomic for the individual to challenge the police case in minor matters, and the ambiguity of the laws on careless and dangerous driving implies an exercise of judgement by the police before the decision to issue a warning (verbal or written) or notice of intended prosecution is taken.

This immediately identifies an area in which little research has been done, the decision tree involved in deciding what action to take when an event which may be an offence against the Road Traffic Acts has been reported or observed. To our knowledge, no statistical analysis has been undertaken of the factors whose presence in the evidence to be presented by the police (or their solicitor) will ensure a better than even chance of conviction. In the absence of such analysis much of the time of senior police officers (sergeant, inspector, chief-inspector and above) is wasted in reviewing evidence on cases which will not be taken to court, or will fail if this is done.

Much research on the role of the police in the system, some of which is reported in a later paper by Freckleton et al.[4], has concentrated on modifying their strategy and tactics, with the effect of increasing the number of serious offences detected, though this has not always been the primary aim of the research. However, to the best of our knowledge no research has been carried out in the U.K. on the effects of successful detection, prosecution and penalties on the behaviour of the driver population. An extensive American experiment (Operation 101 and Operation 500, California Highway Patrol[5,6]) on strict enforcement of their Road Traffic Acts, with withdrawal of licences from those who accumulated twelve penalty points, showed no statistically significant effects on the rate of detection of offences per traffic patrol hour. Again a study by one of us (and others[7]) showed that even at a high level of patrolling the number of offences reported was directly proportional to the hours of observation.

Straight-forward patrolling at the levels normally allowed by economic criteria, ranging from 1 patrol car on duty over 100 km of

major road to at most 1 car to 20 km, imply that the density of police patrol cars is negligible compared with that of other vehicles.
Although with such tactics only a few cars near the patrol car are
directly affected by its presence, there appears to be a subjective
estimate of threat by drivers which maintains reasonable driving behaviour.  A number of studies (see, for example, Munden[8]) show
that if the patrol level is sufficiently high, and patrols are reinforced by radar equipment, then there are significant reductions
in the number of injury accidents, in the percentage of drivers exceeding posted speed limits by a specified amount, and in other parameters.  However, Munden judged that the required levels of patrol-
ling(from  $3^1/2$ to 14 times the normal) were uneconomic.


## 2.2b  The Courts

     The role of the courts is to decide finally who, if anyone, has
committed an offence, what offence, and what the appropriate penalty
should be.

     In the U.K. the courts are very restricted in the types of evi-
dence they can hear, and do not have the power to order a detailed
investigation.  Further, they are restricted in the penalties they
can impose, which for traffic offences other than dangerous driving
are by and large limited by public pressure to fines, licence en-
dorsements and suspensions, and suspended custodial sentences.  Studies of driver reaction following the imposition of such penalties
show that there may be little long term effect.  There is no power
to order errant drivers to undergo a period of retraining, nor are
any facilities provided for such retraining.  (Indeed no assessment
has been made of the benefits associated with a system of driving
licences with speed and lower limitations dependent on the level of
driving skill attained - except for learners, one is either licensed
to drive a motor vehicle or not, though there are further requirements
for public service and heavy goods vehicles.).

     Many of the offences for which a prosecution is brought arise
from involvement in an accident, and a police assessment of the
blame.  Neither the magistrates courts nor the police show great
awareness of the extensive work carried out at the Transport and Road
Research Laboratory[9] on accident causation.  Sabey and others have
made a detailed analysis of the factors involved for 2130 accidents
occurring within a few km radius of the laboratory over a period of
48 months.  The most important factor was found to be driver error,
which was the sole factor in 65% of the cases, and contributory in
all but 5%.  The dominant errors appeared to be lack of care, excessive speed in the circumstances, failure to appreciate external events (i.e. "looked and failed to see"), and being distracted by
other things.  Every one of us who has driven will know that each
of these factors frequently occurs in his own driving behaviour,

fortunately without accident involvement, and it is hard to see how prosecution following assessment of driver error as a contributory factor in a particular accident will assist in preventing the occurrence of errors on subsequent occasions.

The frequent occurence of driver error is confirmed by further observations carried out by TRRL[10,11] and ourselves [12] on conflicts. Serious conflicts are defined as events where two or more vehicles would have collided had not at least one taken rapid evasive action (sudden deceleration or lane change etc.). This research has shown that at non-urban junctions (where 80% of accidents involve more than one vehicle) conflicts are about one thousand times more frequent than accidents. They occur at the same location within the junction, and involve the same vehicle manoeuvres.

It appears that there is a strong case for a national investigation into the form of enforcement system which would be capable of demonstrating a beneficial effect on the road traffic system.

## 2.3 Other Agencies Involved

Apart from the police and the courts at least two other important elements of the national economic system are closely linked to the road traffic system, namely the Health Service and the Insurance Industry. There is no detailed analysis of involvement of either for the U.K. situation, though a detailed investigation has been made of the economics of traffic accidents in Canberra, Australia, where the city can be treated as an isolated system[13]. (Two interesting conclusions were that one third of the premiums paid into the compulsory third-party insurance scheme was taken up in legal expenses, and that half of the accident victims admitted to hospital had no rights to compensation, as they were in some way 'responsible' for their condition).

The involvement of the Health Service is obvious from the accident statistics; in 1974 there were nearly a quarter of a million reported injury accidents, which produced almost 7000 fatalities and over 300,000 non-fatal injuries (Table 3). It can be argued that the Health Service would exist independent of the traffic system, but the levels of staffing and the distribution of staff amongst the various facilities, particularly with regard to the provision of emergency services, would be quite different.

## 2.4 Economics of the System

We are not professional economists, but it is not clear from the published assessments that we are thereby at a disadvantage. The most detailed results are given in Volume 2 of "Transport Policy"[1].

|            | Fatal | Serious | Slight | All |
|------------|-------|---------|--------|-----|
| Accidents  | 6336  | 66738   | 170968 | 244042 |
| Casualties<br>   Children (a) | 719 | 12832 | 40728 | 54279 |
|    Adults (b) | 6153 | 69189 | 194948 | 270290 |
|    All | 6872 | 82021 | 235676 | 324569 |
| Vehicles | 9876 | 103047 | 276580 | 389503 |

Notes: a.  Age below 15 years.
        b.  Age 15 and above, or unknown.
Source:  Road Accidents in Great Britain 1974[14].

Table 3.  Road Accidents and Casualties in 1974.


The costs of reported injury accidents in 1974 are shown in
Table 4, while the costs of all accidents in 1974 are estimated in
Table 5, suggesting a total accident cost of nearly £$10^9$ per annum.
However, the methods of analysis are crude, and estimates ("lost out-
put - £225m") to three figures lend a spurious authenticity to the
data.  The estimation methods used are discussed by Dawson[15,16].

One major element is accurately costed, that is the expenditure
on road construction and maintenance.  This was projected[1] to total £1.2 x $10^9$ for all roads in 1975-76, comparable to the cost of
accidents.

The largest cost is the amount spent on maintaining, replacing
and using vehicles.  Capital expenditure on road transport is detailed in Table 6.  Costs to users are estimated in Table 7, but the
D.O.E. statisticians remark (Transport Policy[1], Vol. 2, p. 10)
"A substantial amount of assumption and interpretation has been necessary in the construction of the estimates."

Thus the estimated overall cost of the system is of the order
of £1.5 x $10^{10}$ per annum.  This is a somewhat frightening figure,
even allowing for uncertainties of a factor of two in the data. The
amount spent on research and development is difficult to establish,
particularly since most vehicle engineering development is commercial and the costs are spread over perhaps twenty years.  However,

|  | Accidents[a] | Cost/accident[b][c] | Total cost[c] |
|---|---|---|---|
| Fatal | 6336 | 44000 | $3 \times 10^8$ |
| Serious | 66738 | 3060 | $2 \times 10^8$ |
| Slight | 170968 | 520 | $1 \times 10^8$ |
| All injury | 244042 | 2360 | $6 \times 10^8$ |

Notes: a.  Source: Road Accidents in Great Britain 1974[14].
       b.  Source: Transport Policy[1], Vol. 2, p.100.
       c.  £'s, January 1976 price levels.

Table 4.  Costs of reported injury accidents 1974 (1976 prices).

|  | £m |
|---|---|
| Property damage | 390 |
| Medical and ambulance costs | 30 |
| Administration costs | 50 |
| Pain and grief | 180 |
| Lost output | 225 |
| Total | 875 |

Source: Transport Policy[1], Vol. 2, p. 118.

Table 5.  Total accident costs 1974 (1975/76 prices).

the budget for TRRL, which deploys the major part of other research
funds is of the order of £7.5 x $10^6$.  We believe the total spent on
R and D is unlikely to be greater than £1.5 x $10^7$ per year (again
within a factor of two), or 0.1% of the total expenditure.  We know
of no other system of comparable size and complexity where so trivial

|                                    | £m   |
|------------------------------------|------|
| Vehicles                           | 2766 |
|   Private cars(a)                  | 1834 |
|   Buses and coaches (b)            |   56 |
|   Other road vehicles(b)           |  876 |
| Plant, machinery and buildings     |   59 |
| Roads                              |  619 |
| Total capital expenditure          | 3444 |

Notes:   a.  Personal expenditure.
         b.  Mainly goods vehicles and business cars.

Source:  Transport Policy[1], Vol. 2, p.26.

Table 6.  Capital expenditure on road transport 1974.

|                            | £m    |
|----------------------------|-------|
| Passenger transport(a)     | 7060  |
|   Motoring                 | 6240  |
| Goods transport(b)         | 7760  |
| Total                      | 14820 |

Source:  Transport Policy[1], Vol. 2.
         a.  P.10.
         b.  P.25.

Table 7.  User expenditure on road transport 1974.

a proportion of the total expenditure goes on research and development, except perhaps the Education system.

It is clear from the amounts involved compared with the U.K. GNP that the central agency in the control system is the Treasury, and the fiscal policies it advises are major control elements. This gives the government the apparent ability to manage in the interests of non-transport objectives[1].


3. CONTROL


3.1  Division of Responsibility

Except insofar as overall control is exercised by the Government acting through fiscal policy, there does not appear to be any central control.  We discuss below some of the more readily identifiable control mechanisms.


3.1a The D.O.E., the Home Office, and the Lord Chancellor's Department

The most important fact is that below the level of overall financial control responsibility for the system is spread between several Departments of State.  The Department of the Environment (D.O.E.) includes the old Ministry of Transport, and is responsible among other things for the provision of roads (in collaboration with the Local Authorities), the code of behaviour for road users ("The Highway Code"), the national research laboratory (TRRL), and for advising the Cabinet on the distribution of resources between road, rail and other forms of transport, as well as having a major say in decisions on traffic control measures.

On the other hand, the Home Office is responsible for the Police but not the Courts, the appointment of magistrates and judges being the responsibility of the Lord Chancellor (i.e. the "head of the Legal Branch").  Within the Home Office, separate divisions or departments handle police matters and the penal system.

In effect the Police operate independently of the Courts, which we do not deny is extremely desirable, but there is no formal feedback mechanism to assist the Police in formulating policy in the light both of the practice of the Courts, and analysis of the results, if any existed, of the sanctions imposed by the Courts.  Thus it is well known to both police and lawyers that the penalty of Disqualification (i.e. withdrawal of licence to drive) is widely ignored. If a disqualified driver is detected driving, the only sanction, other than a custodial sentence, is to fine and disqualify for a further period, and the latter option is usual.  We interpret this,

in the absence of any scientific investigation, as a reaction of the
magistrates courts, staffed as they are by lay magistrates, to a
real or imagined reluctance on the part of the public to accept custodial treatment for road traffic offences.

No independent investigation of the effect of the caution and
prosecution policies of the Police in relation to the conviction
and sentencing policy of the Courts has been permitted by the Home
Office, and if internal analysis has been carried out it has not
been published.  The reaction of the system to Police and Court controls is unknown.


### 3.1b   The Departments of Energy and Industry

Responsibility for other factors lies with the Department of
Energy, since one-third of the oil consumption of the U.K. goes in
fuel for transport[1].  Some attempt at control has been exercised
by imposing reduced speed limits of 50 and 60 m.p.h. on non-motorway
roads,with single and dual carriage-ways respectively, where previously the limit was 70 m.p.h.; in conjunction with pricing policy,
these limits were aimed at reducing fuel consumption.  The success
of these measures is uncertain, though we understand TRRL are studying the situation.

The enforcement of 50 and 60 m.p.h. limits provides substantial
problems for traffic police; severely increased hazards are associated with the usual technique of a "stopper" acting downstream of a
radar meter at these speeds, and the driving population is apparently
aware of the low police activity in this field.  The lack of awareness of the police problem, or the lack of attention to it by the
energy planners, evidences the absence of an overall system model.

Further, the Department of Industry is encouraging the vehicle
manufacturing industry to increase output as rapidly as possible,
mainly with a view to the effect exports have on the balance of payments and on levels of employment, but necessarily implying a growth
in home sales.  The Government[1] anticipate an increase of 25% in
annual road passenger mileage in the period 1975-85, implying at
least a 20% increase in fuel requirements.  The central forecast for
car ownership is of an increase from 0.25 cars/person in 1975 to 0.44
cars/person in 2010, an increase of 75%!  Neither of these forecasts
is compatible with a sane energy policy unless very dramatic improvements in the fuel consumption of vehicles are achieved.


### 3.1c   Local Government

Local Governments' responsibilities in the context of this system include a share in planning the network of major roads and some

executive responsibility for its maintenance, though approved expenditure is recoverable from Central Government funds.  In addition
they have responsibilities, subject to Central Government policy,
for maintenance of minor roads, for street lighting schemes, for
traffic control schemes and parking schemes, and through the local
Police Authority for speed limits.  Thus Local Authorities exercise
a considerable control function, particularly in central urban areas.


### 3.2   The Public


### 3.2a   Preference Patterns

The public are the users of the system, and as such their
individual preferences have an important effect throughout.  For example it is well known that the enforcement of speed limits on urban
trunk roads is only possible where all except a small minority of
users perceive the usefulness of the limit.

At a more important level much evidence indicates that a substantial majority of families wish to own a car, and that they are
prepared to keep this as an important aim over a very wide range of
variation in perceived costs: the projected saturation level of 0.45
cars per person is expected to be reached by 2010[1].  Thus any steps
the Government takes to limit car ownership, by taxation and by limitations on use in congested areas, are likely to be vitiated by public pressure.  Government's aim "to develop a controlled and managed
market in transport" must in a democracy depend on educating and
persuading the voter that the aim is correct, and if correct, desirable.

Public preference is assessed in design studies of urban roads,
in that surveys of current and preferred trip distributions are made
and used in evaluating the benefits of the proposals.  (It is of
interest to note that the mathematical properties of trip distribution and allocation models are such that overall optimization is not
necessarily achieved when individuals optimize their own routes,
and that the provision of new or better roads may cause a degradation in service in some cases[2].)


### 3.2b   Pressure Groups

Official and unofficial pressure groups have an important influence on the behaviour of the system.  We have no detailed model,
but several examples of this process may be noted.  Very recently
public pressure expressed as disruption of a public inquiry into
the proposed route for part of a new motorway (the Aire Valley
route) caused the inquiry to be abandoned, and forced the Ministry

to rethink its proposals.  More subtly applied public pressure has
caused substantial delay in obtaining agreement on the proposals
for the route of the outer London orbital motorway (M25).

Special interest groups also exert pressure and affect control
of the system.  Proposals that dipped headlamps, rather than side-
lights, be made compulsory in urban areas during lighting-up times,
even when street lights were provided, were withdrawn when pressure
from informed independent scientific opinion was applied.  Typically,
for England, this opinion was expressed through letters to "The
Times"!  The point at issue was a technical one, that pedestrians
at night illuminated by street lights, but not by car head lamps,
are perceived by drivers as dark objects against a bright background,
while conversely if illuminated by headlamps and not by street lights,
as bright objects against a dark background; however, when both forms
of lighting are present there is a danger of their effects cancelling.
Many scientists felt that insufficient research had been carried out
to justify the proposed regulation.

Pressure can also be applied by organised labour.  Proposals to
assess certain benefits arising from an individual's use of a company car as income were made in the Finance Bill of 1976, and have
now been amended by the Government.  The effect, as viewed by some
of the Trade Unions in the motor manufacturing industry, would have
been to produce a substantial decrease in the market and a tendency
to increase unemployment in the industry.  It appears likely that
these views influenced the Government.


3.3  Financial Control

The previous discussion indicates how Government decisions on
taxation structure with non-road traffic system objectives can nevertheless have an important influence on the system.  This is only to
say that the traffic system is embedded in the overall economic system.  We have no model of this system (or rather no reliable model),
so that when perturbations are applied their effects on other sub-
systems of the economic system (in particular the traffic system)
cannot be predicted.  Direct financial control can be applied to the
road traffic system, usually by applying it to one of that system's
sub-systems, and the same criticism applies.

The control levels available can be non-specific to the traffic
system, such as general level of taxation or rate of VAT, or specific, such as restrictions on road construction expenditure, on minor
road maintenance, or on pricing of various elements of the system.
The price to the user of access to the whole system can be controlled
by modifying the level of Road Fund Tax, or the taxes on fuel, and
access to limited parts of the system, for example, city centres,
can in principle be controlled by pricing parking, or by licensing.

The Government view appears to be that parking control will remain the best way of limiting congestion in central urban areas, and that this can be achieved through a combination of increased parking charges, reduction in available space, and taxing private parking. The effects of restricting private motor vehicle access to such areas may include a significant change in shopping patterns, where as in the United States much business may transfer to suburban shopping centres, furthering the decay of the inner city. Again, lack of a detailed model means that policies intended to achieve one end may do so at the cost of a disaster elsewhere in the overall system.

## CONCLUSIONS

Central Government claims that it is concerned only with broad general policy matters and does not interfere in detail at the local level. For example, Parliament considers nationally applicable legislation (Road Traffic Acts and amendments to these) but not in general the implementation of these laws by the local Police Force, nor by tradition does it consider the procedures of the Courts, except insofar as the Acts of Parliament set minimum and maximum penalties for offences. Although as in other fields Central Government claims merely to advise Local Authorities and other Statutory Bodies, in practice the Minister often has reserve powers to direct and can in any case apply financial controls over any major local authority scheme.

We conclude that there is no satisfactory mechanism for continuous review and public debate of road traffic policy, and although the recent publication of a discussion document[1] is a helpful step, and is accompanied by some elaboration of the information flows. This lack will not be cured by the proposed establishment of a National Transport Council, unless it is also to be involved in the formulation and not merely the discussion of policy.

### Sub-optimization

The types of study which the discussion document suggests might be undertaken under the auspices of a National Transport Council include, for example, studies of inter-urban passenger services and the return on investment in public transport. These would be valuable, but unless placed in an overall model of the system they are likely to lead to proposals which result in sub-optimization of parts of the system.

The construction of an agreed and accepted, though modifiable, model of the system is a major task, comparable in difficulty with that of constructing a national economic model. The core of such a model would include a clear statement of the many and diverse ob-

jectives of the system, their relationships, and the identification
of those policy parameters which quantify political priorities.  The
primary use of a model would be to assist in understanding the subtle effects in one part of the system caused, often quite unexpec-
tedly, by even relatively small changes in another.  It is apparent
from the available data on growth, for example in car ownership,
that the system is not in a steady-state but is dynamic, and that
without a dynamic model, long-term planning is a hazardous adventure.

The purpose of a model is not to enable detailed quantitative
predictions of response to stimulus to be provided, based on verified relationships, but to assist in understanding, at least qualitatively, the structure and interrelations of the system.


RESEARCH

The situation is not helped by the small amounts expended on
research; much of the research effort, particularly that funded
through TRRL, is of excellent quality, but on too small a scale to
be effective.  The tendency is to direct research towards curing
temporary hot-spots and only secondarily towards establishing a
broad and tested comprehensive understanding of the system.  The
contributions of the Universities have been poor, mainly because few
groups have been able to secure adequate funding on a sufficiently
long term basis, so that expertise, when generated, is neither maintained nor transmitted.  For example, Home Office support of University research is funded on a year-to-year basis, effectively preventing significant career development opportunities for any staff
engaged, but more importantly directing the efforts of the groups
involved towards obtaining rapid results, perhaps with inadequate
testing.

In comparing the way in which research is organized in this
field with say research in a major branch of physics we note several
interesting features.  One, few University libraries can now keep an
up-to-date and comprehensive collection in the field of traffic
studies, while a national laboratory will surely include such a library.  In preparing this paper we have found the TRRL library (and
its staff) extremely helpful.  Two, the expensive equipment required
in some aspects of the subject is more economically provided centrally.  Three, in physics it is usual now to strengthen in-house
laboratory teams with semi-permanent university staff and postgrad-
uate students, thus benefiting both organizations.  The funds to
support the University side are normally provided by the Research
Councils, though occasionally by other agencies (e.g. The Atomic
Energy Authority).  The daily interaction inclines to speed the diffusion of knowledge.  Under the present system, the lack of emphasis
in some quarters of Government science on seminars and publication
and the unwarranted application of the "need to know" principle

severely inhibits progress.

Essentially, we are arguing for the development of TRRL into a
National Laboratory for Transport Studies, on the lines of CERN or
the Rutherford Laboratory.  Indeed, the analogy with CERN is not so
far-fetched: many aspects of traffic science, as with high energy
physics, are independent of frontiers.  The major difference lies
in the scale of funding: several hundred million pounds per year for
CERN, only ten million for TRRL.  The traffic system, with its large
capital investment, high annual cost and direct bearing on the lives
of most of us, deserves a more realistic research appropriation than
it currently receives; we believe that a National Laboratory, with
broad terms of reference, with the ability to investigate all aspects
of the system, and with adequate research funds, is essential if the
road traffic system is to be managed successfully and efficiently to
the benefit of the entire community.

REFERENCES

1.    Transport Policy, A Consultation Document. HMSO, London, 1976,
      two volumes.

2.    LEBLANC, L.J. An algorithm for the discrete network design prob-
      lem.  Transportation Science, Vol. 9, pp. 183-199, 1975.

3.    OECD Road Research Group. Research on Traffic Law Enforcement.
      OECD, Paris, 1974.

4.    FRECKLETON, S., N. FERGUSON and M.E.MONCASTER.  Traffic policing
      effectiveness measurement and resource allocation. NATO Confer-
      ence on Environmental Assessment of Socio-Economic Systems, Is-
      tanbul, Turkey, 4-8 October 1976.

5.    Operation 101 Final Report.  Department of California Highway
      Patrol, State of California.  Phase I, 1966; Phase II, 1969;
      Phase III, 970.

6.    Operation 500 Final Report. Department of California Highway
      Patrol, State of California, 1970.

7.    University of Durham/Durham Constabulary Joint Research Project
      on Trunk Road Patrolling, Second Report, 1969.

8.    MUNDEN, J.M. An experiment in enforcing the 30 mile/h speed
      limit. RRL Report LR24, Road Research Laboratory, Harmondsworth,
      1966.

9.    On-the-spot accident investigation.  TRRL Leaflet LF 392 Issue

2, Transport and Road Research Laboratory, Crowthorne, 1975.

10.  RUSSAM, K. and B.E. SABEY.  Accidents and traffic conflicts at
     junctions. TRRL Report LR 514, Transport and Road Research
     Laboratory, Crowthorne, 1972.

11.  SPICER, B.R. A study of traffic conflicts at six intersections.
     TRRL Report LR 551, Transport and Road Research Laboratory,
     Crowthorne, 1973.

12.  COOPER, D.F. and M.R.C. McDOWELL.  Traffic studies at T-junctions:
     4. Conflict simulation and driver behaviour.  Paper to be pre-
     sented to O.R. Society Conference, Swansea, 21-23 September,
     1976.

13.  TROY, P.N. and N.G.BUTLIN.  The Cost of Collisions.  Cheshire
     Publishing, Melbourne, 1971.

14.  Road Accidents in Great Britain 1974. HMSO, London, 1976.

15.  DAWSON, R.F.F.  Cost of road accidents in Great Britain.  RRL
     Report LR79, Road Research Laboratory, Crowthorne, 1967.

16.  DAWSON, R.F.F.  Current cost of road accidents in Great Britain.
     RRL Report LR 396, Road Research Laboratory, Crowthorne, 1971.

# TRAFFIC POLICING, EFFECTIVENESS MEASUREMENT AND RESOURCE ALLOCATION

S. Freckleton, N. Ferguson; M.E. Moncaster

Plessey Electronic Systems Ltd.; The Home Office

United Kingdom

## INTRODUCTION

The road traffic system is important to the functioning of industrial society as it is today, and provides a means of satisfying many social needs. The structure of this system and its methodology of control is discussed in the paper by McDowell and Cooper. This paper looks in more detail at a sub-set of the control system concerned with aspects of the control of the social costs associated with accidents and congestion. These costs are very heavy - e.g. in the U.K. the costs of accidents was estimated to be in excess of eight hundred million pounds in 1975, with at least as large a figure for congestion. A number of mechanisms are used in order to try and control these costs. Some of these are long term, such as road engineering, improving the performance and safety of vehicles, and raising standards of driver skill and behaviour. On the other hand legislation is provided which is able to produce short term modifications of driver behaviour in order to control the system while it is being modified by these long term mechanisms. That these measures, on the whole, appear to be successful can be gauged by the fact that although traffic in the U.K. has increased by seven times since 1934, the annual number of road fatalities in the U.K. (excluding the war years) has fluctuated only between 4,500 and 8,000. (Ref.1) Clearly the enforcement of traffic legislation, at least in so far as it refers to the safe and free flow of traffic, is fundamental to the control mechanism. It is important, therefore, that ways of making the most effective use of the limited resources of the traffic police, who are responsible for this task, should be found. The objective of the project described in this paper was to develop a methodology for designing traffic policing resource allocations systems and as far as possible to demonstrate the feasibility of implementing the resulting system in a practical operational environment.

The project, which commenced in April 1971 with an initial study, has completed its research phase and currently an operational experimental system is being evaluated in conjunction with the Sussex Police. The paper describes the approach that has been taken towards the development of the design methodology and indicates some of the results achieved.

The work was carried out by a joint project team with members from the Police Scientific Development Branch at the Home Office, the Systems Research Unit of the Plessey Radar Research Centre under contract to the Home Office, and Royal Holloway College, University of London, also under contract. Operational advice, guidance and support were provided by two Traffic Police Officers seconded to the Police Research Services Unit at the Home Office. The full co-operation of the Sussex Police, who also provided an Inspector as liaison officer, and of the East and West Sussex County Councils was generously given.

As initial study of the literature gave clear indications that, while some broad studies had been made in a number of areas concerning the interaction of traffic police, legislation and accident levels, the mechanisms involved were not well understood. It became clear that it would be necessary to develop better understanding of the interactions of traffic flows, road geometry, driver behaviour, legislation, social attitudes, police activity and accidents and congestion. It was felt that in order to have some chance of success it was necessary to start with a relatively simple problem and for this reason the problems studied were those of rural traffic policing rather than the more complex ones of urban traffic policing or the more specialised ones of motorway traffic policing. It may be argued that the greatest problems in accident and congestion occur in the urban and metropolitan areas where the road networks are complex and many pedestrians are involved. However, there seemed little point in tackling this problem if the simpler one could not be solved. The Sussex Police Force was chosen as a suitable one for the project since its area was thought to be representative of many police force areas in England and Wales and provided a broad range of traffic policing problems.

It was realised that trade-offs occur in the system between accidents and congestion - e.g. reducing speeds of traffic will reduce accidents but will also increase congestion - nevertheless, it was felt that the project should concentrate on the reduction of accidents, principally because more work had been done on the interactions between police activity and accidents and a great deal of traffic legislation was aimed at safety. The objective of the work was therefore to develop an overall design approach for the provision of a resource allocation system which had the objective of reducing the number and severity of accidents through the more effective use of available resources.

### Resource Allocation System Requirement

The basic traffic policing resource allocation problem can be stated as follows. The traffic police have a set of resources – men, vehicles, radar meters and so on – which operate in the traffic environment. These resources can be used in various activities such as patrolling, carrying out radar speed checks and carrying out vehicle defect checks. The resource allocation problem is to allocate these resources to particular activities at particular times and places in the traffic environment in such a way as to maximise the success of the system in attaining its objective, which, as stated above, has been taken to be the reduction in the number and severity of road accidents.

The basic mode of operation of such a resource allocation system is illustrated in Figure 1. It is based on a concept of a multi-loop feedback system with different cycle times for each loop.

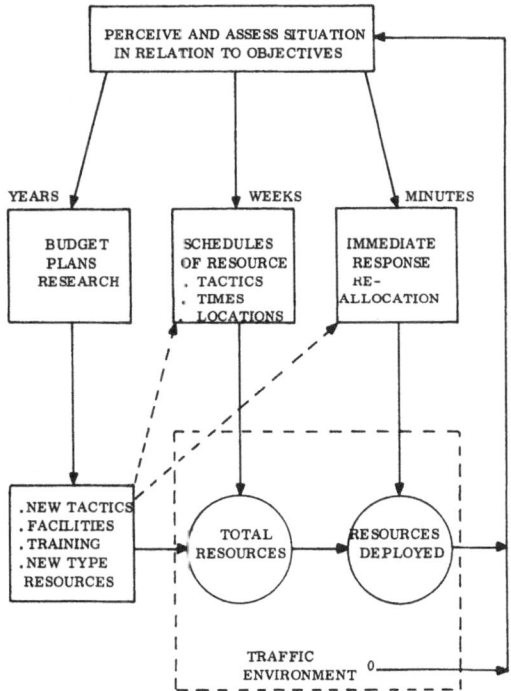

Figure 1. Resource allocation cycles.

The first loop on the right-hand side can be considered as the control loop with a cycle time of a few minutes. In this cycle the following stages occur:

i) Data is collected on the current state of the road traffic environment and the current deployment of the police resources.

ii) This data, when transmitted to a control centre, is processed to determine whether any changes in the deployment of the resources is required as a result of the state of the traffic environment not being as predicted, e.g. an accident may have occurred.

iii) The resources are re-allocated and control instructions transmitted to the deployed resources as required so that the objective achievement of the system can be maximised.

The second loop can be considered as a management loop with a cycle time of one or two weeks where:

1) Analysis of the data collected in past time periods is used to predict times and places of occurrence of situations of high accident risk where police action will be effective in reducing that risk.

ii) The resulting predictions and the knowledge of the effectiveness of different uses of resources in reducing risks are used to produce a schedule of future resource deployments to maximise the risk reduction with the given resources.

iii) The necessary instructions and information required to carry out the scheduled tasks are generated and issued to the resources at the start of their duty.

The third loop is a planning loop with a cycle time in the order of years where decisions are made which determine future resource levels and mixes, new facilities and methods for more effective policing are developed through research and plans made to provide the necessary training, etc. These decisions will be based on analysis of the data collected in the other cycles.


RESEARCH AND DESIGN PROCEDURES

The basic approach adopted to the design of this system involved the following stages:

i) The road traffic system and its interactions were studied

to find ways of predicting situations – times, places and
traffic conditions – where the risk of accident was high.

ii)   The traffic behaviour leading to accidents in these situations was studied to enable police activities to be designed
which would change this behaviour in such a way as to reduce the risk.

iii)  Experiments were then carried out to measure the productivity of these activities in reducing risk.

iv)   Current traffic police operations were studied to find the
constraints that would be imposed on the resource allocation system by police activities outside the objective of
reducing accidents, and the problems that might be encountered in implementing the system.  This work was particularly concerned with the human problems which might arise
from individual police officers' attitude to their job.

v)    Scheduling procedures were then designed to produce allocations of resources to times, places and activities in
such a way as to produce the greatest possible effect on
the basis of known relationships between police activity
and accident occurrence.

vi)   An operational experiment has been set up in which the system has been implemented in an area, so that an assessment
can be made of the effectiveness of the system in reducing
accidents.

Currently, the project has reached the last of these stages –
the operational experiment has been set up and evaluation is now in
progress.

Because of shortage of space, it has not been possible to give
in this paper detailed descriptions of all these stages.  Hence, attention has been concentrated on just one area – the measurement of
the effectiveness of different police activities.


POLICE ACTIVITY MEASUREMENT

The fundamental problem in the measurement of the effectiveness
of police activities in preventing accidents is, of course, the same
problem which has plagued so much accident research – the wide dispersion of accidents in time and space, which means that, although
accidents occur in large numbers over the road network as a whole,
they are very infrequent at any given location.  In Britain a bad
accident "black spot" on a rural road might have fifteen accidents
a year.  This means that to measure the effect of individual police
activities directly by measuring accident frequencies at the sites
where they are used would require either co-ordination of police activities on a very large scale or waiting for a very long time.
Hence, this method was abandoned as impracticable.

The alternative which was adopted was to look for surrogate measure of traffic or driver behaviour which:

    i)   Are capable of easy and preferably automatic measurements.
   ii)   Can be related to accident risk.
  iii)   Will be changed by police activity.
   iv)   Will allow the effect of police activity in changing the measure-parameters relatively easily and quickly.

The number of measure which meets these requirements is relatively small. Among the possibilities are:

a)   Traffic Flow. This is certainly easy to measure and is strongly related to accident risk - so that flow measurement will form an important element in the part of a resource allocation system concerned with predicting high risk situations. This is, therefore, a vital measurement for the allocation system but as considerable difficulties exist in changing flows by police action this measure is not suitable for measuring the effect of police actions on accidents.

b)   Conflicts. Conflicts are near-accident situations - traffic events in which one or more vehicles has to take violent evasive action to prevent an accident occurring. They have been studied extensively by the Transport and Road Research Laboratory - see Ref. (2) - who have found that conflict frequency at a junction is proportional to accident frequency but 2000 times larger. This allows conflicts to be observed directly, by time lapse photography. An experiment was carried out within the Project described here on the use of conflicts for measuring police activity which provided information on the effect of a radar speed check near a junction. Although the results were encouraging and provided useful guidance for the police in reducing junction conflicts, shortage of project resources and time precluded further development of the use of this measure.

c)   Vehicle Defects. Some vehicle defects, particularly defective lights, can be detected easily, and some experimental work was carried out on the effect of police activities designed to reduce them. Such defects, however, only account for a small proportion of accidents (see Ref. (3) ) and these experiments will not be discussed further here.

The measures which were used for the major part of the experimental work within the project were:

d)   Vehicle Speeds.

e)  Gap Acceptance Behaviour.

These will now be discussed in greater detail.

## VEHICLE SPEEDS

Vehicle speed distribution is a measure of traffic behaviour whose relation to accident risk has been extensively studied - see for instance (4).  It is one which extensive efforts have been made to change by legislative action, so the traffic police should be able to have a significant effect on it.  It is also relatively easy to measure and record automatically.  Hence it is an obvious candidate for use in experimental work, and the largest part of the project experiments have been concerned with speed behaviour.

A number of speed measurement techniques were used, including Doppler radar speed meters, but for the majority of experiments the measurements were made using a pair of inductive loops buried in the road surface about 3 metres apart.  The time interval between a vehicle passing over the two loops was recorded automatically.  The recording was done initially by means of a multi-channel data which recorded the data in computer compatible magnetic tape.  Subsequently a specially developed cassette recorder was used.  For the operational scheduling experiment the data will be collected on-line using a microprocessor-based recorder.  Extensive computer software has been developed to process and analyse this data and it is possible to analyse traffic in terms of flow, speed, vehicle length, inter-vehicle gap, direction of travel and time of day.

Despite the amount of work which has been carried out in the past on the relationship between speed and accident risk, it does not appear that a quantitative relationship has been produced which measures the change in risk produced by a given change in speed distribution.  For the purpose of the project it was desirable to be able to measure this change in risk, and hence a "risk index" was devised which it is hoped gives at least a partial prediction of the change in risk produced by a given police action.

This index was based on two sets of observational results obtained from the literature.  The first concerns the relation between speed and accident frequency.  Evidence is quoted in Ref. (5) and (6) for a U-shaped relationship between speed and accident frequency - the probability of a vehicle being involved in an accident increases with the absolute value of the difference between its speed and the traffic average.  Figure 2, taken from (5), illustrates this relationship.  It is possible to fit a quartic curve to the data in Figure 2, of the form:

Accident risk per vehicle = $A + B (v - u)^4$              ---- (1)

where:  v  = vehicle speed
        u  = traffic average
        A, B are constants.

The second set of observational results concerns accident severity.  Ref. (7) gives data on the number of fatalities per accident
against speed which can be fitted by a quadratic function of the form:

$$\text{Severity} = C + Dv^2 \qquad\qquad ----- (2)$$

where C and D are constants.
If we combine (1) and (2) we obtain an accident risk parameter:

$$(A + B (v - u)^4) (C + Dv^2)$$

and if we average this over all vehicles passing through a site we
get a risk parameter:

$$r = \frac{1}{N} \sum_{i=1}^{N} (A + B (v_i - u)^4) (C + D v_i^2)$$

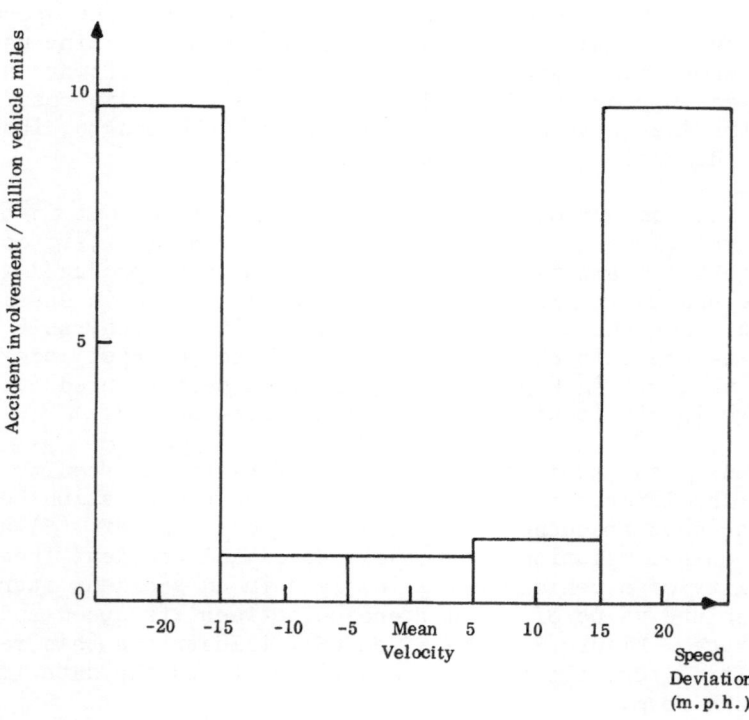

Fig. 2.  The relationship between accident involvement and speed.

where: $N$ = number of vehicles passing through the site in
       given time period

$v_i$ = speed of the ith vehicle

A number of different parameters of the vehicle speed distribution have been used in the experiments to examine the effects of
police activities, such as the proportions of vehicles exceeding
the speed limits. However, this risk index has proved the most
useful in comparing the results of different activities.

## GAP ACCEPTANCE BEHAVIOUR

In Britain some 28% of accidents on rural roads occur at or
near junctions (Ref. (1) ) and the largest single cause of junction
accidents is vehicles turning into or out of a junction into the
path of an oncoming vehicle. Consider for example a vehicle turning
left out of the side road at a T-junction (with a British rule of
the road - vehicles travelling on the left). For the driver of such
a vehicle to make the turn he must merge with the stream of main road
traffic coming from the right, and so he must make a decision to accept a gap in this traffic stream. If he accepts a gap which is too
small, a conflict, as discussed above, or even an accident will occur. Hence the gap acceptance function - the proportion of vehicles
willing to accept a gap of given length - is related to the accident
risk at the junction.

Similarly, a vehicle turning right at a T-junction must accept
two gaps, one in the stream of traffic he is crossing and one in
the stream he is merging with, and will have two gap acceptance functions.

The measurement of gap acceptance behaviour is rather more difficult than that of speed, and no automatic methods were used in
this project. It was assumed that the relevant gap was measured in
terms of time and not distance - that the turning driver based his
decision on the time available before the approaching vehicle reached the junction and not his distance from the junction. The
measurements were made by observing a junction with a TV camera and
a video tape recorder, the gap times available were measured with a
stopwatch subsequently on a TV monitor and noting whether or not the
gaps were accepted.

A considerable amount of work has been reported in the literature on the study of gap acceptance functions in relation to traffic
congestion - see for instance Ref. (8). Less seems to have been done,
however, on their relationship to traffic accidents, and it proved
necessary to devise a method of relating changes in the gap acceptance function to the corresponding change in risk. The method which

was used was to build a computer simulation model of traffic behaviour at a junction. This model predicts the rate of occurrence of
conflicts (as discussed above) at the junction as a function of traffic flow through the junction and gap acceptance behaviour. Assuming the validity of the conflict/accident relation found by TRRL the
model can thus be used to relate the gap acceptance function to risk.
This model has not however yet been validated. The model, which
was developed further by Dr. R. Cooper of Royal Holloway College,
is described in detail in Ref. (9). In addition to its use for these
purposes, the model has also been valuable in helping to study the
relationships between traffic flow and accident sites which are needed for the risk prediction element of the resource allocation system.


## EXPERIMENTS WITH POLICE ACTIVITIES

A large number of experiments have been carried out on different police activities, covering the following areas:-

- different methods of operation of a radar speed check.
- teams of motorcyclists operating in a number of different
  ways.
- the effect on drivers of being prosecuted and cautioned for
  speeding.
- the effects of a number of different methods of communicating
  with the driver, such as signs and newspaper publicity.
- the effect of gap acceptance behaviour at a junction of different police activities.

In the space available here it will only be possible to give
some of the more interesting results in these areas.


## THE EFFECTS OF RADAR SPEED CHECKS

One experiment which was performed on radar speed checks concerned the effect of a repeated check as compared with a single
check at the same site. The mode of operation of the radar check
was that usually adopted by the Sussex Police - the radar meter at-
tended by a police officer at the roadside so as to be visible to
motorists, with two more policemen a few hundred yards down the
road to stop offending motorists. The check was applied at a site
and a time of day (8 a.m. - 9 a.m.) where about 40% of the vehicles
passing through did so regularly. It was expected that, due to association of the position with the radar check, these regular users
would "remember" to reduce speed for a few days after the event.
Furthermore, if an additional "stimulus" was provided before the
effect of the first had worn off the effect would be larger and
would decay more slowly. The experiment to test this hypothesis
consisted of:-

A single application of the radar check, followed by a
period of several weeks to allow drivers to forget.
An application of the check, followed two days later by a
second application.

The results of these applications in terms of the risk index
described previously are given in Table 1.  These results show that:-

   i) The first radar application produced a 12% reduction in
      risk.  This fell to 2% on the day after the check (that
      there was a statistically significant change on the se-
      cond day was found from other parameters of the speed
      distribution) and vanished thereafter.

  ii) The second check produced a 10% reduction, which increa-
      sed to 22% on the third check two days later and a sig-
      nificant reduction in risk was still present ten days
      after this check.  The average reduction throughout the
      thirteen day period was 8%.

Hence the single check produced an average 7% reduction lasting
two days, the double check produced an average 8% reduction over
thirteen days.  Hence, allowing for the fact that the double check
required twice the police effort, the police productivity in reducing risk in the double check was three or four times as high as in
the single check.

It must be emphasised that these are the results of a single
experiment, and that the risk index is subject to large statistical
fluctuations.  This experiment, however, shows the potential for
increasing police productivity.


                        THE EFFECT OF COMMUNICATION

It was expected that driver behaviour would be changed and the
effects of police activities enhanced if means could be found to
communicate to drivers that certain types of behaviour required
modification.  The experiments on the effect of communication included examinations of:-

   i) The effect of newspaper and radio publicity on traffic
      behaviour over an area.
  ii) The effect of a flashing sign warning individual vehi-
      cles that they were exceeding some threshold speed.
 iii) The effect of portable roadside signs warning of the
      presence of speed checks.

The first of these experiments used publicity in local newspapers and on local radio to tell motorists in parts of Sussex firstly

| Day | Risk Level | Percentage Reduction |
|---|---|---|
| Normal (before police activity) | 4.70 | 0 |
| 1st Radar Check (A) | 4.14 | 12 |
| A + 1 | 4.60 | 2 |
| A + 2 | 4.79 | -2 |
| 2nd Radar Check (B) | 4.23 | 10 |
| B + 1 | 4.46 | 5 |
| 3rd Radar Check (B + 2) | 3.67 | 22 |
| B + 3 | 4.23 | 10 |
| B + 4 | 4.24 | 10 |
| B + 5 | 4.48 | 5 |
| B + 6 | 4.30 | 8 |
| B + 7 | 4.42 | 6 |
| B + 8 | 4.68 | 0 |
| B + 9 | 4.25 | 9 |
| B + 10 | 4.57 | 3 |
| B + 11 | 4.36 | 7 |
| B + 12 | 4.46 | 5 |

Table 1.   Effect of Repeated Radar Checks

that the police were concerned about the effects of excessive speed and, secondly, that they would be concentrating attention on two major roads in the area.  This publicity produced small but statistically significant reductions in mean speeds and percentages of vehicles exceeding the speed limits on these roads, but only during "rush hours" when a large proportion of the traffic might be expected to be local rather than long-distance.

The second experiment concerned the effect of a flashing sign which was illuminated with the legend "Police - You are Speeding" when a car passed at a speed above a pre-set threshold.  This sign was placed by the roadside for a period of four weeks, and no police presence was used to support the sign.

The results are given in Table 2.  It will be seen that the sign is consistently effective over the period, producing reductions in risk levels of 25% - 30%, which are comparable with the maximum effect produced by the radar check discussed above, even though the sign presents no actual threat to the motorist.

Two experiments were carried out on the effect of signs warning of speed checks.  The first experiment compared the effect on speed behaviour of a police car stationary by the side of the road with the effect of such a car preceded by a warning sign with the legend "Police Speed Checks Ahead."  It was found that:-

   - the police car on its own produced a reduction in risk
     of 13%
   - the police car with a sign produced a reduction of 49%.

The second experiment concerned the effects of signs on their own, with only occasional police presence.  Temporary signs with the legend "Police Speed Checks Ahead" were placed at either end of a two and a half mile stretch of road, every morning between 8 a.m. and 9 a.m. over a twelve week period.  Feedback, in the form of measurements of the effect being produced was obtained at regular intervals during the experiment and used to decide when police activity was needed to reinforce the signs - the activity being introduced when the effect showed signs of falling off.  Radar speed

| | | Time of Day | | | | |
|---|---|---|---|---|---|---|
| | | 8.00 - 9.00 | 9.00 - 10.00 | 15.00 - 16.00 | 16.00 - 17.00 | 17.00 - 18.00 |
| Normal Risk | | 24.0 | 24.4 | 23.6 | 28.1 | 27.8 |
| 1st Week with sign | Risk | 17.8 | 21.1 | 17.4 | 22.8 | 23.2 |
| | % Reduction | 26 | 14 | 26 | 19 | 17 |
| 2nd Week with sign | Risk | 19.7 | 19.5 | 16.3 | 18.3 | 21.0 |
| | % Reduction | 18 | 20 | 31 | 35 | 25 |
| 3rd Week with sign | Risk | 19.0 | 16.0 | 16.6 | 21.2 | 20.4 |
| | % Reduction | 21 | 34 | 30 | 25 | 27 |
| 4th Week with sign | Risk | 20.2 | 15.4 | 15.7 | 16.3 | 21.6 |
| | % Reduction | 16 | 37 | 33 | 44.2 | 22 |

Table 2.  Effect of Flashing Sign

checks were operated for a total of five times on separate days during this period.  It was found that the signs had a significant effect on the speed distribution throughout the two and a half miles,
though decreasing as the traffic moved away from the signs.  It was
also found that, although the effect was decreasing toward the end
of the period, it was still present at the end, despite the fact
that 35% of the drivers went through the site regularly at the time
the signs were out and so might be expected to know that there was
usually no police presence.  Thus the effectiveness of the feedback
was demonstrated.

This last result was most unexpected by the police.  They had
expected on the basis of previous experience that the signs would
retain their effect for a few days at the most.  It suggests that
there is considerable scope for the use of such signs in police activities.  It is clearly desirable for further experiments to be carried out to confirm the result.  The experiment also illustrates the
value of feedback on effects in the management of police activities.

EXPERIMENTS ON GAP ACCEPTANCE BEHAVIOUR

Two experiments were carried out to measure the effect of police actions on gap acceptance behaviour.  The first concerned the
effect of police presence at a T-junction on vehicles turning left
out of the junction.  The police presence consisted of a stationary
police car placed at the junction so as to be visible to all turning
vehicles.  The results obtained are illustrated in Fig. 3, which
shows the proportion of drivers willing to accept a gap of given
size.  The effect of the police presence was to move this curve to
the right by about one second - when the police car is present all
drivers seem to require a gap of about one second longer before
they are prepared to turn.

The T-junction conflict model gives the result that such a
change in the gap acceptance function would give a reduction of 70%
in conflict frequency.  While, as discussed above, the model is not
yet validated, this suggests that this type of police presence may
well be very effective in reducing accidents of this type.

The second experiment concerned the effect of signs on vehicles
turning right at a T-junction.  A series of four signs was placed
on a 200-yard stretch of road up to the junction.  In addition to
the word "Police", their legends were "Fast Traffic Ahead", "Allow
Time to Turn", "Fast Traffic" and "Have you Time to Turn?".  The
signs were used over a number of days at the same site from 8 a.m.
to 9 a.m.  The effect found was that:-

i)  For the gap acceptance behaviour in the traffic stream
    from the right - that which right-turners have to cross -

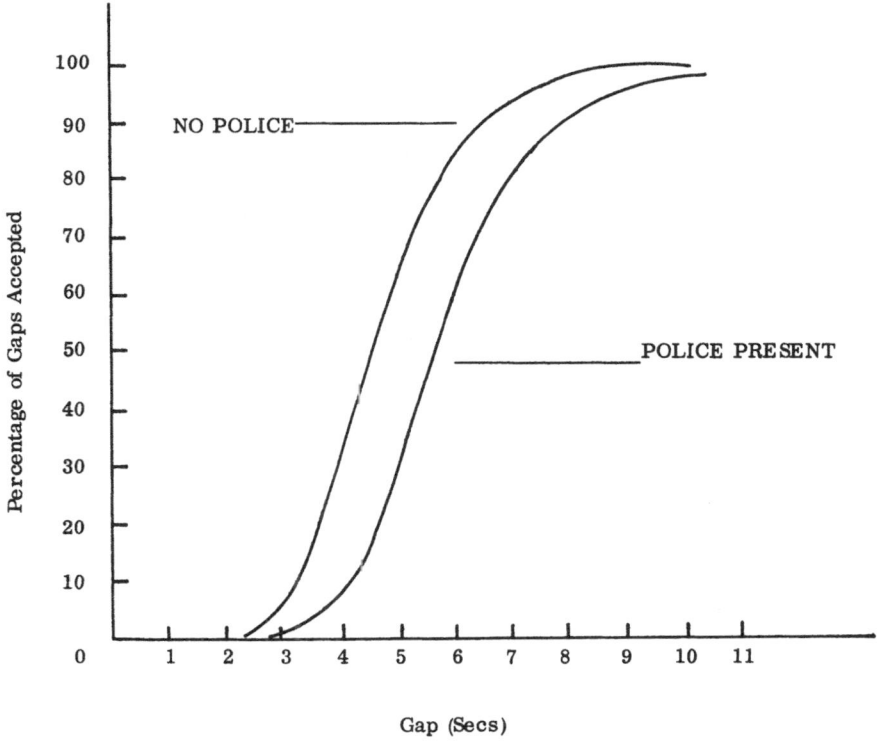

Figure 3. Fitted Gap Acceptance Functions

the effect was similar to that for the previous experiment. All vehicles seemed to require a gap about one second longer.

ii) For gap acceptance behaviour in the stream from the left, which right-turners have to merge with, there was no detectable effect.

These results have not yet been processed by the conflict model, but it seems likely that, since only half the gap acceptance decisions have been affected, the effect of this activity on right-turning accidents will be considerably less than the effect found in the previous experiment on left-turning accidents.

CONCLUSIONS

Some conclusions can be drawn from the work already carried out in this project. It appears from the experimental word discussed above, for instance, that the productivity of police activities can be enhanced by the use of slightly more sophisticated modes of operations, such as the use of repeated radar checks.  There is evidence, particularly, that the use of warning signs enhances police effectiveness, and that these signs have larger and longer-lasting effects than was expected.

The main conclusions, however, will emerge in the future.  The results of the research work discussed above has been incorporated into a resource allocation system for the traffic police, incorporating on-line collection of traffic data and production by computer of optimal resource schedules.  This system has been implemented in part of the Sussex Police Force area, and will be evaluated in operation over the next year.  This should allow the measurement of the effect of the system on accident frequency – which is, after all, the final objective.

The authors thank the Home Office and the Plessey Company Limited for their permission to publish this paper.  The views and conclusions expressed, however, are those of the authors and do not necessarily express the policy of the Home Office.

## REFERENCES

(1)      Road Accidents Great Britain 1973, Department of the Environment – HMSO 1974.

(2)      Spicer, B.R., A Pilot Study of Traffic Conflicts at a Rural Dual-Carriageway Intersection TRRL Report LR 410 Transport and Road Research Laboratory, Crowthorne, 1971.

(3)      TRRL Leaflet LF 392 Issue 2, Transport and Road Research Laboratory, Crowthorne.

(4)      Smeed, R.J., The Influence of Speed and Speed Regulation on Traffic Flow and Accidents. Road and Road Construction. 1960, 38 (456).

(5)      Research Triangle Institute.  Speed and Accidents Vol. 1. Research Triangle Institute 1970.

(6)      Munden, J.M., The Relation between a Driver's Speed and his Accident Rate. TRRL Report LR 88, Transport and Road Research Laboratory, Crowthorne.

(7)      Joksch, H.C., An Empirical Relation between Fatal Accident Involvement and Speed.  Accid. Anal. and Presention. Vol. 7 No. 2, June 1975.

(8)     Bottom, C.G. and Ashworth, R.A., Driver Behaviour at Priority Type Intersections First Int. Conference on Driver Behaviour, Zurich 1973.

(9)     Cooper, D.F., and Ferguson, N., Traffic Studies at T-junctions 2.  A Conflict Simulation Model. Traffic Engineering and Control July 1976.

A SUGGESTED FRAMEWORK FOR EVALUATING THE ASSESSMENT

OF A TECHNOLOGICAL SYSTEM

D. W. Fischer, R. W. Keith

University of Waterloo

Ontario, Canada

## INTRODUCTION

Few places can be said to escape the present concerns over the complex interaction of energy with the socio-environmental milieu. Wherever one looks, a variety of interests and actions concerning energy, environment and society preoccupy the attentions of individuals, organizations and nation states.

The uncertainty about physical, biological and social systems is world-wide. Time and space configurations transcend nation states and regions within them. As one consequence of these conditions it has been suggested that there exists in society a growing disenchantment with technology and especially the consequences of technology over time. (1) Coupled with this is a growing concern for the quality of environments. In addition, and as an outgrowth of both these broad areas of concern, there is increasing attention being given to the study of human institutions which are involved in and reflect technological and environmental issues, their assumptive worlds and value orientations. (2)

Manifestations of the concern for technology, environment and society has taken a number of forms. For the purposes of this paper the development of broadly conceived notions of "assessment" is central. While it might be argued that assessment of technology, environment and institutions is not at all novel, it is suggested here that the focus of such analyses has shifted from technical feasibility studies, biological systems analysis and institutional or organizational productivity investigations respectively, to more broadly conceived frameworks which at once encompass physical, biological and human parameters and their interactions. (3)

The emphasis is upon assessment as comprehensive information gathering and evaluation processes. Illustrative of this focus is a definition of technology assessment as

"an activity to provide information about and systematic analyses of the internal and external consequences (short, medium and long term) for a society of the application and diffusion of a technological capability into its physical, social, economic and political systems. (4)

Comprehensive assessments, whether of energy, environment or institutions introduce the important dimensions of "By whom, using what methods and for what purposes?". Thus, it is necessary to examine not only the information components of assessments but in addition, those social systems that are involved in such assessments and the nature of their assessment activities. In a study designed to identify and analyse the nature of assessment activities and processes surrounding off-shore petroleum development in Eastern Canada, Gibbons and Voyer focused their analysis on the concept of a "technology assessment system" which they define as

"... those social groups which are (or should be) concerned with developing a given technological program. The elements which make up this system may or may not be bound together by any formal arrangements: coupling is effected by their mutual interest in the development and diffusion of a given technological capability." (5)

Taking together the notions of "assessment" and "assessment systems," a number of important concepts are identified. These include:

. a development program
. social groups or actors
. actors' objectives and roles
. actors' levels of involvement
. relationships among actors
. strategies of actors
. consequences of development
. information bases
. decisions; kinds, over time
. emergent issues

For purposes of a comprehensive analysis it is important that the concept of actors be construed broadly to include not only those who create and develop a particular program but also those who are ultimately affected by it, as such developments become embedded in society over time. This too emphasizes the need to consider consequences or impacts other than direct ones. Attention must be given to secondary, tertiary, etc. or induced effects which may appear in the short, medium or long term. Assessment system analysis also requires consideration of convergent and divergent interests, co-

operative and competitive actions, values and motives, categorization of actors, relationships or linkages among actors, information
bases and their sufficiency, relevance and accessibility to actors,
kinds of decisions and their consequences and the issues or problems
which emerge.

By examining an assessment system in this way one can observe
"multi-party and multi-dimensional" processes.  The "technology assessment system" approach as outlined above provides one framework
from which to undertake such an analysis.  In addition to identifying possible logical outcomes, assessment system analysis provides
an opportunity to examine activities in selected time frames. Through
a consideration of the kinds of decisions taken, the sequence in
which they occur (logical or otherwise) and an identification of
those factors which are seen to affect decisions, it may be possible
to gain some understanding of the pacing of activities and their consequences.

Perhaps the major strength of the assessment system approach is
in its capability to provide some indication of the extent to which
the assessment system as a whole has achieved a sense of comprehensive overview, a holistic perspective of the program and issues at
hand.  Where there are important voids in the information base, where
there is evidence that actor relationships preclude the flow of essential information and where control over important aspects of the
development program is highly centralized and those in control have
limited terms of reference, the likelihood that such a system of
actors can achieve a comprehensive overview may be markedly reduced.

The purpose of this paper then, is to show from a framework of
a technology assessment system how one can approach some aspects of
a reference projection for global energy systems.  This system by
denoting the actor hierarchy and the issue dimensions can also be
a basis for evolving a normative strategy set for accomplishing given aims, either from within the system or external to it.  Its basic strength, however, is its usefulness for attempting to evolve
strategic and operational plans for achieving institutional change.
On the basis of research presently in progress on northern development assessment systems in Canada, it is hoped that such concepts
as actors, information, decisions, strategies and issues cost in a
dynamic assessment system framework will further our understanding
of the nature of global multi-party and multi-dimensional energy
systems.

It is essential to understand that the approach used is not an
attempt to model the petroleum assessment system.  Rather it is designed to create a framework for analyzing that assessment system,
i.e. to assess the assessors.  This distinction is vital.  Our attempt has been not to model actor behaviour but to assess actor assessments in the light of a comprehensive view capable of being

adopted at both national and international levels.  Whether or not
this approach is capable of assessing the global energy assessment
system is open to further research.  The framework espoused in this
paper is one proposal for furthering and sharpening the collective
research thrust on aspects of the global energy system.


TECHNOLOGY ASSESSMENT SYSTEM: A MODEL FOR ANALYSIS OF A GLOBAL
                PETROLEUM DEVELOPMENT PROGRAM

A number of important concepts which are integral to our analysis of an assessment system were identified.  These include actors,
issues, information base, information flows, relationships among actors, strategies and decisions.  For schematic purposes and as a representation of the sequence of analysis Figure 1 depicts a simplified model of a technology assessment system analysis.


                    Petroleum Development Program

The initial step in the analysis is the identification and description of the various components of, say, a petroleum development
program.  In addition, the sequencing or phasing of the various components is identified because of their importance to the time frames
which exist in actors' minds.  It is not necessary at this time to
describe in detail the nature of the petroleum program. Suffice it
to say that eight broad development activities have been identified.
These are:

1. Reconnaisance
2. Exploration
3. Development
4. Production
5. Transportation
6. Processing
7. Sales
8. Abandonment

Although these eight phases represent a sequential development
process they are clearly overlapping in time, a characteristic which
is influenced by many factors (e.g. size of proven reserves, regulation processes, financial capability, markets, etc.). Nonetheless,
the general sequential characteristic of the petroleum development
program does permit one to sense the time perspectives and the pacing
of development.

## Supporting Infrastructure

The ease with which petroleum development is conducted is in large measure related to the availability of a host of services and facilities often provided by governments as they seek to encourage resource development.  Examples of infrastructure support systems include:

1. government geological surveys
2. air transport facilities
3. highway and road systems
4. navigation and harbour facilities
5. community development
6. communication services
7. incentive programs

While the importance of a well-developed infrastructure for a petroleum program will vary depending on the location and the nature of the program itself, the infrastructure is seen by many as a very important element.  In the north of Canada where transportation costs alone may account for 60 percent of an exploration budget, its importance is very substantial.  In addition, an understanding of the infrastructure may aid in analyzing likely time frames.

## Actor Analysis Framework

Based upon the identification and description of the petroleum development program and the supporting infrastructure it becomes possible to identify actors.  It must be emphasized at this point, however, that in addition to simply the "technical" aspects of a petroleum program and the supporting infrastructure one must include the consequences or impacts of such activities, for it is only when these are also taken account of that we can have a reasonable assurance that the list of actors is a comprehensive one.

Actors refer to those who are in one way or another participants in an assessment system; i.e. they are "involved". Such involvement may take several forms and may include decision-making, information generation, seeking and utilization, exercising regulatory functions, coalition or sub-system formation, etc.  In addition to describing the nature of an actor's involvement it is useful to identify the degree of involvement.

Degree of involvement is central to the notion of an assessment system.  By indexing levels of involvement and categorizing actors on this basis it becomes possible to locate certain actors at the centre of the system, those who are intensively and continuously involved.  It is important in this regard to remain cognizant of the various phases of the petroleum development program, for as the nature of activities change so may the configuration of actors change.

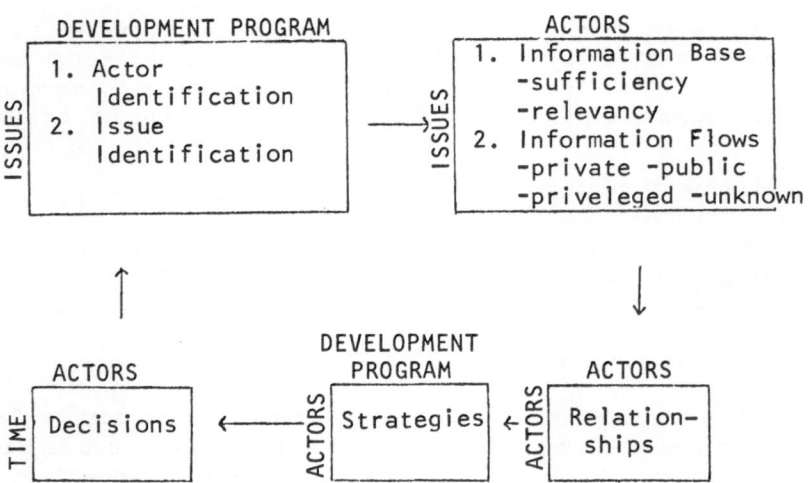

Figure 1. Resource Assessment System Model: Assessing the Actors
in Development

        In addition to involvement, it is useful to identify actors in
terms of their favourability or value orientation to the development
program and issues at hand.  From a knowledge of such attitudes it
is possible to identify cooperative and competitive or rival relationships among actors.  Moreover, such attitudes may account for
information-related behaviour and subsequent strategizing.

        In their study of the Canadian east coast off-shore petroleum
assessment system Gibbons and Voyer identified three categories of
actors. (6) These were:

1.    Core actors: those actors who were intensively and continuously
      involved in the technology assessment system.

2.    Supporting actors: those actors who were less intensively and/
      or intermittently involved in one or another aspect of the tech-
      nological program, but whose decisions and actions nonetheless
      could at least potentially have a significant effect on the
      overall program.

3.    Should be actors: those actors who will be influenced in some
      way by the development program but for one or another reason
      are not involved (e.g. unaware, not participating by choice,
      don't know how to participate or become involved).

This classification is based solely upon the concept of involvement. In this paper the additional factor of attitudes toward the development of petroleum resources permits further distinctions among actors, distinctions which it is felt provide greater explanatory power when attempting to analyse information, decision, strategy and issue concepts. Thus, for this paper a two-dimensional actor classification framework has been created. While admittedly a much simplified representation of actors, it nonetheless permits certain distinctions not previously made in technology assessment system analysis. In particular it allows for a certain degree of differentiation among what Gibbons and Voyer refer to as "supporting actors." Figure 2 depicts the actor classification framework.

The right-hand column of the framework represents actor groups with varying levels of involvement but all being positively or favourably disposed to the development program. As the degree of involvement decreases the specificity of the supportive relationship will likely change from intimate knowledge of the development activities on the part of core actors to more generalized supportive value orientations on the part of "supporting exogenous actors."

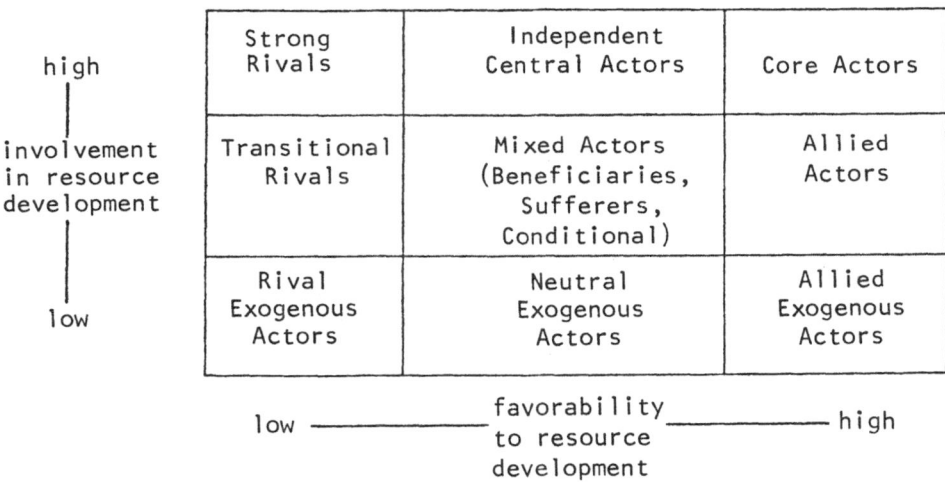

Figure 2.  Actor Classification Framework.

The centre column area of the framework represents a "mixed"
or middle range attitude or value orientation.  Actor types included in this segment of the framework would include

1.  neutral actors
2.  "don't knows" - the should be actors
3.  independent actors - who theoretically are without an a priori
    value position.  Such actors may conduct "objective" analyses.
    Regulatory agencies whose concern is the "national interest"
    would be classified as independent actors,
4.  conditional actors - depending on decisions of other actors
    (as well as own decisions) may be either favourably or unfavou-
    rably disposed to a development program at some point in time,
5.  certain beneficiaries and sufferers.

The left hand column of the framework consists of those actors
opposed to the development program, the adversaries and rivals.
Reasons for opposition will vary among such actors and be related
to:

1.  their own particular objectives
2.  the situation or context in which they are located

Rivals and adversaries may demonstrate varying degrees of involvement; however, even those peripheral to the technology assessment system may significantly affect the system and the petroleum
development program.

In order to clarify the actor classification framework a description of each actor type follows.


## Core Actors

This group have continuous and intensive involvement in the
technological development program.  Though other actor types may
make fundamental decisions it is usually the core actors who initiate a program via one or more fundamental decisions.  Core actors
are usually fewer in number than other actor types though core actors may change as the development program shifts through a sequence
of major phases (e.g. exploration, production, distribution, etc.).


## Allied Supporting Actors

These actors are characterized by a positive or favourable
orientation to the development program.  Their activities enhance
the development program (e.g. complementarity, facilitation, prior
enabling actions, legitimation, etc.).  Suppliers of goods and services, provision of infrastructures, enabling decisions, passage of
legislation, etc., are examples of allied supporting actors' activities.

## Independent Central Actors

Actors of this type are seen to have a degree of independence or autonomy from both the proponents and adversaries of a given development program. Their autonomy may be either constitutionally or legally based or a function of an "objective" information position, i.e. performing their own research, information gathering and interpretation.

## Middle Range Actor Cluster

This type of actor has moderate involvement in the development program and attitudinally may have both pro and con orientations to the development or may be neutral. An actor of this type may be so classified because of:

1. own choice - wishes only moderate involvement
2. expertise is not central to program
3. legal criteria - may have limited terms of reference
4. others' decisions - because of lack of information especially from other key actors, middle range actors' positions may not be sufficiently well developed to qualify as another actor type
5. conditional - waiting for decisions by others to determine or declare position.

## Transitional Rivals and Adversaries

This group of actors consists of those who for reasons of expertise, power, resources and information have only moderate involvement and are declared rivals or adversaries of the technological development program. Not only may actors be emerging as rivals and adversaries but also, given a fundamental decision against the prevailing program, former "core" or "allied supporting actors" may shift to rival or adversary status while those who were formerly rivals and adversaries assume "core" or "allied supporting actor" status.

## Strong Rivals and Adversaries

Strong rivals are characterized for the most part by having developed viable alternate technological programs. To the extent that two or more development programs are recognized (one of which is that of the core actors) and are similarly feasible, then strong rivals will exist. Strong adversaries might include political organizations who may be central to the development program in terms of power but ideologically opposed to the particular technological-economic program mix.

## Exogenous Rival Actors

This group of actors are seen to be outside the technology assessment system, certainly in its day-to-day, week-to-week functioning.  Exclusion may be on a geopolitical basis (e.g. another country).  Their position is in opposition to the prevailing development program and is often based on an alternative program which may or may not differ technologically.

## Exogenous Independent Actors

These actors are seen as exogenous to the assessment system usually for geopolitical reasons.  Like "independent central actors" they have an independent or autonomous role due to constitutional or legal factors.

## Exogenous Supporting Actors

Two types are suggested:

1.  those actors who have definite links to "allied supporting" and "core" actors and who though geopolitically distinct from members of the technology assessment system may indirectly and significantly control their actions (e.g. multinational corporations),
2.  those actors who are characterized both as marginal to the development program and having supportive or favourable attitudes to the program.  Usually they can only be identified through solicitation or indirect representation of their views (e.g. other countries with similar development programs).

## Issue Analysis Framework

Two basic issue orientations have been identified.  The first of these might be referred to as substantive issues.  By this is meant that set of issues which emerge from the very nature of petroleum development.  The second aspect of issue analyses concerns that which is at issue in the assessment system processes.  Such issues will reflect the inadequacies of the assessment system in terms of involvement of actors, quality and flow of information, decisions, and strategies and their resultant effects on the nature of the assessment itself.

Substantive issues are many and may be categorized in a variety of ways.  For purposes of the northern development projects five substantive issues types are identified.  They are:

1. Technological
2. Environmental
3. Social
4. Economic
5. Political

Technological issues are seen as those problems which arise when the limits of a particular technology in its interaction with environment or man have been reached or surpassed. Such situations often give rise to technological adaptation or innovation, though in some instances substantial technological difficulties may bring about consideration of alternate technologies or even abandonment of the program.

Environment, social and economic issue frameworks are broadly conceived to encompass significant time and space configurations. Impacts of technological development are considered in the short, medium and long term. Direct and indirect (or induced) consequences in regional, national and international contexts are examined. Political issues refer to jurisdictional or multi-party conflicts which may be localized or extend through regional and national spheres to the international scene.

Assessment system issues derive from the structure and functions of those social groups or actors which comprise the system. Thus, such issue types are anticipated as:

1.  Actor issues; an example of which would be the absence of certain actors from the system i.e. "should-be" actors.

2.  Information issues; examples of which might include barriers to the information flows and limitations in accessibility as well as deficiencies in the kind and quality of information available.

3.  Decision issues; examples of which might include the existence of highly centralized decision-making perogatives, over-lapping actor decision-making loci, the absence of formalized decision-making authority or responsibility for certain issues and the relevance of information to decisions.

4.  Strategy issues; examples might include instances of cooptation, cooperation and coercion.

5.  Overview comprehensiveness as an issue; which refers to the extent to which the collective assessments represent all the major dimensions of the program or problems which are at issue.

INFORMATION ANALYSIS FRAMEWORK

Information seeking, evaluation and utilization is conducted
from a variety of viewpoints.  For the most part actors information
activities bear a close relationship to their primary objectives or
mandates.  In some instances,however, certain deviations from this
norm can be noted.  Where an actor is induced either by regulatory
requirements or changes in the nature of its objectives the actor's
information activities can lead to information which is in addition
to or tangential to primary needs.

Of particular importance to an analysis of information is an
assessment system is its anticipatory nature.  Inasmuch as ideal
assessments are a priori to development projects, the identification
of information and decisions in a temporal frame is useful.  Extended time frames provide some indication of consideration of medium
and long term consequences of development.  Thus, information analysis ought to encompass external and nth order consequences in time
and in the nature of the information base pertinent to them.

An essential characteristic of an information analysis framework is that of actors' perceptions of information.  Consideration
of such concepts as relevance, sufficiency, accessibility and time-
liness are important in understanding information seeking, evaluation and utilization.

For purposes of the northern development studies a matrix format has been utilized to organize actor perceptions of information.
Matrices have been extensively used in environmental impact assessments (7).  A matrix is primarily a checklist which is designed to
show possible interactions between elements of a system.  In the
case of information analysis, a matrix which arrays perceived issues against actors is developed.  Cell entries reflect actors perceptions of information, more specifically the relevance, accessibility and flow of information. (Refer to Figure 1.).

A number of entries are possible for each cell.  First is an
indication of the perceived sufficiency of information by an actor
group about a selected issue.  By sufficiency is meant both the
kinds and amount of information, in other words, an index of perceived comprehensiveness.  While there are undoubtedly a number of perceptual filters operating some of which may not be recognized, the
information analyst's task is to report the perceptions of actors.

In addition, an attempt is made to record perceptions of accessibility.  Information may be thought to have four accessibility conditions; 1) public; 2) private; 3) privileged; 4) unknown.  Public
information is that which exists in the public domain and is accessible assuming one is aware of its existence and has the resources
necessary to acquire it.  Private information is that which may be

shared among a limited number of actors and/or may be divulged with a confidentiality rider attached. Privileged information is that which remains in the possession of the source and will not be divulged to other interests. Unknown information consists of those areas where knowledge is lacking for want of further research and development.

Where no cell entry is made the assumption is that an actor perceives as irrelevant information concerning the particular issue. Thus, wherever sufficiency and accessibility are indexed one can assume some degree of perceived relevance.

## Strategy Analysis Framework

Policy and program decision-making is subject to a number of forces which may influence the direction or pace of activities. Perceptions of uncertainties, rewards and costs are important factors which affect the composition of an assessment system and its decision-making structures. The distribution and control of technological, informational, financial and management resources are also important aspects of decision-making systems. The nature of the issues at hand and the relationships among actors may have important effects. (8).

Strategies are those purposive actions designed to influence some set of the above-mentioned factors; including the decisions of actors in the assessment system. Some strategies may be intended to alter perceptions of uncertainties, costs and benefits. Other strategies may focus upon control and utilization of resources to influence decisions. Still other strategies may attempt to redefine issues or relationships among actors. Strategies imply a sense of normativeness, that which ought to be, albeit that a narrow perspective of the normative condition may be pursued.

Establishing a focus for policy-making or program decisions can be a strategic consideration. For some actors, their best interests are seen to be served by narrow problem or issue definition which limit the scope of ensuing information and decision fields. For example, emphasis on a specific project may deflect attention from the broader issues in which a project is embedded, preclude participation of certain actors and obscure alternative proposals for goal attainment.

On the other hand, broadly defined problems tend to be complex, often amorphous, making delineation of their dimensions and relationships a monumental task. In such situations policy or program development may become mixed in a tangled mix of interests and priorities the result of which is often inaction. Of course, such is the strategic intent of some actors in some situations.

Strategists are also sensitive to the locus of decision-making or control, the degree of access to such points and the influence they can exert over them. To the extent that an actor is unable to identify the appropriate "arena" then strategies may not yield the desired results. This becomes particularly important when complex problems are at issue and rather than there being a single decision-making point, a cluster of actors must make one or several decisions. In the context of arctic petroleum development actors, both pro and con, the developments are faced with several complex regulatory processes involving many other actors and decision points. The complexity of the bureaucracy has clearly been difficult for a number of actors to comprehend, with the result that their efforts have yielded few desired results.

An analysis of the nature and development of relationships among actors is very much at the heart of a great deal of "strategizing." Two general strategies are identified; 1) consent building and, 2) conflict containment, both of which represent interactive modes of planning. (9)

Consent among actors may be achieved by command, bargaining or persuasion. Command-based consent usually has its basis in a coercive, end-means control relationship between actors. Legislative action is one example of consent-building through command.

Bargaining and persuasion are common strategies in consent-building operations. Where the effective power base of an actor is insufficient to achieve the desired kind and level of influence attempts may be made to broaden the support base by forming coalitions. Coalitions may form to achieve desired levels of financial, informational, technological and management resources. In so doing, any number of bargains and tradeoffs may be mutually arrived at. In the case of large scale technological developments or broad policy issues the formation of several coalitions or consortis may be observed resulting in "giant armies against each other in a struggle for the majority." (10)

In addition to coalitions in which all members are more or less active we may find alignments whose basis is "mutual non-interference." In such cases agreement is reached among some number of actors who do not share a common interest but agree not to jeopardize the actions and purposes of others.


Decision Framework

Technology assessments are forms of analysis, the purpose of which is to aid decision-makers. In the case of multi-party and multi-dimensional assessment systems one can expect many decisions from many sectors. The focus of analysis of assessment system decisions is on identification of the directions and pace of the technological program as evidenced by the kinds of decisions, their importance to the program and when they occurred.

Various approaches to analysis of decision-making are reviewed by Gibbons and Voyer and Doern and Aucoin. (11) Rational and disjointed incrementalism models are discussed, particularly in terms of their limitations with respect to empirical observation of policy decision processes. Neither model is especially suited to explaining decisions taken under conditions of incomplete information or decisions which represent major changes in policy or program.

To compensate for these limitations the above-mentioned analysts utilized the "mixed scanning model" advocated by Etzioni which distinguishes between "fundamental" and "incremental" decisions. (12)

Fundamental decisions are characterized by major in shifts in policy relative to an actor's primary goals or major investments in projects or programs which leads to substantially increased activity on the part of various actors in an assessment system. It may also be useful to characterize fundamental decisions as general or somewhat abstracted decisions (rather than particularistic or detailed) which affect a relatively large organizational space over a considerable period of time.

Incremental decisions are those of a more specific and limited nature which are both antecedent and consequent to fundamental decisions. As actors "scan" their environments for opportunities and maintain on-going programs, decisions of an incremental nature may be made. Subsequent to a fundamental decision a host of new or different actions may be required which derive logically from the fundamental decision. These too, may be characterized as incremental decisions.

By distinguishing fundamental from incremental decisions, it is possible to account for major changes in policies and programs, which the disjointed incremental model is not suited to do. Moreover the distinction also provides a basis for explaining how actors, having scanned their environment for opportunities and allocated certain resources to do so, make fundamental decisions without complete information.

In the context of the northern petroleum development studies the framework evolved attempts to cast decisions in an actor-time matrix (see Figure 1). In this fashion it is possible to display decision chronologies for actor groups. Cell entries would consist of the decisions taken by actors in a given time frame. Inspection of the columns would then reveal the succession of decisions over time of the various actors.

The matrix format for decisions also permits one to compare and contrast decision profiles of various actors. For example, it may be instructive to identify enabling decisions and thus ascertain "initiating" and "reactive" roles. Moreover, the matrix would allow visual inspection of changing role patterns. Again the initiator-reactor patterns would be examined to identify when the "lead is changing hands." Reasons for these patterns may be associated with the characteristics of the technology, the nature of regulatory functions or important exogenous factors.

Inspection of the matrix can also provide evidence of the presence of fundamental decisions. If one accepts the premise that fundamental decisions bring about substantially increased levels of actor activity, of which decisions are a part, then the clustering of decisions within a particular time frame may suggest that a fundamental decision has occurred.

Any actor can use any of these possible operations carried out in combination to achieve whatever aim is desired. The strategy set employed is a function of that actor's perceptions of the other actors in the energy assessment system and their strategies as well as the perceived issues pertinent to those actors. Thus, the strategies followed by each of the actors form and reform the issue in the energy system. As the issues shift changes in both actors and strategies will change. Such shifts, of course, can occur within any one phase of the petroleum program as well as throughout the entire petroleum development program. Thus, an entire circuit can be followed throughout the petroleum technology assessment system which begins with an actor-issue configuration and carries on through information, actor relations and interactions and decisions to the strategies which re-shape the multi-actor and multi-dimension matrix.

## NOTES

(1) Francois Hetman, Society and the Assessment of Technology, OECD, Paris, 1973.

(2) Ibid., and Raphael G. Kaspar (ed.), Technology Assessment: Understanding the Social Consequences of Technological Applications, N.Y., Praeger, 1972.

(3) Harvey Brooks, Technology Assessment as a Process, Int. Soc. Sci. J., Vol. XXV, No. 3, 1973, pp. 247-256; and Joseph F. Coates, Technology Assessment: The Benefits, the Costs, the Consequences, The Futurist, Vol. 5, No. 6, December 1973, pp. 225-231; and Martin V. Jones, The Methodology of Technology Assessment, The Futurist, Vol. 5, No. 6, December 1973, pp. 19-26; and I.L. Whitman, et. al., Final Report on Design of an Environmental Evaluation System, Columbus, Ohio, Battelle Memorial Institute, 1971.

(4) M. Gibbons and R. Voyer, A Technology Assessment System: A Case Study of East Coast Offshore Petroleum Exploration, Background Study No. 30, Science Council of Canada, Ottawa, March 1974, p. 14.

(5) Ibid., P. 14-15.

(6) Ibid., pp. 71-81.

(7) D.W. Fischer and G.S. Davies, "An Approach to Assessing Environmental Impacts," Journal of Environmental Managements, Vol. 1, July 1973, pp. 207-227.

(8) H.M. Ingram, Patterns of Politics in Water Resource Development: A Case Study of New Mexico's Role in the Colorado River Basin Bill, No.79, University of New Mexico, Alberquerque, December 1969, pp. 1-96; and E.W. Erickson and L. Waverman, The Energy Question: An International Failure of Policy, Vol. 1, The World, University of Toronto Press, Toronto, 1974, pp. vii-376.

(9) H.M. Ingram, op. cit.

(10) Ibid.

(11) Ibid., and B. Doern and P. Aucoin, (eds.), The Structure of Policy-Making in Canada, Toronto, Macmillan, 1971.

(12) A. Etzioni, The Active Society: A Theory of Societal and Political Processes, Free Press, 1968.

ASSESSMENT OF ALTERNATIVE ENERGY/ENVIRONMENT FUTURES: A COMPARATIVE
CASE STUDY OF WISCONSIN (U.S.A), THE GERMAN DEMOCRATIC REPUBLIC,
AND RHONE-ALPES (FRANCE)

W.K. Foell, F. Buehring, W. Buehring, R. Dennis,
K. Ito; R. Keeny and B. Lapillonne

IIASA, Austria, Osaka University, Osaka, Japan
Woodward-Clyde Consultants, San Francisco, Ca. USA

I. INTRODUCTION

A companion paper at this conference has described a conceptual
and methodological framework used at IIASA for long-term policy assessment of energy/environment systems (1). This current paper provides a complementary overview description of research results for
three comparative case studies conducted within that framework,
namely, the state of Wisconsin (USA), the German Democratic Republic
(GDR), and Rhone-Alpes (France).

Section II provides a brief description of some relevant energy-
related characteristics of the regions. Section III lays out the
main assumptions and constraints for the scenarios. Section IV presents some of the scenarios themselves. An application of multiattribute decision analysis, an approach to evaluation and communication of the energy/environment policies, is presented in Section
V. Conclusions derived from the case studies are given in Section
VI, and the final section provides some comments on future directions.

II. DESCRIPTION OF REGIONAL ENERGY/ENVIRONMENT SYSTEMS IN GDR,
RHONE-ALPES, AND WISCONSIN

Table 1 gives a comparison of the size, population, and population density in the regions. The contrast between the overall density of Wisconsin and the heavily populated GDR is striking. The
GDR at present has zero population growth, in contrast to continuing
though modest growth rates in Rhone-Alpes and Wisconsin (currently
approximately 1% and 0.8%, respectively). The contrasting popula-

tion dynamics had a strong influence on the scenarios written for the regions.

Both the GDR and Rhone-Alpes are more industrialized than Wisconsin and their infrastructures differ significantly. Table 2 gives a cross regional comparison of fractional industrial activity by sector. The greatest relative differences occur in the food and chemical sectors. Wisconsin relies heavily on the automobile; however time-series show that auto ownership in the GDR is increasing at an annual rate of 12% in comparison with a 4% growth in Wisconsin. Also striking is the heavy GDR reliance on mass transit. Table 3 gives a cross regional comparison of motor vehicles.

The comparison of primary energy use in Table 4 shows that, although the per capita energy use is greatest in Wisconsin, the density of use is by far the greatest in the GDR. The primary energy sources for the three regions differ significantly. The GDR relies heavily on coal (mainly domestic strip-mined lignite), whereas Rhone-Alpes is dependent on petroleum and hydropower (Figure 1). Wisconsin, although having no naturally occurring fuel resources, has a diverse supply mix comprised mainly of petroleum, natural gas, and coal; nuclear is providing a rapidly growing portion of its energy.

This brief description provides only a glimpse of the three energy systems, but gives an indication of the diversity of the three regions. The natural and environmental characteristics are discussed in some detail in a forthcoming publication (2).

## III.  CHARACTERISTICS OF THE SCENARIOS

As described in Reference (1), scenario building was employed as a device for analyzing alternative energy and environment policies and strategies in the regions. They were constructed for Wisconsin, Rhone-Alpes, and a composite region ("Bezirk X"), typical

|  | Population ($10^6$ people) | Area ($km^2$) | Density (people/$km^2$) |
|---|---|---|---|
| GDR | 17.0 | 108,178 | 157 |
| RHONE-ALPES | 4.7 | 43,634 | 108 |
| WISCONSIN | 4.5 | 145,370 | 31 |

Table 1.  Cross-Regional Comparison of Population and Area 1972.

| Industrial Activity | GDR % of Net Industrial Product | RHONE-ALPES % of Industrial Value Added | WISCONSIN % of Industrial Value Added |
|---|---|---|---|
| Food | 11.6 | 8.7 | 15.8 |
| Building Materials | 2.1 | 3.5 | 1.3 |
| Primary Materials | 4.7 | 5.8 | 5.6 |
| Machinery,(Mech. Elec., & Transp. Equipment) | 42.0 | 44.5 | 49.0 |
| Chemicals and Rubber | 17.0 | 14.7 | 6.0 |
| Light Industry | 22.6 | 22.8 | 22.3 |
| | 100.0 | 100.0 | 100.0 |

Table 2.  Cross-Regional Comparison of Fractional Industrial Activity by Sector (1972).

| | GDR Total Per Capita ($10^6$) | | RHONE-ALPES Total Per Capita ($10^6$) | | WISCONSIN Total Per Capita ($10^6$) | |
|---|---|---|---|---|---|---|
| Autos | 1.400 | 0.082 | 1.259 | 0.270 | 1.969 | 0.436 |
| Motorcycles | 1.373 | 0.081 | 0.502 | 0.106 | 0.070 | 0.015 |
| Buses | 0.018 | 0.001 | 0.007 | 0.001 | 0.010 | 0.002 |
| Trams & Trollies | 0.0048 | 0.00028 | 0.0003 | 0.00007 | -- | -- |
| Trucks | 0.256 | 0.015 | 0.328 | 0.069 | 0.376 | 0.083 |
| Tractors | 0.203 | 0.012 | 0.011 | 0.002 | 0.230 | 0.051 |

Table 3.  Cross-Regional Comparison of Motor Vehicles (1972).

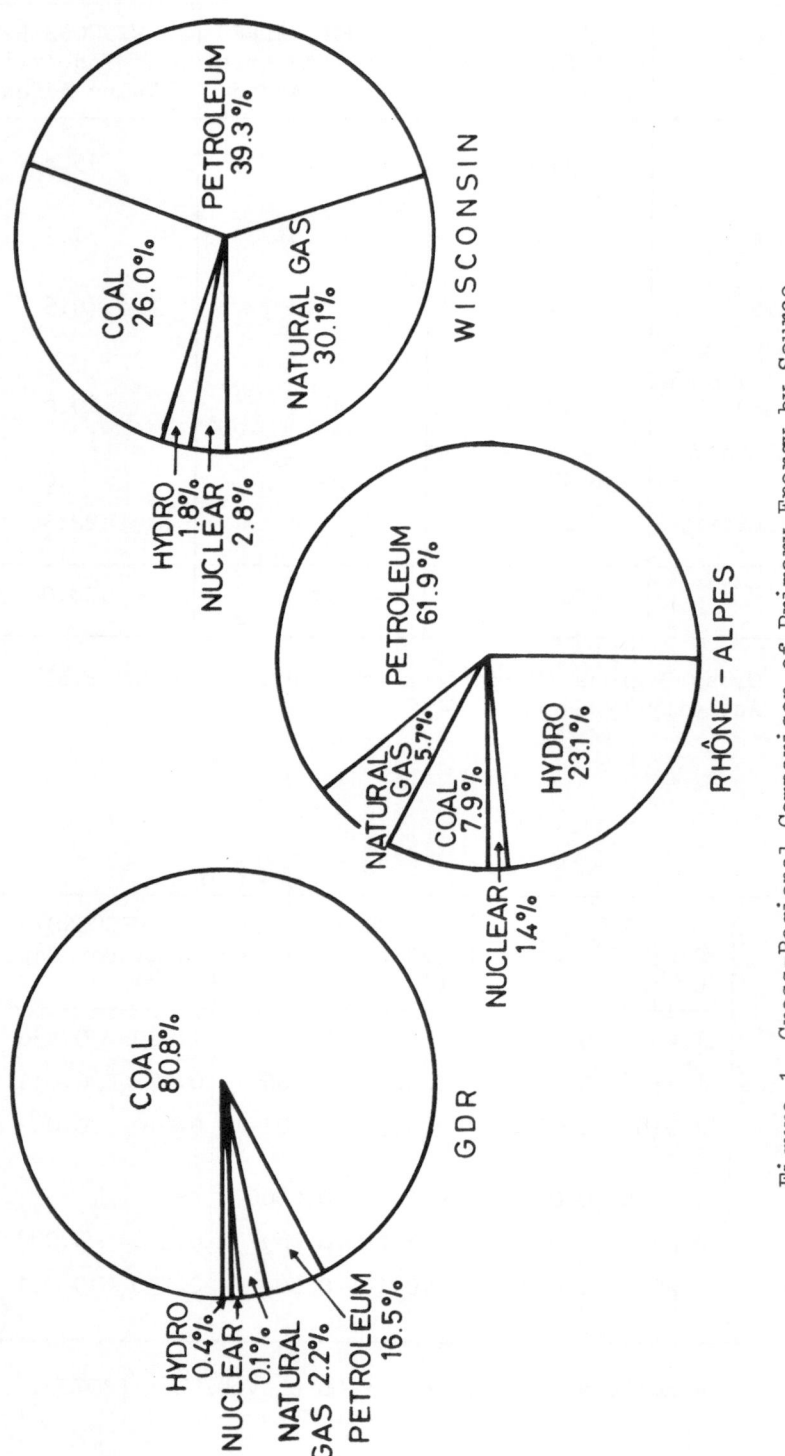

Figure 1. Cross-Regional Comparison of Primary Energy by Source
(1972)

|            | Annual Energy Use $(10^{15}cal/yr)$ | Annual Energy Use Per Capita $(10^{9}cal/p/yr)$ | Density of Annual Energy Use $(10^{12}cal/km^{2}/yr)$ |
|------------|------------|------------|------------|
| GDR        | 749        | 44.0       | 6.9        |
| RHONE-ALPES | 168       | 35.7       | 3.8        |
| WISCONSIN  | 319        | 70.9       | 2.2        |

Table 4.  Cross-Regional Comparison of Primary Energy Use
(1972-73 Data)

of the heavily industrialized southeastern GDR.  The resulting scenarios were not developed as predictions.  They are intended to help test and compare the consequences of different policy choices.  It is obvious that each of the regions has many alternate energy futures open to it; the scenarios chosen for the research highlight only three of them in order to improve our understanding of energy and environmental management.

It should be emphasized here that the assumptions underlying the scenarios were not chosen arbitrarily by the IIASA research team. They were developed after lengthy and repeated interactions with collaborating specialists in the respective regions; in some cases exogenous inputs to the scenarios were supplied directly by the collaborating institutions.  Whenever possible these were tested by reference to other economic or technical studies, e.g. the Ford Foundation Energy Policy Study (3), the GDR Long Term Energy Plan (4), and national energy assessments in France (5).  Where feasible, the Wisconsin Energy Model was used to construct sectoral energy demand descriptions, based upon data and parameters from the respective regions.  However, the 50 year time span of the scenarios clearly introduces major uncertainty into many of the underlying assumptions and parameters.

As pointed out earlier, regional rather than national scenarios were studied because of the former's value in addressing environmental issues -- often regional in nature.  However, because none of the regions are institutionally or economically autonomous, it is not possible to discuss regional scenarios while ignoring the evolution of the national systems.  The scenarios were based upon policies which could be implemented, albeit in varying degrees, both nationally and regionally; the resulting contrasting energy patterns and environmental consequences were then evaluated for the regions.

The policy areas explicitly addressed in the scenarios are:

- Urban Form
- Energy-use Technology
- Transportation

| | Scenario | | |
|---|---|---|---|
| | S1 | S2 | S3 |
| Urban Form | .Present patterns | .Same as S1 | .Same as S1 |
| Technical | .Large decrease in industrial energy intensiveness<br>.High penetration of district heating | .Smaller decrease in industrial energy intensiveness relative to S1<br>.Medium penetration of district heat | .Industrial energy intensiveness for electricity decreases faster than S1<br>.Very high penetration of district heat plus penetration of solar-thermal and solar-electric |
| Transportation | .Low automobile use<br>.Efficient freight mode | .Greater auto use than S1<br>.Much less **efficient freight** relative to S1 | .Lower auto use than S1 with more mass transit<br>.Freight same as S1 |
| Energy Supply | .Electricity almost all coal, some nuclear after 2000 | .Must import electricity due to increased direct demand for coal | .Less nuclear than S1 and solar-electric penetrates |
| Environmental | .Present trends in control of particulates and $SO_2$ | .Low control of particulates and no $SO_2$ control | .High control of particulates and $SO_2$ |

Table 5. Policies for Bezirk X (GDR) Scenarios, 1970-2025

- Energy Supply
- Environmental Control.

Contrasting policy alternatives were formulated in each of these areas. These alternatives formed the basis for the three scenarios, chosen to correspond to a consistent rationale.

The scenario characteristics are summarized by the following:

| | Scenario | | |
| --- | --- | --- | --- |
| | S1 | S2 | S3 |
| Urban Form | .Continuation of current patterns | .Dispersal<br>.Emphasis on single family homes (80% of post 1975 homes) | .Growth in small compact cities<br>.Multifamily buildings |
| Technical | .High level of fuel prices inducing:<br>- high penetration of electricity in industry<br>- High decrease in energy intensiveness<br>.Emphasis on elec. heating | .Low fuel prices smaller decrease in energy intensiveness<br>.Low penetration of electricity in industry | .Energy conservation<br>.Low penetration of electricity in industry<br>.High reduction of energy intensiveness |
| Transportation | .Predominance of auto<br>.Slight increase of intercity mass transit | .Same as S1 | .Increase of mass transit<br>.Efficiency gains for autos and trucks<br>.Some freight shift from trade to rail |
| Energy Supply | .Electricity generation from nuclear | .Mix of nuclear oil and coal for electricity generation | .Max. feasible hydro-electric generation<br>.Solar electric<br>.No new nuclear plants |
| Environmental | .Present trends of increasing controls | .Low controls | .Stringent controls |

Table 6.  Policies for Rhone-Alpes Scenarios, 1970-2025

|  | Scenario | | |
|---|---|---|---|
|  | S1 | S2 | S3 |
| Urban Form | .Suburban extension<br>.25% apartments | .Exurban dispersal<br>.50% apartments | .Small compact cities<br>.50% apartments |
| Technical | .Almost constant energy use per unit value-added in service and industry | .Increasing energy use per unit value-added<br>.Emphasis on electricity | .Declining energy use per unit value-added<br>.Conservation measures |
| Transportation | .Auto gain efficiency | .No auto efficiency gain | .Large auto efficiency gain |
| Energy Supply | .Synthetic fuel from coal<br>.Mix of coal and nuclear for electricity | .Synthetic fuel from coal<br>.Mostly nuclear for electricity | .Solar for electricity and heating<br>.No new nuclear<br>.Synthetic fuel from coal |
| Environmental | .Present trends of increasing controls for $SO_2$ and particulates | .Low controls of $SO_2$ and particulates | .Stringent controls of $SO_2$ and particulates |

Table 7.  Policies for Wisconsin Scenarios, 1970-2025

Scenario S1: The "Base Case"

- No dramatic changes in the current policies (e.g. in the GDR, it follows objectives of the 20 year plan); this therefore represents a continuation of current socio-economic trends.

Scenario S2:

- Results from combinations of policies favoring higher growth in energy use than in S1;
- Based on presumption of low or moderate energy costs and few or no special incentives for improved efficiency of energy use.

Scenario S3:

- Based on presumption of high cost of energy and on desirability of implementation of energy-saving measures;
- Emphasis is placed on conservation of energy resources,
  leading to development of solar energy.
-Pollution control is stressed.

Tables 5, 6 and 7 show the policy sets for the three scenarios.

The building of the scenarios within the policy frameworks of
Tables 5, 6 and 7 enabled us to systematically examine the energy
and environmental consequences.  In addition several "sensitivity
studies", in which only a single parameter is varied, were carried
out.

A most important similarity among the three scenarios for a
given region is that they are all based on essentially the same
overall economic and population growth rates.  Macro-economic studies of the Ford Foundation Energy Policy Project indicated that
the U.S., within the range of the above policy frameworks, the coupling between energy use and economic growth rate is loose, although
there can be significant sectoral interactions.  We have followed
this approach in our studies.  The population and overall economic
growth developments, as provided by the collaborating institutions,
are discussed in detail in Reference (2).


IV.  SELECTED SCENARIO RESULTS

Space limitations permit the presentation here of only a few
results of the scenarios.  Because of the focus of this conference,
energy use and environmental impact are emphasized.


A. Primary Energy Use

Cross-regional comparisons of primary energy use per capita,
excluding exports, for Scenarios 1 and 3 are shown in Figures 2 and
3.  As pointed out earlier, the total energy use shown there is built
up from sector and subsector components.  The Base Case (S1) in
Fig. 2 shows primary energy use in each region growing steadily, but
at rates under the historical trends of the past few decades.  The
per capita annual growth rates over the 50 year period in the three
regions are 2.4% (Bezirk X), 2.5% (Rhone-Alpes), and 2.3% (Wisconsin).
Caution must be exercised in drawing conclusions from these comparisons because of both the differing initial conditions and the differing assumptions related to the policy constraints of Tables 5,6,
and 7.  For example, in Wisconsin much of the growth in energy use
occurred in the service sector, whereas in Bezirk X, the growth was
dominated by the industrial sector.  These differences were discus-

sed in detail at a Workshop in November, 1975, and are described in
forthcoming publications (2). For comparison, the 1970 per capita
use in the U.S. (10.8Kw), the industrialized countries of the world
(4.4Kw), and the developing countries ( < 1Kw) are also shown in Fig.
2. The Scenario S3 depicted in Fig. 2, reveals much lower energy
growth rates, though the population and economic growth rates are
identical to S1. The per capita annual energy growth rates are
1.5% (Bezirk X), 1.2% (Rhone-Alpes), and 1.5% (Wisconsin).

As an example of the sectoral distribution of energy use, Fig.
4 shows the time dependent percentage of total end-use energy in
each demand sector for Wisconsin Scenario S1. End-use energy includes only energy consumed in end-use processes; conversion losses,
such as in electricity generation, are excluded. The service sector
increased its share of end-use energy from 13 to 31% over the 55
year period, while the residential sector's share dropped from 30
to 15 percent. Transportation maintains approximately the same
fraction of the total only because freight energy increases in relation to economic activity; personal transportation energy grows
at a much lower rate than freight energy in Scenario S1. The supply
scenarios corresponding to these demand figures are described in
detail in (2).

B. Environmental Impacts

The analysis of environmental consequences was one of the major
objectives of scenario-building. A wide range of environmental indicators is used to characterize the environmental implications.
Some environmental indicators used were associated with "quantified"
human health and safety impacts. Quantified here refers only to
those impacts explicitly included in the Envrionmental Impact Models,

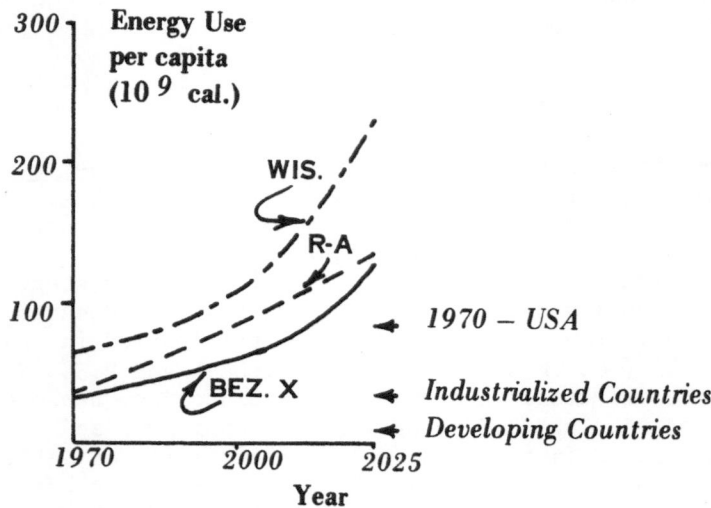

Figure 2. Cross-Regional Comparison of Primary Energy Use Per Capita
- Scenario 1.

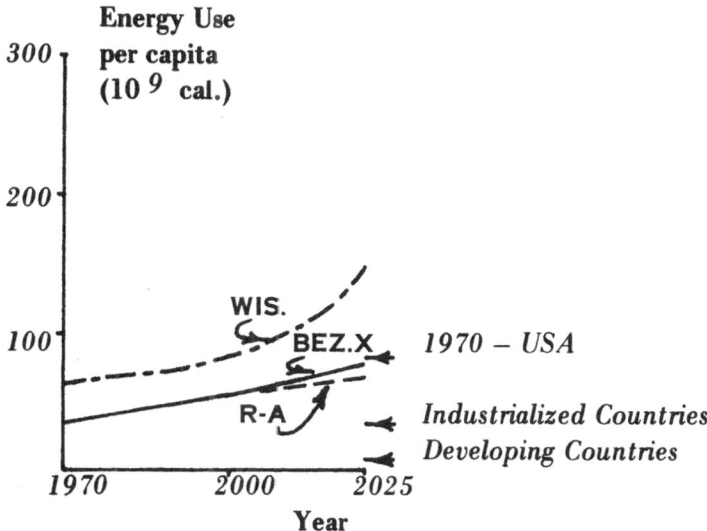

Figure 3.   Cross-Regional Comparison of Primary Energy Use Per Capita
                 - Scenario 3.

(e.g. 1,6, 7) used in this research.  The choice of this set of impacts is to some extent subjective; in addition, some degree of uncertainty (and perhaps controversy) is associated with some of the
impact factors.  There are also many indicators which are recognized
but remain unquantified; there are others that are unrecognized and
hence unquantified.  An approach to coping with this uncertainty and
subjectivity is described in Section V.  The impacts described below
are only a fraction of those studied with the methodology described
in Reference (1).  Emphasis is given here to a cross-regional comparison of the impacts.

    1. Carbon Dioxide Emissions.  Carbon dioxide emissions are of
concern on a global scale since the atmospheric concentration of
$CO_2$ affects average global temperature (8,9,10).  Burning of fossil
fuels has produced enormous quantities of $CO_2$, e.g. about 10,800
million metric tons in 1960 (8), for which there are three main reservoirs: the oceans, the biosphere (defined as the mass of living
and non-living organic matter), and the atmosphere.  About half of
the $CO_2$ liberated by fossil fuel combustion has remained in the at-
msophere (8,9).  Calculations have shown that $CO_2$ concentration in
the atmosphere may increase from about 320 parts per million (ppm)
by volume in 1970 to 370-380 ppm in 2000; the resulting global temperature increase may be nearly one degree Celsius.  Global temperature changes of this magnitude may have serious implications for
agriculture, global sea level, and global precipitation patterns.
Thus, $CO_2$ emissions that result from burning of fossil fuels may involve a significant long-term risk to future generations.

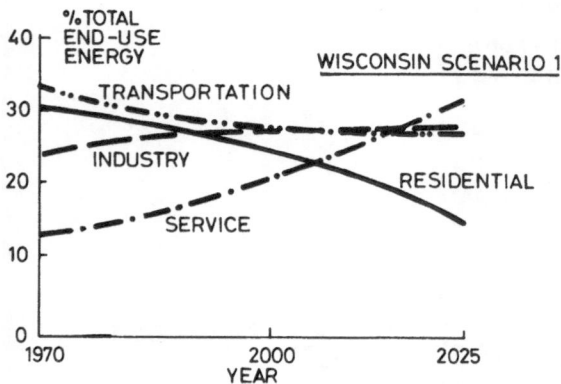

Figure 4.   Percentage of Total End-Use Energy by Sector for
            Wisconsin - Scenario S1


     The <u>per capita</u> $CO_2$ emissions in the year 2025 for the three
scenarios in each region are shown in Figure 5 along with the 1970
emission levels.  All three regions have greater $CO_2$ emissions in
2025 for all three scenarios than they had in 1970.  The total $CO_2$
emissions in Wisconsin in 2025 for Scenario S1 are more than a factor of five greater than the 1970 emissions (population increases
by about 50 percent over this period).  Bezirk X relies on coal for
a large fraction of its energy in all scenarios; the total emissions
for the high energy scenario (S2) in the Bezirk are nearly five
times greater than the 1970 emissions.  In Rhone-Alpes the availability of fossil fuels is more limited and a significant factor of
the energy comes from other sources, such as hydro or nuclear.

     The emissions resulting from Wisconsin's energy use in 2025 are
approximately four percent of the total emissions for the world in
1960.  If all regions of the world were to increase their $CO_2$ emissions as in these scenarios, methods for reducing $CO_2$ emissions to
the atmosphere, e.g. Marchetti's disposal method (11), or even $CO_2$
removal from the atmosphere, may be required, or the fossil fuel
option might be unacceptable.

     Other potential long-term problems, such as radioactive waste
production and nuclear safeguards, must also be considered in the
evaluation of alternative energy futures.  These types of impacts
are in contrast to the usual localized impacts; cooperation among
nations will be needed to avoid "the tragedy of the commons" (12)
on a global scale.

     2. $SO_2$ Emissions.  Sulfur dioxide ($SO_2$) is generally thought
to have unfavourable effects on human health, vegetation, and struc-

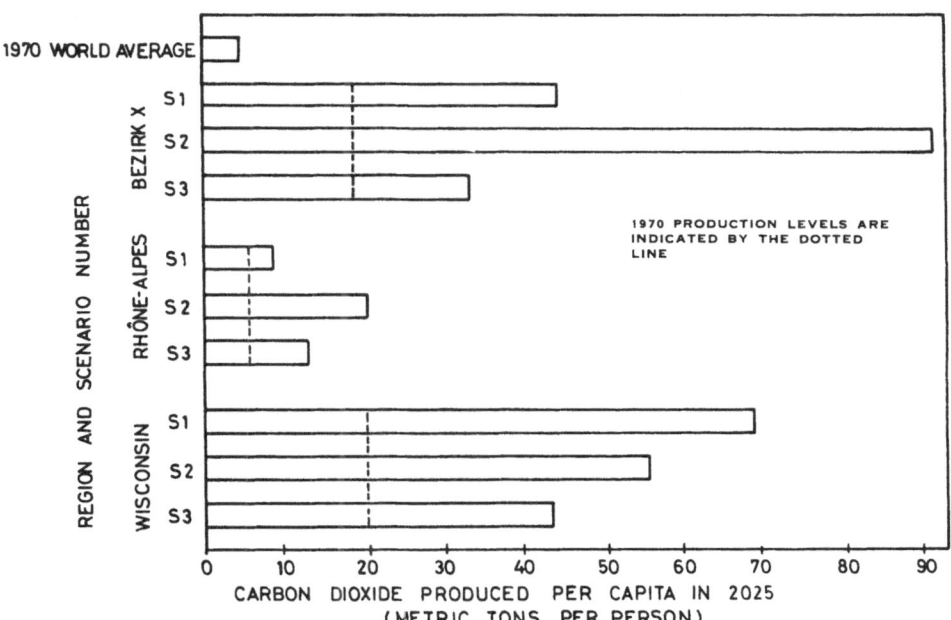

Figure 5. Carbon Dioxide Production Per Capita in 1970 and 2025.

tures.  The $SO_2$ emissions for Scenario S1 in the year 2025 are compared for the three regions in Figure 6.  Four different indices
for the emissions are shown: total absolute emissions, emissions
per unit area, emissions <u>per capita</u>, and emissions per unit of primary energy use.  Wisconsin has the largest total emissions, but
Bezirk X has the largest emission impact on the other three scales.
Bezirk X is a highly industrialized region that uses lignite for
most of its energy.

    <u>3. Human Health and Safety</u>. The total "quantified" human health
and safety impacts in the year 2025 for Scenario S1 are shown in
Figure 7, which is similar in format to the previous figure.  Person days lost (PDL) are used to combine the effects of mortality and
morbidity.  The quantified totals shown in the figure include health
and accidental impacts on the general public and those people employed throughout the energy system, from resource extraction to waste
disposal.  In contrast to $SO_2$ emissions, Bezirk X has the largest
impact only on the scale showing PDL per unit area. On the other
three scales Wisconsin has the greatest impact.  Quantified impacts
of air pollution are a major share of the total PDL for Wisconsin.

    To provide some perspective on these numbers, let us compare
the PDL <u>per capita</u> in Figure 7 with the PDL <u>per capita</u> that result
from all accidental fatalities in the U.S.  The risk of fatality
from all accidents (autos, falls, burns, drowning, firearms, etc.)

was 49 per 100,000 in the U.S. in 1974 (Source: <u>1974 U.S. Statistical Abstract,</u> Bureau of Census). This is equivalent to 2.9 PDL per person per year.* The quantified PDL <u>per capita</u> in 2025 that result from Wisconsin energy use are about seven percent of the PDL <u>per capita</u> from all fatal accidents at current incidence levels.

The quantified health and safety impacts are not spread evenly over all population groups. For example, nearly 30 percent of the quantified PDL in 2025 for Wisconsin Scenario S1 are imposed on less than one percent of the total population, namely 53,000 elderly people who live in the industrialized Milwaukee area and have preexisting heart or lung disease.

The total PDL per 1000 population in the region are displayed in Figure 8 for the years 1970 and 2025 for all scenarios. It is interesting to note that there is at least one scenario in each region that has fewer PDL <u>per capita</u> in 2025 than occurred in 1970. Also, it is clear from the figure that a significant fraction of the total PDL are imposed in regions other than where the energy is consumed. Impacts in other regions result from consuming fuels that must be mined elsewhere and transported into the region, or from the expected global health effects from radioactive releases, etc. The quantified health impacts of air pollution are an important consideration in all scenarios considered.

It may be desirable to consider occupational health and safety impacts separately from the public impacts because occupational risks are presumably taken voluntarily, while the public is exposed to risks involuntarily. The fraction of the total quantified PDL that is occupational is shown in Figure 9 as a function of time for the Wisconsin scenarios. The occupational impacts include activities throughout the fuel supply system, from resource extraction through waste disposal. A particularly hazardous occupation per unit of energy produced is underground coal mining, which is assumed to supply between 25 and 50 percent of the coal for the Wisconsin scenarios. One reason that the occupational fraction increases for S3 but not for the other two scenarios is that relatively low levels of air pollution impact on the public occur in S3, thereby reducing the major contributor to total PDL from the other two scenarios. The total PDL in S3 for 2025 are more than a factor of three less than the total PDL in S1 for that year; in fact even the occupational PDL in S1 are considerably greater than in S3.

4. Land Use. The low energy scenario (S3) for Wisconsin has greater impacts in some categories than the other scenarios. For example, land disturbed per unit of primary energy is considerably

---

* One accidental fatality is equivalent to 6,000 PDL in this accounting system (13).

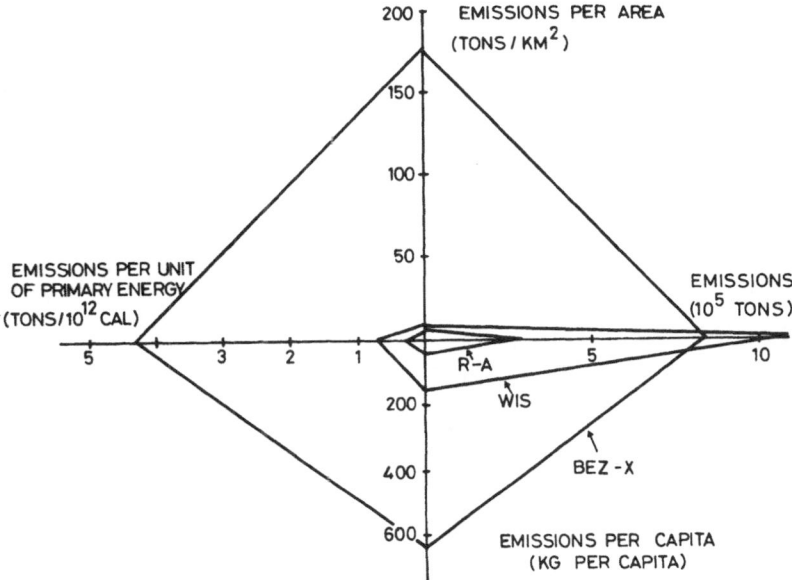

Figure 6.   $SO_2$ Emissions in Year 2025 for Scenario 1.

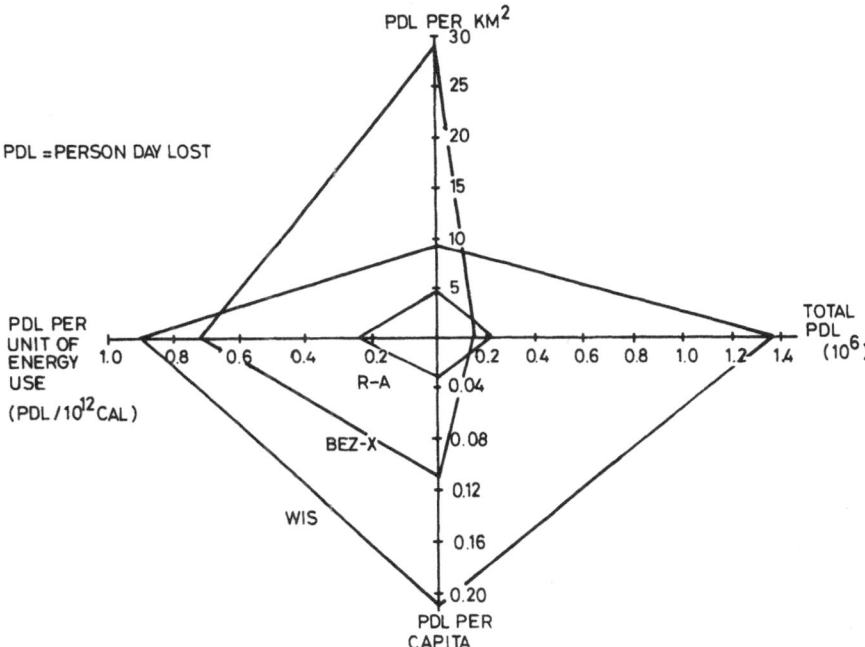

Figure 7. "Quantified" Human Health and Safety Impacts; Year 2025
fcr Scenario 1.

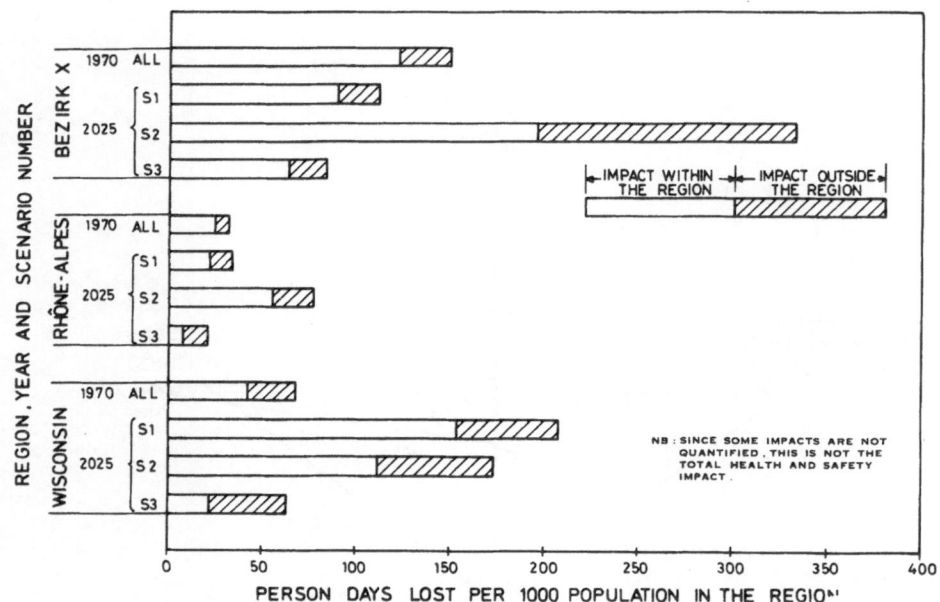

Figure 8.   Cross-Regional Comparison of Quantified Human Health
Impacts.

greater in S3 because of the solar electrical generation plants.
The land use shown in Figure 10 is the sum of all land disturbed,
except for electricity transmission, in order to produce the energy
in the year shown; this includes land used for sites for power plants
and fuel system facilities, for mining, and for storage of various
waste products.  The land use in 2025 for Scenarios S2 and S3 is
about 0.3 percent of the area of Wisconsin; however, not all land
use shown here is in Wisconsin.  The solar land use by 2025 in S3
is 240 km$^2$; solar electrical generation accounts for 30 percent of
the total in that year.  The annual land use in S2 is slightly greater than for S3; however, the primary energy requirement is approximately twice as large for S2 as for S3.

5. Heat Discharge from Power Plants.  Waste heat from electrical
generation presents a problem that varies in magnitude among the
three regions.  Three categories of cooling options have been considered for power plants: once-through cooling on rivers; evaporative cooling, such as provided by wet cooling towers; and dry (non-
evaporative) cooling towers.  A high control and a low control case
for each region were based on different allowable temperature increases for rivers.  Since Rhone-Alpes and Wisconsin have sizeable
water resources, once-through cooling and wet cooling towers are
the only options considered.  For Bezirk X the water resources are
very limited, and once-through cooling is not an option.  Therefore,
as indicated in Figure 11, the high control case for Bezirk X has

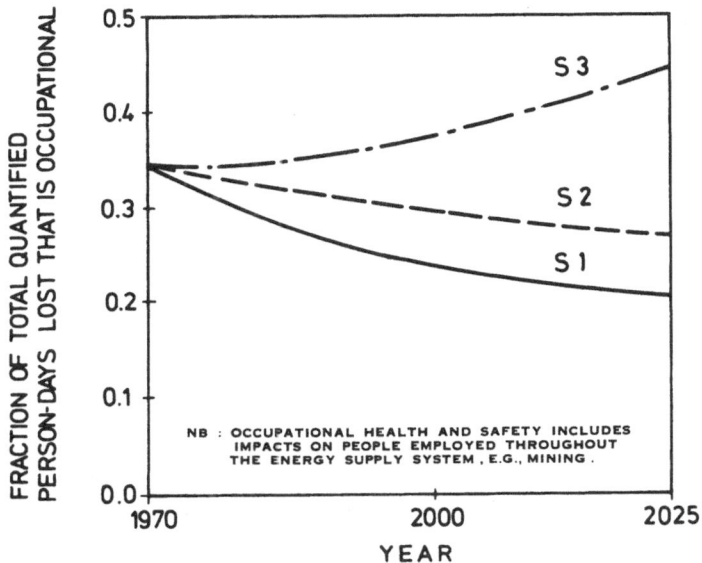

Figure 9.   Occupational Fraction of Total Quantified Person-Days
Lost for Wisconsin Scenarios

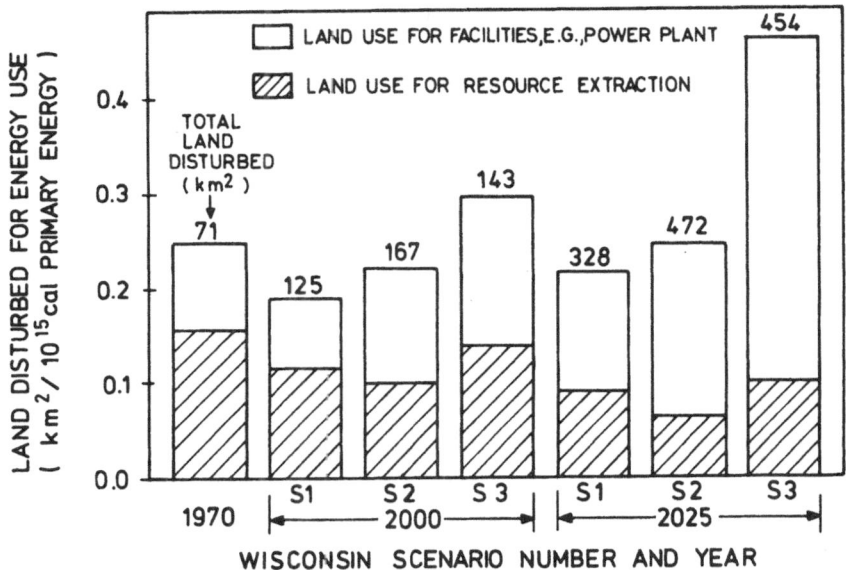

Figure 10. Land Disturbed Per Unit of Primary Energy for the Wisconsin Scenarios

less evaporation (more dry cooling), while the high control case for Rhone-Alpes and Wisconsin means more evaporation (less once-through cooling). If strict environmental standards are imposed, dry cooling may also eventually be required in Rhone-Alpes and Wisconsin. Current designs for dry cooling systems are avoided because of large efficiency penalties and costs.


## V. APPLICATION OF DECISION ANALYSIS TO ENERGY/ENVIRONMENT POLICY ASSESSMENT

The previous section pointed out the difficulty of coping with uncertainty and subjectivity in the assessment of energy-related environmental impact. The companion paper (1) described a framework for the application of Multiattribute Decision Analysis to this problem; it also argued that decision analysis was potentially a valuable tool for embedding energy/environment models and their outputs into the decision and policy making apparatus of a region. This section summarizes an initial application of the tool to the IIASA research program.

This application was based upon a set of policies related to the choice of electricity generation systems for Wisconsin. Because of space limitations, only a highly simplified version is presented here as an example. The electricity Impact Model was used to generate the following four attributes (shown in Table 8) of a set of scenarios based upon alternative policies. The ranges are representative of the cumulative impacts and electrical generation that may

| Attributes | Units | Range |
|---|---|---|
| $X_1$ = Total Quantified Fatalaties | Deaths | 100 - 700 |
| $X_2$ = $SO_2$ Pollution | $10^6$ Tons | 5 - 8 |
| $X_3$ = Radioactive Waste | Metric Tons | 0 - 200 |
| $X_4$ = Electricity Generated | $10^{12}$ kWh | 0.5 - 3.0 |

Table 8. Attributes and Ranges Used for Utility Measurements.

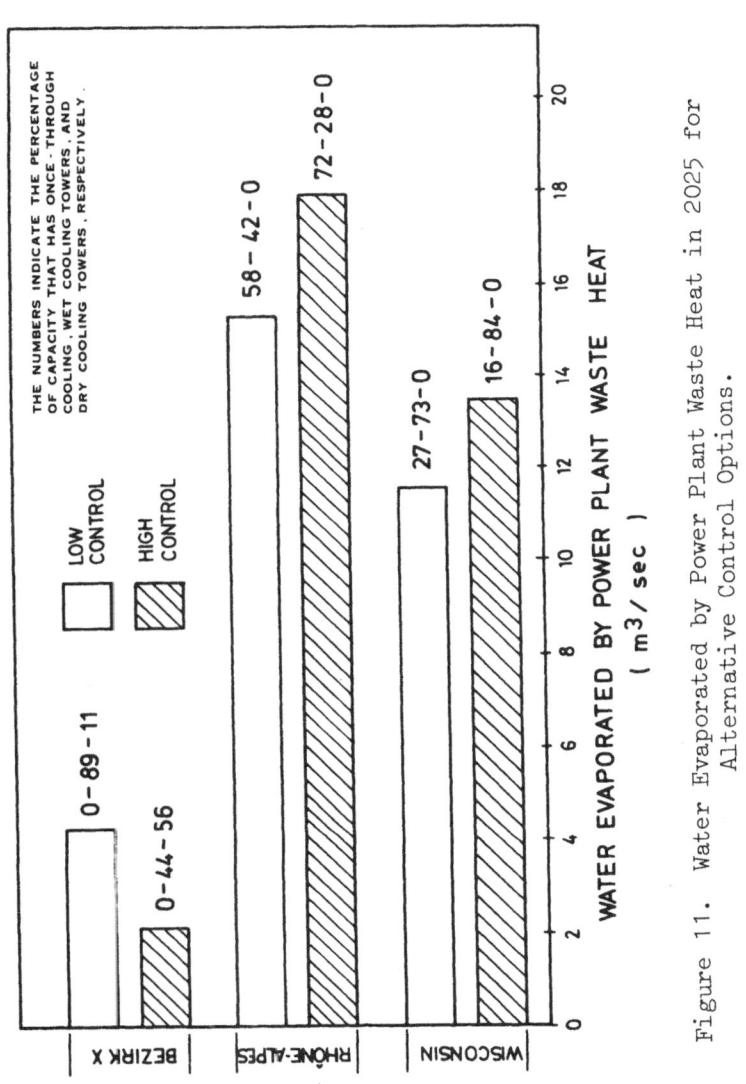

Figure 11.  Water Evaporated by Power Plant Waste Heat in 2025 for
Alternative Control Options.

occur for a variety of Wisconsin scenarios over the period 1970
through 2000. The set of scenarios is not identical but similar to
the set S1, S2, and S3. Preliminary utility assessments were completed for five individuals from Rhone-Alpes, the GDR, and Wisconsin. The group included a mixture of decision makers and energy/
environment specialists; the non-Wisconsin individuals were famili-
arized with current trends in Wisconsin electricity use so that they
would understand the ranges of the attributes. Details of the assessments formed the basis of a preference model for each of the individuals (see Fig. 5 in Ref. 1).

A utility function $u_i$ over attribute $X_i$ is set equal to zero
at the least desirable level of $X_i$ in the range and set equal to one
of the most desirable level of $X_i$ in the range. The results for one
of the assessed individuals are shown in Fig. 12. The assessments
also provided scaling constants for each of the attributes shown in
Figure 12; comparison of these scaling constants for an individual
indicates the relative importance of each of the attributes for the
specified ranges. Total quantified fatalities had either the
largest or second largest scaling constant for all five individuals.

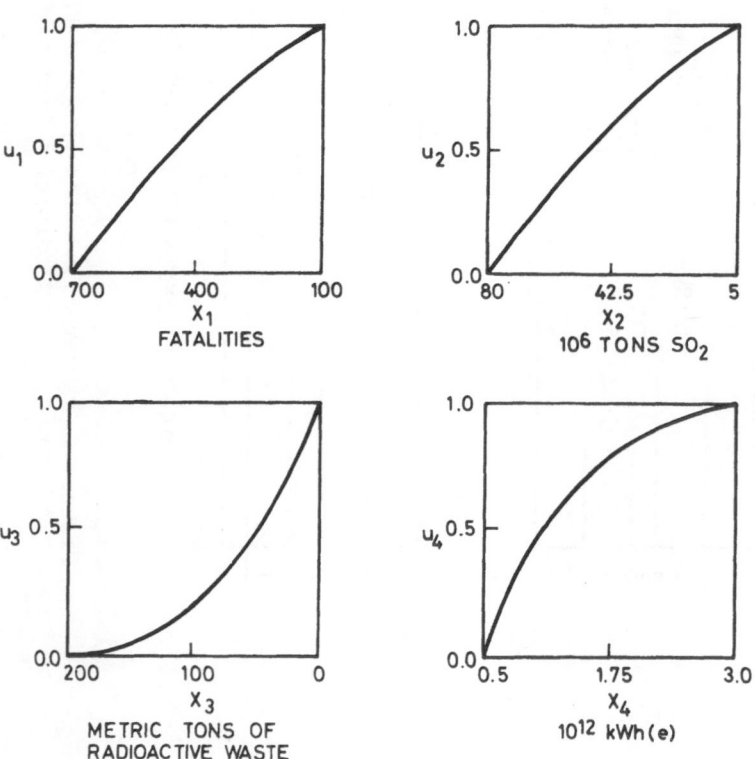

Figure 12. Utility Functions for One Individual.

| Attributes and Expected Utilities | | Reference Case: Attributes at Extreme Le- | Policy 1: Mostly Coal, Good Pollution Control | Policy 2: Mostly Nuclear | Policy 3: Low Sulfur Coal from Distant Mines & Some Nuclear | Policy 4: Mostly Coal with less Electricity |
|---|---|---|---|---|---|---|
| Total quantified fatalities | | 100 | 380 | 240 | 680 | 280 |
| $SO_2$ pollution ($10^6$ tons) | | 5.0 | 12 | 8.0 | 8.6 | 9.5 |
| Radioactive waste (metric tons) | | 0.0 | 61 | 160 | 110 | 54 |
| Electricity generated ($10^{12}$kWh(e)) | | 3.0 | 1.7 | 1.7 | 1.7 | 1.3 |
| Expected utility for individual | A | 1.00 | 0.53 | 0.66 | 0.14 | 0.65 |
| | B | 1.00 | 0.56 | 0.63 | 0.14 | 0.65 |
| | C | 1.00 | 0.76 | 0.83 | 0.64 | 0.41 |
| | D | 0.92 | 0.62 | 0.66 | 0.24 | 0.73 |
| | E | 1.00 | 0.65 | 0.72 | 0.31 | 0.74 |

Table 9.  Expected Utilities for Five Individuals for Several
Policies

The functions $u_i(x_i)$ and the scaling factors specify completely
the multiattribute utility function $u(x_1,x_2,x_3,x_4)$.  These five preliminary utility functions were used to evaluate expected utilities
associated with the set of alternative policies for electrical generation in Wisconsin.  The levels of the four attributes and the expected utilities for each of the individuals are listed in Table 9.
If it is assumed that the individuals expressed their true preferences
and that they act in a logically consistent manner, the expected
utilities can be used to indicate their overall preferences.  Under
these conditions Table 9 shows that all five individuals should prefer one or more of the other policies to policy 3.  This is primarily the result of the large number of fatalities expected for policy

3 and the relatively high scaling factor each of the individuals placed on fatalities.

In applying this technique to an actual policy study, the attribute list must be expanded to include other impacts and conventional costs (eleven impacts have been used in another Wisconsin study (16) ).  Clearly, considerations other than environment play a major role in the policy design.  Our initial efforts with this approach have convinced us that the process of assessing the utility function has many benefits in itself.  It can be a substantial aid in identifying and sensitizing individuals to important issues, generating and evaluating alternatives, isolating and resolving conflicts of judgement and preference among members of the decision making team, communicating among the several decision makers, and identifying improvements needed in the impact model.  Because of these and other benefits experienced during the initial use of this approach we are incorporating it into our future energy/environment research both at IIASA and at the University of Wisconsin.

## VI.   CONCLUSIONS AND SUMMARY

This paper and its companion report (1) have presented an overview of the framework and initial results of a research program on long-term policy assessment of regional energy/environment systems. Some major results and conclusions of the study to date are:

1.   A quantitative description and comparison of energy/environment systems has been developed for the three regions; the comparative descriptions have provided insight into the relationships between energy and the regional socio-economic patterns.

2.   Alternative energy/environment scenarios were written for the three regions as a vehicle for analysis of selected long-term policy issues; indications are that these scenarios are playing a role in energy/environment planning in the regions.

3.   A significant socio-technical interaction of specialists and decision makers took place at an IIASA workshop in 1975 during which the energy/environment scenarios were discussed and analyzed.

4.   A set of energy/environment models were tested for their relevance and validity by application to the greatly differing regions.

5.   A decision analysis approach was developed and applied to energy/environment policy analysis.

6.   A significant transfer of models and analytic methodology occurred among the collaborating institutions in the three region study.

## VII. FUTURE WORK

One of the most important outputs of the 1975 research has been creation of a network of research institutions coordinated by IIASA. This has provided IIASA with encouragement in its role as a catalyst and coordinator of policy-oriented research in the international scientific community. The three collaborating institutions will continue to pursue research during 1976, and in addition IIASA will extend the studies to other regions, again with very different socio-economic, geographic, and institutional characteristics. Specifically, one of them will be located in a less industrialized country to allow the IIASA team to further generalize their models and methodologies. Although we realize that there can never be a universal energy/environment model, our long-range goal is generalization of the approaches into a coherent and sound process for resource management in all regions of the world.

## ACKNOWLEDGEMENTS

The authors wish to acknowledge the research contributions of the following individuals:

IIASA: J. Bigelow, J.-P. Charpentier, A. Hoelzl, H. Stehfest, J. Weingart and R. Yorque.

Institut für Energetik, Leipzig, GDR: W. Haetscher, P. Hedrich, W. Kluge, D. Ufer.

Institut Economique et Juridique de l'Energie, Grenoble, France: B. Chateau, D. Finon, and J.-M. Martin.

University of Wisconsin-Madison: M. Hanson, J. Mitchell and J. Pappas.

## REFERENCES

1) Foell, W.K., "Long-Term Policy Assessment of Energy/Environment Systems: A Conceptual and Methodological Framework", A paper presented at NATO Conference on Environmental Assessment of Socio-Economic Systems, Istanbul, Turkey, October 4 - 8, 1976.

2) Proceedings of a Workshop on Integrated Management of Regional Energy/Environment Systems held November 10-14, 1975. International Institute for Applied Systems Analysis, Laxenburg, Austria, forthcoming.

3) Energy Policy Project of the Ford Foundation, A Time To Choose: America's Energy Future, Ballinger, Cambridge, Mass. 1974.

4) Personal Communication, Institute für Energetik, Leipzig, German Democratic Republic, 1975.

5)   Chateau, B., and B. Lapillonne, "Prévisions à Long Terme de la
      Consommation d'Energie: pour une Nouvelle Approche Méthodo-
      logique", Synthetical report 74p, IEJE, Grenoble, 1976.

6)   Buehring, W.A., and W.K. Foell, Environmental Impact of Electri-
      cal Generation: A Systemwide Approach, RR-76-13, Interna-
      tional Institute for Applied Systems Analysis, Laxenburg,
      Austria, 1976.

7)   Dennis, R.L., Regional Air Pollution Impact: A Dispersion Metho-
      dology Developed and Applied to Energy Systems, RM-76-22,
      International Institute for Applied Systems Analysis,
      Laxenburg, Austria, 1976.

8)   Matthews, W.H., W.W.Kellogg, and G.D.Robinson, eds., Man's Im-
      pact on the Climate, MIT Press, Cambridge, Mass, 1971.

9)   Broecker, W.S., "Climate Change: Are We on the Brink of a
      Global Warming?" Science, 189 (1975), pp. 460-463,

10)  Niehaus, Friedrich, A Nonlinear Eight Level Tandem Model to
      Calculate the Future $CO_2$ and C-14 Burden to the Atmosphere
      RM-76-35, International Institute for Applied Systems Ana-
      lysis, Laxenburg, Austria, 1976.

11)  Marchetti, Cesare, On Geoengineering and the $CO_2$ Problem,
      RM-76-17, International Institute for Applied Systems Ana-
      lysis, Laxenburg, Austria, 1976.

12)  Hardin, Garrett, "The Tragedy of the Commons", Science, Vol.
      162, pp. 1243-1248.

13)  American National Standards Institute, "The American National
      Standard Method of Recording and Measuring Work Injury
      Experience", ANSI Z16.1 - 1967.

14)  Keeney, R.L., Energy Policy and Value Tradeoffs, RM-75-76,
      International Institute for Applied Systems Analysis,
      Laxenburg, Austria, 1976.

15)  Buehring, W.A., W.K. Foell and R.L.Keeney, Energy/Environment
      Management: Application of Decision Analysis, RR-76-14,
      International Institute for Applied Systems Analysis,
      Laxenburg, Austria, 1976.

16)  Buehring, W.A., "A Model of Environmental Impacts from Electri-
      cal Generation in Wisconsin." Unpublished PhD dissertation,
      Department of Nuclear Engineering and Institute of Environ-
      mental Studies, University of Wisconsin, Madison, 1975.

Session V

CASE STUDIES: ASSESSMENT OF

SOCIAL SYSTEMS

# THE DISTRIBUTION OF ENVIRONMENTAL QUALITY: SOME CANADIAN EVIDENCE

Brian A. Emmett, Linda E. Perron, Paolo F. Ricci*

Office of the Science Advisor, Environment Canada

Ottawa, Canada

## INTRODUCTION

This paper represents an attempt to define the natural environment in a way which is operational in economic terms and then examine the influence the use of the natural environment has on socio-economic disparities within society. Its purpose is twofold. First it contains a brief outline of the approach taken and the evidence assembled at Environment Canada on the distribution of environmental costs and benefits among individuals and households in society, some of which have previously been reported elsewhere.[12] Second, and in more general terms, it discusses the importance of distributional considerations in understanding the behavioural nature of environmental problems, and how such considerations are a concern in implementing effective environmental management programs.

In section 1, the problems of evaluation of environmental costs and benefits and their relationship to distributional considerations are discussed. A definition of the environment and the services it provides is introduced as a basis for the following empirical and theoretical discussion. In section 2, consumption patterns for commodities which are relatively heavy users of indirect environmental services are examined to capture the distribution of environmental quality, appropriated as a factor of production among income groups. In section 3, the direct consumption of environmental resources by individuals is investigated. In view of the theory and evidence

---

* Current address: Department of Geography and Regional Planning, University of Ottawa, Ontario, Canada.

discussed in sections 2 and 3, the probable distributional impacts
of hypothetical environmental managemnet programs are outlined in
section 4.

Finally, section 5 discusses the role distributional considerations might play in contributing to effective environmental regulation and management and makes suggestions regarding possible directions for future research.

## 1. DEFINITION OF THE ENVIRONMENT

For the purposes of this study, the term environment is defined
in a way which highlights the comparability of services provided to
man by the environment to services provided by other natural resources and economic goods. The natural environment may be considered
as an asset or non-reproducible capital good which provides a variety of services to productive sectors and to households which are
analogous to those provided by commodities which are exchanged in
the marketplace, bear positive prices and are more traditionally
thought of as economic goods. These environmental services "are all
economic goods in the sense that people are willing to pay to receive
more or to avoid a reduction in the quantity or quality of a service.
From this perspective, environmental quality means the level and composition of the stream of environmental services that people receive,
and in principle at least, the ultimate measure of quality is the
value that people place on it."(6)

Services of the natural environment are consumed directly by
individuals or as intermediate inputs by firms. Direct environmental services are analogous to final products in the economic sense.
They contribute directly to the well-being of individuals. Such
services include recreation, amenity (that is, the entire range of
rather ill-defined aesthetic services provided by the environment)
and, importantly, life support. Intermediate environmental services
include the use of air, water and land by firms as a factor of production for the removal, dispersion, storage and degradation of waste
or thermal effluent.

If the natural environment provides services like other resources, it follows that the use of the natural environment should ideally be valued and account kept of changes in environmental quality in
assessing the performance of an economic system in contributing to
the well-being of individuals.

Such values are difficult to establish, however, because of the
common property nature of environmental resources. Common property
resources are not subject to the usual market mechanisms for alloca-
ting resources into their most productive uses when buyers and sel-
lers exchange privately owned resources. Values cannot be establi-

shed because the environment cannot be owned, bought and sold to establish its value relative to other goods and services. The common property resources must come under some form of collective management to avoid overuse and congestion.

Environmental management is an attempt to reconcile various competing uses of environmental resources. It is necessary, therefore, to determine the value of alternative uses of the natural environment in order to allocate environmental resources in such a way as to maximize the net benefit to society arising from their use. Consequently, a large body of literature has grown up around the complex problems of valuing environmental resources, particularly the difficulties of establishing values in monetary terms which can be directly compared to marketed commodities.[5] [8] [10]

Less attention has been paid to determining the distribution of valuable environmental services although it would be desirable to determine the value of environmental services received by individuals in order to arrive at a clearer understanding of the relative well-being of individuals in society. Values in dollar terms would be ideal for comparison with standard measures of well-being such as money income or wealth.

Valuation and distribution are closely related problems:

"While it is true that the study of the welfare effects of the distribution of the environment must logically await the successful valuation of environmental services, paradoxically it is also true that valuation cannot proceed without first settling two very important distribution issues. These are the assignment of property rights to the common property resource services of the environment and the question of the acceptability of the existing distribution of income and the market prices associated with it.[6]

A discussion of these fundamental problems is beyond the scope of this paper. Further detail and references are available in Fisher and Peterson.[8]

Because of the theoretical and practical difficulties of establishing the value of the various services of the environment, recourse is made here to approximate measures of use for both direct and indirect service flows arising from the use of the natural environment. Consequently, nothing can be said about the value of the environment in alternative uses nor about the ultime impact of the distribution of environmental quality on socio-economic disparities. Nevertheless, the measures provide an indication, admittedly rough, of the likely direction of influence of the distribution of environmental quality.

## 2. DISTRIBUTION OF INDIRECT ENVIRONMENTAL SERVICES

The definition of the natural environment introduced in Section 1 highlights the comparability of the natural environment to other factors of production. Residuals production is regarded as the use of environmental resources as an input to the production process.[3]

Given that environmental services are analogous to other factor inputs, it follows that the demand for environmental services by firms is similarly derived from the underlying demand for the commodity which is produced. Input demand functions are derived from the firm's profit maximizing behaviour, and input demands are determined by the price of the factor in question, the price of other factors of production and the price of the product output.[7]

Unlike other factors of production, however, environmental resources can generally be appropriated by firms at zero cost because of their common property nature. This by no means implies that the contribution of environmental resources to production is zero, nor does it imply that the appropriation of environmental resources to the production process is without cost to society. Rather, it means that the use of the natural environment will not be accounted for in the firm's calculation of cost minimizing inputs and consequently, the use of scarce common property resources will not be reflected in the selling price of commodities.

The contribution of environmental resources to the production process is, therefore, significant but unpriced. Conceptually, the value of the environment as a factor of production could be determined by assuming that firms are denied the use of environmental resources entirely. For example, a regulatory body could establish a price for the use of common property environmental resource and could raise the price high enough so that firms are effectively denied its use (outright prohibition is equivalent). A reduction in the use of environmental resources to zero implies increased use of other factors of production - all of which bear positive prices. Cost curves are shifted in a way which resembles the imposition of an excise tax. The ultimate result will be an increase in the price of commodities to the consumer and a reduction in quantities demanded. [6]

This, of course, is a highly simplified treatment of a complex process. In reality, the exact impacts of such a regulatory measure are largely unknown. The effect on output, prices and employment for individual industries and the economy as a whole would depend upon the complex interaction of interindustry relationships,* elasticities of supply and demand for commodities, the external economic environment and so on.[4] Further, this treatment mentions only

---

*   see over

the loss in production caused by a hypothetical environmental regulation - it ignores the associated improvements in environmental
quality which have positive value.

Despite the simplifications, however, the point remains that
non-marketed environmental resources make significant contributions
to the production process which are generally unpriced, contributions
which ultimately benefit individual consumers who pay less than they
would if environmental resources bore appropriate prices.  The distribution of indirect benefits to consumers is determined by consumption patterns for goods using environmental services as inputs.
Expenditure patterns for commodities can therefore be regarded as a
rough proxy to capture the indirect consumption of common property
environmental resources.

Assuming that the residual load per dollar of expenditure on
a given type of commodity remains constant, unaccounted environmental
benefits will increase with increasing consumption of commodities
which impose high residual loads on the environment.  Consumption
patterns in turn are affected by the distribution of income and individual tastes and preferences.  Therefore, a priori expectations
are that the consumption of indirect environmental resources embodied in commodities will be positively associated with wealth or
money income.

Numerical results for Canadian data derived at by Environment
Canada tend to support these a priori expectations.  Engel curves,
which relate expenditures on different classes of commodities to income levels, were fitted to data on expenditure for commodities which
make relatively heavy use of indirect environmental services.[13]
The commodity classifications chosen were food, paper products, power
and fuel, gasoline for vehicle operation, intraurban transport, and
interurban transport.

Income elasticities of demand, which are an indication of the
responsiveness of demand for specific commodity classifications to
changes in income, were calculated directly from the estimated Engel curves and used to classify the distribution of indirect environmental benefits as either pro-rich or pro-poor.  As per Freeman,
if income elasticity is greater than zero, consumption of indirect
environmental services increases with income and is therefore characterized as pro rich.  Conversely, if income elasticity is less
than zero, the distribution of indirect environmental services is
termed pro poor. [6]

---

* Dick [3] and Freeman [6] contain more detailed mathematical and
  geometric treatments along similar lines, allowing for different
  production technologies, and different forms of damage function.

Table 1 gives elasticities for each commodity classification
per urban resident.*  All results were significant at least at the
10% level of confidence.  Consumption expenditure for all commodity
classification which are thought to impose high residual loadings is
significantly positively related to income.  Highest elasticities
were observed for paper, interurban transport and gasoline for vehicle operation.  Given that consumption can be considered a rough
indicator of the use of indirect environmental services, results
indicate that the benefits of these services are distributed in a
pro rich manner.

| Commodity Category | Income Elasticity |
|---|---|
| Total food | 0.59 |
| Total paper | 0.71 |
| Total power and fuel | 0.53 |
| Gasoline for vehicle operation | 1.36 |
| Intra urban transportation services | 0.34 |
| Inter urban transportation services | 0.88 |

Table 1.  Income Elasticities of Demand Calculated from Engel
          Curves for Average Expenditures by Families and unat-
          tended Individuals, Major Urban Centres, Canada, 1969.

---

* Engel curves were also estimated allowing for the impact of family
  size and for finer commodity breakdowns within each classification.
  See Ricci, Perron and Emmett. (12)

## 3. DISTRIBUTION OF DIRECT ENVIRONMENTAL SERVICES

Direct environmental services include air quality, open space, recreation, amenity, and life support.

In this section of the paper, a simple theoretical framework is outlined which establishes first how the demand for direct environmental services is likely to vary between income classes and second, how variations in demand for such services can, to some extent, be realized through an individual's choice of residential location.

If the services the environment provides to individuals are similar to services provided by other goods, and if environmental quality is a normal good, it can be expected that demand for direct environmental services will increase with income. Baumol[1] has used a simple indifference curve analysis to illustrate this point under the simplifying assumption that a single level of environmental quality such as atmospheric purity is shared in common by all residents.

This simple treatment brings out two important points  First, demand for direct environmental services can generally be expected to increase as income increases. Second, as all residents share the same level of environmental quality, some compromise will be reached which Baumol assumes will involve a level of environmental quality higher than the level desired by poorer residents, and lower than that desired by the better off. Consequently, the rich will value any increments of environmental quality more highly than the poor.

Underlying the indifference curve analysis are a number of important assumptions: individuals do have preferences defined over environmental quality; if so, environmental quality is a normal good; preferences maps of the poor and the rich are roughly similar; and that there is a fixed price for environmental quality which does not vary with income. A major question is whether individuals actually do have preferences regarding environmental quality. For some important environmental attributes this may not be the case. There may be lack of knowledge regarding the benefits of improved air and water quality, for example, especially the link between environmental deterioration and the incidence of disease. Even if the risks are known, there is a serious difficulty in establishing values for low probability events. How can the consumer place a value on, say, a reduction in the risk of dying from an air pollution-related disease from 1/1,000 to 1/1,050? There is also the problem of making decisions which affect others, particularly children and future generations.[9]

The simple indifference curve analysis assumed a single level
of environmental quality was shared by all members of the community.
In reality, environmental quality varies considerably from place to
place because of the type of man's activities and the complex interaction of residuals with the natural environment.  The quality of
environmental resources which accrue to individuals, such as air
quality, recreational opportunities and access to open space is
largely determined by the choice of location within an urban area.
Simple indifference curve analysis cannot hope to explain the large
number of variable factors which is involved in the choice of residential location.  A useful approach to take, following Freeman, is
Lancaster's goods characteristics theory.(6) (9) In the goods characteristics model utility functions are defined over characteristics rather than commodities.  Commodities are consumed in order to
produce the set of characteristics desired by the consumer.

Goods characteristics theory is helpful in analyzing commodities, such as land, which have a variety of characteristics.  From
the point of view of goods characteristics, each site is seen as
producing a set of characteristics over which individual utility
functions are defined.

The consumer's problem in choosing a residential location can
be stated in terms of characteristics theory.  The individual chooses
a residential location in order to maximize his utility as defined
in characteristics space and subject to budget or endowment constraints.

Characteristics, however, may come from a variety of sources,
which can complicate the analysis greatly in the case of environmental attributes.

Given the overall level of environmental quality in a region
or an urban area, an individual can achieve improved environmental
quality in a number of ways, by choice of residential location, by
consumption of private commodities, or some combination of the two.

"Some (characteristics) are associated with, or available from,
both consumer goods and environmental sites, indicating a
potential for substitution between sites and goods in satisfying the demands for these characteristics.  For example, urban residence and air-conditioning may substitute for rural
residence in shady areas." (6)

A major factor influencing the distribution of direct environmental services, and a primary reason for expecting that consumption
of environmental quality will be positively associated with income
is the extent to which the consumption of complementary commodities,
which are private goods, is required in order to receive high levels
of environmental quality.  If, for example, vacation homes are

required to utilize an environmental resource, then the ability of
an individual to enjoy this service is limited by his ability to purchase the necessary complements.  This analysis also applies to residential location and air quality.  If high air quality is indeed
a desirable environmental characteristic, then sites with higher air
quality can be expected to command a premium and will be rationed
to those individuals willing and able to pay.

     A second consideration highlighted by the goods characteristics
approach is the difficulty of distinguishing between housing environment and the ambient environment.  Levels of environmental quality
enjoyed by individuals and households can be altered both by changing
the micro environment as well as the macro environment.  Thus it becomes difficult to state conclusively that those who reside in areas
characterized by lower ambient air quality actually receive lower
levels of environmental quality. They can substitute privately produced services for those of the environment.

     This discussion provides only the briefest outline of a possible
framework for understanding the way in which environmental benefits
could be distributed amongst income groups.  It is by no means complete.  Rather than elaborate on a theoretical framework, the approach
taken by this paper has been to set theory aside and employ multiple
regression techniques to investigate the distribution of environmental quality amongst income groups using various proxies for environmental quality, socio-economic characteristics of households, and
variables which are generally considered to represent other aspects
of the urban environment and which influence individual well-being.

     Data was obtained from a cross-section of observations by census
tracts for the Montreal urban area. (14)  The independent variable
chosen was median family income.  The explanatory variables included
air quality (AIRQIDX) measured by a biological index of atmospheric
purity which was extracted for each census tract from a map of the
Island of Montreal; (11) household crowding (CROWD) in terms of persons per room was chosen to indicate the environmental and social
adequacy of housing; (POPDENS), population density, in terms of persons per square mile, was assumed to capture congestion and noise;
DISTCBD measures distance from the Central Business District under
the assumption that residential areas further removed from the urban
core will be characterized by higher levels of comfort and security;
(DISTPARK) measures distance to nearest urban parkland; (VACHOM) indicates vacation home ownership which is seen as a complementary commodity to the enjoyment of recreational activities and amenities.
Finally, three variables (OCCUP), relative number of dwellings which
are owner occupied, (AVGRENT), expenditure on tenant occupied dwellings, and (MEDNVAL), expenditure on owner occupied dwellings, are
used to capture the quality of individual dwellings and more generally, the quality of a neighbourhood. AIRQIDX, VACHOM, AVGRENT,
DISTCBD and MEDNVAL were hypothesized to be positively associated

with income while CROWD, POPDENS, and DISTPARK are expected to display a negative relationship.

Results of the multiple regression analysis both for cumulative income groups and all income groups were mixed.

The estimation of the model for all median income groups yields:

$$
\begin{aligned}
\text{MEDINCM} = 12596.219 \quad &- \quad \underset{(1.179)}{10.876 \text{ AIRQIDX}} \quad - \quad \underset{(11.950)^{**}}{11146.495 \text{ CROWD}} \\
&+ \quad \underset{(5.290)^{**}}{87.037 \text{ VACHOM}} \quad + \quad \underset{(8.426)^{**}}{45.784 \text{ OCCUP}} \\
&- \quad \underset{(1.623)}{137.188 \text{ POPDENS}} \quad + \quad \underset{(6.389)^{**}}{838.291 \text{ DISTCBD}} \\
&- \quad \underset{(2.218)^{*}}{376.643 \text{ DISTPARK}} \quad + \quad \underset{(4.518)^{**}}{5.99 \text{ AVGRENT}} \\
&+ \quad \underset{(7.951)^{**}}{0.044 \text{ MEDNVAL}}
\end{aligned}
$$

$$
\begin{aligned}
N &= 414 \\
F &= \text{computed } 189.602^{**} \\
R^2 &= 0.80
\end{aligned}
$$

The value in parantheses is the computed-t statistic (one tail). N is the number of observations, F is the joint test on the overall equation, and $R^2$ provides the measure of explanatory power of the overall equation.

The level of significance of statistical tests is shown by:
** is 1.0 per cent level of significance
* is 5.0 per cent level of significance

Regressions were also estimated for six different income groups and results are available from the authors on request.

For all statistically significant cases, the examination of residuals indicates that the fundamental assumption underlying the ordinary least squares are not grossly violated.

For the all inclusive group, the air quality variable is not significant at the level of significance chosen for the analysis. The variable (AIRQIDX) is significant in one of the regression models of the cumulative group of census tracts but the sign of the coefficient is not as hypothesized.

The variable which measures ownership of vacation homes (VACHOM) yields mixed results. For some sub groups, the sign is not as hypothesized. Factors other than income, such as amount of leisure time available or tastes and preferences, influence individual choice with respect to owning a vacation home.

The signs of the coefficients of locational variables (DISTCBD) and (DISTPARK) are significantly related to income in such a way as to support the contention that higher incomes enable urban residents to have greater accessibility to more spacious and quiet residential areas and to parks or open spaces.

In contrast to the mixed results obtained for both neighbourhood and environmental characteristics, results for variables thought to reflect housing quality were generally significant and consistent with expectations.  For example, the variable POPDENS, measuring neighbourhood density, was significant for only one subgroup, while CROWD, the measuring household density, was highly significant in all areas.  The two housing expenditure variables (AVGRENT) and (MEDNVAL) were also highly significant.

The results indicate households tend to allocate increments in income to improve the quality of their dwelling, or enhance the micro environment rather than seek locations characterized by higher ambient environmental quality and lower population density.

## 4. DISTRIBUTIONAL CONSIDERATIONS IN ENVIRONMENTAL MANAGEMENT

Sections II and III have assessed the distribution of direct and indirect environmental services amongst income groups in a static sense.  They have been an attempt to assess "who gets what" in terms of environmental costs and benefits at a point in time.  More interesting from the point of view of environmental management is "who gets what" when measures are implemented to improve environmental quality.  In this section, some inferences are made about the incidence of the economic costs of meeting environmental regulations on different income groups.  Because no data currently exists linking residuals discharge, ambient environmental quality, and impact on individual receptors, nothing can be ventured in even the most tentative way regarding the associated impact of environmental management on the distribution of direct environmental services.

As Baumol and Oates remark, "To offer a prediction on so broad a subject is hazardous ... Nevertheless, by making some reasonable assumptions and exploring the available evidence, ... some inferences (can be reached) concerning the likely pattern of incidence of these costs," (2)  Suppose, for example, that the use of the environment for waste disposal in production of commodities were to impose direct environmental costs on individuals.  A management authority attempts to reduce these damages by imposing emission standards or effluent charges.  The firms involved must incur costs to meet these standards and the effects of these costs will ultimately result in higher product prices and reduced consumption.  As argued in section

2, the precise impacts of regulatory measures are largely unknown.
Ultimately, however, it can be expected that the real costs of environmental regulation will be borne by:

1) consumers of those commodities whose production and consumption generates relatively large environmental disruption;

2) labour and other factors specialized to the production of
   these commodities;

3) holders of equity in these industries.

Costs of implementing environmental regulations can be roughly
categorized as transitional costs, those of achieving an improvement
in environmental quality standards, and continuing costs, those of
maintaining a desired level of environmental quality.  Of these categories, transitional costs are likely to be the most disruptive,
uneven, and difficult to predict. (2)

Integration of pollution abatement into the production process
may alter the firm's labour requirements.  The pattern of cost incidence may be distributed differently across different occupational
groups. Employment-restricting effects of environmental programs
will be unevenly distributed across different geographical areas.

The effect of the continuing costs of maintaining environmental
quality, however, relates to changes in the structure of prices and
consumption of goods and services which use indirect environmental
services and are more easily handled in the framework of this paper.
"(The) expectation here is that there will be a rise in the relative
price of those goods and services whose production involves substantial external costs (at least where techniques of production that
reduce destructive emissions are significantly more costly than
'free' dumping of wastes into the atmosphere or local waterways,)"(12)

If higher commodity prices do, in fact, result from environmental regulation, the distribution of costs among income groups will
depend upon consumption patterns for residuals producing goods,
which were introduced in section 2.  Such price increases are roughly
the equivalent of:

(i)   a proportion tax when expenditures upon such commodities
      vary proportionately with income;

(ii)  a progressive tax when such expenditures account for a
      larger percentage of income as income increases;

(iii) a regressive tax, when such expenditures account for a
      smaller percentage of income as income increases.

In table 2, expenditures for those commodities which were sing-
led out in section II are expressed as percentages of income.  When
calculated as a percentage of income, expenditures on food, paper,

power and fuel, intra urban transport and inter urban transport show a general decline as income increases, indicating price increases arising from environmental regulation would exhibit a regressive pattern of incidence. Expenditure patterns for gasoline show that increased prices for gasoline would impact relatively more severely on middle income groups. For some finer commodity breakdowns (like air travel within inter urban transports) the general pattern of regressivity may not be exhibited, but in general environmental regulation is likely to have generally regressive impacts. These results are in broad agreement with other studies. (2) (6)

| Income Class | Total Food | Total Paper | Power & Fuel | Gasoline for Vehicle Operation | Intra-urban transport services | Inter-urban transport services |
|---|---|---|---|---|---|---|
| - 3000 | 32.05 | 1.97 | 4.65 | 0.82 | 2.04 | 0.91 |
| 3000 - 3999 | 25.46 | 1.71 | 3.29 | 1.25 | 2.32 | 1.66 |
| 4000 - 4999 | 24.48 | 1.44 | 3.08 | 2.23 | 1.88 | 1.07 |
| 5000 - 5999 | 21.55 | 1.42 | 2.58 | 2.18 | 1.88 | 8.98 |
| 6000 - 6999 | 21.54 | 1.42 | 2.73 | 2.75 | 1.37 | 0.79 |
| 7000 - 7999 | 20.60 | 1.26 | 2.48 | 2.43 | 1.29 | 0.80 |
| 8000 - 8999 | 19.85 | 1.35 | 2.45 | 2.78 | 1.14 | 0.77 |
| 9000 - 9999 | 18.41 | 1.23 | 2.44 | 2.54 | 0.98 | 0.78 |
| 10000 - 10999 | 17.39 | 1.33 | 2.21 | 2.55 | 0.81 | 0.85 |
| 11000 - 11999 | 17.36 | 1.23 | 2.42 | 2.42 | 0.91 | 0.97 |
| 12000 - 14999 | 15.12 | 1.15 | 1.86 | 2.29 | 0.85 | 0.72 |
| 15000 and over | 12.00 | 0.97 | 1.47 | 1.70 | 0.58 | 0.89 |

Table 2. Expenditure on Selected Commodities per families and unattached individuals expressed as a percentage of income by family income, major urban centres, 1969.

## 5. THE ROLE OF DISTRIBUTIONAL CONSIDERATIONS IN ENVIRONMENTAL MANAGEMENT

It would be ideal, as Stevens and Kalter point out, to establish a functional relationship which would describe both the efficiency and the equity considerations associated with projects which affect the environment. As it is, it is not even possible to assess the existing distribution of the environment in a sophisticated way.(15 The complex economic interactions which determine how costs of environmental management are transmitted eventually to individuals have barely begun to be examined. No values have yet been established for the consumption of direct and indirect environmental services by individuals and firms. Even if such values were established and functional relationships estimated, a social welfare function would still be required to rank various alternatives.

No clear cut evidence has been developed in this paper that the distribution of direct environmental services will be either pro-rich or pro-poor. However, evidence strongly suggests that the economic costs of environmental regulation are likely to affect the poor relatively _more_, while simple economic theory suggests that the poor will value increments in environmental quality relatively _less_ than the rich. It has not been argued, however, either that environmental management should be undertaken in order to redistribute income, or that environmental regulation should not be undertaken solely because they are likely to accentuate existing inequities in the distribution of income. Rather, proper attention to distribution enhances understanding of environmental problems and could aid considerably the effective design of environmental regulations.

First, society has ethical preferences about the distribution of income, if vaguely articulated, as evidenced by such redistributive measures as progressive income taxation.

"Within an economic framework, social welfare depends in part upon both the level of income, where income is used as a proxy for well-being, and its distribution ... Thus, evaluation of alternatives by economic efficiency analysis alone is inadequate to determine that set of investments which will make maximum contribution to social welfare." (15)

Therefore distributive impacts of environmental regulations are of interest in policy formulation. This is particularly the case when a given level of environmental quality may be achieved by a variety of different regulatory mechanisms, possibly financed by a number of different instruments.

Information on the distributional impact of environmental measures can play an important role in developing effective programs. Political acceptability of various programs is likely to be heavily dependent on the distribution of their costs and benefits. Information on distributional impacts can help in the design of effective and equitable financial arrangements.

Finally, and perhaps most importantly, this paper emphasizes that the natural environment is an important scarce natural resource which contributes substantially to the welfare of individuals in society, the management of which is an important collective decision in a society which is characterized by concern about equity and social justice.

## REFERENCES

1. Baumol, W.J., "Environmental Protection and the Distribution of Incomes," in Problems of Environmental Economists, Paris, OECD, 1972.

2. Baumol, W.J., and Oates, W.E., The Thoery of Environmental Policy, Englewood Cliffs, N.J. Prentice Hall Incorporated, 1975.

3. Dick, D.T., Pollution, Congestion and Nuisance, Lexington, Mass., Lexington Books, 1974.

4. Evans, M.K., "A Forecasting Model Applied to Pollution Control Costs" American Economic Association: Papers and Proceedings, 1973, pp. 244-252.

5. Fisher, A.C., and Peterson, F.M., "The Environment and Economics: A Survey," Journal of Economic Literature, Mar, 1976, pp. 1-33.

6. Freeman, A.M., "Distribution of Environmental Quality," in Kneese, A.V., and Bower, B.T., Environmental Quality Analysis: Theory and Method in the Social Sciences, Baltimore, Johns Hopkins Press for Resources for the Future, 1972.

7. Henderson, J.M., and Quandt, R.E., Microeconomic Theory, New York, McGraw-Hill Book Company, 1971.

8. Kneese, A.V. Management Science, Economics and Environmental Science," Management Science, Volume 19, No. 10, pp. 1122-1137.

9. Lancaster, K.J., "A New Approach to Consumer Theory", Journal of Political Economy, Vol. 74, No. 2, 1966, pp. 132-157.

10. Lave, L.B., "Air Pollution Damage: Some Difficulties in Estimating the Value of Abatement," in Kneese, A.V., and Bower, B.T., Environmental Quality Analysis: Theory and Method in the Social Sciences, Baltimore, Johns Hopkins Press for Resources for the Future, 1972.

11. LeBlanc, F., and de Sloover, J., "Relations between Industrialization and the Distribution of Epiphytic Lichens and Mosses in Montreal," Canadian Journal of Botany, Vol. 48, No. 7, 1970, pp. 1485-1496.

12. Ricci, P.F., Perron, L.E., and Emmett, B.M., "An Analytical In-
    vestigation of the Distributional Impact of Environmental Qua-
    lity on Socio-economic Groups in Canada," paper prepared for the
    40th Annual Educational Conference, National Environmental Health
    Association, June 26 - July 1, 1976, Nashville, Tenn.

13. Statistics Canada, Family Expenditure in Canada: Volume III,
    Major Urban Centres, 1969. Catalogue 62-537, Occasional, Ottawa,
    Information Canada, 1973.

14. Statistics Canada, Population and Housing Characteristics by
    Census Tracts, Catalogue 95-704, Ottawa, Information Canada,
    1974.

15. Stevens, T.H., and Kalter, R.J., "Evaluation of Public Invest-
    ments: Distributional Impacts of Water Resource Projects,"
    Search: Agriculture, Vol. 2, No. 12, 1972, pp. 1-39.

SOCIAL STABILITY ANALYSIS - CAN CARRYING CAPACITY PROVIDE AN

ANSWER TO PUBLIC POLICY IN CANADA?


Paolo F. Ricci

University of Ottawa

Ontario, Canada


## INTRODUCTION

The usage of the phrase "carrying capacity" in recent planning and policy statements requires a discussion of the meaning and transferability of the concept to policy-making.  Recently, the concern with population size and its probable growth over time has increased to the point that specific policy issues may well turn out to be based on some arbitrary measure of a "carrying capacity."  For this reason policy-making, as distinct from decision-making, becomes the focus of this paper, while the emphasis is not on providing measures of carrying capacity for any area; rather it is on developing an operational definition.

The objectives of this paper thus are:

1.  To review the concept of carrying capacity, and

2.  To attempt a synthesis of critical elements that an operational definition of carrying capacity, amenable to simulation, ought to incorporate to enhance the policy-making process.


## A REVIEW OF THE CONCEPT

In this paper, policy-makers are defined as those making the policy decisions and often are elected or appointed officials.  The decision-makers, on the other hand, are concerned with providing operational inputs to the policy-makers.  Broadly, the differentiation between the two groups is given by the level of detail breadth, and scope of the societal issues being addressed.

In current policy-making usage, the maximum number of individuals an area can support in terms of a limiting factor, normally
food availability becomes the "carrying capacity" of the area.  This
supportive capacity is the quantification of a population ceiling
which can be rationalized as that which the future population should
not be encouraged to exceed.

The facility with which the phrase has recently been used –
outside of its traditionally technical area – leads to the obvious
assumption that a rational population figure can be established on
the basis of the total availability of a limiting factor and the
per capita consumption for that limiting factor.  The policy implication of the concept, is gravid with consequences, in that it leg-
itimizes, by lending "scientific credibility", projections in which
a single variable, say agricultural output over time, may yield an
optimum population ceiling for an area, a region, or a nation.

While the simplification inherent to assuming that a limiting
factor (i.e., caloric intake) establishes a technical carrying capacity, when it is done at the exclusion of social, economic, institutional, and political factors it is not acceptable for policy
making.

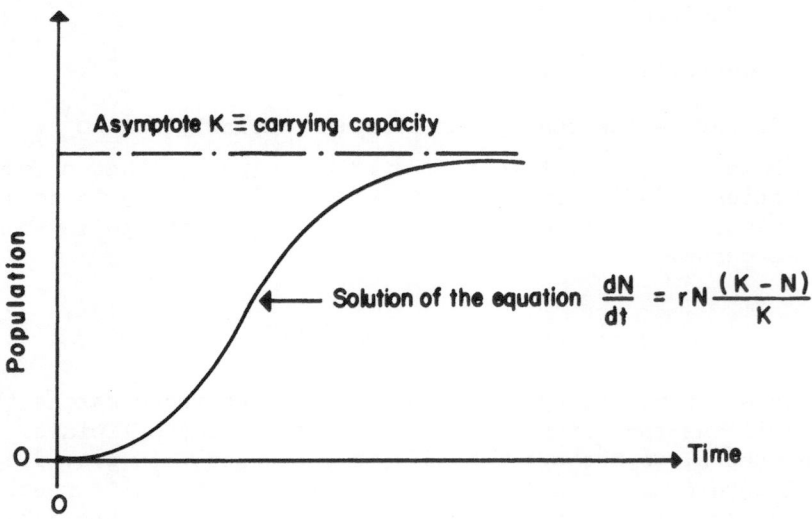

Figure 1.   Representation of the carrying capacity model shown in
Equation 1.

Carrying capacity is a well defined concept in ecology[a] and describes the upper asymptote of the rate of population growth,[b] as shown in Figure 1.  An example that typifies the misuse of carrying capacity and gives a practical meaning to the concept is provided by solving the logistic equation (1) under the assumption that fresh water is the limiting factor to population growth.  We can simulate the fresh water budget of Canada by assuming it to be a given percentage of the fresh water budget of the world. Figure 2 shows some of the time paths that water consumption take under the assumption of varying growth rates, for specific values of the carrying capacity asymptote, i.e., the theoretical amount of fresh water.  When per capita water consumption is known then the population that can be supported, if fresh water is the limiting factor, is easily found. Thus, if consumption of water occurs at a 3 per cent growth rate, the water budget of Canada is $23 \times 10^{11}$ cubic meters, one tenth of the world's budget, the population in the year 2030 consumes 3000 cubic meters of water per year, then the actual population of Canada would turn out to be :

$$\frac{2.65 \times 10^{11}}{3000} = 8.83 \times 10^{7} \text{ people (at the 2030 horizon)}$$

while its potential (i.e., the carrying capacity) is $7.67 \times 10^{8}$ people.

---

[a]The rate of population growth, dN/dt, with a carrying capacity asymptote is:

(1)   $\frac{dN}{dt} = rN \frac{(K - N)}{K}$     where: K is the upper asymptote, carrying capacity
                                                            N is the population
                                                            t is time
                                                            r is the growth rate

This differential equation is interpreted as:

| The rate of population growth over time | equals | an exponential growth rate | adjusted by | how much growth can actually occur |

[b]The logistic growth rate is shown by equation 1, and, upon solution, yields the path for consumption of water over time, shown in Figure 1:

$$V(t) = \frac{K}{1 - e^{a - rt}} \qquad (2)$$

in which a is the constant of integration, V(t) is the amount of water used at any time t, K is the total fresh water budget, and r is the growth rate.

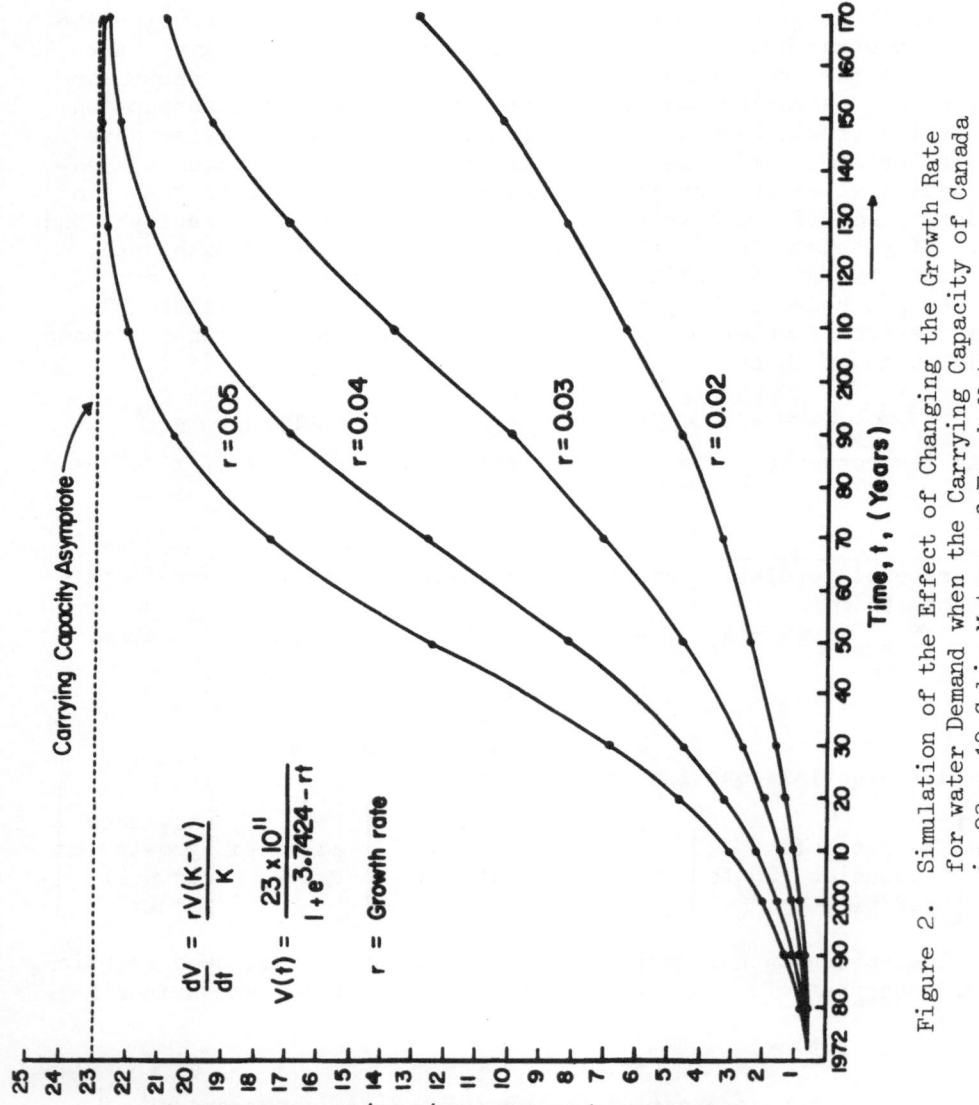

Figure 2. Simulation of the Effect of Changing the Growth Rate
for Water Demand when the Carrying Capacity of Canada
is 23 x 10 Cubic Meters of Fresh Water.

An obvious flaw with this simplistic reasoning is that not all the fresh water is available consumption and this estimate becomes unrealistic!

Yet the concept is appealing since it provides concise and terminology for defining a problem vexing the policy makers of many countries[2] as:

> "...the upper level, beyond which no major increase (in a population) can occur ... is the upper asymptote carrying capacity."

Ecologists have developed more complex approaches which range from the seminal work of Lotka[2] and Volterra[3] to the recent work by May[4] and others[c]. Models of population dynamics now are available and include such factors as unlimited growth rate, self-crowding effects, interspecific competition, and environmental effects by utilizing systems of differential equations that seek to reproduce realistic growth patterns for the populations undergoing study. Greater similarity with the uncertainty of biological life is analytically reproduced by assuming population growth.[d]

As an example of including randomness in the model, Figure 3 describes the probability-dependence of population growth when the solution of the model is stochastic. It can be seen that the effect of imparting random behaviour may result in the completely different time paths which the assumption of determinism would not necessarily capture.

The conclusion that can be reached from the viewpoint of ecological population dynamics is that increasing the level of mathematical sophistication, although it may provide a more accurate description of species behaviour, also creates almost absurd complexity when the analytical effort is transferred to model policy-making. Indeed, part of the possible inaccuracy involved in the transfer of the ecological concept of carrying capacity to policy-making lies in the assumption that a set of boundary conditions leads to a neat description of societal growth and thus determines the carrying capacity of a country or other are given that the parameters of the

---

[c]Lotka[2] and Volterra[3] model populations in terms of predator-prey interaction by coupling two differential equations that describe different growth rates.

[d]May[4] provides a discussion of the effect of deterministic and stochastic environments on biological population growth and stability. In particular, May provides generalizations from the diffusion equation to the modeling of populations whose growth rate is affected by a randomly fluctuating environment.

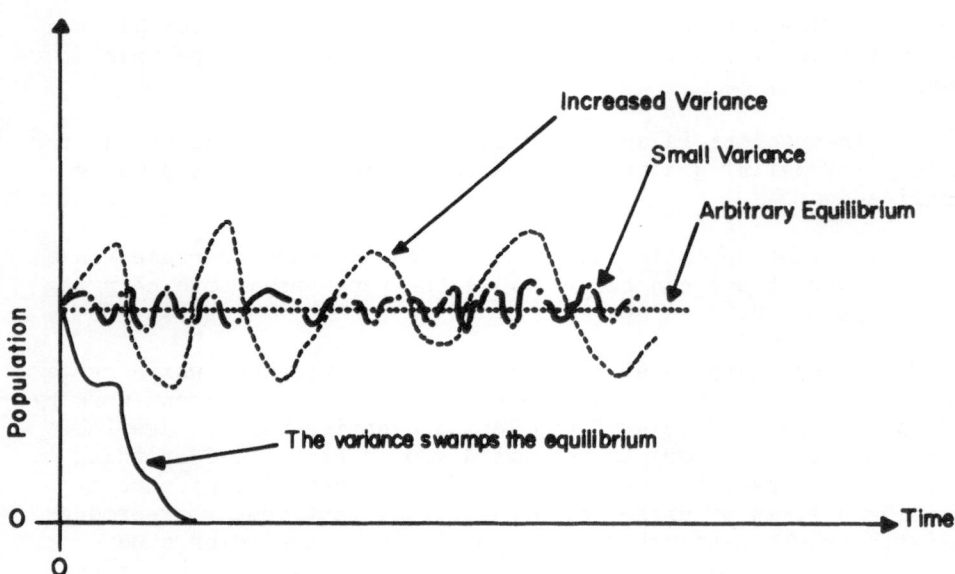

Figure 3.  Changes in the fluctuation of population size over time.
          As the variance increases the population fluctuates around the
          stable equilibrium with increasing amplitude,when the variance
          is too large the population extinguishes rapidly.[*]

[*] Adapted from May[4].


system modeled are known or can be approximated.  Additional insight
with the preoccupation inherent to how much and how far can a population expand is provided by anthropology and geography.

     The approach taken by modeling the "carrying capacity" of a
region consists of calculating sustainable agricultural yields and
from those yields establish how many individuals can live in the
area being studied.  The definition that has the greatest usage
among anthropologists is:[5]

          "The maximum number of people that a given land will main-
          tain in perpetuity under a given system of usage without
          land degradation setting in."

     Within the anthropological definition "perpetuity" and "land
degradation" show the futility of trying to come to grips with an

estimate of a single measure of carrying capacity. The assumption of technological and gastronomic constancy, uniformity of human needs and wants make anthropologic carrying capacity akin to the asymptotic definition just reviewed and the magic figure easy to reach. For policy-making, the simplifying assumptions of constant technology and taste leave too many facets, the political, institutional and social dimensions inevitably bound with societal activities. Little help, at least so far as expanding the concept of carrying capacity, is forthcoming when the attention is turned to geography.

Geographers also have been found to utilize carrying capacity in terms of the maximum theoretical population that can be supported by a given resource base. In most cases the process common to this approach consists of calculating simple ratios, a variation of which is:

$$\frac{(available\ land)}{(utilized\ land)} = \frac{(theoretical\ population)}{(actual\ population)} \qquad (3)$$

The approaches to achieve a measure of "carrying capacity" tantalize into realizing the potential value of the concept, but the gap from technical definitions to one appropriate for policy-making needs to be bridged since we cannot escape the fact that:[6]

> "The basic problem facing organized society today boils down to determining in some objective manner when we are getting "too much of a good thing." This is a completely new challenge to mankind because, up until now, he has had to be concerned largely with too little rather than too much."

Meaningful answers to the "getting too much of a good thing" are not satisfied by assuming that population levels can be determined from agricultural food yields divided by some measure of consumption. Society, through its public representatives, cannot avoid realizing that simple answers to complex problems may not obtain and that future actions need to be viewed in a systematic and comprehensive manner. Policy-making can be systematized and an expanded concept of carrying capacity does provide the means to conceptualize the state-of-the-system, but further work is needed. The concept should be amplified to include specific recognition of factors which the disciplinary definitions leave out. Only recently some explicit efforts in this direction have begun to appear in the literature; for example, EPA, defines carrying capacity[7]

> "Carrying capacity (as) the level of human activity (including population dynamics and economic activity) which a region can sustain as acceptable quality of life in the long term."

This is in the nature of a goal and does not suffice, since it provides a "motherhood and apple pie" statement, which is at the other extremity of the spectrum of definitions reviewed. Indeed, EPA adds interdeterminacy to the concept of carrying capacity by including in it the "quality of life." Although this attempt reflects concern with societal well-being, it appears that the need for carrying capacity as a means by which the goal can be approximated - is not recognized to be the essential function of the concept.

The transfer of carrying capacity to policy-making is facilitated by realizing the relationship that exists between the human and the natural environments in terms of responses to stimuli of temporal and spatial nature, as shown in (4):

Human Environment ⇄ Natural Environment

In this scheme, the two "environments" affect one another actively and reactively, as well as they affect each other.

The ecological, socio-economic, institutional, and policy spaces inherent in (4) can be integrated into a single operational definition under the carrying capacity umbrella. The term loses the strict (i.e., asymptotic) connotation and becomes a tool for describing supply, maintenance, and sink potentials of policy area. These potentials may possess various tendencies (away, toward, oscillating or bifurcating from an equilibrium). The realization that explosion and implosion may occur leads to the need for modeling the outcoming of policy-making.

If policy-making is to be systematized to provide more enlightened understanding of the possible effects of policy-making then, as Hollig notes:[13]

> "A management approach based on resilience, on the other hand, would emphasize the need to keep options open, the need to view events in a regional rather than a local context and the need to emphasize heterogeneity... The resilience framework can accommodate this shift of perspective, for it does not require a precise capacity to predict the future, but only a qualitative capacity to devise systems that can absorb and accommodate future events in whatever unexpected form they may take."

The utility and validity for maintaining - as an operational concept - carrying capacity is that it allows modelling policy options to portray the possible outcomes that may result from enacting and implementing a policy. Hence, carrying capacity provides a ready-made, appealing, dynamic frame of reference in which the boundaries of the policy domain are not sacrosanct, rather vary with the degree of aggregation and level of detail required to understand

the implications of policy options.  These boundaries, which may not
overlap, identify the time, space, kind and scope attributes of the
policy as shown in Figure 4.

Dorfman's[8] statement provides the setting for the level of detail that carrying capacity should provide when used for modelling
policy issues:

> "We can often say whether things will go up and down when
> we cannot say how far or how quickly.  But these merely
> qualitative appreciations - predictions of the probable
> direction of change - are important.  They are frequently
> all that are needed for informed policy decisions."

The implication of wanting to determine not the "how much" but
the "where" is that policy-makers can utilize the concept to determine the resiliency - as the ability to rebound from shocks - of
the system as it becomes perturbed by a policy action.[13]  In other
words, an operational carrying capacity provides the generalized,
systematic approach from which we can approximate the behaviour of
key factors affected.

The interrelatedness of policy-induced effects upon society,
and the paramount objective of maintaining well-being, cannot be
solved by concentrating only on narrow objectives (i.e., food).
For example, it is not the "no-growth" option that ought to be examined, rather it is the type and mix of goods and services that
should be studied in terms of bottlenecks and the social implications of scarcity, wants, and needs.  Typically, a "no-growth" policy would ignore the expectations and choices of segments of the
population who face the uncertain struggle of day-to-day living,
and presumably have little inclination, or the time, to pursue and
accept abstract theories of self-sufficiency:[9]

> "The self-indulgent, materialistic, mass-consuming society
> ... proclaims every man a "king" and every woman a "queen".
> But the monarchial prototype is neither sober, hard-working
> Frederick William of Prussia, nor urbane, philosophical
> Marcus Aurelius, it is the "I'll take mine now and to hell
> with the consequences" Bourbon Louis XV who, however, had
> the prescience to declare - après moi le déluge."

## TOWARD AN OPERATIONAL DEFINITION OF
## CARRYING CAPACITY FOR POLICY-MAKING

Carrying capacity requires the determination of critical factors and resources which are the principal determinants of social

and economic life.[e] These determinants include renewable and non-renewable resources, population dynamics as well as the feedback mechanisms which relate these.  In this vein the approach to describe an operational form of carrying capacity, as well as the initial schematization at the broadest level of analytical detail and generalization, is:

$$(\text{carrying capacity})_{t+1} = f((\text{carrying capacity})_{t-k}, (\text{carrying capacity})_t,$$
$$(\text{random and non-random disturbance}))    (5)$$

This relationship indicates that the dynamic nature of carrying capacity is affected by its previous, present values, non-linearities thresholds and the unexpected disturbances imparted on the entire system by the real world.  The lagged dependency includes the effect of past policy-making upon present population growth, lags, leads and discontinuities to the policy-making process.

The model shown by (5) is the first step needed to develop a set of relationships that capture the gross behaviour of key components of a carrying capacity system.  Although the final development of the model is outside the scope of this paper, the process which expands the functional relation shown as (5) may be one which includes:

1.  Representations of rate of change over time, where change is a function of the change itself at any time (t), modified by the difference between actual and expected occurrences.

2.  Representations of the rate of change of incentives existing within the system.

3.  Representations of multiplier effects inherent to the system, as analytical functions of endogenous and exogenous variables that make up the system.

4.  Choosing the output state variable through experts in the specific areas and the policy-makers.

5.  Establishing the functional relationships with the help of the experts and policy-makers.

---

[e] The quantification and qualification of carrying capacity for Canada, even if feasible, require models which to date have not been formulated or constructed.

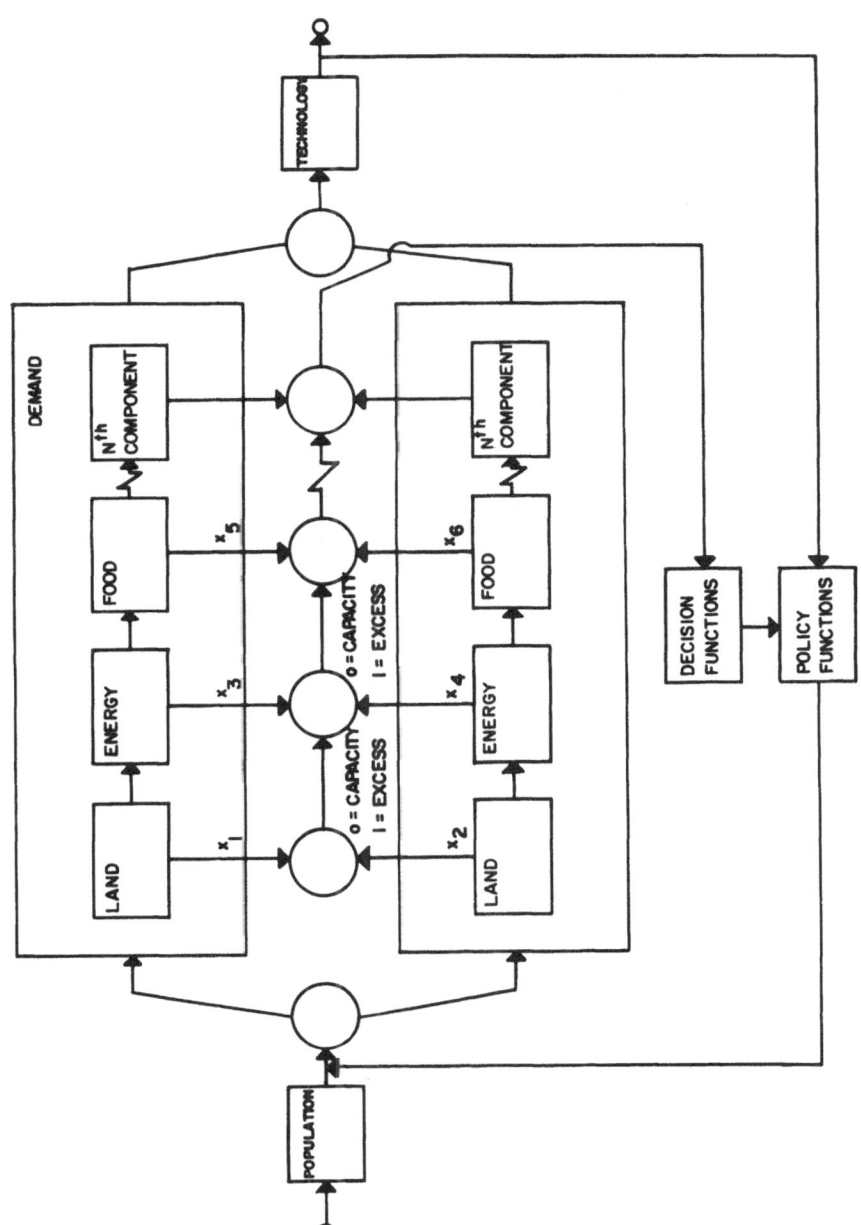

Figure 4. Components Chosen for an Operational Carrying Capacity

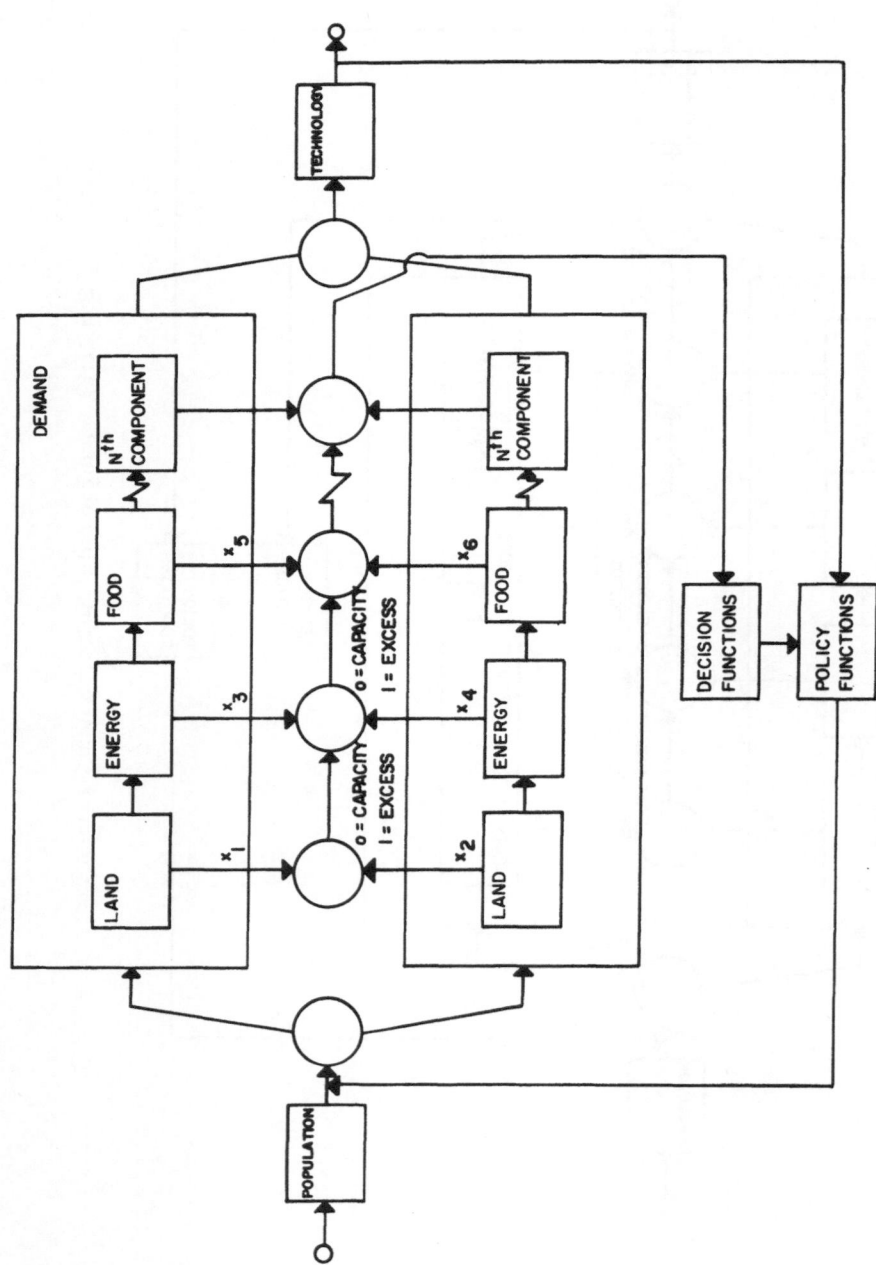

Figure 5. Heuristic Model of Carrying Capacity

The management process necessary to rationalize the implications
of policy acts (which should conceptually precede the specification
of the operational carrying capacity model) would consist of:

1. Bounding the problem;
2. Choosing the key components;
3. Determining the level of detail needed;
4. Describing the processes either qualitatively
   or quantitatively;
5. Analyzing the responsiveness to the "policy questions".

As an example, the strategies that can be selected to establish
levels of carrying capacity - given specific policy alternatives -
range from the "Principle of Insufficient Reasons" to the "Minimax
Regret Criterion,"[f]

The immense number of relationships that make a society and
that ought to be included, requires the heuristic model shown in
Figure 5 as a synthesis of key factors and their active-reactive[g]
nature. In this figure, the policy "actuator" provides for management actions and interfacing with the components of the carrying capacity system. The components, shown as black boxes, represent various rate relationships, multipliers, and incentives which would
allow the policy-maker to trace out the effect of policies and observe changes that result from management probes. The operational
carrying capacity can be expanded and schematized as shown in Figure 6. In this last elaboration, the operating rules of the model
are not specified, while internal consistency among the state variables of the system is assumed to be feasible at a future stage of
development. The policy and decision relationships are also not
specified, but are shown as two "black boxes", with the understanding that the operating rules are state variables also, and subject
to modeling.

---

[f]For discussion of these strategies, see 14.

---

[g]For example, one can investigate the environmental carrying capaci
ty of an air basin in terms of the concentration of pollutants from
human and animal sources, and determine the environmental carrying
capacity of the system. In this case, carrying capacity obviously
does not include land-use or other parameters affected by and affecting the carrying capacity of the air-land-water system.

## SOME INITIAL ATTEMPTS

The few studies concerned with developing operational measures of carrying capacity have been regional or urban in character and with marked qualitative-prescriptive emphasis. The applications that have appeared in the literature identify an "optimum growth" for the population of a specific area. The "optimum population growth" concept is exploded by introducing environmental quality, assumed levels of quality of life, and sustainable production yields under demand by population and economic growth in the Pacific North-west Study.[10] This approach provides the basis for defining the initial values of carrying capacity since it recognizes that if we progress from satisfaction of basic physiological needs to societal security, society has to continue to "invest" in the production of goods and services. In turn, "growth" is characterized both by population increase and pent-up demand for resources.

Carrying capacity, as a ceiling or "optimal population growth" of the region, is approximated through a series of steps, the out-come of which is in units of population under specific assumptions as to the technology and policies, recycling and population abate-ment. Any relaxation of the constraints causes a downward estimate of population carrying capacity.

A second approach to the regional carrying capacity has been developed for the State of Florida.[11] The concepts developed in the study, i.e., the environmental relationships, are based on ener-gy flows, while the relationships themselves portray the dynamic be-haviour of the environmental characteristics being modeled. This approach indicates that theoretically, at least, the assumption of a steady-state carrying capacity would be quite inadequate since the combination of (1) historical, (ii) spatial, (iii) social and eco-nomic, and (iv) structural and environmental parameters produce dy-namic, non-steady state behaviour in the system's stability and re-silience.

In the U.S.A., in some instances, carrying capacity has been legislated into action by the executive branch of the state govern-ment. For example, Godschalk-Parker[12] report that the legislature of Hawaii mandates that the governor's office develop:

> "Criteria for defining the state's optimum carrying
> capacity as related to its environmental systems ...
> to include ... population, air quality, water quality
> and supply, energy supplies, transportation systems,
> and land-use capabilities."

The same authors also report that the State of Maine and the State of Vermont mention carrying capacity when determining feasi-bility of large-scale projects.

## CONCLUSIONS AND SUMMARY

The concept of carrying capacity, in the decision-making context, requires the development of an operational formulation to reflect the policy-making purpose for which the concept has been used
in this research.  By reviewing the technical-traditional meaning
of carrying capacity, a mechanistic definition, borrowed from academic disciplines, is found to be lacking since it does not capture
the full vitality of society.  In this paper, a suggestion is made
that an operational definition of carrying capacity is necessary to
enhance policy-making.

The implication of the discussions entertained in this paper
provides a heuristic framework from which the carrying capacity can
be adapted, without undue loss of information, to public policy-
making.

Since the interest of this paper is to view carrying capacity
as a direct aid to the formulation of policies, the concept must be
understood not solely in a national framework, but should also be
studied in terms of the relationships with the rest of the world.
The spectrum of activities necessary to provide full realization of
the concept, given the host of social, economic, institutional and
policy constraints, should be amplified to include external affects
such as formation of cartels, food demands from areas with recurring
famine, external trade, etc.  It is obvious that the open character
of most economic systems requires that global data be available so
that national responses can be accurately formulated.

In summary, the operational definition of carrying capacity is
a valid addition to policy-making since:

1.    It provides bounding for the problem of managing technical as
      well as societal needs and wants.  The operational definition
      reflects the spatial, temporal and topical bounds of carrying
      capacity.

2.    It provides an understanding of the goal-oriented policy im-
      plications of the definition.

3.    It establishes grounds for discussion, given that objectives
      have been formulated.

4.    It provides a conceptual model which can serve as a basis of
      developing a simulation methodology for use by the policy-
      maker.  This is meant as a vehicle to eventually qualify spec-
      ific policy options, as general indications of where the sys-
      tem is going - up or down - not necessarily by how much it is
      going to vary.

5.    It promotes the understanding that the components of (3) above
      must be developed with the direct input for the users.

      The expanded concept of carrying capacity, shown in the prev-
ious sections, provides a concise statement of various constraints
on the growth and development of populations, since:

1.    It points to critical limits that manifest themselves in the
      life support system of populations.

2.    It allows to model the future implications of existing policies
      and trace out unstated or absurd assumptions.

3.    It allows comparisons to be made with past policies.

4.    It allows the integration of disciplinary research from many
      areas into an integrated framework.

5.    It recognizes that even though the inclusion of constraints is
      enhanced, it does not obviate the need for political choices
      on the how, where and for whom.  These choices involve more than
      mere compatibility with environment capacity; they must reflect
      social values and preferences.

6.    It permits identification of a carrying capacity which may pro-
      vide the option of using or filling excess capacity.

7.    It allows eventual transferability of the concept to simulate
      regional carrying capacity in terms of resource shifts, sub-
      stitutability,cultural shifts, scientific innovation and tech-
      nological innovations.

8.    It modulates population growth on the basis of alternative pol-
      icy options.


REFERENCES

1.    Odum, D.P., Fundamentals of Ecology, Third Edition, W.B.Saunders
      Company, Philadelphia, Pa., 1971.

2.    Lotka, A.J., "Growth of Mixed Populations," Journal of Washing-
      ton Academy of Science, Vol. 22, pp. 461-469, 1932.

3.    Volterra, V., Lecon sur la Théorie Mathématique de la Lutte pour
      la Vie, Gauthier-Villars, Paris, France, 1931.

4.  May, R.M., <u>Stability and Complexity in Model Ecosystems</u>, Monograph in Population Biology, No. 6, Princeton University Press, Princeton, New Jersey, 1973.

5.  Street, J., "An Evaluation of the Concept of Carrying Capacity", <u>Professional Geographer</u>, Vol. 21, No. 22, pp. 104-107, March, 1969.

6.  Odum, E.P., "The Strategy of Ecosystem Development," <u>Science</u>, 164(3877): 262-270, April 18, 1969.

7.  EPA 600/5-74-021, <u>Carrying Capacity in Regional Environmental Management</u>, Socio-economic Environmental Studies Series, February, 1974.

8.  Dorfman, R., "Discussion," in <u>American Economic Review</u>, May, 1973, pp. 253-256.

9.  Utton, A.E., and D.H.Henning, (eds.), <u>Environmental Policy</u>, <u>Concepts and International Implications</u>, Praeger Special Studies in International Economics and Development, Praeger Publishers, New York.

10. Peterson, E.K., 1973, <u>Ecology and Economy</u>, Pac. N.W., River Basins Comm., Vancouver, Wash.

11. Shadix, Jerry A., 1974, <u>Interface Four: Urban Design Studio Report Four</u>. Depart. Arch., University of Florida, Gainesville.

12. Godschalk, D.R., and F.H.Parker, "Carrying Capacity, A Key to Environmental Planning?" <u>Journal of Soil and Water Conservation</u>, Vol. 3, July-August, 1975.

13. Holling, C.S., "Resilience and Stability of Ecological Systems", <u>RR-73-3, IIASA Research Report</u>, September, 1973, Schloss Laxenburg, Austria.

14. Intriligator, M.D., <u>Mathematical Optimization and Economic Theory</u>, Prentice-Hall, Inc., Englewood Cliffs, N.J., 1971.

ENVIRONMENTAL QUALITY CHANGE IN SEVERAL SOCIO-ECONOMIC CONTEXTS IN

FRANCE : 1970 - 1995

Jacques Theys

French Ministry for the Quality of Life

Neuilly sur Seine, France

## INTRODUCTION

The economic crisis has temporarily put the public concern about environment into the background. But, as a paradox, the crisis, by setting the question of "an other growth", has also led to emphasize the part that natural resources and qualitative goods might play in the future development. One fact has been clearly pointed out: it will not be henceforth possible to deal with environmental problems in isolation, without referring to a precise socio economic project.

This assumption is the starting point of a study which has been engaged for the past two years on the future of the environment in France.

The research originally aimed at developing a methodology to assess the efficiency of different environmental control strategies in a fluctuating context. But the main conclusion of this first step was that reliance on control alone, without fundamental changes in the industrial or consumption structures and without strong modifications in the localization trends would not be the most effective way of reducing pollution and saving natural resources substantially.

As drastic improvements of the environment seemed to be in the long term closely linked with basic transformations of some socio economic key variables, it became apparent that an "a strategic environmental assessment process" had to be based on a brief analysis of some possible "futures" of the french society.

The strategic environmental assessment progress itself con-

sists of three other major components, each of which is a separate
model:

- a quantitative estimation of some selected "futures" of the
  environment;
- a qualitative evaluation of the human, social and natural
  impacts associated with each of these environmental situations.
- a selection of management options available for improving
  the environmental quality involved by each scenario.

1. A first model (SIMULATION MODEL) is used to estimate the
variations of selected environmental parameters such as concentration and discharge of pollutants, density of population, consumption of water and land, depletion of natural resources, involved by
a composite set of assumptions on demography, rate of growth, patterns of social needs, territorial distribution of the activities,
international trade and technology (SCENARIOS).

The simulation step results in a range of fourteen different
pictures of the environment in 1995 by region and for the whole
country.  They are based on the assumption that development is realized without any further reinforcement of specific environmental
policies.  This reference background given, it is then possible to
compare the efficiency of three levels of control strategies more
and more stringent and costly.

The main output of this first part is a better knowledge of
the relative influence of factors controlled or not controlled by
the environmental protection administrations, of the "winning cards"
and limits of their action in each socio economic context.

2. A second model (EVALUATION MODEL) translates the quantitative data obtained in the previous step in a more global and expres-
sive appreciation of the environmental situation in France.  For
this purpose the full range of the impacts of residuals is taken into account: effects of non-economic or economic nature on human
populations, on nature and on the man-made goods.  The geographic
specificity of the impacts and the synergetic effects are given an
important place by first dividing the whole country in areas similar as regards to their reactions to environmental stresses and then
treating the pollution problems in each of these subsystems.  In the
fifteen blocks - previously clustered - the damage values are roughly
estimated for different classes of separate receptors.  The problem
of no bridge between the different receptors - man, nature, economic
goods ... implied by this functional analysis is solved by introducing several systems of value - a key concept of the study.

Finally, the decision makers are given five criterias to appreciate the environmental situation:

1. The total of damages corresponding to each of the scenarios.

2. A ranking of pollutants per system of value.

3. A hierarchy of polluting activities and sources.

4. A ranking of areas and regions depending both on their carrying capacity and on the environmental stresses.

5. The damage levels for each of the receptors.

This last information may easily be connected with a definition of long term objectives or standards - such as the protection of man's health or the maintenance of natural ecological balance.

3. A third step (PLANIFICATION PROCESS) is at last necessary to define investment priorities and programs corresponding to each of the three levels of policy previously used in the simulation stage. It consists mainly in an assessment and optimisation of several actions and incentives "generated" through a systematic and morphological approach.

4. Those three parts must be considered as a framework for a holistic approach not only to some of the potentially serious long term environmental problems but also to a more precise definition of future social needs and demands relating to nature protection, human settlement and conditions of life in France. The results themselves are very provisional and should be frequently revised as soon as new datas are available; however it is already possible to conclude that according to the socio economic contexts and the strategies applied the environmental quality index will vary, in France (or at least in some exposed areas), in a proportion of one to five or six.

In the axis of the conference program, it will be given a large part to some methodological aspects of the study. The different points will be treated according to the basic structure drawn in the following scheme (exhibit 1).

᙭ ᙭ ᙭

(1) The author apologizes for any possible mistakes (or misprints) introduced in the text. This one was written directly (without a translator).

(2) The paper presents some elements of a study just now undertaken by a team composed mainly of R. BARRE, J.P. BORDET, A. DRACH, P. MIREMOWICZ and J. THEYS.

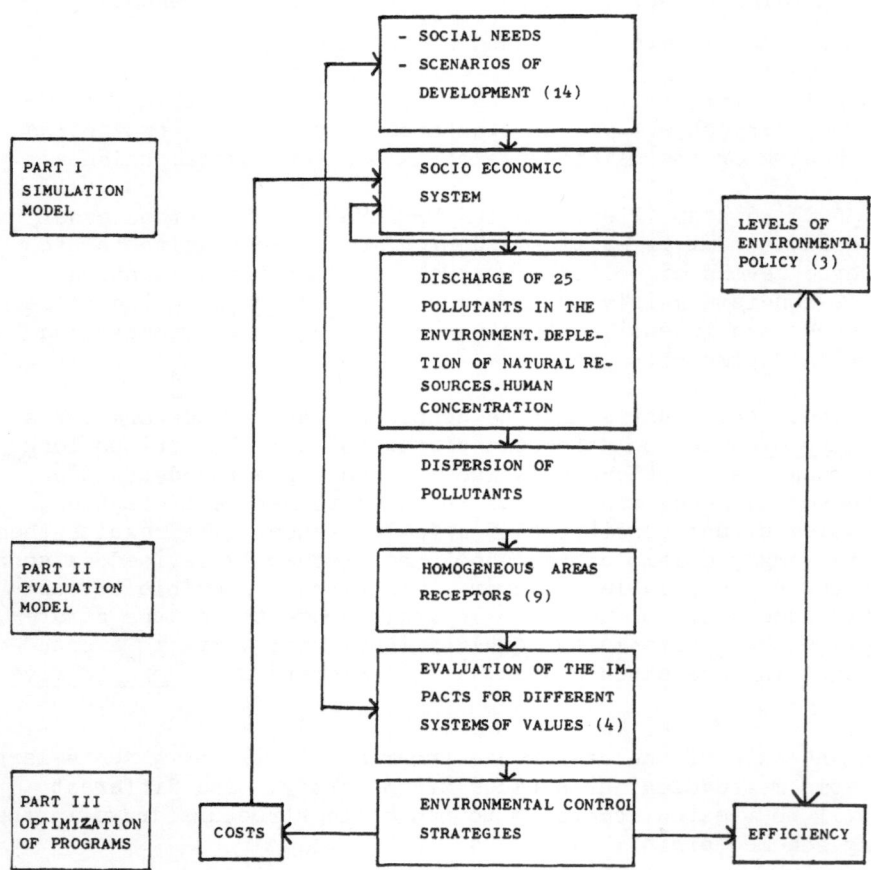

Exhibit 1.   Basic Structure of the Study

## I. SOME BASIC ORIENTATIONS OF THE STUDY

Industrial societies have discovered that they have an "Environment" which is for them both a wealth to protect (or to exploit) and a constraint to respect.  In the last ten years policies have been implemented or reinforced in several countries (in France a Ministry of Environment was created in 1971), but it has become apparent that neither the problem nor the solutions were simple. In fact the environmental quality is the result of manifold actions and processes interacting in a complex unmanageable system.

Between a _laissez faire_ policy and the temptation of a new "enlightened despotism", a broad spectrum of interferences between man and his natural environment is imaginable.  Actually ecologists, engineers, architects, politicians are concerned with different aspects of the problem, and each has his own view on the way the environment is to be controlled.  The ecologist will focus on the limits of the self adaptation of the eco-systems; the conservationist will look for the most effective ways to lead a campaign against the destruction of _local_ natural resources, the engineer will concentrate on the management of energy and raw materials through an industrial process ... This study especially lays emphasis on the relationship existing between the socio-economic system and the environment.  This set purpose determines four major options:

- (1) a broad definition of the environmental issues
- (2) a long term prospect
- (3) a structural analysis
- (4) a dialectical approach of goals and means (a "scenario" approach)

### 1 - The Definition Problem

There is a notable lack of agreement about a definition of the environmental issues, the scale of the problems, the elements to take into consideration.  Administrative divisions and a regard for efficiency tend to favor the approach in specialized fields, reducing the complex socio-environmental system in a series of elementary problems.  So water pollution, air pollution, solid waste disposal, land use, congestion... have been for a long time studied in isolation.  This strategy has resulted in unanticipated side effects - the improvement of air quality leading, for instance, to a solid waste problem.  Furthermore, this fragmented analysis has generally neglected to fully consider the feedbacks between pollution, supply of natural resources, and functioning of ecosystems, forgetting:

- that everything in an ecosystem is related to everything else in the system
- that total waste output ought to be equal to total waste input though the form of the waste may be altered.

In the study a less specialised approach is implemented:

- a broad range of pollutants and natural resources is taken
into consideration.

- an identical process is applied to all pollutants thus providing a rationale for a comparison of investment decision not only
within air, land and water but also among the media.

Furthermore, the concept of pollution is extended to elements
such as noise and congestion

- the synergies between pollutants are fully taken into
account.

Such purpose requires the study to permanently ply between
a national and regional approach as it is not then possible to ignore the geographic specifics of the environmental linkages.


## 2.  A Long Term Prospect

Many reasons plead in favour of a long term evaluation of environmental issues:

- (1) The time required for environmental damages to become
apparent is often important owing to significant delays between the
generation and the effects.

- (2) Once the threshold is exceeded, reversibility is often
very slow.  A forest may be restored in one hundred years, an
eutrophic lake in thirty years.  The Mediterranean sea is renewed
in eighty years and it takes nearly three hundred years for the
Black sea.

- (3) Though emphasis has been, in the past, layed upon flows
of pollutants problems of stocks might become the worry in the
future.  These two problems are quite different (*).

- (4) Implementation of significant environmental changes is
certainly to require adjustments in population, technology, consumption, production, or even modifications of social values.  Results

---

*) The costs necessary to cope with the problems of stocks might be
   ten times higher than the one expended to reduce the flows.

**) Source: Environmental problems: their causes, cures and evolu-
    tion. Using southern California smog as an example (A.KARLIN
    and G.E. KOLMER) The Rand Corporation, May 1971.

are not to be expected without a continuous long term policy.  It
has been estimated for instance, that fifty years will have been
necessary to cope with the problem of Californian Smog (✱✱).  The
same order of magnitude is usual in water planning.

    - (5) The generation of pollutants is very dependent on long
term structures and decisions: a power plant is built for thirty
years, towns or motorways are expected to last from half a century
to at least one or two, some radio-active wastes have a half life
of two hundred centuries.

    Every study in the environmental field has to cope with the
difficult problem of choosing the relevant term.  In that case it
has been estimated that <u>thirty years</u> were necessary to evaluate the
impacts of some foreseen major changes such as growing urbanization,
agricultural modifications, pressures on land use, adaptations to
nuclear energy, increasing needs of water...

## 3. A Structural Analysis

    Since 1970 several efforts have been conducted to develop dif-
ferent policies aimed mainly at controlling flows of pollution through
regulatory approaches.  Many parallel management models have been
built to relate residuals generation and discharges with production
activities, or to assess the relative changes in environmental quality associated with a range of specific actions in terms of cost
benefit analysis.  However, these evaluations have neglected to
consider all relevant factors: social, economic, physical in a systematic way, thus distorting the choices.

    Actually the tools useful to control the quality of the media
in the short term are not necessarily effective to deal with the
environment-economic-population complex, more especially as the efficiency of epuration techniques is limited by increasing costs in
energy and themodynamic constraints.  The building of multisectorial
models seems to be the best way to introduce the relationship and
feedbacks among resources allocation, production, consumption, pollution discharge and demography in a dynamic system.

    However, it must be pointed out that such models are not very
useful to give the decision makers precise quantified results.  As
in system dynamics the emphasis is layed upon structures, relationship  trends more than upon figures.  The study follows basically
a <u>structural approach</u> in so far as it aims mainly to assess the relative importance of some variables assumed to be determinant for
the environmental state.  The model is finally a black box used to
link two categories of variables (socio-economic and environmental)
as follows:

| CHANGES IN | DETERMINE | CHANGES IN |
|---|---|---|
| . DEMOGRAPHY<br>. TECHNOLOGY<br><br>. RATE OF GROWTH<br><br><br>. PATTERNS OF<br>  CONSUMPTION<br><br>. INTERNATIONAL<br>  TRADE<br><br>. LAND USE<br>  PLANNING<br><br>. LEVELS OF<br>  ENVIRONMENTAL<br>  POLICY | | . DISCHARGE OF<br>  POLLUTANTS (25)<br><br>. AMBIANT AIR<br>  AND WATER<br>  QUALITY<br>. USE OF NATURAL<br>  RESOURCES (WATER)<br><br>. LAND USE<br><br><br>. CONGESTION |

4. Dialectics of Goals and Means: the Scenario Approach

       There is no scientific or political evidence of what should
be the goals in the environmental field.  The only principle on
which it is now possible to rely is the standstill principle which
suggests that the present level of pollution should be looked upon
as maximum for the future.  This principle does not say anything
on the needs for a better environment in the long term.

       Since targets are not clearly fixed, the use of optimization
models doesn't seem adequate in spite of the fact that they are
generally more credible for the decision makers (*),  as simulation
techniques lead to an unmanageable number of situations and results.

_____

*)
  A good comparison between the simulation models and the optimi-
  zations models has been done by A. V. KNEESE in "Management
  Science Economics and Environmental Science" R.F.F. September
  1973.

The SCENARIO APPROACH (*) appears to be a transitory compromise between these two methodologies.  It allows the user to select the
different futures which will be examined in considerable detail and
therefore to assess several assumptions about long term environmental objectives.  In the study a special emphasis was set on the opportunity for different users (pressure groups, associations, decision makers ...) to build their own scenario and hence to realize
the dialectical relationship between goals and means.


## 5. What the Study does not Include

In its present form, the study is far from being as systematic
as is theoretically necessary.  Many feedbacks are still ignored:

- the dynamic functioning of the ecosystems is not explicitly
  introduced,

- the macro economic impacts of the environmental investments
  or regulations are only very roughly taken into account(**),

- the global impact of pollution on demographic trends is not
  estimated.

Researches will be undertaken in these different fields but
their results are not expected to be included in the existing models.  A global system including all these elements does not seem
at the present time to be manageable.(***).

---

(*) The scenario approach has been developed in France as a long
    term planning technique by the D.A.T.A.R. (the regional action
    group for land use planning).

(**) By now, it is only assumed that some environmental strategies
    are not consistent with some of the scenarios.  A more serious
    way to assess the macro economic impacts is under construction.

(***)
    One may remember that the number of feedbacks in a complex sys-
    tem is approximately equal to the square of the number of ele-
    ments of the system (n) and that the number of possible states
    of the system equals $N^{n^2}$ (if N = the number of possible states
    for each element).

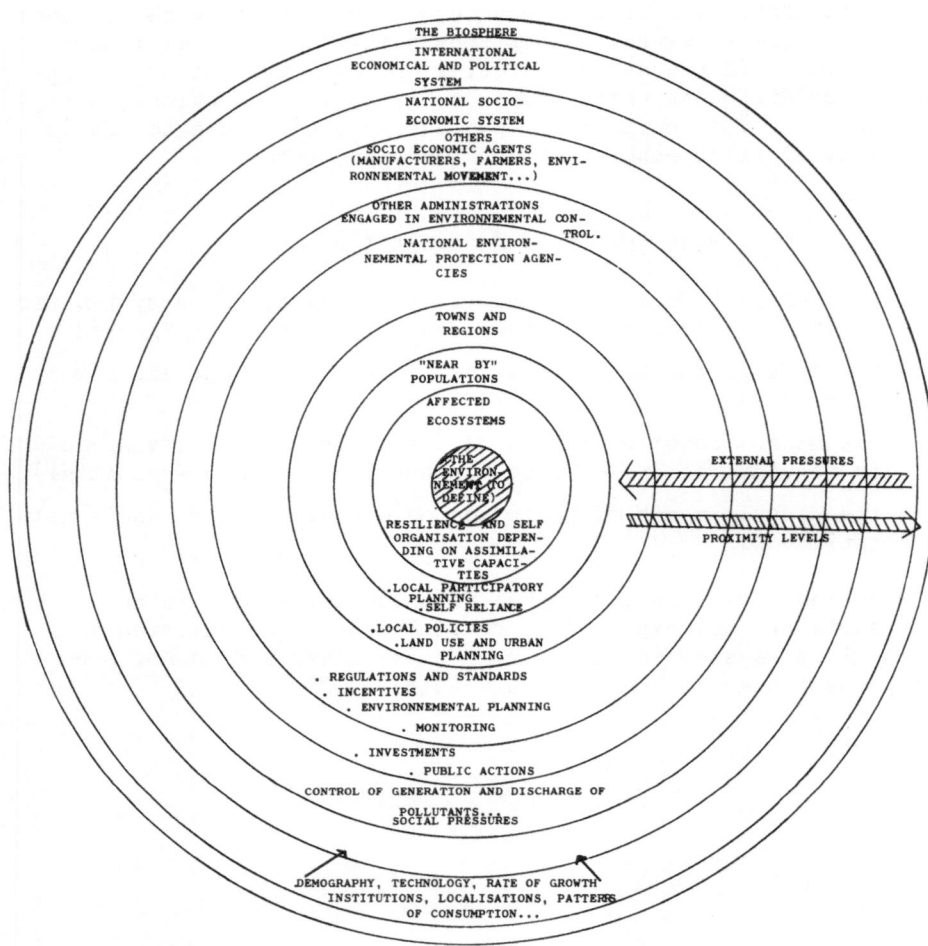

THE BIOSPHERE

INTERNATIONAL
ECONOMICAL AND POLITICAL
SYSTEM

NATIONAL SOCIO-
ECONOMIC SYSTEM

OTHERS
SOCIO ECONOMIC AGENTS
(MANUFACTURERS, FARMERS, ENVI-
RONNEMENTAL MOVEMENT...)

OTHER ADMINISTRATIONS
ENGAGED IN ENVIRONNEMENTAL CON-
                                    TROL.
NATIONAL ENVIRON-
NEMENTAL PROTECTION AGEN-
CIES

TOWNS AND
REGIONS

"NEAR BY"
POPULATIONS

AFFECTED

ECOSYSTEMS

THE
ENVIRON-
NEMENT (TO
DEFINE)

RESILIENCE AND SELF
ORGANISATION DEPEN-
DING ON ASSIMILA-
TIVE CAPACI-
TIES
.LOCAL PARTICIPATORY
PLANNING
.SELF RELIANCE

.LOCAL POLICIES
.LAND USE AND URBAN
PLANNING

. REGULATIONS AND STANDARDS
. INCENTIVES
. ENVIRONNEMENTAL PLANNING
. MONITORING

. INVESTMENTS
. PUBLIC ACTIONS

CONTROL OF GENERATION AND DISCHARGE OF
POLLUTANTS.
SOCIAL PRESSURES

DEMOGRAPHY, TECHNOLOGY, RATE OF GROWTH
INSTITUTIONS, LOCALISATIONS, PATTERNS
OF CONSUMPTION...

EXTERNAL PRESSURES

PROXIMITY LEVELS

2. TIME CONSTRAINTS

A case study : the problem of photochimical smog in California.

|  | APPROXIMATE DATE |
|---|---|
| 1. BEGINNING OF THE PROBLEM | 1940 |
| 2. FIRST PUBLIC REACTIONS | 1943 |
| 3. OFFICIAL RECOGNITION OF THE PROBLEM | 1943 |
| 4. THEORETICAL UNDERSTANDING OF PHENOMENON AND DISCOVERY OF THEORETICAL SOLUTIONS | 1951 |
| 5. PUBLIC ACTION AND PUBLIC DEMAND FOR EFFECTIVE SOLUTION | 1968 |
| 6. APPROVED SOLUTION BY GOVERNMENT | 1968 |
| 7. EFFECTIVE SOLUTION ACCEPTED BY THOS MOST DIRECTLY AFFECTED | 1970 |
| 8. SETTING OF SOLUTIONS | 1975 to 1980 |
| 9. RESOLVED PROBLEM | 1985 to 1990 |

Exhibit 2.   Levels of Control and Time Constraints in Environmental
Policies

## II - THE SIMULATION MODEL

Environmental problems have not hitherto formed a part of a long term development plan of economy or society. This is possibly due to a lack of public interest or to less severe situations in the past, but another reason is the existence of many unknown factors such as the relationship between the amount of pollutants and the human activities. The simulation model used here tries to bridge the gap by connecting different patterns of social needs to both concentrations of pollutants and use of natural resources.

The model includes five steps:

- a first step consists of a quick analysis of the future social needs,

- a second step simulates the ways those needs may be satisfied through the socio economic system.

It results in several distributions of human activities (consumption, production, employment, level of population) on the whole territory, depending on several socio economic scenarios.

- A third step describes the factors influencing the generation, modification and final discharge to the environment of residuals. It links the regional levels of activities previously determined to emissions of pollutants and consumption of water, land and renewable natural resources - using mainly the level of employment as an indicator of production.

- A fourth step estimates the quantities of residuals really discharged in the different media taking in consideration three levels of environmental policy more and more stringent.

- A last step gives some idea of the concentrations of pollutants related to each of the scenarios considered, mainly for water residuals. For this purpose an hydrographic model is tested on a national scale.

The emphasis will be layed upon the four steps as the diffusion model has not yet been completed.


## 1. Some Trends in Social Needs

The objective functions of the model are not explicitly formulated but they may be qualitatively expressed in terms of social need functions: quantities of specific goods that the society wishes to get in a fixed period of time among which environmental and non-market goods are to be included.

The general form of this function is:

$$FSN_t = \sum_{k=1}^{n} (\beta_k \cdot P_o \cdot e^{pt})^{1+\sigma}$$

were

$FNS_t$    = the quantity of goods wished for the period $t$

$\beta_k$    = the quantity of the good k individual and unit of time

$P_o$    = population in time 0 (1970)

p    = rate of population growth

$\sigma$    = a coefficient taking in account the growth and diffusion of the consumption induced by the societal
development

n    = the number of different goods included in the pattern.

α/ In a simulation model it is not necessary to specify a
unique function of social needs.  Nevertheless it remains extremely
useful to have in mind some realistic patterns of consumption and
growth expected in the future as they are the best baseline for
the scenario making.  In the study the investigations were limited
to <u>six basic situations</u>:

. ASSUMPTION 1: High rate of growth without change in the con-
                sumption trends.
. ASSUMPTION 2: High rate of growth with some diversification
                of the consumption pattern:
                        more social goods, less energy use.
. ASSUMPTION 3: Economic instability due to a decrease in the
                productivity rate and to acute social conflicts.
. ASSUMPTION 4: Social and economic instability due to a con-
                tradiction between a desire for a high level
                of consumption and (i) the development of hed-
                onistic values (ii) a decrease in the rate of
                demographic growth or in the rate of activity.
. ASSUMPTION 5: Low rate of growth and regression in the con-
                sumption trends due to severe economic crisis $^{(*)}$
. ASSUMPTION 6: Low or moderate rate of growth with important
                modifications of the consumption pattern due
                to a lasting change in the system of value or
                in the international situation.

Such an assumption may include: a better adaptation to ecological constraints, a better use of natural goods and energy, less
use of fuel, less international trade, less concentration of the
population.

---

$^{*)}$ An ecological catastrophy is not expected to happen in the next
thirty years.

Actually those assumptions are not equally probable in the time and may be distributed along the period as follows:

| periods of time \\ A | 1970-1980 | 1980-1990 | After 1990 |
|---|---|---|---|
| possible assumptions | $A_1$ | $A_1$ $A_2$ $A_3$ $A_4$ $A_5$ | $A_2$ $A_3$ $A_4$ $A_5$ $A_6$ |

$\beta$/  . The demand for environmental goods and services is naturally given an important part in this analysis.  However, this demand presents characteristics which make every estimate very hazardous:

- the demand is very dependent on the local and social context,
- the nature of the goods desired is badly known,
- the demand is not expressed on a market,
- it does not seem that the need for a defined quality of en-environment follows a rational law (statistically expressed by a gaussian curve).  This demand might be rather characterized by the coexistence of two basic behaviours: from one side, a minimalist behaviour characterized by the fact that people are satisfied by low levels of quality; from the other side, a maximalist behaviour characterized by a desire for very high levels of quality.

These two structures of needs may be easily represented by the curves in Exhibit 3.

This consideration is very important on a political point of view, as the strategies and the costs implied by the two situations are quite different.  In the most realistic case (case II) the problem will be to satisfy either the minimalists (by avoiding a severe disruption) or the maximalists (by offering a high quality environ-

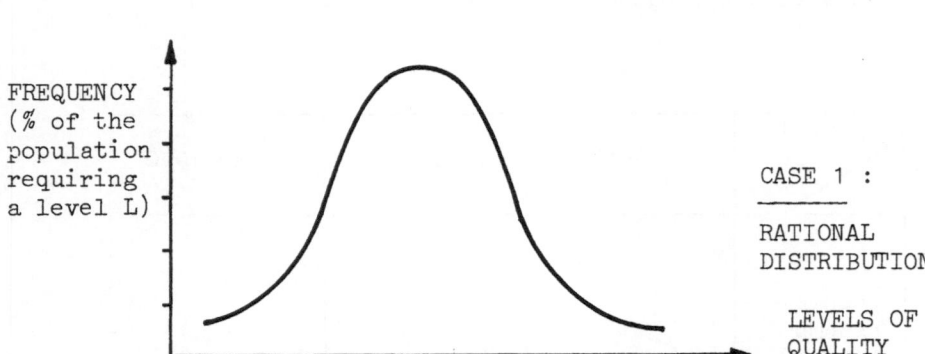

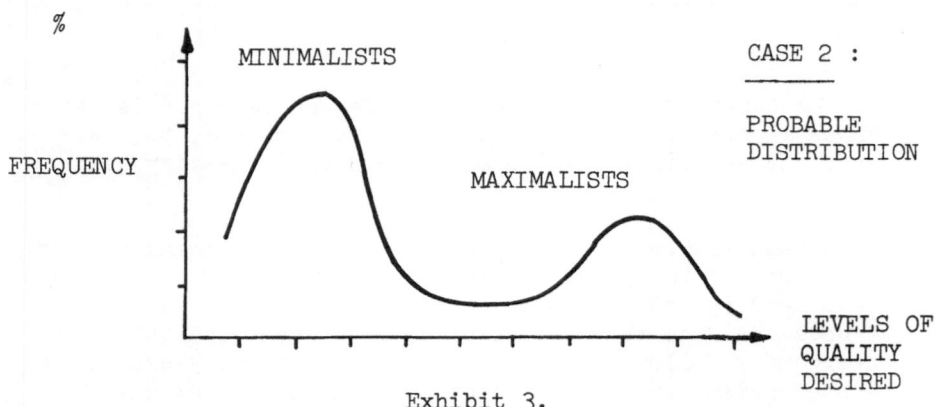

Exhibit 3.

ment).  The problems to resolve are not at all in the same order of magnitude: if it is, for instance, rather easy to prevent an eutrophic lake from biological death, it is extremely long and costly to make this lake come back to an oligothrophic state.  This dysmetry in the needs for a better environment presents naturally some important social aspects (*).

_____

(*) The demand for the environmental goods is  much influenced by the revenue, the social group and the employment.  Environmental objectives may be largely affected by redistributive aspects.

## 2.  The Socio-Economic Module

The socio-economic module (or model) constitutes the very heart of the study.  It consists mainly in the simulation of an iterative process of equilibration between the offer and the demand of employment on a national and regional level.  It results in a distribution of populations and activities on the whole territory depending on the following variables. (termed: "scenario variables in the study").

- consumption patterns
- structures of import and export        } determining the
- rate of growth                              final demand
- rate of investment
- energy and technological trends        } determining the in-
  and structures                              put-output coeffi-
- input of raw materials                      cients
- demographic trends and structures
- internal and external migration
- rate of activities                     } determining the
- projected localization                   distribution of
- land use policy                          activities
- urbanization process

The functioning of the model is described by the sequence drawn in the next exhibit (3).

Three such models have been built as parallels for energy, agriculture and touristic activities.

α/ The input-output matrix is classically used to get the gross production levels corresponding to different sets of final demand.

For this purpose the inter-industry coefficients have to be extrapolated using the R.A.S. technique.  It is possible to introduce in this phase substitution of products or technologies, and change in the consumption of raw materials.

β/. The process of equilibration between the offer and the demand of employment on a regional level is simulated using, first a demographic model, and second an economic model which estimates the future distribution of the employment by sector of activity given the demographic trends and the past distribution (a gravitational technique is applied).  In this process, internal and external migrations, rate of activity, local productivities, and even the national level of production may be modified.  Urbanization trends and land use policies are also, taken into account.

Finally the model may be characterized by its ability to assess

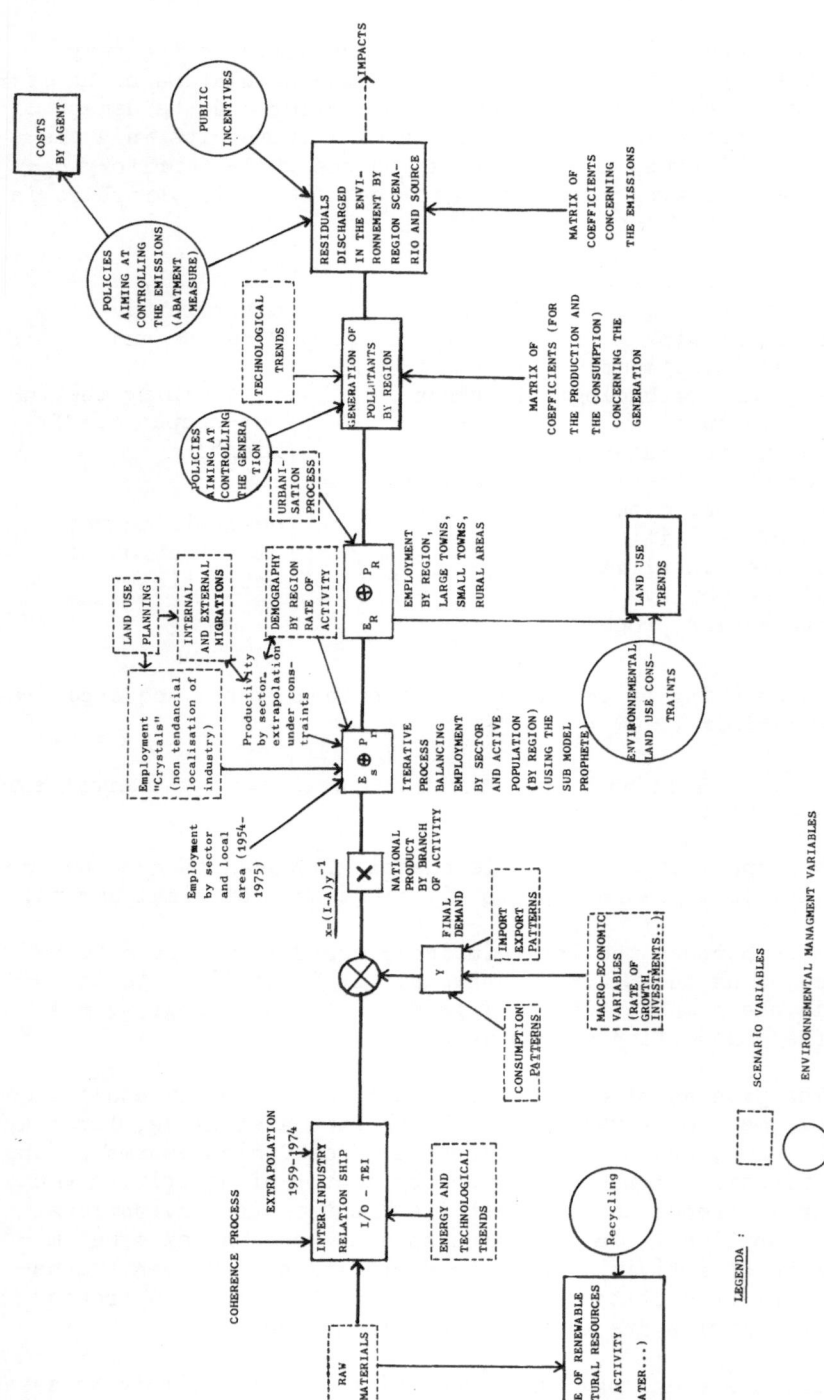

Exhibit 4.  General Scheme of the Simulation Model

a broad range of structural modifications – sometimes marginal for the socio-economic system but of a great significance for the environmental quality (*).


3. Estimating Quantities of Residuals Generation and Discharge

Using the levels of activities as data it is not theoretically difficult to estimate the residuals generation.

In practice, however, three main problems hampered the task:

– From a scientific point of view there is no rational process to choose the relevant pollutants (**). In the study the choice was made according to the present importance of the residuals – what might be quite unsatisfying for the future. (see exhibit 4).

– From a statistical point of view it is difficult to get accurate information on generation processes and abatement systems – as data on each plant or even industry are lacking. In the study the changes in the residuals discharges due to modifications of the industrial structures has been taken into account at only a regional level (***).

– Residuals generation coefficients were not available in France for every production or consumption process. One important task of the study was to estimate systematically all these coefficients both for combustion, consumption and production processes.

Finally residuals discharge was considered as a function of the following variables:
- the employment level (for industrial activities)(E)
- the population (for consumption)(P)
- the use of energy and raw materials (oil, coal, electricity) (RM)
- the regional industrial structure (age of the plants)(IS)
- the characteristics of the product output or product specifications (PS)

_____

(*) For instance modifications in the intermediate products used by the consumers to satisfy a similar need.

(**) Two millions of chemical products are now in use in the western countries.

(***) As the level of employment is used to estimate the pollutions (through a residual generation coefficient by employee), the regional level of productivity may reflect differences in the industrial structure.

| POLLUTANTS | CODING | POLLUTANTS | CODING |
|---|---|---|---|
| . ARSENIC        (*) | ARS | . NITROGEN OXIDES | $NO_x$ |
| . ASBESTOS       (*) | ASB | . NOISE RESIDUALS      (**) | NOIS |
| . BIOCHEMICAL OXYGEN DEMAND AND CHEMICAL (IN WATER) | MO | . ODORS     (*) | ODOR |
| . CARBON MONOXIDE | CO | . OIL AND PETROLEUM (IN WATER AND SOIL) | OIL |
| . COLIFORMS      (*) | COL | . ORGANIC AND NON ORGANIC MATTERS IN THE SEA. | OMINS |
| . CONCRETE      (**) | CONC | | |
| . FLUORINE, HCL | FLU | . PARTICULATES | PRT |
| . HEAT | HEAT | . PESTICIDES (EXCEPT INS) | PES |
| . HEAVY METALS | HEVM | . PHOSPHATE | PHO |
| . HYDROCARBONS (IN THE AIR) | HC | . RADIOACTIVE WASTES | RADW |
| . HYDROCARBONS (IN THE SEA) | HYDS | . RADIONUCLIDES | RAD |
| . INSECTICIDES (ORGANO-CHLORINES) | INS | . RESIDUALS FROM TREATMENT FACILITIES AND SEWAGE | RESE |
| . IRON, CHROMIUM, ZINC (*) | IRON | . SULFUR OXIDES AND SULFA-TES | $SO_x$ |
| . LEAD | LEAD | . SUSPENDED SOLIDS AND FLOA-TING MATERIALS (WATER) | MES |
| . NITRATE | NIT | . SOLID WASTE | WAST |
| | | . TOXIC ORGANIC CHIMICAL COUM-ROUNDS | CTOX |

(*) Not included in the simulation model.
(**) Treated separately in the simulation model.

Exhibit 5.   Coding of Pollutants

- the technology of residuals modifications (TRM)
- the discharge constraints (DC)

Such a function may be reduced as follows:

$$R\,D\,(I) = f\,(\underbrace{E,\ RM,\ IS,\ PS,}_{\text{GENERATION}}\ \underbrace{TRM,\ DC}_{\text{EMISSIONS}})\ (\ast)$$

The main interest of such a formulation is to establish a clear distinction between the generation – depending on industrial structure or technologies of consumption and production – and emissions – directly related to an environmental policy.

At the end of the two proceeding steps the simulation model results in:

a) – quantities of pollutants by scenario, by region and by source;
b) – consumptions of water, wood, sand, phosphate and land by scenario – on a national level;
c) – indicators of congestion by region and scenario.

The results presented in this paper will concern mainly the discharge of pollutants.

## 4. Applying the Model

Thirteen socio-economic scenarios have been simulated, each combined with three different environmental policies ($EP_1$, $EP_2$, and $EP_3$).

In this manner the possible combinations of the numerous variables were in fact reduced to some of the most likely and significant "futures"'.

In each of these scenario the variables are evolving in accordance with the following trend:

---

$(\ast)$ Such a function was developed in the United States for the national residuals discharge inventors (January 1976).

| Scenario | Land use Planning | I/O technologies | Rate of growth | consumption pattern | Import Export | Y | Generation | | |
|---|---|---|---|---|---|---|---|---|---|
| | | | | | | | 1970 | 1995 Prod | 1995 Cons |
| $A_1$ | T | VII Plan | 5.7% | T | T | $Y_1$ | 70.1 | T | T |
| $A_2$ | T | VII Plan | 5.7% | T | T | $Y_1$ | 70.2 | T | T |
| B | T | I/O 1974 | 5.7% | T | T | $Y_1$ | 70.1 | T | T |
| $C_1$ | T | NT | 5.7% | T | T | $Y_1$ | 70.1 | T | T |
| C | T | NT | 5.7% | T | T | $Y_1$ | 70.1 | T | T |
| D | NT | VII Plan | 5.7% | T | T | $Y_1$ | 70.1 | T | T |
| E | T | NT | 5.7% | T | NT | $Y_2$ | 70.1 | NT | T |
| F | T | VII Plan | 3.6% | T | T | $Y_3$ | 70.1 | T | T |
| $G_1$ | T | VII Plan | 5.7% | NT | T | $Y_4$ | 70.1 | T | NT |
| $G_2$ | T | VII Plan | 5.7% | NT | T | $Y_4$ | 70.1 | T | T |
| H | | NT | 5.7% | NT | NT | $Y_5$ | 70.1 | NT | NT |
| I | NT | NT | 5.7% | NT | NT | $Y_5$ | 70.1 | NT | NT |
| J | T | NT | 3.6% | NT | NT | $Y_6$ | 70.1 | NT | NT |

with $\begin{cases} \text{T = extrapolated trend} \\ \text{NT = non tendencial evolution ($\ast$)} \\ \text{VII}^{\text{th}} \text{ Plan = trend expected by the VII the French national} \\ \qquad \text{plan.} \end{cases}$

The discharge of pollutants induced by these 27 scenarios (13 multiply by 1, 2 or 3 abatement policies), are summarized in the exhibit 5.

But in fact the main conclusions of this simulation process are not to be found in these figures.

---

($\ast$) It is not possible in this paper to give an accurate description for each of the variables considered, of the so-called "non-tendencial evolutions" (NT). In the case of land use planning for instance, a non-tendencial evolution would signify:

- a best balance between the east and the west part of the country
- a higher increase of the small towns
- some constraints on the carrying capacity of some old industrial areas.

All these modifications may be taken into account by the simulation model.

Much more information may be extracted from a <u>structural ana-lysis</u> of the system.

The tendencial scenario ($A_1$ + $Ep_1$) is thus assumed to lead to an increase of 125 percent of the total pollution between 1970 and 1995. Applying the most severe policy ($Ep_3$) will reduce this increase to 25 percent - but at a very high cost. A non-tendencial evolution <u>of all the</u> variables would induce, on the contrary, an important decrease in the total amount of residuals (fifty percent). The scenarios A,B,C,D,E,F are expected to give rise to worse environmental situations in 1995 ( /1970) and the scenarios G,H,I,J to better situations. On a more detailed scale, the land use planning, for instance would have the same impact that a middle range abatement policy ($Ep_2$). A systematic analysis of all these trade offs is now in progress.


III - THE EVALUATION MODEL

The previous calculations provide series of <u>numbers</u> that are of a highly questionable value as a measure of actual impacts on nature and human welfare. A next step is then necessary to set priorities between the different environmental issues and define the outer limits of the economic growth. It is the purpose of an evaluation model.


1. Developing the Model

The classical premise of this model is that pollution damage occurs when a pollutant adversely affects a receptor of pollution. Naturally equal amounts dispersed in different places will not cause even roughly identical damage. It is therefore assumed that the damage functions which link explicitly pollutants to receptors have no significance except for small geographic homogeneous areas. This involves a preliminary division of the whole territory in sub-regions presenting similar pollutant and receptor profiles.

Within each of these areas, the evaluation procedure consists in 1) defining the elements or receptors affected, 2) evaluating their presence or density over the area, 3) estimating the impact of each pollutant upon each receptor and 4) evaluating the total damage by pollutant by summing the impacts to the individual receptors.

The national priorities are finally set by summing the regional results: each scenario is associated with a global indicator of risk, a hierarchy of pollutants and a ranking of polluting activities.

From what has already been said, it is clear that the geogra-

Unit : T. $10^3$
C. $10^2 - 10^9$ ther./year

| | CO | SOZ | NOX | PRT | HCA | LEAD | RAPA | NIT | PHOS | MO | MES | HCW | RADW | HEAT | EQTOX | PES | WAST |
|---|---|---|---|---|---|---|---|---|---|---|---|---|---|---|---|---|---|
| 70R | 6238 | 3083 | 1313 | 2340 | 932 | 10,2 | 80 | 451 | 83 | 3820 | 3512 | 126 | 800 | 89,3 | 34,8 | 30 | 16752 |
| 70Ø | 4336 | 2739 | 1306 | 1883 | 923 | 10,2 | 80 | 236 | 64 | 2461 | 2568 | 126 | 800 | 89,3 | 34,8 | 0 | 9573 |
| 80R | 7246 | 3474 | 1718 | 2501 | 799 | 12,6 | 1258 | 510 | 107 | 3372 | 3012 | 199 | 5411 | 186,4 | 31,4 | 41 | 16201 |
| 80Ø | 4945 | 3049 | 1706 | 1973 | 789 | 12,6 | 1258 | 219 | 81 | 2249 | 2222 | 199 | 5411 | 186,4 | 31,4 | 0 | 9133 |
| ARAA 1 | 21092 | 5431 | 3606 | 3109 | 1958 | 28,7 | 3505 | 597 | 230 | 4442 | 3354 | 557 | 22217 | 399,3 | 54,8 | 55,1 | 12453 |
| 2 | 12131 | 3722 | 2414 | 1909 | 1316 | 28,7 | 1753 | 557 | 191 | 2488 | 1743 | 557 | 11284 | 255,2 | 27,4 | 55,1 | 9572 |
| 3 | 4961 | 2606 | 1484 | 1530 | 802 | 28,7 | 350 | 495 | 93 | 1332 | 1015 | 557 | 2538 | 139,9 | 6,8 | 55,1 | 7139 |
| A ØAA1 | 14764 | 4697 | 3594 | 2571 | 1941 | 28,7 | 3505 | 202 | 194 | 2576 | 2383 | 557 | 22217 | 399,3 | 54,8 | 0 | 6884 |
| BRAA 1 | 22047 | 7309 | 4114 | 3368 | 2323 | 28,7 | 4708 | 458 | 217 | 3927 | 3173 | 561 | 34242 | 575,8 | 50,4 | 35,7 | 12452 |
| CRAA 1 | 21082 | 4422 | 3363 | 2930 | 1863 | 28,7 | 3089 | 497 | 221 | 4121 | 3223 | 557 | 18054 | 348,8 | 52,5 | 41,1 | 12453 |
| CRBA 1 | 21082 | 1954 | 3045 | 2959 | 1622 | 28,7 | 3089 | 497 | 221 | 3359 | 2873 | 552 | 18054 | 277,3 | 42,3 | 41,1 | 12453 |
| 2 | 12121 | 1483 | 1906 | 1772 | 980 | 28,7 | 1544 | 456 | 182 | 1845 | 1467 | 552 | 9203 | 193,9 | 21,2 | 41,1 | 9572 |
| 3 | 4951 | 1130 | 1003 | 1410 | 465 | 28,7 | 309 | 394 | 84 | 1011 | 878 | 552 | 2122 | 127,1 | 5,3 | 41,1 | 7139 |
| DRAA 1 | 21075 | 5433 | 3604 | 3085 | 1958 | 28,7 | 3505 | 597 | 223 | 4391 | 3275 | 557 | 22217 | 399,3 | 54,8 | 55,1 | 13806 |
| ERBA 1 | 20946 | 1901 | 3035 | 2797 | 1620 | 28,7 | 2950 | 497 | 221 | 3221 | 2817 | 552 | 16667 | 264,8 | 41,5 | 41,1 | 12453 |
| 2 | 11985 | 1449 | 1898 | 1707 | 977 | 28,7 | 1475 | 456 | 182 | 1756 | 1431 | 552 | 8509 | 187,2 | 20,7 | 41,1 | 9572 |
| 3 | 4815 | 1115 | 996 | 1350 | 463 | 28,7 | 295 | 394 | 84 | 967 | 859 | 552 | 1983 | 125,1 | 5,2 | 41,1 | 7139 |
| FRAA 1 | 12475 | 4239 | 2393 | 2451 | 1268 | 16,1 | 2580 | 597 | 161 | 3668 | 3027 | 315 | 12967 | 283,3 | 33,8 | 55,1 | 12451 |
| 2 | 7438 | 2848 | 1700 | 1474 | 907 | 16,1 | 1290 | 557 | 136 | 1974 | 1524 | 315 | 6659 | 194,5 | 16,9 | 55,1 | 9570 |
| 3 | 3408 | 2052 | 1171 | 1181 | 618 | 16,1 | 258 | 495 | 75 | 1076 | 906 | 315 | 1613 | 123,6 | 4,2 | 55,1 | 7137 |
| GRAB 1 | 4466 | 4865 | 1488 | 2960 | 685 | 6,5 | 3181 | 597 | 144 | 3991 | 3224 | 68 | 18979 | 359,6 | 48,8 | 55,1 | 8729 |
| 2 | 3837 | 3351 | 1362 | 1747 | 640 | 6,5 | 1591 | 557 | 123 | 2198 | 1653 | 68 | 9665 | 234,9 | 24,21 | 55,1 | 6712 |
| 3 | 3334 | 2431 | 1286 | 1484 | 603 | 6,5 | 318 | 495 | 70 | 1187 | 970 | 68 | 2214 | 135,1 | 6,1 | 55,1 | 5000 |
| HRBB 1 | 4301 | 1675 | 945 | 2646 | 396 | 6,5 | 2811 | 497 | 135 | 2809 | 2700 | 64 | 15279 | 253,0 | 35,2 | 41,1 | 8730 |
| 2 | 3672 | 1298 | 870 | 1543 | 351 | 6,5 | 1406 | 456 | 114 | 1509 | 1356 | 64 | 7815 | 181,1 | 17,6 | 41,1 | 6713 |
| 3 | 3169 | 1039 | 817 | 1300 | 315 | 6,5 | 281 | 394 | 61 | 843 | 822 | 64 | 1844 | 123,7 | 4,4 | 41,1 | 5001 |
| IRBB 1 | 4282 | 1672 | 942 | 2618 | 395 | 6,5 | 2811 | 496 | 130 | 2758 | 2621 | 64 | 15279 | 253,0 | 35,2 | 41,1 | 9680 |
| JRBB 1 | 3065 | 1406 | 764 | 2211 | 296 | 3,5 | 1489 | 496 | 101 | 2329 | 2444 | 35 | 7724 | 179,2 | 22,2 | 41,1 | 8728 |
| 2 | 2725 | 1096 | 716 | 1256 | 272 | 3,5 | 745 | 456 | 86 | 1219 | 1205 | 35 | 3960 | 142,4 | 11,1 | 41,1 | 6711 |
| 3 | 2454 | 906 | 684 | 1038 | 252 | 3,5 | 149 | 394 | 52 | 698 | 746 | 35 | 949 | 113,1 | 2,8 | 41,1 | 4999 |

Exhibit 6.   Quantitative Results of the Simulation Model-
Discharge of Pollutants.

phic level of analysis is quite different from that which is used
in the simulations model.  So, the evaluation procedure begins with
the building of a new data base.

   a). The data base.  The inventory process consists of collec-
ting and organizing all datas necessary for the following computa-
tions.

   Small areas, or CELLS, are the units used to record all the
data on pollutants and on physical and socio demographic characteristics of receptors.  Actually the information has only been systematically gathered for the starting baseline of the scenarios (1970
to 1975).  For the future it is assumed that the cells belonging to
the same regions (22) are evoluting according to the same profiles.

   To fit with available data the whole territory was divided into approximately 400 cells corresponding either to an intermediate
French administrative division (the "Arrondissements") or to the
main urban districts.  Unfortunately, it was not possible to take
into consideration ecological areas, which is one of the main limits
of the model.

   All the data collected on each pollutant and receptor were
coded according to levels 0,1,2,3,4.  From these data are extracted
fifty computer-maps which give a rough description of the physical
and socio-economic characteristics of the country.

   b). The determination of homogenous areas.  For easier mani-
pulations the data base may be compressed with minimum loss of information since some of the cells may be considered as having the
same environmental problems.  The technique used for this aggregation is a cluster analysis. (*).

   The process includes three states: a definition of homogeneous
areas based on level of pollutants (1), a definition based on levels
of receptors (ecological and social structure) (2), a combination
of (1) and (2).

   Clustering (data analysis) is a key feature of the methodology
for several reasons.  First, it reduces the data base to a manageable
set of entities.  Second, clustering considerably lessens sensitivity
to bad information.  Furthermore, it can help spot individual cells
which might justify further local investigations.  Third, clustering
determines a hierarchy of areas where problems are similar in nature
and magnitude.

_____
(*) Clustering defines groups of cells which are the most alike among
themselves and the most dissimilar vis à vis other groups through an
iterative process using a special type of factor analysis. Clustering
is performed by computer.

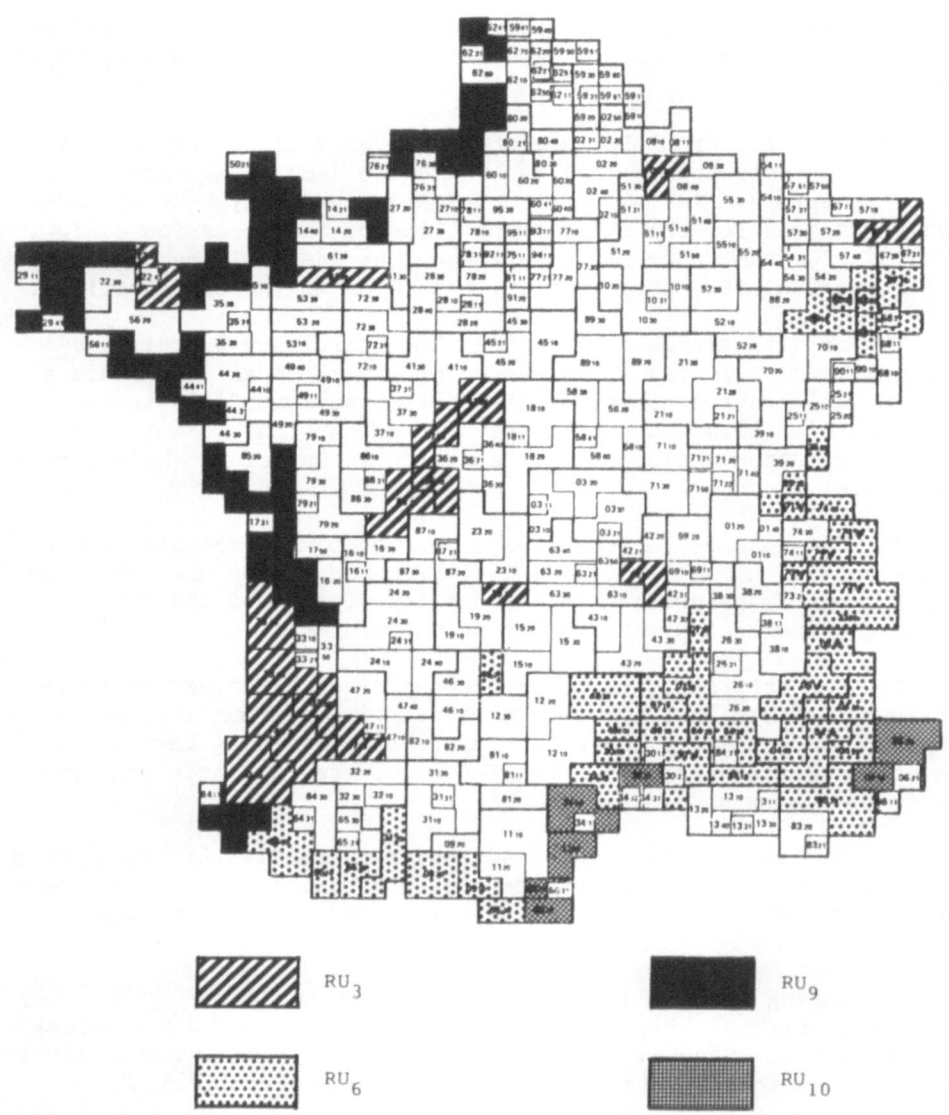

Exhibit 7.  Cell Clustering According to Pollutants and Preindicators
(Rural Areas)

For these reasons, the "socio-ecological regions" generated by the cluster analysis are a very practical and effective policy tool. They are themselves an important result of the study (see exhibit 6).

c). Definition of receptors. Another original feature of the model consists in a systematic and functional definition of the receptors. These receptors are formed by combining five different elements (or "populations"): Human population (H), natural resources (N), man-made elements (M), space (S) and Time (T) with four value functions: economic (E), use (U), patrimony, affectivity and quality (A). All these elements are thus not concrete physical entities, but concepts used to describe the total environment. They are themselves the result of an aggregation in which all the major elements of damage are taken in account. (see exhibit 7).

Actually the description has been temporarily limited to nine of those entities, corresponding to combinations of three sets of "populations" and of value functions, as follows:

The Matrix of Receptors

|  |  | POPULATIONS | | |
|---|---|---|---|---|
|  |  | H | N | M |
| VALUE | E | HE | NE | ME |
| FUNCTIONS | U | HU | NU | MU |
|  | P | HP | NP | MP |

d). Evaluation of the presence of receptors. Each of the nine receptors has to be assigned a number (or indicator) which indicates its level of presence in the area considered. Nevertheless, the indicator figures cannot be defined directly since receptors are only abstract concepts. They are in fact computed from more elementary components of the receptors terms PREINDICATORS.

Two criterias determine which of these preindicators will be taken into consideration for the evaluation process: major elements of damage to receptors and availability of data. It is set by approximation that a linear relation can be formulated between receptors and preindicators.

An example would help to make clear the form of these rela-

| SYSTEMS OF VALUATION / RECEPTORS | ECONOMIC VALUES | USE VALUES | PATRIMONIAL VALUES | SENSORIAL AND AFFECTIVE VALUES |
|---|---|---|---|---|
| MAN | . ABSENTEEISM AND DE-CREASE IN PRODUC-TIVITY<br>. ECONOMIC COST OF HEALTH | . PHYSICAL AND MEN-TAL HEALTH IMPAI-REMENT<br>. DECREASE OF NON MARKET ACTIVITIES | . GENETIC DAMAGES | . IMPAIREMENT OF SENSORIAL PERCEP-TIONS<br>. NUISANCE<br>. FEELING OF INSECU-RITY |
| NATURAL GOODS AND RESOURCES | . DAMAGES IN TERMS OF: DECREASE OUT-PUT DUE TO ALTERA-TION OF COMMERCIAL-LY USED NATURAL GOODS : CROPS, WATER, FOREST IN VALUE<br>. COST OF REPAIRING THE NATURAL GOODS | . LOSS OF AMENITY DUE TO ALTERATION OF NATURAL GOODS<br>. DECREASE IN RECRE-ATIONAL OPPORTUNI-NITIES<br>. MISUSE OF NATURAL RESOURCES | . DETERIORATION OF BIOLOGICAL PATRI-MONY<br>. DECREASE IN THE DIVERSITY OF THE ECOSYSTEMS<br>. LONG TERM MODI-FICATIONS OF THE ECOSYSTEMS, CLI-MATE LANDSCAPES | . AESTHETICAL AND INTELLECTUAL AL-TERATION OF NATU-RE<br>. DECREASE OF THE FEELING FOR NATU-<br>. |
| MAN MADE GOODS | . DECREASED OUTPUT DUE TO ALTERATION OF MAN GOODS IN VALUE<br>. CLEANING COSTS<br>. CORROSION OF PIPES STRUCTURES, MATE-RIALS | . LOSS OF AMENITY OF THE URBAN SET-TING<br>. DECREASE IN USE AND DURABILITY OF MAN MADE GOODS | . DAMAGES TO ELE-MENTS OF HISTO-RIC, CULTURAL AND ARCHEOLOGI-CAL VALUE<br>. LONG TERM MODIFI-CATIONS OF THE URBAN ECOSYSTEMS<br>. DECREASE IN THE DIVERSITY OF MAN MADE GOODS | . AESTHETICAL DETE-RIORATION OF THE MAN MADE ENVIRON-MENT (CITIES...)<br>. DIRTINESS |
| SPACE AND LAND | . DECREASE IN LAND VALUE<br>. COSTS OF SOIL MAINTENANCE | . MISUSE OF LAND<br>. EROSION<br>. OVER DENSITY | . IRREVERSIBLE LAND USE<br>. DECREASE IN THE OPPORTUNITIES OF MAN SETTLMENT (DESERTS...) | . IMPACTS ON :<br>. AFFECTIVE SPACE<br>. INDIVIDUAL TERRI-TORY<br>. INTIMACY |
| TIME | . DECREASE OF THE PRODUCTIVE TIME | . DECREASE OF THE RELEASE TIME<br>. LIFE DURATION | . LOWER OPTION VA-LUE FOR THE NEXT GENERATIONS | . PERCEPTION OF TI-ME :<br>. PESSIMISTIC OPI-NION ON THE PRE-SENT, THE PAST OR THE FUTURE |

Exhibit 8.   Matrix of Damages by Receptor.

tions: the impact of pollutants on the use of natural goods. It is
assumed that density of natural goods (use function) is proportional
to: agricultural potential (AGP), forestry potential (FOP), natural
recreation sites (NRS) and accessability (ACC). The higher the level of presence of these elements, the more use of natural resources.
Hence the following equation is used:

$$DNU = AGP + FOP + 4 \ NRS + ACC$$

The model includes nine of these equations - one for each
receptor.

Such formulations, of course, are the result of an arbitrary
compromise. However, in the present stage, there are no undeniable
starting points for this kind of evaluation.

e). Evaluation of damages to receptors. To evaluate damage
to receptors (i.e. to preindicators) requires a set of damage functions which can indicate the levels of impact in each receptor related to different concentrations of each pollutant.

At this stage the specific vulnerability of preindicators and
synergistic effects between pollutants can be considered. (*)

The approach assumes that each area is really homogeneous in
terms of pressure on the environment and of assimilative capacity -
what is only very roughly observed in reality.

It should be apparent that available information falls far
short of that needed for such evaluation. Epidemiological studies
don't yet lead to evident conclusions (**) and researches on interactions between human activities and ecosystems are far away from
reaching to undeniable measurements. Even less is known for other
receptors. However, it is possible to use a "crude weighting" of
qualitative information. Two empirical techniques were used in the
model:

---

(*) For instance, thermal pollution increases the damages associated
with MO and toxics. Particulates and $SO_x$ have also synergetic
effects.

(**) Such a study has been undertaken in France for two years - to
assess air pollution impacts.

GEOGRAPHIC INDICATORS

1. Farm land
2. Pasture land
3. Forest land and potential
4. Coastal line
5. Rivers
6. Lakes and marshlands
7. Erodability
8. Slope
9. Groundwater potential and vulnerability
10. Categories of ground
    - 1
    - 2
    - 3
    - 4
    - 5
11. Agricultural potential
12. Precipitation (rainfalls)

 - Socio-economic Indicators

13. Resident population
14. Per capita income
15. Level of energy consumption
16. Old and young population
17. Amount of workers (percent)

Functional Indicators

18. Agricultural productivity
19. Agricultural population
20. Wild life density
21. Recreational potential and sites
22. Recreational revenues (in rural areas)

 - Level of tourist activities (in urban areas)
23. Forestry revenue
24. Water use
25. Level of medical consumption
26. Accessibility.

---

Exhibit 9.  Preindicators of Receptors


     - a systematic analysis of the dose-response relationships
quantified in the literature has been completed and the results
transformed to be included in the impact matrix.
     - A large survey of experts in the environmental field was
simultaneously conducted using a DELPHI procedure.

As a result of these two investigations all pollutant damage were ranked, per area and per receptor on a 0-5 or 0-10 scale - which gives a rationale to weight concentration or discharge data by a numerical factor (SPECIFIC SEVERITY INDEX).

It must be pointed out that <u>such functions are non-linear</u> as damages are supposed to increase exponentially ($D = e^f$ with $f$ = impact figures).

<u>f). Evaluation of total damage by area.</u> Given the level of presence of the different preindicators or pollutants and the damage functions the model computes the total damage by area. However, for two basic reasons, this cannot be done directly.

First, the damage functions have different units of measure across receptors (it is not possible or even wise to aggregate effects on human health and on natural patrimony ...) and, second, even if the unit of measure were the same, the relative importance of each receptor is, because of its conceptual nature, basically dependent upon a <u>VALUE SYSTEM</u>.

These difficulties have been however alleviated by using weighting factors and defining a preference function ranking the nine receptors.

In this case also two different methods were applied:

- in a first step, a multicriteria analysis (✻) provided an adequate framework to simulate fourteen kinds of weights, out of which four were finally extracted as the most interesting for the decision makers. Such distributions of weighting factors are termed "PHILOSOPHIES" in the model.

- In the second step a more empirical approach was conducted with the help of professionals in the environmental field and of representatives from different socio-economic groups. It was asked to this panel to distribute among the nine receptors 1000 damage points according to their own preferences. The answere were used to compute the two average patterns which follow. (✻ ✻).

---

(✻) The model used was developed in France by the Society of Economics and applied Mathematics under the name of <u>ELECTRE II</u>.

(✻ ✻)
The survey yielded various patterns of answer which were classified in function of the different backgrounds of the people interviewed (using clustering method and data analysis).

| RECEPTORS | HE | NE | ME | HU | NU | MU | HP | NP | MP |
|-----------|-----|-----|-----|-----|-----|-----|-----|-----|-----|
| Reference pattern $\Sigma$ = 1000 | 91 | 59 | 35 | 159 | 121 | 81 | 176 | 226 | 52 |
| More "economist" pattern | 151 | 89 | 65 | 199 | 201 | 81 | 56 | 136 | 22 |

These two distributions could have been applied as non-monetary functions of preference without any reference to social groups, regional objectives, systems of value or even time. In fact it seemed much more interesting to keep different sets of weights representing alternative "philosophies" and so to reintroduce the idea of fluctuating environmental needs. Theoretically each scenario has so to be associated with one system of value.

Actually the model takes into consideration five value systems, each tending to emphasize one of the potential objectives of an environmental and social strategy:

- VS 0: corresponding to the reference pattern,
- VS 1: emphasis on economical factors and on human health (anthropocentric philosophy),
- VS 2: emphasis on nature conservation on use.
- VS 3: emphasis on quality of life (use of man-made goods and of the urban environment, emphasis on visible effects of pollutants),
- VS 4: emphasis on long term considerations (conservation of the genetical, ecological and cultural patrimony, high option value for the future generations).

For this purpose, the model uses different patterns of answer given by the questionary to calibrate the figures selected in the multi-criteria analysis. Finally, the total damage by area is obtained through the following process:

Average concentrations (c) of pollutant for the area (A)

$\downarrow$

Average presence of the preindicator P

$\downarrow$

Average presence of the receptor R = f(P)

$\downarrow$

Impact on the receptor R = f (A, R, C)

$\downarrow$

Aggregation of the impacts on the nine receptors according to the philosophy Ph:

I = f (A, R, C, Ph)

Global damage per area - per philosophy and scenario (synergistic effects of all the pollutants).

g). Global evaluation of the scenarios. The summing up of the preceding results on the geographic dimension leads to three kinds of classification which provide a global evaluation of the most plausible scenarios, for the whole territory:

1 - a ranking of scenarios per value system - function of the total damage.
2 - a ranking of pollutants per value system and scenario.
3 - a ranking of areas per value system and scenario (*).

These results may be analysed in many different ways but the main interest of the approach is to test the stability of the hierarchies associated with several socio-economic contexts and value systems and with different valuations of the impacts.

As a matter of fact, the numerous assumptions made in the model rise some suspicions about the validity of the results.

An important work was undertaken to show how sensitive the damage figures are to these choices: the conclusions are that even with rather arbitrary measures of impact it is possible to reach to a stable partition of pollutants, areas, or scenarios, in at least two categories, inducing severe or not severe problems of environment.(**).

---

(*) It is possible to divide the "quantities" of damage by the population, the surface area, the revenue...which provides other sets of useful indicators.

(**) For instance, because of the uncertainties associated with damage functions, it was necessary to take into account the sensitivity

h). Evaluating the monetary costs of damages? The main idea
underlying the preceding functional approach is that the market determined costs do not necessarily capture the whole value of the elements affected; indeed, some do not have a market value or even a
use value in the economical sense of this term.  Under those conditions the notion of monetary cost of damages is somewhat contradic-
tory with the general philosophy of the model.

However, some economic policy makers would prefer to have information on damage functions put into monetary value terms in a
form which can easily be used in benefit cost calculations.  It may
also be useful to have only a rough idea of the costs implied in the
environmental disruptions as a ratio of the gross national product.

Therefore a test has been undertaken in this direction so as
to balance the global costs and the global benefits of some forecast
strategies.  In the state of knowledge this aspect of the study must
be considered only as an exercise.

Actually the method applied is very simple.  In the last step
of the evaluation process, the pollutants were defined by a special
quantity of points reflecting the total damage, caused on each receptor.  An economic expression of these figures may be obtained by
affecting to each point a monetary value.  In this case it was possible to calculate a rough estimate of this point of damage using
different monographs established in France for some selected pollutants such as particulates, carbon monoxyde and noise (*).

The main difficulty of this approach was in fact to ensure a
minimum of homogeneity between the monographs and the model as regards the definition and weight of receptors affected: an iterative
process was necessary to get a satisfactory average value of each
damage point.

The results are comparable with those obtained, by different
ways, in some west countries.


2. Applying the Model

The model was applied to assess the reference scenario. The
damages were estimated for 1970 and 1995 only.

_____

of the results to change in the damage figures.  Two sets of functions were prepared: the base case representing the generally prevailing view and a more "conservative" case attributing a higher
mark to INCERTAIN effects, such as impacts of nuclear radiations.
*)Reference: the social cost of noise residuals due to ground trans-
portation. MQV 1976.

The necessary information was recorded in 386 cells (292 rural cells and 94 urban cells). It took nearly 20,000 data to describe the situation of the environment in France in 1970. The cluster analysis defined six urban homogeneous areas and ten rural based on the simultaneous presence of pollutants and preindicators (*). Using the methodology summarized above, the damages associated with twenty-five pollutants were evaluated for the five value systems. Some of the main results (ranking of pollutants, damages, functions...) are given in the following exhibits (9, 10). Finally, the monetary cost of the damages was estimated as an exercise (see exhibit 10).

As shown in exhibit 9, the ranking order of pollutants is some-what sensitive to changes in value system. However, some pollutants remains in the first group whatever the philosophy adopted: it is the case for solid wastes, particulates and nutrients. The situation is quite different, for instance, for noise or radiations, which are very dependent on the value system chosen. Some pollutants will be-come a more serious problem in the future if no supplementary measure is taken: noise, nitrogen oxides, heat, radiation, residuals from treatment plants.

---

(*)

The clustering method divided the whole country into nine categories of rural areas and six categories of urban areas.

However it is possible to schematically reduce these divisions in four great kinds of region:

1. The Mediterranean coast, Alpine and Pyrennean "Piedmont" characterized by a strong pressure of the activities and a great ecological vulnerability,

2. The North-west, characterized by a very strong pressure of the activities but a moderate vulnerability (high population and intensive agriculture.

3. The North-east characterized by a large concentration of industrial activities and localized vulnerable areas.

4. The West and Center parts of the country characterized by a low pressure and a moderate vulnerability (the coastal zone excepted.).

| | 1970 | | | | 1995 | | | |
|---|---|---|---|---|---|---|---|---|
| | | | SYSTEMS | OF | VALUE | | | |
| | VS1 | VS2 | VS3 | VS4 | VS1 | VS2 | VS3 | VS4 |
| 1 | PRT | NIT | WAST | PRT | CTOX | PES | NOIS | CTOX |
| 2 | PES | INS | PRT | INS | NOIS | NIT | WAST | HEVM |
| 3 | CTOX | WAST | CONC | CTOX | PES | WAST | CONC | PES |
| 4 | $SO_x$ | PES | NOIS | PES | COL | HEAT | RESE | HEAT |
| 5 | COL | PHO | ODOR | WAST | $NO_x$ | PHO | PRT | RAD |
| 6 | WAST | HYDS | MO | PHO | WAST | CONC | ODOR | RADW |
| 7 | NOIS | OMINS | $SO_x$ | NIT | PRT | CTOX | MO | HYDS |
| 8 | MO | MO | HYDS | HEVM | MO | HEVM | $NO_x$ | NIT |
| 9 | $NO_x$ | LEAD | PHO | HYDS | RESE | INS | $SO_z$ | RESE |
| 10 | NIT | $SO_x$ | MES | OMINS | $SO_x$ | HYDS | HEAT | WAST |
| 11 | MES | FLU | OIL | $SO_x$ | HEVM | MO | CTOX | PHO |
| 12 | HEAT | HEUM | CTOX | LEAD | MES | OMINS | PHO | $NO_x$ |
| 13 | PHO | CTOX | CO | RAD | NIT | RADW | HC | PART |
| 14 | FLU | CONC | NIT | $NO_x$ | PHO | $SO_x$ | NIT | LEAD |
| 15 | HYDS | HEAT | FLU | COL | OMINSQ | LEAD | HYDS | IRON |
| 16 | OMINS | MES | OMINS | MO | HYDS | RESE | OMINS | INS |
| 17 | OIL | PRT | $NO_x$ | CONC | HEAT | FLU | CO | $SO_x$ |
| 18 | INS | IRON | HEAT | HEAT | RAD | RAD | OIL | MO |
| 19 | LEAD | OIL | PES | IRON | LEAD | IRON | COL | OMINS |
| 20 | HEVM | RADW | HC | RADW | HC | PRT | RAD | HC |
| 21 | CO | HC | RESE | FLU | RADW | MES | MES | CONC |
| 22 | CONC | RAD | LEAD | NOIS | CONC | $NO_x$ | PES | FLU |
| 23 | HC | $NO_x$ | ASB | RESE | INS | NOISE | HEVM | MES |
| 24 | ODOR | NOIS | HEVM | MES | FLU | HC | FLU | OIL |
| 25 | RAD | CO | RAD | HC | OIL | ARS | RADW | NOIS |
| 26 | RADW | ARS | RADW | OIL | IRON | OIL | LEAD | COL |
| 27 | IRON | ASB | INS | ASB | CO | ASB | ASB | CO |
| 28 | ASB | RESE | COL | ARS | ODOR | CO | INS | ASB |
| 29 | RESE | COL | IRON | CO | ASB | COL | IRON | ARS |
| 30 | ARS | ODOR | ARS | ODOR | ARS | ODOR | ARS | ODOR |

Exhibit 10.   Ranking of Pollutants per value system.

For the reference pattern the evaluation leads to a rough classification of the pollutants in five categories as follows: (in 1973).

| CATEGORIES | POLLUTANTS | IMPACT FIGURES |
|:---:|---|:---:|
| 1 | NOISE, SOLID WASTE | 180 to 130 |
| 2 | DISSOLVED OXYGEN, NITRATED PARTICULATES, NITROGEN OXIDES, SULFUR AND SULFATES | 85 to 67 |
| 3 | PESTICIDES, PHOSPHATES (NUTRIENTS) HYDROCARBONS, HEAT, SUSPENDED MATTER | 52 to 37 |
| 4 | CADMIUM -BERRYLIUM, MERCURY, RADIATIONS COLIFORMS LEAD, FLUORINE, DISCHARGE OF HYDROCARBONS IN WATER AND SEA | 30 to 14 |
| 5 | CARBON MONOXIDE, CHROMIUM, ASBESTOS, CYANIDE, PHENOLS, ARSENIC | 7 to 1 |

It is clear that this classification does not take into account the potential severe problems caused by high concentrations of toxic products in specific conditions of time and location: it does not tell anything about the severity factors peculiar to each pollutant.

. As a final result it is possible to estimate that the social monetary costs of the pollution associated with the damages recorded would represent in 1970 - according to the most plausible assumptions not far from three to four percent of the gross national product. (*)

Some supplementary work will be necessary to know how this "burden" is distributed among the regions and social groups.

_____

(*) According to the same calculations the damages would increase less fast than the national product. This global damage represented 770 francs per inhabitant in 1970 and 980 in 1973.

UNITE: Millions de Francs 1973

| POLLUANTS | Note D'Impact relatif en 1970 | Emissions en 1970 (en Tonne Curies Milliard de Thermies | Note de gravité spécifique (1/(2)) | Emissions en 1973 (estimation) | Dommages en 1973 Hypothèse Haute (M F) | Dommages en 1973 Hypothèse Basse (M F) | Note d'Impact relatif en 1973 |
|---|---|---|---|---|---|---|---|
| PARTICULES | 95 | 1.539.000 | $61,7 . 10^{-6}$ | 1.172.000 | 3515 | 2812 | 72,3 |
| CO | 6 | 6.006.000 | $0,99 . 10^{-6}$ | 6.504.000 | 315 | 252 | 6,5 |
| NOx | 59 | 1.3>5.000 | $43,9 . 10^{-6}$ | 1.600.000 | 3412 | 2730 | 70 |
| SO2 - SULFATES | 60 | 2.990.000 | $20,1 . 10^{-6}$ | 3.372.000 | 3305 | 2644 | 67,8 |
| HC - (AIR) + ALDEHYDES... | 32 | 920.000 | $34,9 . 10^{-6}$ | 1.120.000 | 1895 | 1516 | 39 |
| FLUOR | 11 | 23.000 | $488 \cdot 10^{-6}$ | 29.000 | 690 | 552 | 14,1 |
| AMIANTE CHLORE (PVC) NH+4 | 2 | - | - | - | 116 | 92 | 2,4 |
| SOUS-TOTAL "AIR" | - | - | - | - | 13248 | 10598 | - |
| CADMIUM-BERRYLIUM | 27 | 246 (Cad) | $109756 \cdot 10^{-6}$ | 278 (Cad) | 1480 | 1184 | 30,5 |
| PLOMB | 17 | 10.300 | $1652 \cdot 10^{-6}$ | 12.900 | 1040 | 832 | 21,4 |
| MERCURE | 25 | 60 | $416666 \cdot 10^{-6}$ | 72 | 1458 | 1168 | 30 |
| CHROME | 4 | 2.300 | $1792 \cdot 10^{-6}$ | 2.700 | 235 | 188 | 4,8 |
| SOUS- TOTAL METAUX LOURDS | - | - | - | - | 4213 | 3372 | - |
| RADIATIONS (Tritium...) + Déchets R.A | 20 | 88.000 | $227 \cdot 10^{-6}$ | 113.900 | 1170 | 936 | 23,8 |
| POLLUTION THERMIQUE | 29 | 110.000 | $264 \cdot 10^{-6}$ | 145.000 | 1852 | 1484 | 38,2 |
| SOUS-TOTAL | - | - | - | - | 3022 | 2420 | - |
| MATIERES OXYDABLES | 79 | 2.331.000 | $34 \cdot 10^{-6}$ | 2.525.000 | 4180 | 3344 | 85 |
| MATIERES EN SUSPENSION | 34 | 2.656.000 | $11,9 \cdot 10^{-6}$ | 3.100.000 | 1798 | 1436 | 37 |
| HYDROCARBURES (eau + mer) | 12 | 283.000 | $42,4 \cdot 10^{-6}$ | 340.000 | 695 | 556 | 14,5 |
| CYANURES | 2 | 58 | $34.000 \cdot 10^{-6}$ | 67 | 112 | 88 | 2,3 |
| PHENOL | 2 | - | - | - | 112 | 88 | 2,3 |
| ARSENIC | 1 | - | - | - | 54 | 44 | 1,1 |
| AUTRES TOXIQUES | Non Estimé | - | - | - | - | - | - |
| COLIFORMES - VIRUS | Non Estimé | - | - | - | - | - | - |
| SOUS-TOTAL "EAU" | - | - | - | - | 6951 | 5556 | - |
| NITRATES (Engrais) | 70 | 626.000 | $112 \cdot 10^{-6}$ | 714.000 | 3880 | 3104 | 79 |
| PHOSPHATES(Eng+ Détergents) | 41 | 90.000 | $455,5 \cdot 10^{-6}$ | 104.000 | 2300 | 1940 | 47,3 |
| PESTICIDES | 46 | 30.000 | $1536 \cdot 10^{-6}$ | 36.000 | 2508 | 2004 | 51,6 |
| SOUS-TOTAL | - | - | - | - | 8688 | 6948 | - |
| DECHETS SOLIDES | 123 | 21.000.000 | $5,85 \cdot 10^{-6}$ | 22.260.000 | 6324 | 5044 | 130 |
| BRUIT (Transports) | - | - | - | - | 8645 | 8645 | 177 |
| BETON | (11) | - | - | - | - | - | - |
| TOTAL | 960 | | | | 51091 | 42583 | |

Exhibit 11. Provisional Evaluation of the Damages Related to Residuals Discharge.

## IV - OPTIMIZATION OF ENVIRONMENTAL MANAGEMENT STRATEGIES

Three kinds of works are now in progress to assess the long term strategies available to cope with the environmental problems previously stressed:

. A systematic investigation and review of a broad range of programs corresponding to the three levels of abatement policy tested in the simulation model (by sector of activity or consumption)

. The shaping of a methodology aiming at optimizing combination of abatement strategies and structural modifications.

. The development of optimization models dealing with specific problems such as management of water investments or industrial pollution (*).

Results are not yet available.

### CONCLUSION

A great amount of work will be still necessary to improve the quality of the results and explore the scenarios in a more detailed way. What may be already said is that environmental policy will soon come to a great turning point where it will appear that reinforcement of environmental control techniques must be associated with structural modifications of the economy and way of life. However, it cannot be expected that the strategies aiming at preventing residuals discharge or depletion of natural resources will lead to effective results in the short or even the middle term. In this respect the present decade may be considered as a transitory period.

---

(*) Such optimization models are developed separately.

# ENVIRONMENTAL CONTROL: A METHODOLOGY FOR PLANNING CHANGE IN PUBLIC HOUSING

Richard M. Fenker

Texas Christian University

Fort Worth, Texas, U.S.A.

## INTRODUCTION

The dilemma of large scale public housing is that on the one hand it is perhaps the only reasonable option available to cities for sheltering people in high density, low income areas while on the other hand it represents an unhappy solution which is costly both from an economic and social standpoint and which often exaggerates many of the problems it was intended to solve. The present paper represents, in part, a report[1] of the activities of a group of architects, social scientists and housing officials concerned with the problem of improving the physical and social environment of a housing project in Dallas, Texas. Although work on the Dallas project is far from finished the majority of the planning process has been completed. The present paper examines this process and other related experiences as a basis for improving planning strategies with future projects.

The paper is organised into the following four sections: (1) A description of the West Dallas Project, the proposed changes, and my involvement with the planning team; (2) A discussion of the theoretical basis, that of "control", which provides a common framework for interpreting resident behaviors and evaluating the impact of physical and social changes; (3) A brief outline of a methodological approach for the measurement of residents' "control" needs and the evaluation of whether changes in the physical and social environments effectively satisfy these needs; and, (4) A discussion of how recent social changes within the housing community (initiated primarily by the project's resident council) and changes in the physical structures (as a result of the architects and planning team) influence the resident's control of their environment.

## THE WEST DALLAS HOUSING PROJECT

During the next two years approximately ten million dollars will be spent in an effort to "modernize" the West Dallas Housing project. The project consists of conventional one and two story apartment buildings constructed in the early 1950's. Although the housing is open to all races, and a balanced mixture was desired, at present almost 100% of the 15,000 residents are Black. Approximately 5,000 of the residents are single mothers or elderly individuals and the remaining 10,000 residents are children. The project has a high crime rate, garbage disposal problems, rats, poor property maintenance and a variety of other difficulties associated with low income public housing.

The modernization program is unusual in that the Housing Authority and Planners are concerned not only with improving the physical environment but also with using the physical changes as a basis for modifying the present social systems. The planners or "Design Team" for the project consisted of the project architects, representatives of the housing authority and members of the project's resident organization. Although the modernization program is intended to bring about certain desirable social changes (e.g., elimination of rats, crime reduction, improved property maintenance) through modification of the physical environment, the program's primary focus is on "nuts and bolts" repairs to present dwellings plus some rerouting of streets and landscaping work. My intended role as an environmental psychologist was to assist in evaluating the social systems present in the project, primarily those concerned with resident needs, and then to provide specific input to the design team on the direction physical changes should take. Evaluation of the effectiveness of the changes was a second proposed area of responsibility.

## PHYSICAL AND SOCIAL ENVIRONMENTAL CONTROL

To a project resident almost all important issues depend on the question of control. To what extent can residents regulate the physical, social and economic environments in which they exist? West Dallas project residents and residents in similar housing arrangements in urban settings nationwide generally suffer as a consequence of having very little control over their environment. Residents perceive themselves as having almost no influence over the flagrant crime, as not being able to improve the general cleanliness of the project or eliminate rat populations, as not having much effect on the general quality and maintenance of their facilities, and as not being able to regulate the privacy and security of their grounds and apartments.

A number of researchers who have studied public housing projects have identified various aspects of the control problem. Altman's (1975) concept of privacy as "a changing self/other boundary-regulation process in which a person or a group sometimes wants to be separated from others and sometimes wants to be in contact with others" succinctly summarizes the issues related to social control. Newman's (1972) analysis of primary and secondary territories and crime statistics illustrates the importance of territorial control. Other research which demonstrates the relationship between the residents' ability to successfully modify or alter their living environment (make necessary repairs, add furniture, paint, change facilities) and both their satisfaction with the environment and how well they maintain it suggest the necessity of physical control. In some instances (the Brownsville Project in New York is a good example) simply providing residents with certain kinds of territorial controls brought about a dramatic reduction in crime. In other cases some outside incentives are necessary for control mechanisms to become effective.

Although there are perhaps other theoretical perspectives that would prove useful for communicating, planning, and evaluating I chose the concept of control because: (1) It seems to include the entire range of problems and needs associated with public housing, from the user's perspective; (2) It helps translate these problems and needs into explicit, behavioral, operational language; and, (3) It provides a convenient, interpretive framework within which the influence of both physical and social changes can be assessed. Discussions with project residents and planners plus the results of a "need" survey (conducted by the resident council) suggested that "control" was deficient in at least seven areas for many residents. These areas are:

1. Personal security. This includes the ability to protect oneself from theft, assault, vandalism and other types of crime. An important component of security is a change in psychological attitude about the project from fearful and guarded to comfortable and safe.

2. Control of secondary areas. This type of control is obviously closely related to personal security. It encompasses Newman's concept of "defensible space" in that it refers to control of secondary areas such as hallways, sidewalks or yards. In the most general sense this form of control represents the ability to regulate the behavior of others in areas outside of the home. Thus, whether residents use the streets, the playgrounds, or any areas surrounding the home will depend to a large extent on the effectiveness of secondary controls in delimiting the activities that occur in these areas.

3. Accessability of needed services. Despite the availability of policemen, community services, information or assistance for home repair, various training schools, public transportation, grocery stores, shopping areas and parks many of these kinds of "services"

are not <u>accessible</u> to project residents.  Fear (in some areas a
walk to or from the bus stop can be a harrowing experience), location (because a supermarket is not within walking distance many
people buy from extremely overpriced sidewalk vendors) and ignorance
(a relatively small percentage of people benefit from the training
programs or use the child care center which is available) often prevent residents from utilizing these services.  Thus, project residents have very limited control over many kinds of facilities which
are taken for granted in other parts of the city.

    4.  <u>Privacy</u>.  Because of crowded conditions within the home and
the large percentage of children (10,000 of the 15,000 project residents) privacy regulation both within the home and in surrounding
areas is often difficult.  As Altman (1975) points out, "privacy"
depends on one's ability to control the access of others and is
therefore not necessarily related to population density or crowding
per se.

    5.  <u>Correction and repairs control</u>.  This type of control con-
cerns the residents' ability to fix, repair, correct (or have someone else do these things) anything that goes wrong.  Normally such
control would apply specifically to problems with the apartment,
the management or neighbors where someone with special skills or in
a particular administrative position is required.

    6.  <u>Economic control</u>.  The economic problems of project resi-
dents, their incentives for being on or off welfare and their strategies for dealing with these issues would fill a book.  Needless to
say one of the first actions of a resident gaining some measure of
economic control has typically been to move out of the project.

    7.  <u>Control of self-image</u>.  A typical resident's self-image is
influenced to a considerable degree by the social mores of the project and by the fact that he or she lives in the project (which connotes various economic, social and personal characteristics) rather
than somewhere else in Dallas.  The project is perceived by residents
to be a very distinct community, with its own laws and behavior patterns, and <u>not actually part of the city of Dallas</u>!

    As I mentioned above, this list is not necessarily exhaustive
but it does provide a fairly specific set of guidelines, <u>from the</u>
<u>perspective of a project resident</u>,for assessing the impact of vari-
ous changes in the physical environment.


                    A METHODOLOGICAL APPROACH

    The methodology described below was in large part developed on
the basis of my limited experience with the Dallas project.  It represents an attempt to describe the "ideal" approach to be followed

by a planning team responsible for changing the physical and social
environment of public housing (or any other large socio-economic
system). The steps are perhaps obvious to anyone familiar with systems or organizational analysis. Unfortunately they have seldom
(to my knowledge) been applied to the domain of public housing.

1. <u>Assessment of the control needs of project residents</u>. This
includes defining the different areas of control and answering
through the use of questionnaires, interviews, or resident consultants the following questions: (1) Which areas of control are of
primary importance; (2) What deficiencies exist in each area; (3)
How are the different types of control interrelated; and, (4) What
changes (physical or social) will help improve controls? It will
be important to carefully examine the perceived needs of residents
to insure that needs unrelated to "control" are not overlooked.

2. <u>Determination of specifically how each area of control will
be affected by the possible changes in the physical environment</u>.
Social changes which result as a consequence of the physical changes
should also be considered.

To anyone concerned with housing issues, Step 2 sounds difficult
if not impossible. There is very little systematic knowledge relating specific types of physical changes to behavioral changes in
housing project residents. Newman's work on control of secondary
areas is an excellent example of an attempt to collect such knowledge.
The best a group of planners can do is to utilize information on the
impact of particular physical changes if available; but if the needed information is not available then use residents or specialists
to make "good guesses." If the results of these guesses are evaluated (as is recommended below) eventually a catalog of "design-
behavior" knowledge can be constructed. In practice, I think planners will find that if they can agree on a theoretical framework
which breaks the complex array of resident behaviors into a comprehensible system of needs and strategies for achieving these needs,
then hypotheses about the impact of physical changes are likely to
be somewhat correct and can at least be evaluated.

3. <u>Implementation</u>. Implement the design changes with the best
judged potential for improving the resident's control in needed areas.

4. <u>Evaluation</u>. Evaluate the results of the physical changes.
This is also an incredibly difficult task because of the confounding
of physical changes, social influences, normal behavioral fluctuations and the dependent measures such as control related behaviors
or estimates of need satisfaction. Multivariate evaluation methods
are obviously needed to separate the various components which can
influence behaviors. Such evaluation methods are available, however,
and have been used in industry for many years despite their complexity.

The key to the success of the above methodology is the willingness of the planners to focus at the level of specific, measurable
behaviors, to hypothesize about the effect of design changes on
these behaviors, and to evaluate the results in terms of these behaviors.  In the next section where specific changes are discussed
one can see more clearly how the theoretical issue of control and
the willingness to consider specific behaviors resulted in a concrete
set of design proposals.


DESCRIPTION OF THE PHYSICAL AND SOCIAL CHANGES AND THEIR JUDGED
                 IMPACT ON THE AREAS OF CONTROL

In order to be consistent with the previous discussion the
changes are discussed in the context of the seven types of control.
Also note that in many cases particular changes may influence more
than one type of control.

1.  Personal security and secondary area control.  Approximately
half of all the proposed project changes were directed toward increasing personal security and control of secondary areas.  Because a
development which improves security in one of these areas concomit-
ently improves security in the other area both types of control are
discussed together.

a.  The back yards of many apartments will be fenced to limit
access to residents living in adjacent apartments.

b.  The elderly will be moved to one of two areas of the project.
Both areas have public transportation and other services in close
proximity.

c.  A number of low fences will be used to designate front yards,
while other fences will divide the project into discrete areas of
approximately 100 units.

d.  The 100 unit areas called Porteros will be organized around
a central service center containing a laundromat, vending machines
and possibly some form of food service or craft sales.  Individuals
in the project will be permitted to run business-like operations in
conjunction with the service center.  The major business, run by a
"resident manager," will be the laundromat and its related services.

e.  A number of through streets will be converted to dead ends
in order to reduce the amount of casual traffic and auto theft which
typically occurs.

f.  An armed, resident patrol trained and approved by the Dallas Police, will be on 24 hour duty throughout the project.

g. Current areas of high density will be broken up by replacing apartments with playgrounds, parks and other public facilities.

In summary, these changes are intended to divide the large project into a collection of 100 unit residential communities. A variety of physical changes are intended to reinforce the boundaries of these communities and permit residents to limit access to outsiders and to regulate behavior within the communities. The use of resident police will hopefully help resolve some of the current problems between residents and the Dallas Police (at present a relatively small percentage of crimes, threats, problems are reported to the Dallas Police).

2. Access to services.

a. One of the major goals of the portero concept with its resident manager was to provide some important services within each small community. A well cared for laundromat and collection of vending machines was judged to be very important by project residents.

b. Negotiations are underway concerning the construction of a shopping center in a vacant area near the project.

c. The city bus routes will be changed to conform to the new pattern of streets and communities within the project. Plans for a special intra-project transportation service are also under consideration.

3. Privacy. Because the population density of the project is comparable to middle and upper class projects throughout the city (all the apartments are one or two-story buildings) the planners made no special proposals for the purpose of enhancing personal privacy regulation. It would appear that many of the changes discussed in (1) could, however, have a favorable impact on privacy control.

4. Correction and repairs. The possibility of a resident-run repair center for apartment problems within each portero is being considered. The newly formed Resident Council has, however, already demonstrated its effectiveness as a community organization capable of obtaining needed services and repairs for the residents. The council has also proved valuable as a liason between the project and the housing authority, the police, and other community officials. A number of policy changes and recommendations for action in specific cases involving project residents have resulted from the resident council's requests.

5. Economic changes. At present there is little incentive to take a low-salaried job rather than exist solely on welfare since this would result in an immediate rent raise! A rent scale that creates positive instead of negative incentives for employment is

being discussed. Also the possibility of a craft center, open to
the city, where residents can sell their work is being considered.

6. <u>Self-image</u>. In a weak sense it is hoped that by improving
the physical makeup of the project, residents will take more pride
in the facilities and regard the project as "not such a bad place
to live." The above objective is weak in that one cannot reasonably
hope that a few new fences and painted rooms will erase the social
stigma the City of Dallas connotes to project residents and which
is obviously part of the resident's self-image. The planners' goal
in this area was through physical means to let people both inside
and outside the project recognize the residents of the project as
simply citizens of West Dallas rather than as members of a special
community. Changes in bus routes, locating services outside the
project in the center of the West Dallas community and opening the
project to public activities (with a park, golf course, or stadium)
are directed at this purpose.


## CONCLUSION

Although the types of control described above are probably necessary (in varying degrees) for most Western cultures, it would be
unwise to interpret "how necessary" or "to what extent" these controls are important to the project residents on the basis of our
typical life experiences. Altman's discussion of privacy perhaps
best illustrates this point. It is not that an individual lives in
a densely populated area or has a home with six children and two
rooms that determines whether his or her privacy needs are being
satisfied. It is the issue of control that matters. If the person
selected these "crowded" conditions by choice and has the potential
to change them then some measure of control exists and pathological
behaviors associated with uncontrolled crowding are not likely to
occur. Similarly, it is not the economic condition, crime rate,
state of repair of the apartments, self-image, etc. which define
the quality of life, per se, for project residents but their potential to exert some measure of control over these circumstances if
desired.

The first two steps of proposed methodology are completed. Data
on specific needs and behaviors was collected. A variety of relationships between the needs and the proposed physical changes were
hypothesized. The previous section of the paper summarized some of
these hypotheses within the theoretical framework of "control." At
present many of these changes are being implemented in the Dallas
project. Evaluation hopefully will follow.

---

1 The author accepts full responsibility for the opinions and
ideas presented in this paper.

ENVIRONMENT AND PUBLIC POLICY IN THE UNITED STATES: A DIFFUSION

ANALYSIS

Vijay Mahajan, Manoj Agarwal

State University of New York at Buffalo

New York, U.S.A.

## INTRODUCTION

Comparative research on legislative behaviour in the United
States has focused on the characteristics of states that are useful
in predicting their behavior. These characteristics range from
socio-economic conditions (Crittenden, 1967; Hofferbert, 1966) to
political organizations (Jacob and Lines, 1965; Sullivan, 1973)
while the state behavior being analyzed has ranged from welfare expenditures (Dawson and Robinson, 1963; Dye, 1966) to policy outputs
(Crew, 1970; Hofferbert, 1966). This paper presents a continuation
of that tradition. In this case the behavior we are analyzing represents the state adoption of new policies to protect its environment. From the perspective of the adopting state, the policy is an
innovation, not meaning that the state originated it but that the
policy is new for that state.

Adoption of an environmental policy by a state represents a
response by its legislature to harmonize its social system and its
environment to ensure an 'optimum' quality of life. Whether the environmental damage is caused by technological development (Commoner,
1971), population growth (Ehrlich and Holdren, 1972; Meadows, 1972),
economic growth or scarcity of resources (Hines, 1973), adoption of
an environmental policy represents an action by the individual state
to minimize social losses generated by the impairment of the total
physical environment (air, water, and land use), and the deterioration in social environment (work environment, living environment,
recreation, food, consumer protection, etc.).

This paper studies this response or action of the United States
over time. More specifically, we study the diffusion (spread) of

441

twenty-eight environmental policies (physical and social) among the
forty-eight states (mainland). The policies selected satisfy the
following criteria:

(a) they are non-mandatory -- i.e. the states have the option of adoption and (b) they are equally applicable to all states -- e.g. strip
mining policy is not relevant to all states and such policies are
not included in the analysis. A list of these policies is given in
Table 1 (see Annex). We have selected thirteen policies in the area
of health, education and welfare, five in consumer rights and ten
in physical and natural environment. Table 1 gives also the number
of states in each of the ICPR (inter-University Consortium for Political Research) regions which have not adopted these policies. The
adoption time span for these policies varies from four years for auto
vehicle safety (adopted by forty-four states) and waste treatment
agency (adopted by thirty-five states) to eighty-three years for
conservation of oil and gas (adopted by forty states). Approximately
65% of these policies have been adopted by forty-four or more states.
Ten policies, in Table 1, have been adopted by all the forty-eight
states. The time of adoption of these ten policies for all the states varies from nine years (for soil and water conservation districts)
to seventy-three years (for welfare agencies) which indicates that
rate of diffusion for individual policies is different. However,
which American states are more likely to adopt environmental policies
earlier than the others? What are the underlying socio-economic and
political dimensions that facilitate the adoption of these policies?
This paper will concentrate on these two questions. More specifically, based on the twenty-eight policies selected, this paper will
(a) develop an aggregate innovation index for the American states,
(b) identify adopter categories (innovators, laggards, etc.) based
on the innovation index using clustering techniques and, (c) identify the underlying socio-economic and political factors that may
facilitate the adoption of these policies using multidimensional
scaling techniques.


DIFFUSION OF INNOVATIONS:  A BRIEF REVIEW

The process by which an innovation spreads is called diffusion;
it consists of the communication of a new idea in a social system
over time. There are more than eighteen research disciplines in
the social sciences sharing this concept with each having different
applications.[1] For example, anthropologists have been concerned
about how new ideas and practices diffuse from one culture or society to another culture or society (Barnett, 1953). Most early socio-

---

[1] For a comprehensive bibliography on these disciplines, see Rogers
and Shoemaker (1971).

logists traced the diffusion of a single innovation over a geographical area (Bowers, 1937; Tarde, 1903).  Rural sociologists have studied the spread of new agricultural technology among farmers (Katz,
Levin and Hamilton; 1963).  Educationalists have studied school adoption of new teaching methods and equipment (Carlson, 1968; Mort,
1964).  The innovation studied by medical sociologists consisted of
(a) either new drug or medical techniques, where adopters are doctors, or (b) polio vaccine, family planning methods, where adopters
are clients or patients (Coleman, Katz, Menzel, 1966). Economists
have been concerned with the imitation process by which firms adopt
new technological and production processes (Mansfield, 1971).  Communication researchers have focused on the diffusion process to better understand the dynamics of persuasion and propaganda (Rogers and
Shoemaker, 1971).  Marketing researchers have been concerned about
understanding the diffusion process of new products to help the marketing managers to launch new products most efficiently (King, 1966;
Robertson, 1971).  Researchers in geography have dealt with the spatial aspect of diffusion process (Brown, 1968; Hagerstrand, 1965).
In very recent years political scientists have been concerned about
the diffusion of public policies among the American states (Gray,
1972, 1973; Walker, 1969, 1973).  Two research disciplines which
usually are not reviewed in the diffusion literature are: engineering
and natural sciences, and organization theory and design.  For some
time engineers have been studying the transfer process, typified by
a diffusion process, of momentum, heat and mass in physical entities
(Bird, Stewart and Lightfoot, 1960).  Examples of diffusion studies
in natural sciences include the studies on population growth in ecology (Peilou, 1969) and studies on the spread of epidemics (Bailey,
1957).  It is interesting to note that the diffusion process models
used in the social sciences have been modeled after physical or biological diffusion processes, such as heat transfer, the spread of
epidemics and the reliability engineering theory (Robertson, 1971).
Scholars in organization theory and design have been concerned with
the relationship of the organization to its environment.  Two main
problems seem to be of concern to these scholars: (a) what type of
organizations innovate and (b) how do organizations develop different
organization designs and structures to cope with changes in environment (see for example, Burns and Stalker, 1961; Galbraith, 1973;
Lawrence and Lorsch, 1967; Woodward, 1965; Zaltman, Duncan and Holbek,
1973).  In almost all the diffusion studies where the adoption unit
is an organization, the effect of organization structure on the diffusion process has been ignored (Betly and Rogers, 1973).  This may
be because of the fact that in all these studies, researchers are
concerned with the macro aspect rather than the micro aspect, and
the adopter, whether an individual, an institution or a country is
treated as the unit of analysis.  On the other hand, scholars in organization theory and design have narrowed their scope to the micro
aspect of the diffusion process and have examined the introduction
of change or organizational change.  This paper examines diffusion
and not the introduction of change of environmental policies among
the American states.

## Adoption over Time

The temporal behavior of the spread of an innovation may be represented by the following equation (Mahajan and Haynes, 1975).

$$n(t) = \frac{dN(t)}{dt} = g(t) \, (\overline{N} - N(t))$$

$$N(t = t_o) = N_o \qquad\qquad (1)$$

where

$n(t)$  = the number of adopters at time t.
$N(t)$  = the cumulative numbers of adopters at time t;
    $N(t) = \int_o^t n(t)dt.$
$\overline{N}$  = the ceiling on the ultimate number of adopters in the population of M potential adopters (in case of the American states M = 50).
$g(t)$  = growth 'constant'.

Equation (1) gives the rate of diffusion at time t, i.e. the differential change in cumulative adoption in the time period (t, t + dt); and at any time t this rate is obtained by multiplying the number of nonadopters, $(\overline{N} - N(t))$, by a growth or diffusion 'constant' $g(t)$. Solution of equation (1) will yield cumulative adopters distribution $N(t)$ and the first derivative of $N(t)$ with respect to time t will provide noncumulative adopters distribution $n(t)$. As can be observed from equation (1), the rate of diffusion or number of adopters at time t is controlled by 'constant' $g(t)$ which will vary from innovation to innovation, the social system in which it is diffused, the channel, the change agents and other relevant diffusion elements, i.e,

$$g(t) = f \text{ (innovation, social system, channels, change agents, etc.)} \qquad (2)$$

Hence, $g(t)$ is 'constant' only when the above characteristics of the diffusion process are delineated.

A number of empirical studies have suggested that the cumulative adopters distribution, $N(t)$, resembles an S-shaped or logistic curve (Rogers and Shoemaker, 1971). Rogers and Shoemaker (1971) have hypothesized that noncumulative adopters distribution, $n(t)$, follow a bell-shaped curve over time and approaches normality. In fact, assuming a normal distribution, they have proposed a classification of 'adopter categories' from innovators (first to adopt) to laggards (last to adopt) by using mean and standard deviation (see Figure 1). Innovators, for example, are those lying beyond two standard deviations to the left of the mean year of adoption (2.5 percent of adopters); laggards are those falling beyond one standard deviation to the right of the mean year of adoption (16 percent of adopters). This categorization is, of course, arbitrary and furthermore, though

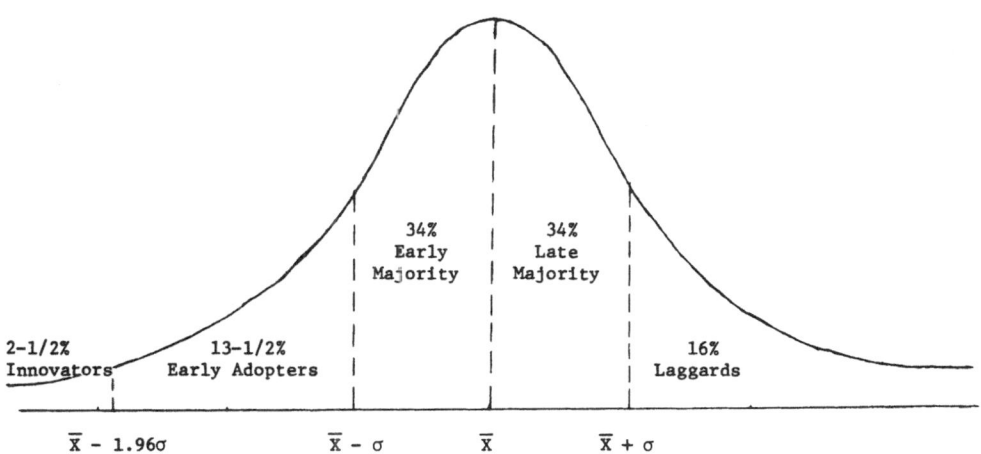

Figure 1. Adopter Categories

a number of studies in rural sociology have confirmed that noncumulative adopters distribution is normal, a number of researchers
have questioned the validity of this assumption especially in industrial or consumer marketing where the competition and "bandwagon"
feeling among the firms may lead to a distribution more peaked and
skewed than is normal (Peterson, 1973). Peterson (1973) has proposed the use of a unidimensional clustering algorithm to determine
the optimal number of categories without making normality assumptions. This paper will use this approach to determine the optimum
number of adopter categories for the adoption of environmental policies.

What are the factors or underlying dimensions that result in a
logistic cumulative adopters distribution, N(t), normal or non-normal
cumulative adopters distribution, n(t) or affect the rate of diffusion in equation (1). There is no coherent theory to answer this
question. Rogers and Shoemaker (1971), for example, explain the logistic curve arguing from psychological research on learning curves
and social interaction. In the case of diffusion of public policies
among the American states Gray (1973) and Mahajan and Haynes (1975)
have fitted interaction model (which assumes that adoption takes

place because of communication between the states which have adopted
a policy and states which have not adopted the policy) with good results.  Recalling equation (1), it is noticed that adopters' distribution is controlled by the diffusion constant which is a function
of diffusion elements.  Mansfield (1971) studying the adoption of
industrial innovations, identified size of investment, investment
policy and the behaviour of firms as the relevant diffusion elements
in the development of diffusion constant g(t), affecting the rate of
diffusion or adopters' distribution function.  Walker (1973), for
example, studying the diffusion of eighty-eight public policies among
the American states suggested certain socio-economic and political
variables as the correlates of the diffusion of public policies. In
this paper we will characterize socio-economic and political factors
underlying the diffusion of public policies among the American states which may affect the development of g(t) by showing the application of a new technique, multidimensional scaling, to the diffusion
literature.

## THE TECHNIQUE OF MULTIDIMENSIONAL SCALING

Nonmetric multidimensional scaling is a relatively new multivariate technique which was developed at Bell Telephone Laboratories
by Shephard and Kruskal (see Shephard, Romney and Nerlove, 1972;
Shephard, 1974).  This technique involves the development of a similarity matrix between a set of points (in our case similarity matrix
between forty-eight states across the twenty-eight policies on the
time of adoption) and determining a concise, invariant and assimi-
lable representation of the essential pattern of structure that lies
more or less hidden in the data or determining the number of dimensions and the configuration of the set of points along these dimensions which best summarizes the information contained in the data.

The ability of the technique to reconstruct a configuration
from ordered pairs of objects was first demonstrated by Shephard
(1966) and has been further explored in a series of Monte Carlo studies (Spence and Ogilvie, 1973).  A visual demonstration of the technique has been provided by Neidell (1969) who used the ordered distances between fifteen cities to construct a map of the United States.  Other applications have included such subjects as nations
(Wish, Deutsch and Beiner, 1972), colors (Indow, 1963), artistic
drawings (Skager, Schultz and Klein, 1966), tastes (Gregson, 1968),
sounds (Copperman, 1970), decision strategies (Rigby and Debow,
1967), common stocks (Green and Maheshwari, 1968) and cigarette
brands (Klahr, 1969).

Most of these studies attempt to develop a perceptual map,
which is a graphical representation of a person's or a group's view
of a set of objects.  Each dimension of the map, when identified or
characterized, represents an attribute of the set of objects, while

the position of an object along a dimension represents the extent to
which the object is perceived to possess that attribute.

In our case, each policy produces an ordered set of the forty-
eight states based upon the time each state adopted the policy. The
objective is to reconstruct the ordering or configuration of states
across all the twenty-eight policies in minimum, meaning possible
number of dimensions. TORSCA, a multidimensional algorithm suggested by Young and Torgerson (1967) is used for this purpose. Incorporated in this algorithm is the calculation of "stress" which is
Kruskal's (1964) measure of goodness of fit of the reconstructed
configuration to the initial data input. The stress curve can be
used to determine the most suitable number of dimensions.

ADOPTION ANALYSIS

Identification of Social, Economic and Political Factors

An examination of the data on the date of adoption on each of
the twenty-eight policies by the states indicates that cumulative
adopters distribution is a logistic curve but the noncumulative adopters distribution is not normal.[2] In order to characterize the
dimensions underlying the diffusion process, thirty-four explanatory
variables representing the social, economic and political characteristics of the states are selected and are listed in Table 2 (see
Annex). A factor analysis of the states' characteristics on the
thirty-four explanatory variables yield eight significant factors
explaining 82% of the variance. After examining the varimax rotated
factor matrix (Table 3 - see Annex) these factors are characterized
as follows.

---

[2] For example, the ten policies which have been adopted by all the
states - Board of Health, Child Labor Standards, Compulsory School
Attendance, Library Extension System, Welfare Agencies, Workman's
Compensation, Air Pollution Control, Soil and Water Conservation
Districts, Urban Renewal, Zoning in Cities, the Kurtosis is - .274,
3.246, -.933, .789, -.1032, 5.084, 1.863, .827, 1.443 and 24.944
respectively and Skewness is .750, .589, -.014, 1.138, -.219,
2.297, .525, 1.330, 1.178 and 4.439 respectively. (See Snedecor
and Cochran (1967, p. 84) for normality tests.).

[3] The data on these variables is collected from Jacob and Vines
(1973).

Factor 1: Gubernatorial Power-Legal Professionalism-Urbanization
Factor 2: Citizen Political Involvement and Inter-party
          Competition
Factor 3: Welfare Expenditures
Factor 4: Natural Resources
Factor 5: Legislature Professionalism
Factor 6: Industrialization
Factor 7: Interest Group Pressure
Factor 8: Party Cohesiveness

These factors represent the political, social and economic inputs to the states legislative adoption process.


Innovation Index

The development of an innovation index for the states based on their time of adoption of the twenty-eight policies is now discussed.

Consider the matrix $\underline{X}$ = $(X_{ij})$

$i$ = 1,2...48

$j$ = 1,2...28

Here $X_{ij}$ is the number of years that state 'i' has taken to adopt policy 'j'; the time period being counted from the first adoption date of policy 'j'. Hence if a state 'i' is the first state to adopt a particular policy 'j', $X_{ij}$ will be zero.

The nature of data is such that a lot of policies have not been adopted yet by the states at the time of data collection there are missing cells in the matrix $(X_{ij})$. Before an innovation index can be developed, these cells have to be filled with some logically derived figures.

Various alternatives present themselves.

1.  Arbitrarily choose a particular year in which the policy will be adopted. Walker (73) has used this methodology. The logic is less than clear. Walker (73) gives a score of 1.00 to states which have not yet adopted a particular policy. Thus he implicitly assumes that the policy would have been adopted by the state in the same year as that of the last known state's adoption year.

The arbitrariness is thus twofold.

(a) If the data was collected in 1965 and say a state has not yet adopted policy X. The last state to adopt to policy X was say in 1955. Then Walker's procedure implicitly assumes that all the nonadopting states would have adopted in 1955, when in fact we know for certain that they have not adopted it until 1965. Thus we are disregarding available information. The least that should have been done was to have assumed at least an adoption date of 1966.

(b) But why 1966? Why not 1984? This question is still not answered satisfactorily at all, although the missing value which is assumed will have a major impact on the innovation score.

2.  Find the mean for each state i across all policies 'j' which have been adopted, i.e., $\bar{X}_i$ and use this to fill the missing cells. One problem with this approach is that different kinds of policies may have taken vastly varying number of years to diffuse. For example, while soil and water conservation districts program has taken nine years to be adopted by all the states, the welfare agencies policy has taken seventy-three years. The mean may thus be completely out of line with the average number of years that may have been taken for a particular policy to diffuse. Thus the mean may be ten years while the maximum number of years for the policy which has not been adopted may be seventy years. In the calculation of the innovative index, this state though a laggard will show up as an innovator. This approach is thus unacceptable.

3.  Use Walker's method of calculating the index on each policy that has been adopted and then use the mean of the index to fill the missing cells. However, this procedure does not guarantee that the implied year of adoption will necessarily be later than the date of data collection or later than the last date of adoption of that policy.

4.  A completely different method, which is believed to be more valid than those discussed above is now proposed. In the first step the data in the column entries is standardized to zero mean and unit variance.
    Let the standardized matrix be denoted by $(S_{ij})$ where $S_{ij}$ now denotes the number of standard deviations before or after the mean adoption time of the policy 'j' that state 'i' has adopted the policy. Hence $S_{ij}$ gives an idea of how innovative the state is. For states which adopt the policy earlier than the average, $S_{ij}$ will be negative and it will be positive for the laggards. Most of the $S_{ij}$'s will lie between $\pm 3$.

    In the second step we fill the missing entries. The procedure is to first estimate for each state 'i', the means $s_i$ and the variance $\sigma_i^2$. across all the policies that the state has adopted.
    The missing entries are then filled with $m_{ij} = s_{i.} + 3\sigma_{i.}$

The logic of the above procedure is as follows. All states
are not innovators for all policies. Some policies they may
adopt relatively early and some they may adopt relatively late.
We find out on the average how late or early they adopt policies (given by $S_j$.) and then look at the variance from the average in their adoption behavior. Our best expectations based
on past behavior then are that, if the $S_{ij}$'s are distributed
normally, the state should adopt a policy not yet adopted such
$S_{ij} = S_i + 3\sigma$.

The advantages of the above approach are two fold: (a) there
is no arbitrariness and (b) all the past information that is
available is used.

If this method has validity, then the year of adoption generated by this procedure should definitely be later than the last
date of known adoption of any particular policy (true in 95%
of the filled missing values) and also later than the year of
data collection, i.e. 1976 (true in 50% of the cases). At
first sight the 50% result implies no better results than the
other methodologies criticized earlier. But we think that the
results are mixed not due to the method but due to the small
sample size. Because only twenty-eight policies are used, it
does not provide a wide enough range of the state's adoption
history. This results in the variance 0 being low, thus not
making $m_{ij}$ as large as it would be otherwise. It is anticipated that if the same procedure is followed with a much larger
set of data, the results will be much better.

One modification which could have been tried was to replace all
adoption dates earlier than the 1976 (the date of data collection)
with the year 1976.

In order to calculate the innovation index, the mean $S_{ij}$ should
be used. But since the sample was not very large it was decided to
use the median of the distribution of $S_{ij}$'s as an index of the inno-
vativeness of the state. Thus for innovative states the median is
expected to be negative as most of the laws are adopted relatively
early. The opposite would be true for the laggards.

Table 4 (see Annex) shows the rank order of the states obtained.
The northeastern states have the higher rankings (Massachusetts,
Ohio, New Hampshire, Michigan, Connecticut) while the southern sta-
tes have the lower rankings (Arkansas, Wyoming, South Carolina,
Mississippi, Nevada).

## Adopter Categories

The unidimensional clustering algorithm suggested by Peterson (1973) produced four optimal adopter categories. The input to this algorithm is the innovation index reported earlier. The clustering algorithm produces optimal categories by maximizing the ratio of between groups to within groups variance. The percentage of the states grouped in each category are given in Table 5 (see Annex). The labeling of these groups is consistent with that suggested by Rogers and Shoemaker (1971).

## The Dimensions of Innovation

In order to identify the dimensions underlying the diffusion process first a one way analysis of variance is performed across the four adopter categories. The results, given in Table 6 (see Annex), suggest that three factors; welfare expenditures, legislature professionalism and interest group pressure distinguishes these groups. The group means of the important variables loading on these factors are given in Table 7 (see Annex). The data suggests that per capita welfare expenditures are the highest for innovators and the lowest for laggards. Interest group pressure is very strong in laggard states, strong to medium for late adopters and medium to weak for the early adopters. In the case of the innovative states the percentage of states having strong, moderate or weak interest group pressure is evenly distributed. On the factor legislature professionalism, innovators tend to have highly professional legislative bodies as compared to late adopters which tend to have more professional legislature than early adopters and laggards. Early adopters and laggards tend to have less or non-professional legislature.

The other approach to elicit the underlying dimensions which may help explain similarities in states adoption behavior is to use multidimensional scaling.

The starting data matrix was the 48 x 28 matrix with the entries $X_{ij}$ denoting the number of years from the first recorded enactment of the legislative program, that state i took to adopt program j. All missing cells were filled as described in an earlier section.

This matrix was standardized across all 'i' to get the $(S_{ij})$ matrix, which was then input to the DISTRS program to get interstate euclidean distances. The output from DISTRS is a 48 x 48 matrix with the cells being the euclidean distances between the states. This distance provides a measure of the similarity between the adoptive profiles of the different states taking into consideration all the policies simultaneously. Hence, if two states have exactly the same adoptive profile, i.e., they adopted the laws in the same years,

the distance between them will be zero.  As the profiles become more
and more disparate, the distance between them increases.

    The distance matrix was then input to the multidimensional sca-
ling program TORSCA (Young and Torgerson, 1967).  This program tries
to find a spatial configuration of the states in a specified number
of dimensions such that the rank order of interstate distances best
reproduces the original rank order of the input data.

    In order to determine the minimum number of dimensions while
still retaining a good fit, Kruskals (1964) stress Form I was used.
Although a two dimensional solution is visually appealing, we decided not to use this representation as the stress was unacceptably
high at 17.65%.  A three dimensional configuration satisfied our
requirements of a good fit.  Stress was 12.92%, which is in the fair
range according to Kruskal.  The optimality of three dimensions was
also confirmed by the existence of an elbow at three dimensions, when
the number of dimensions versus stress was plotted.

    Once the dimensionality was decided, the next step was to inter-
pret the dimensions.  This process is the most difficult and one of
the popular "scientific" methods in the use of "head scratching and
eye-balling".  This as well as some correlational analysis was used.

    The spatial configuration output provided by TORSCA is invarient
under orthogonal axial rotation, i.e., the axis can be rotated without changing the derived distances between the states.  This strategy was followed in order to find out the best orientation of the
axis.  The dimensional coordinates of each state were correlated
with the factor scores of the state on each of the eight factors.
Different angular rotations were then tried and the orientation
which provided the best interpretative solution was used.

    Figures 2 and 3 are a plot of the states in this rotated space.
The four groups based on the cluster analysis done earlier are also
shown in the figure.

    An examination of the correlations between the dimensions and
the factors leads to the following conclusions: (a) Axis I seems to
represent the interest group pressure - legislative professionalism
dimension, (b) Axis II the political involvement - inter-party competition dimension and (c) Axis III the welfare expenditure dimension.

    Except for the factor political involvement, the other three
factors which characterize the scaling dimension are the same as obtained in the analysis of variance results (Table 6 - see Annex).
An examination of the variables that load on the factor political
involvement suggest that in addition to the profiles developed earlier the early adopters tend to have the highest turnouts in both
presidential and gubernatorial elections and also the highest per-

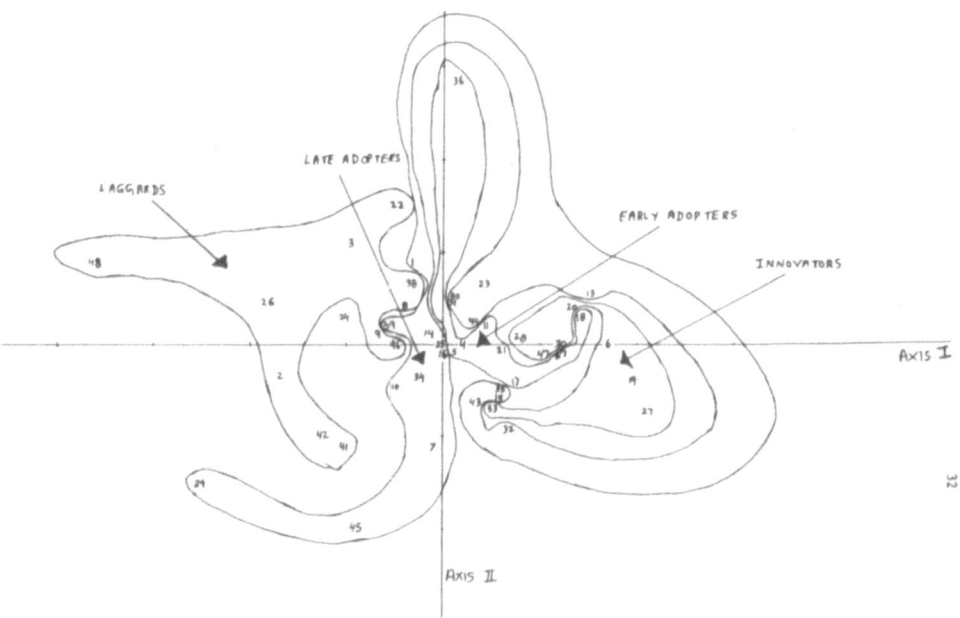

Figure 2. Two Dimensional Plot of the TORSCA Solution – Axis I vs.
Axis II (The 48 states are numbered alphabetically)
See Table 4.

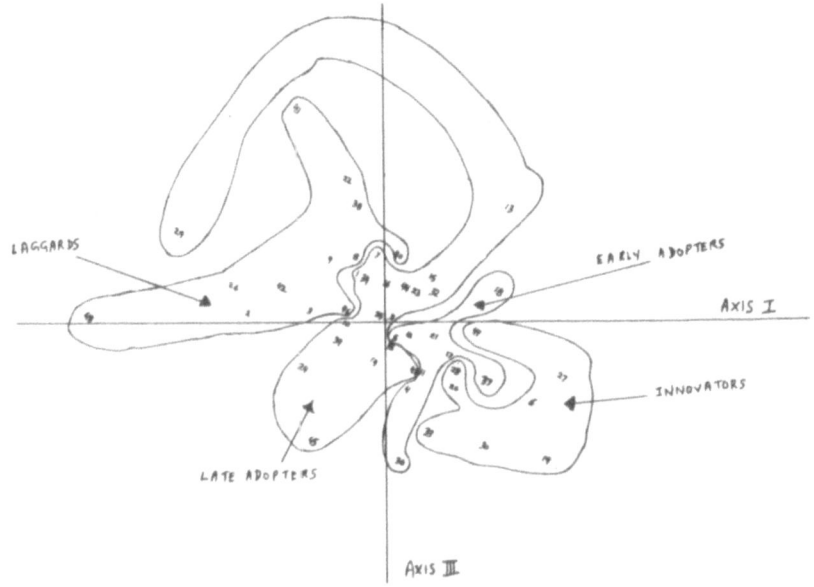

Figure 3. Two dimensional Plot of the TORSCA Solution – Axis I vs.
Axis III (The 48 states are numbered alphabetically)
See Table 4

centage of non-negro population. The laggards and the late adopters have the lowest turnouts in the elections and have a higher inter-party competition index.

To summarize, analysis of variance and the multidimensional scaling techniques have suggested similar dimensions underlying the adoption of environmental policies by the American states. In addition, as depicted in Figures 2 and 3, multidimensional scaling results in geometrically mapping the similarity in the adoption pattern of environmental policies among the American states.

## SUMMARY

There is an increasing concern (Aaker and Day, 1970) about the quality of our social and physical environment at the local, national and international levels. The American states have been responding to this critical issue by enacting legislative policies to ensure a harmony between the environment and the quality of life.

This paper has attempted to study this response or action of the American states to cope with this problem. It has suggested a different methodology for calculating the innovation index, has determined adopter categories of states based on this index and has introduced the new technique of multidimensional scaling to the diffusion literature. This study of adoption of policies over time provides a better understanding of the state legislature behavior in terms of social, economic and political characteristics in that factors which facilitate the adoption are identified. Also, the adopter categories of the states and their profiles may serve as a device to predict their behavior.

Table 1.   Policy Innovation Adoption Among the American States

| Policy Innovation | ICPR Regions*-No. of Non-Adopters | | | | | | | | Total Adopters | Non-Adopters | Time Span |
|---|---|---|---|---|---|---|---|---|---|---|---|
| | I | II | III | IV | V | VI | VII | VIII | | | |
| **A. HEALTH, EDUCATION & WELFARE** | | | | | | | | | | | |
| 1. Alcoholic Beverage Control | 1 | 0 | 1 | 1 | 0 | 2 | 1 | 1 | 42 | 6 | 1926-1959 |
| 2. Board of Health | 0 | 0 | 0 | 0 | 0 | 0 | 0 | 0 | 48 | 0 | 1869-1919 |
| 3. Child Labor Standards | 0 | 0 | 0 | 0 | 0 | 0 | 0 | 0 | 48 | 0 | 1901-1925 |
| 4. Compulsory School Attendance | 0 | 0 | 0 | 0 | 0 | 0 | 0 | 0 | 48 | 0 | 1852-1918 |
| 5. Defense and Disaster Supervision | 0 | 0 | 0 | 0 | 1 | 0 | 0 | 0 | 47 | 1 | 1949-1959 |
| 6. Equal Pay for Females | 1 | 1 | 1 | 1 | 6 | 0 | 2 | 0 | 36 | 12 | 1919-1974 |
| 7. Library Extension System | 0 | 0 | 0 | 0 | 0 | 0 | 0 | 0 | 48 | 0 | 1890-1949 |
| 8. Minimum Wage Law | 0 | 0 | 0 | 2 | 4 | 1 | 0 | 0 | 41 | 7 | 1915-1975 |
| 9. Mental Health Standards | 0 | 0 | 0 | 1 | 4 | 1 | 4 | 1 | 37 | 11 | 1955-1969 |
| 10. Supervision of Public Institutions | 0 | 1 | 0 | 0 | 0 | 0 | 0 | 0 | 47 | 1 | 1813-1891 |
| 11. Teachers Certification, Secondary | 1 | 0 | 1 | 3 | 1 | 0 | 1 | 0 | 41 | 7 | 1896-1956 |
| 12. Welfare Agencies | 0 | 0 | 0 | 0 | 0 | 0 | 0 | 0 | 48 | 0 | 1863-1935 |
| 13. Workman's Compensation | 0 | 0 | 0 | 0 | 0 | 0 | 0 | 0 | 48 | 0 | 1911-1948 |
| **B. CONSUMER RIGHTS** | | | | | | | | | | | |
| 1. Auto Vehicle Safety Compact | 0 | 0 | 0 | 0 | 3 | 1 | 0 | 0 | 44 | 4 | 1962-1965 |
| 2. Home Solicitation Sales Law | 0 | 0 | 0 | 0 | 1 | 0 | 2 | 0 | 45 | 3 | 1966-1974 |

Table 1 (Cont.)

| Policy Innovation | ICPR Regions* —No. of Non-Adopters | | | | | | | | Total Adopters | Non-Adopters | Time Span |
|---|---|---|---|---|---|---|---|---|---|---|---|
| | I | II | III | IV | V | VI | VII | VIII | | | |
| 3. Slaughter House Inspection | 0 | 0 | 0 | 0 | 0 | 1 | 0 | 0 | 47 | 1 | 1891-1970 |
| 4. Unfair and Deceptive Trade Laws | 0 | 0 | 0 | 0 | 1 | 1 | 0 | 0 | 46 | 2 | 1957-1974 |
| 5. Utility Regulation | 0 | 1 | 0 | 0 | 0 | 0 | 0 | 0 | 47 | 1 | 1839-1915 |
| **C. PHYSICAL AND NATURAL ENVIRONMENT** | | | | | | | | | | | |
| 1. Air Pollution Control | 0 | 0 | 0 | 0 | 0 | 0 | 0 | 0 | 48 | 0 | 1943-1973 |
| 2. Conservation of Oil and Gas | 3 | 2 | 0 | 2 | 0 | 1 | 0 | 0 | 40 | 8 | 1892-1974 |
| 3. Fish and Game Agencies | 1 | 1 | 0 | 1 | 0 | 1 | 2 | 0 | 42 | 6 | 1878-1932 |
| 4. Forest Agencies | 0 | 0 | 0 | 0 | 0 | 0 | 2 | 0 | 46 | 2 | 1885-1961 |
| 5. Nuclear Supervision | 5 | 3 | 3 | 6 | 0 | 0 | 1 | 0 | 30 | 18 | 1960-1969 |
| 6. Park Agencies | 0 | 0 | 0 | 0 | 0 | 0 | 2 | 0 | 46 | 2 | 1885-1937 |
| 7. Soil and Water Conservation Districts | 0 | 0 | 0 | 0 | 0 | 0 | 0 | 0 | 48 | 0 | 1937-1945 |
| 8. Urban Renewal | 0 | 0 | 0 | 0 | 0 | 0 | 0 | 0 | 48 | 0 | 1941-1969 |
| 9. Waste Treatment Agency | 0 | 0 | 0 | 2 | 4 | 2 | 4 | 1 | 35 | 13 | 1967-1970 |
| 10. Zoning in Cities | 0 | 0 | 0 | 0 | 0 | 0 | 0 | 0 | 48 | 0 | 1913-1965 |

xInter-University Consortium for Political Research (ICPR) Regions include the following states:
I. New England: Connecticut, Maine, Massachusetts, New Hampshire, Rhode Island, Vermont. II. Middle Atlantic: Delaware, New Jersey, New York, Pennsylvania. III. East-North Central: Illinois, Indiana, Michigan, Ohio, Wisconsin. IV. West-North Central: Iowa, Kansas, Minnesota, Nebraska, Missouri, North Dakota, South Dakota. V. Solid South: Virginia, Alabama, Arkansas, Florida, Georgia, Louisiana, Mississippi, North Carolina, South Carolina, Texas. VI. Border States: Kentucky, Maryland, Oklahoma, Tennessee, West Virginia. VII. Mountain: Arizona, Colorado, Idaho, Montana, Nevada, New Mexico, Utah, Wyoming. VIII. Pacific: California, Oregon, Washington. From Walker. (1973).

Table 2. Dictionary of Explanatory Variables

| Variable | Variable Index | Variable Definition |
|---|---|---|
| Social and Economic Variables | $x_1$ | Value added by manufacture per capita. |
| | $x_2$ | Percentage employed in manufacturing |
| | $x_3$ | Population per square mile. |
| | $x_4$ | Percentage foreign workers. |
| | $x_5$ | Total population. |
| | $x_6$ | Percentage urban population |
| | $x_7$ | Median school years completed. |
| | $x_8$ | Percentage non-negro. |
| | $x_9$ | Personal income per capita. |
| | $x_{10}$ | Total motor vehicle registration. |
| | $x_{11}$ | Telephones per 1,000 population. |
| | $x_{12}$ | Rural road mileage. |
| | $x_{13}$ | Fishing licenses. |
| | $x_{14}$ | Hunting licenses. |
| | $x_{15}$ | Percentage highway expenditures to all state and local expenditures. |
| | $x_{16}$ | Total per capita welfare expenditures. |
| | $x_{17}$ | State per capita welfare grants. |
| | $x_{18}$ | Federal per capita welfare grants. |
| | $x_{19}$ | Welfare expenditure as a percentage of general expenditure. |
| | $x_{20}$ | Number of welfare recipients per 1,000. |
| | $x_{21}$ | Average monthly payment per family (aid to families with dependent children, AFDC). |
| | $x_{22}$ | Average percentage turnout in Gubernatorial and Senatorial elections. |
| | $x_{23}$ | Average percent turnout in Presidential elections. |

Table 2. (Cont.).

| Variable | Variable Index | Variable Definition |
|---|---|---|
| Political Variables | $x_{25}$ | $x_{25}$ = 1 if legislative party cohesion is strong.<br>= 0 otherwise. |
| | $x_{26}$ | $x_{26}$ = 1 if legislative party cohesion is moderate<br>= 0 otherwise. |
| | If $x_{25}$ = $x_{26}$ | = 0, legislative party cohesion is weak (see Jacob and Vine, 1973, p. 113) |
| | $x_{27}$ | $x_{27}$ = 1 if the strength of pressure groups is strong.<br>= 0 otherwise |
| | $x_{28}$ | $x_{28}$ = 1 if the strength of pressure groups is moderate<br>= 0 otherwise |
| | If $x_{27}$ = $x_{28}$ | = 0, the strength of pressure groups is weak. (See Jacob and Vine, 1973, p. 127). |
| | $x_{29}$ | Index of citations of states from the 1970 state legislative program of the Advisory Commission on intergovernmental relations. |
| | $x_{30}$ | $x_{30}$ = 1 if state legislators are highly professional.<br>= 0 otherwise. |
| | $x_{31}$ | $x_{31}$ = 1 if state legislators are professional.<br>= 0 otherwise. |
| | $x_{32}$ | $x_{32}$ = 1 if state legislators are less professional.<br>= 0 otherwise. |
| | If $x_{30}$ = $x_{31}$ = $x_{32}$ | = 0, the state legislators are non-professional. (See Jacon and Vine, 1973, p. 194). |
| | $x_{33}$ | Index of formal powers of the governors (See Jacon and Vine, 1973, p. 232). |
| | $x_{34}$ | Index of legal professionalism in the state. (See Jacob and Vine, 1973, p. 292). |
| | $x_{24}$ | Ranney's inter-party competition index (see Jacob and Vine, 1973, p. 87). |

Table 3

Factor Analysis of the Explanatory Variables

Varimax Rotated Factor Matrix

| Variable # | Factor 1 | Factor 2 | Factor 3 | Factor 4 | Factor 5 | Factor 6 | Factor 7 | Factor 8 |
|---|---|---|---|---|---|---|---|---|
| $x_1$ | .74147 | -.02249 | .06635 | .52601 | -.08112 | -.31904 | .11437 | -.06409 |
| $x_2$ | .26651 | -.12665 | -.05589 | -.02923 | .00524 | -.79987 | -.04614 | .06934 |
| $x_3$ | -.60179 | .12722 | -.07300 | -.10243 | -.15079 | .66084 | -.16301 | -.15842 |
| $x_4$ | .05234 | .15600 | .11461 | -.05710 | -.12896 | .72316 | -.18350 | -.10345 |
| $x_5$ | .65341 | -.03856 | -.06628 | .05211 | -.10818 | -.14572 | -.04011 | .04483 |
| $x_6$ | .67036 | .02747 | .03347 | .02257 | .28043 | .09743 | .06471 | .03847 |
| $x_7$ | .28606 | .71596 | -.10794 | -.02034 | .17832 | .39768 | .11262 | -.01911 |
| $x_8$ | -.04593 | .85267 | -.02929 | -.03221 | -.08641 | .30394 | -.03976 | -.01546 |
| $x_9$ | .72711 | .42685 | -.04198 | -.05220 | .26549 | -.03084 | .23971 | .17805 |
| $x_{10}$ | .68578 | -.13985 | .20185 | .58425 | .05455 | -.07289 | .05501 | -.06776 |
| $x_{11}$ | .51878 | .42316 | -.00840 | -.01232 | .35791 | .17186 | .21944 | .13570 |
| $x_{12}$ | -.02218 | -.05112 | -.01697 | .87897 | .07451 | .23115 | -.10828 | .09163 |
| $x_{13}$ | .21745 | .04159 | .18116 | .40853 | .23881 | .02757 | .28667 | -.15584 |
| $x_{14}$ | .26450 | -.14756 | .03527 | .75955 | -.03230 | -.23446 | -.18122 | .00976 |
| $x_{15}$ | -.66349 | .13206 | -.23783 | -.19187 | -.16041 | .21433 | .10356 | -.35727 |
| $x_{16}$ | .11145 | .10718 | .98562 | .01922 | .01242 | .09433 | -.04495 | .00708 |
| $x_{17}$ | .30977 | .38747 | .77144 | -.10880 | .08611 | -.01417 | .02936 | -.00662 |
| $x_{18}$ | -.05042 | -.18392 | .91281 | .11844 | -.04177 | .13402 | -.11699 | -.00102 |
| $x_{19}$ | .20043 | -.07239 | .85522 | .09053 | -.01510 | -.07557 | -.02047 | .11510 |
| $x_{20}$ | -.12339 | -.44494 | .77317 | .01749 | -.14044 | -.00300 | -.05904 | .06529 |
| $x_{21}$ | .52761 | .74059 | .06158 | -.03783 | -.01203 | -.04390 | .07127 | .03711 |
| $x_{22}$ | .12612 | .91589 | -.00474 | -.15014 | -.07165 | .06271 | .13334 | .00337 |
| $x_{23}$ | -.08812 | .82863 | .03071 | -.07209 | .02035 | -.10342 | .16177 | .00084 |
| $x_{24}$ | -.08048 | -.85409 | .21059 | -.02198 | .21121 | -.02291 | -.15090 | .18107 |
| $x_{25}$ | .28009 | .38564 | -.00064 | -.06999 | .02112 | .43532 | .07391 | .58451 |
| $x_{26}$ | -.06710 | .30923 | -.17515 | -.05090 | .02639 | .19102 | .28684 | -.68109 |
| $x_{27}$ | -.15124 | -.38385 | -.10385 | .25832 | .05485 | .15295 | -.71046 | -.03084 |
| $x_{28}$ | .12364 | .18541 | -.06726 | -.11372 | -.04864 | -.07517 | .79411 | -.20705 |
| $x_{29}$ | .48815 | .18067 | -.00040 | .08875 | -.03276 | -.41735 | -.09026 | .40220 |
| $x_{30}$ | .80783 | .11083 | .17851 | .14087 | -.24703 | -.15735 | .09670 | -.09497 |
| $x_{31}$ | -.02832 | -.22631 | -.05283 | .09714 | -.89635 | -.18529 | -.10706 | -.00256 |
| $x_{32}$ | -.32075 | -.10638 | -.05856 | -.05901 | -.46462 | .05129 | .02425 | .43796 |
| $x_{33}$ | .51260 | .32217 | .15534 | -.05871 | .08451 | -.03057 | .06144 | -.19061 |
| $x_{34}$ | .78677 | .40905 | .14138 | .01956 | .00816 | -.01212 | -.03998 | -.03808 |

Table 4.  Innovation Index for the American States

| Alphabetical Order | Rank Order | State | Innovative Median Score |
|---|---|---|---|
| 19 | 1 | Massachusetts | -.9660 |
| 33 | 2 | Ohio | -.6863 |
| 27 | 3 | New Hampshire | -.6619 |
| 20 | 4 | Michigan | -.6348 |
| 6 | 5 | Connecticut | -.6204 |
| 28 | 6 | New Jersey | -.5978 |
| 30 | 7 | New York | -.5939 |
| 47 | 8 | Wisconsin | -.5408 |
| 4 | 9 | California | -.4976 |
| 37 | 10 | Rhode Island | -.4471 |
| 36 | 11 | Pennsylvania | -.4105 |
| 11 | 12 | Illinois | -.3844 |
| 5 | 13 | Colorado | -.3312 |
| 18 | 14 | Maryland | -.3088 |
| 12 | 15 | Indiana | -.3051 |
| 21 | 16 | Minnesota | -.2909 |
| 17 | 17 | Maine | -.2707 |
| 35 | 18 | Oregon | -.2228 |
| 43 | 19 | Vermont | -.1591 |
| 45 | 20 | Washington | -.1274 |
| 15 | 21 | Kentucky | -.1235 |
| 14 | 22 | Kansas | -.1126 |
| 31 | 23 | North Carolina | -.1109 |
| 13 | 24 | Iowa | -.0707 |
| 7 | 25 | Delaware | -.0442 |
| 23 | 26 | Missouri | -.0331 |
| 32 | 27 | North Dakota | -.0219 |
| 29 | 28 | New Mexico | .0010 |
| 44 | 29 | Virginia | .0132 |
| 24 | 30 | Montana | .0171 |
| 16 | 31 | Louisiana | .0222 |
| 1 | 32 | Alabama | .0268 |
| 25 | 33 | Nebraska | .0278 |
| 10 | 34 | Idaho | .0305 |
| 42 | 35 | Utah | .0404 |
| 39 | 36 | South Dakota | .0445 |
| 40 | 37 | Tennessee | .0759 |
| 46 | 38 | West Virginia | .1322 |
| 9 | 39 | Georgia | .1956 |
| 3 | 40 | Arizona | .2035 |
| 34 | 41 | Oklahoma | .2152 |
| 8 | 42 | Florida | .2437 |

Table 4. (Cont.)

| Alphabetical Order | Rank Order | State | Innovative Median Score |
|---|---|---|---|
| 41 | 43 | Texas | .2791 |
| 2 | 44 | Arkansas | .3589 |
| 48 | 45 | Wyoming | .3737 |
| 38 | 46 | South Carolina | .4284 |
| 22 | 47 | Mississippi | .6173 |
| 26 | 48 | Nevada | .7482 |

Table 5.   Optimal Adopter Categories.

| Group * | Number of States | % of Total Number of States |
|---|---|---|
| 1. Innovators | 8 | 16.67 |
| 2. Early Adopters | 9 | 18.75 |
| 3. Late Adopters | 19 | 39.58 |
| 4. Laggards | 12 | 25.00 |

* Innovators include Massachusetts, Ohio, New Hampshire, Michigan, Connecticut, New Jersey, New York, Wisconsin.

Early Adopters are California, Rhode Island, Pennsylvania, Illinois, Colorado, Maryland, Indiana, Minnesota, Maine.

Late Adopters are Oregon, Vermont, Washington, Kentucky, Kansas, North Carolina, Iowa, Delaware, Missouri, North Dakota, New Mexico, Virginia, Montana, Louisiana, Alabama, Nebraska, Idaho, South Dakota, Utah.

Laggards are Tennessee, West Virginia, Georgia, Arizona, Oklahoma, Florida, Texas, Arkansas, Wyoming, South Carolina, Mississippi, Nevada.

Table 6.  One-Way Analysis of Variance.

| Factor | Wilk's Lambda | F-ratio |
|--------|---------------|---------|
| 1 | 0.9490 | 0.7888 |
| 2 | 0.9530 | 0.7238 |
| 3 | 0.8491 | 2.6065* |
| 4 | 0.9703 | 0.4484 |
| 5 | 0.8395 | 2.8044** |
| 6 | 0.9859 | 0.2098 |
| 7 | 0.8571 | 2.4462* |
| 8 | 0.9390 | 0.9527 |

* Level of significance less than .1 at 3, and 44 degrees of freedom

** Level of significance less than .06 at 3 and 44 degrees of freedom

Table 7

Significant Characteristics of Adopter Categories

| Characteristic | Innovators | Early Adopters | Late Adopters | Laggards | All Groups |
|---|---|---|---|---|---|
| **Interest Group Pressure** | | | | | |
| % of states with strong interest group pressure | 37.50 | 22.22 | 57.89 | 75.00 | 52.08 |
| % of states with moderate interest group pressure | 25.00 | 44.44 | 42.11 | 8.30 | 32.25 |
| % of states with weak interest group pressure | 37.50 | 33.34 | 0.00 | 16.70 | 16.67 |
| **Legislature Professionalism** | | | | | |
| % of states with: | | | | | |
| a.  highly professional legislature | 37.50 | 22.22 | 5.26 | 8.33 | 14.58 |
| b.  professional legislature | 12.50 | 0.00 | 52.63 | 33.33 | 31.25 |
| c.  less professional legislature | 37.50 | 55.56 | 26.31 | 25.00 | 33.33 |
| d.  non-professional legislature | 12.50 | 12.22 | 15.80 | 33.34 | 20.84 |
| **Welfare Expenditures** | | | | | |
| a.  Total per capita expenditures | 40.59 | 39.23 | 31.71 | 27.42 | 33.53 |
| b.  State per capita expenditures | 16.07 | 15.62 | 11.64 | 10.94 | 12.95 |
| c.  Federal per capita expenditures | 24.52 | 23.62 | 20.06 | 17.95 | 20.94 |
| d.  Welfare expenditures as % of general expenditures | 14.87 | 14.09 | 10.69 | 9.86 | 11.82 |
| e.  Number of recipients per 1,000 population | 50.19 | 38.41 | 36.48 | 34.69 | 38.66 |

Table 7 (continued).

| | Adopter Categories | | | | All Groups |
|---|---|---|---|---|---|
| Characteristic | Innovators | Early Adopters | Late Adopters | Laggards | |
| f. Average monthly | 148.62 | 155.89 | 130.89 | 138.08 | 140.33 |
| Citizen Political Involvement | | | | | |
| a. Average % turnout in gubernatorial election | 45.57 | 52.58 | 43.00 | 43.14 | 45.26 |
| b. Average % turnout in presidential election | 63.32 | 66.22 | 60.99 | 61.02 | 62.37 |
| c. Ranney's inter-party competition index | 0.54 | 0.54 | 0.65 | 0.61 | 0.60 |
| d. % non-negro | 89.97 | 94.24 | 89.60 | 91.30 | 90.96 |

## REFERENCES

1.  Aaker, D.A. and G.S. Day, 1970, "A Guide to Consumerism,"
    Journal of Marketing, 34: 12-19.

2.  Bailey, N.T.J., 1957, Mathematical Theory of Epidemics. New
    York: Hafner Publishing Co.

3.  Barnett, H.G., 1953, Innovation: The Basis of Cultural Change.
    New York: McGraw Hill.

4.  Betty, S.A. and E.M. Rogers, 1973. "The Diffusion of Innova-
    tions and Change in Health Systems," in Quality Assurance of
    Medical Care, monograph, Regional Medical Program Services,
    U.S. Dept. of HEW, pp. 403-424.

5   Bird, R.B., W.E.Stewart and E.N. Lightfoot, 1960. Transport
    Phenomena. New York: John Wiley and Sons.

6.  Bowers, R.V., 1937. "The Direction of Intra-Social Diffusion",
    American Sociological Review. Vol. II, pp. 826-36.

7.  Brown, L.A. 1968 Diffusion Dynamics: A Review and Revision of
    the Quantitative Theory of the Spatial Diffusion of Innovation.
    Gleerup, Lund: Lund Studies in Geography.

8.  Burns, Tom and G.M.Stalker, 1961. The Management of Innovation.
    London: Tavistock Publishcations.

9.  Carlson, R.O. 1968. "Summary and Critique of Educational
    Diffusion Research," Paper presented at the National Conference
    on the Diffusion of Educ. Ideas, East Lansing, Michigan.

10. Coleman, J.S., E. Katz and H. Menzel, 1966. Medical Innovation:
    A Diffusion Study. Indianapolis: The Bobs-Merrill Company Inc.

11. Commoner, Barry. 1971. The Closing Circle. New York: Alfred
    A. Knopf.

12. Copperman, N. 1970. "Multidimensional Analysis of Similarity
    Judgements of Multicomponent Tones," Journal of the American
    Acoustical Society, Vol. 38, pp. 95-98.

13. Crew, R.E., Jr. 1970, State Political System and Public Policy:
    An Analysis of State Commitment to the Conservation and Develop-
    ment of Natural Resources, Ph.D. Dissertation, University of
    North Carolina at Chapel Hill.

14. Crittenden, J. 1967. "Dimension of Modernization in the Ameri-
    can States," American Political Science Review, 166: 989-1001.

15. Dawson, R.E. and J.A. Robinson, 1963. "Interparty Competition,
    Economic Variables, and Welfare Policies in the American
    States" The Journal of Politics, 25: 265-289.

16. Dye, T.R. 1966. Politics, Economics and Public Policy Outcomes
    in the American States. Chicago: Rand McNally,

17.    Ehrlich, P.R. and John P. Holdren, 1972. Environment, XIV,
       p. 24.

18.    Galbraith, J. 1973. Designing Complex Organizations. Reading:
       Addison-Wesley Publishing Company.

19.    Gray, V.H. 1972. Theories of Party Leader Strategy and Public
       Policies in the American States, Ph.D. Dissertation, Washington.

20.    Gray, V.H. 1973. "Innovation in the States: A Diffusion Study,"
       American Political Science Review, 67: 1174-1185.

21.    Gray, V.H. 1973. "Rejoinder to 'Comment' by J.L. Walker,"
       American Political Science Review, 67: 1192-1193.

22.    Green, P.E. and Arun Maheshwari. 1969. "Common Stock Percep-
       tion and Preference: An Empirical Application of Multidimen-
       sional Scaling," Journal of Business, Vol. 42, pp. 439-57.

23.    Gregson, R.A.M. 1968. "Simulating Perceived Similarities Be-
       tween Taste Mixtures Having Mutually Interacting Components,"
       British Journal of Mathematical and Statistical Psychology,
       Vol. 22, pp. 117-30.

24.    Hagerstrand, T. 1965. "A Monte Carlo Approach to Diffusion,"
       Archives European de Sociology, Vol. 6, pp. 43-67.

25.    Hines, L.G. 1973. Environmental Issues. New York: W.W.Norton
       & Company, Inc.

26.    Hofferbert, R.I. 1966. "The Relation Between Public Policy
       and Some Structural and Environmental Variables in the Ameri-
       can States," American Political Science Review, 60: 72-82

27.    Hofferbert, R.I. 1968. "Socio-economic Dimensions of the Ameri-
       can States: 1890-1960,""Midwest Journal of Political Science,
       12: 401-418.

28.    Indow, T. 1963. "Two Kinds of Multidimensional Scaling Methods
       as Tools for Investigating Color Space from the Macroscopic
       Point of View," Acta Chromatica, 1,pp. 60-71.

29.    Jacob, H. and K.H. Vines. 1971. Politics in the American Sta-
       tes.  Boston: Little Brown.

30.    Johnson, S.C. 1967. "Hierarchical Clustering Schemes,"
       Psychometrika, 32(3): 241-254.

31.    Katz, E., M.L. Levin and H. Hamilton. 1963. "Traditions of
       Research on the Diffusion of Innovation." American Sociolo-
       gical Review, Vol. XXVIII, pp. 237-52.

32.    King, C.W. 1966. "Adoption and Diffusion Research in Marketing:
       An Overview." In Proceedings of the American Marketing Associ-
       ation. R.M. Hass, ed. Chicago: American Marketing Association.

33.    Kahr, D. 1969. "Study of Cognitive Structure of Cigarette

Brands", Journal of Advertising Research, Vol. 9, pp. 39-46.

34. Kruskal, J.B. 1964. "Non-Metric Multidimensional Scaling by Optimizing Goodness of Fit to a Non-Metric Hypothesis," Psychometrica, Vol. 29, pp. 1-27.

35. Lawrence, P.R. andJ.W. Lorsch. 1969. Organization and Environment. Homewood, Illinois: Richard D. Irwin, Inc.

36. Mahajan, V. and K. Haynes, 1975. "Modeling the Diffusion of Public Policies in the American States," Studies in the Diffusion of Innovation, Discussion Paper No. 34, Department of Geography, The Ohio State University, Columnus, Ohio.

37. Mansfield, E. 1971. Research and Innovation in the Modern Corporation, New York: W.W. Norton and Company, Inc.

38. Meadows, D.H. 1972. The Limits to Growth. New York: Universe Books.

39. Mort, P.R. 1964. "Studies in Educational Innovation from the Institute of Administrative Research: An Overview," In Innovation in Education. M.B. Miles, ed. New York: Columbia University.

40. Neidell, L.A. 1969. "The Use of Non-metric Multidimensional Scaling in Marketing Analysis," Journal of Marketing, Vol. 33, pp. 37-43.

41. Peilou, E.C. 1969. An Introduction to Mathematical Ecology. New York: Wiley Interscience.

42. Peterson, R.A. 1973. "A Note on Optimal Adopter Category Determination," Journal of Marketing Research, Vol. X.

43. Rigby, J.W. and E.H. Debow. 1967. "Multidimensional Scaling Analysis of Decision Strategies on a Threat Evaluation," Journal of Applied Psychology, Vol. 51, pp. 205-10.

44. Robertson, T.S. 1971. Innovative Behavior and Communication. New York: Holt, Rinehart and Winston.

45. Rogers, E.M. and F.F. Shoemaker. 1971. Communication of Innovations: A Cross-Cultural Approach. New York: The Free Press.

46. Rose, D.D. 1973. "National and Local Focus in State Politics: The Implications of Multi-Level Policy Analysis," The American Political Science Review, 67: 1162-1173.

47. Shepard, R.N. 1966. "Metric Structures in Ordinal Data," Journal of Mathematical Psychology, Vol. 3, pp. 287-315.

48. Shepard, R.N., A.K. Romney and S.B. Nerlove (eds.). 1972. Multidimensional Scaling: Theory and Application in the Behavioral Sciences, Vol. I and Vol. II, New York: Seminar Press.

49. Shepard, R.N. 1974. "Representation of Structure in Similarity Data: Problems and Prospects," Psychometrika, 39(4): 373-421.

50.   Skager, R.W., C.G. Schultz and S.P.Klein, 1966. "Multidimen-
      sional Scaling of a Set of Artistic Drawings: Perceived Struc-
      ture of Scale Correlates," Multivariate Behavioral Research 1,
      pp. 425-36.

51.   Spence, I. and Ogilvie, J.C. 1973. "A Table of Expected Stress
      Values from Random Rankings in Non-metric Multidimensional
      Scaling," Multivariate Behavioral Research, Vol. 8, pp. 511-17.

52.   Sullivan, J.L. 1973. "Political Correlates of Social, Econo-
      mic, and Religious Diversity in the American States," The
      Journal of Politics, 35, 1: 70-84.

53.   Tarde, G. 1962. The Laws of Imitation. Gloucester, Mass,:
      Reprinted by Peter Smith.

54.   Van Gigch, John. 1975. Applied General Systems Theory, New
      York: Harper and Row.

55.   Walker, J.L. 1969. "The Diffusion of Innovation Among the
      American States," American Political Science Review, 63: 880-899.

56.   Walker, J.L. 1973. "Comment: Problems in Research on the Dif-
      fusion of Policy Innovations," American Political Science
      Review, 67: 1186-1191.

57.   Wish, M., M. Deutsch and L. Beiner, 1972. "Differences on
      Perceived Similarity of Nations," in Multidimensional Scaling:
      Theory and Applications in the Behavioral Sciences, ed. R.N.
      Shephard, Al Romney and S. Herlove, New York: Academic Press.

58.   Woodward, Joan. 1965. Industrial Organization: Theory and
      Practice. London: Oxford University.

59.   Young, F.W. and W.S.Torgerson. 1967. "TORSCA, A FORTRAN IV
      Program for Shepherd-Kruskal Multi-dimensional Scaling
      Analysis," Behavioral Science, 12: 498.

60.   Zaltman, G., R. Duncan and J. Holbek, 1973. Innovations and
      Organizations. New York: John Wiley and Sons.

RESPONSES TO CHANGING URBAN SYSTEMS: AN ANALYSIS

OF PUBLIC EDUCATION

B. Anderson, J. Mark

Washington University, St. Louis, USA

Some years ago, Howard Becker (1952) studied the career of the Chicago public school teacher. In this study, Becker found that teachers seemed to leave teaching at high rates, and that those who remained tended to move from inner city locations towards the outer suburbs. In an era when the number of both schools and students were expanding, these career patterns had the effect of producing a school system with a large pool of young, comparatively inexperienced and low cost teachers at one extreme, and another large pool of older, more experienced "career" teachers at the other.

Becker's study, with its pioneering insights about the career patterns of teachers in large urban areas, is particularly important as school systems begin to prepare themselves for the declines in enrollment and revenue which have recently begun and which we expect will become more intense in the near future.

The distribution of teaching experience, newly trained teachers, and administrative assistance is not random across an urban area. These factors are, in part, the collective sums of individual career lines, and the career pattern of teachers is shaped not only by the numbers and kinds of students to be taught, but also by the state of development of the urban area in which the school system is built.

In this paper, we present data showing how career patterns in the metropolitan St. Louis area vary from one portion of the metropolis to another. We then attempt to show how these changes in the career pattern might account for certain characteristics of the area's school systems.

Data are presented on four important aspects of the education

system's manpower situation.  First, we discuss the pattern of mobility in the metropolitan area.  Second, we deal with drop-out behavior among the areas' certified personnel.  Third, we discuss the
shift of personnel towards positions in administration as opposed
to teaching.  Last, we discuss the side effects of these changes in
salary payments in the system.  We conclude with a frankly speculative section dealing with the effects of these four aspects on the
school systems of large urban areas.


## DATA SOURCES

Through the cooperation of the Missouri State Department of
Education, we have obtained data which describe all of the certified
personnel in the public schools of the five counties of the Missouri
portion of the St. Louis SMSA: St. Louis City, St. Louis, St. Charles,
Jefferson and Franklin Counties.  The data cover the period 1968 to
1974 and are in a form which allows us to use individuals, schools,
school districts, or mixtures of all three as our unit of analysis.
There are from seventeen to twenty thousand people in any one year of
the data file, and thirty thousand appear in the file over the seven
year period
Using the annual report for the schools of Missouri, as well
as annual compilations of each district's statistics, we have been
able to add district level variables having to do with tax base,
number of students and total numbers of personnel to the file. More
such variables are required.  Notably absent are records dealing
with sources of teachers' training, recency of training, the budget
of each district, and the external causes of teachers' movement --
such as retirement, death, movement to another area, and so forth.
We hope to add these data to our file at a later point in time.


## TEACHER MOBILITY PATTERNS

As mentioned earlier, Becker found that teachers' mobility
within the school system was not random.  Usually, those teachers
who moved did so by seeking out what they considered to be "better"
jobs, and those happened to be towards the outer edge of the city.
Greenburge and McCall (1973) observed the same phenomenon in San
Diego.  This movement can be placed in a broader context -- one that
we believe useful in understanding what happens to schools as an
urban area develops.

The Missouri portion of metropolitan St. Louis may be viewed
as a large horseshoe, with the aging City of St. Louis in the middle
of the open end and with much new residential and commercial development taking place all along the outer edge.  In the metropolitan
area, this is accompanied by a process called neighborhood succession in which waves of people move from the urban core towards the
fringes.

The effect of this expansion was described by Park, Burgess and McKenzie in their work on the social ecology of Chicago (1925). Burgess described it as

> ... a series of concentric circles which may be numbered to designate both the successive zones of urban extension and the types of areas differentiated in the process of expansion. This ... represents an ideal construction of the tendencies of any town or city to expand radially from its central business district ... Encircling the downtown area there is normally an area of transition which is being invaded by business and light manufacture. A third area is inhabited by the workers in industries who have escaped from the area of deterioration but who desire to live within easy access of their work. Beyond this zone is the residential area of high-class apartment buildings or of exclusive ... single family dwellings ... This brings out clearly the main fact of expansion, namely the tendency of each inner zone to extend its area by the invasion of the next outer zone. This aspect of expansion may be called <u>succession</u>, a process which has been studied in detail in plant ecology. (Burgess, 1961).

We feel that the movement of people from inner city to outer areas of the metropolis must be accompanied by a movement of teachers. In order to examine this possibility, we divided the Missouri portion of the SMSA into six rings, which are shown in Figure 1. The inner Ring (I) is formed by the City of St. Louis. Rings II and III are formed by older residential areas with some light industry. Generally speaking, these communities have had no physical room for expansion in the last eight years. Thus, they are relatively stable, except that succession has been more rapid in Ring II, where a Black population increase is evident. Rings IV and V are bounded on their inner edges by well-developed extensions of the city. Their outer fringes are comparatively open and have been growing in recent years. Ring VI, which surrounds the built-up portion of the city, is predominantly rural, although it is occupied by some small towns.

Examining this distribution of the area's population over the last thirty years led us to conclude that teachers were likely to flow from the center of the area to the fringes, and that rates of entry for new teachers would decline as areas stabilized and rise as they expanded.

Table I shows net flows of teachers across the various rings for the period 1968-1974. Over the seven year period, the City of St. Louis experienced a net loss of 284 teachers to the outer rings. Of these, 189 went into the adjacent Rings (II and III). Rings II and III experienced a net loss of 179 teachers -- to Rings IV, V and VI. The pattern of mobility, such as it is, is plain enough from this. Teachers are moving away from the center of the urban area and towards the suburbs.

DROP-OUT PATTERNS

There have been many studies of teacher turnover, and it is conventional wisdom that the profession is characterized by large numbers of people who teach for a few years and then quit, or else who teach only sporadically. Generally speaking, dropping out occurs more frequently among young, inexperienced teachers, and more frequently among females than males. At a district level then, the biggest explanation for teacher drop-out lies in the proportion of teachers who are young, inexperienced and female.

The career patterns of individuals, aggregated over districts and time, are then important if we wish to see what the age structure of a districts' teaching staff will look like. One way to examine this is to study the survival rates of people who enter teaching. Figure 2 shows those rates for the cohorts who entered teaching in metropolitan St. Louis in the period 1968 to 1973. It is clear that the survival rate for each cohort is increasing. Moreover, Table II shows that not only is the survival rate increasing, but the differences between men and women appear to be decreasing. Table III shows drop-out rates ring by ring for the six of the years in our sample. Drop-out and survival are not the same thing, but since so many young teachers do quit, they are highly related.

These things do not vary randomly across the area. An examination of Table III shows that the drop-out rate for Ring I ranges from 14% to 8%, while for Ring IV it ranges from 19% to 11%. The same trend is apparent across the Table. The percentage of drop-outs is declining everywhere over time, but is lower in the inner rings than the outer rings for any given year. Indeed, the outer rings are now at about the same state as was Ring I in 1968.

In addition to changes in retention and drop-out rates, there have been changes in the rate at which new teachers enter the force. Such a change is to be expected in view of declining birthrates and the correlated decline in school enrollments, but like drop-out and retention rates, the rate of new entries to teaching varies systematically across the rings.

Table IV shows the percentage changes in staff size and enrollment from year to year, by ring; as well as the proportion of the staff that is new to teaching in each year and ring. The Table shows that the growth of staff seems unrelated to growth of enrollments, at least over the short period of time we are examining. It also shows that the rate at which new entrants enter the force seems to be most responsive to enrollment growth or reduction.
enter the force seems to be most responsive to enrollment growth or reduction.

The Table also shows that enrollment growth has turned negative

in the inner rings, but remains positive in the outer portion of
the metropolitan area, although at a much slower rate than in the
inner three rings. No doubt these slowed positive growth rates mask
a probable decrease in enrollments as the number of young students
entering schools in these outer areas continues to decline.


## ADMINISTRATIVE EXPANSION

Many teachers who remain in the profession try to move up a
career ladder, as well as towards better teaching positions. In
public school systems such movement means promotion out of the classroom into some sort of administrative or support role -- librarian,
counselor, principal and so forth. A sizeable subset of teachers
prepare themselves for promotion within or across systems (see Griffiths, et al., 1965), and of these many are promoted.

Promotion is made possible by two factors: simple expansion of
the system and by expansion of the administrative component within
the system. There is considerable debate over what seems to occur
in schools, however. On the other hand, theorists argue that the
administrative component of a system can be smaller in relation to
the rest of the system in large systems. Their view is supported
by some empirical evidence from schools in the work of Holdaway and
Blowers (1971). On the other hand, there is also evidence that the
administrative component of an organization grows independently of
the rest of the organization (see Parkinson, 1957, ch. 1). To our
knowledge, there have been no studies of schools which examine the
growth of school administration. Indeed, the data presented by
Holdaway and Blowers in regard to administrative size could equally
well have been used by Parkinson to make his point that administration
seems to grow at about 5-6% per year. This is because schools in
their sample were expanding enrollments each year.

Table Va shows the relationships between system size as measured
by number of professional staff and enrollments and also shows the
percentage of the staff in administrative positions. The Table shows
that the administrative component of the system does increase in size,
but at less than the five percent predicted by Parkinson. Particularly interesting is the increase in the number of administrators
per pupil and per teacher as a result of this growth. In part,
this finding replicates the work of Gittel and Hollander (1968) who
found that the number of administrators per 1,000 pupils rose steadily over the ten year period of their study in two out of four districts studied. (They examined New York, Detroit, Chicago, St. Louis,
Baltimore and Philadelphia.)

Parkinson makes the point more strongly in his study of the
British admiralty and Foreign Service. For example, he cites a decline of 31.5% in the number of officers and men in the Royal Navy

accompanied by a 78.5% increase in the number of men at the Admiralty Offices (1957, 21).  Upon examining Table Va, we are inclined to
agree with Parkinson, in that the growth of the administrative segment of school systems in these rings appears to be proceeding at a
rate which cannot be accounted for by growth of the systems being
administered.

Since errors in predicting the size of the administrative component are largest in the fast-growing rings, we computed a regression equation to correct for changes in enrollment and also for
changes in staff size, and then applied Parkinson's rule that the
administrative component would grow at no less than 5.17% per year.
That rule, in conjunction with the correction for changes in enrollment and staff size, produced a prediction of the size of the administrative component, and the rule accounts for 20.5% of the
variance in size of the administrative component over and above the
effects of enrollment and staff size.

Generally speaking, the data seem to support the notion that
increasing proportions of school resources are going into administrative rather than teaching functions.  As yet, this should not be
judged, since we know of no thorough treatment of the relationship
between such developments and student outcomes.  However, we view
this development with some apprehension, since it is of substantial
magnitude, its effects are unknown, and it seems to be developing
without the conscious attention of school boards and other agencies
responsible for setting school policy.

COSTS

One result of the foregoing three factors is increased costs
for education.  While the overwhelming determinant of costs is enrollment, particularly enrollment which causes a need for capital
expansion, the second most important factor is probably salary expense per pupil.  It is in this area that schools in the older,
slow growth or negative growth areas are in a difficult position.
These systems have little need to add new personnel, the people who
are on staff are older, have acquired more training, and therefore
are more expensive.  Moreover, the schools have no way to control
this salary expense, for teachers gain experience and training at
their discretion, not at the discretion of the school system which
must then pay for them.

In the past, when large numbers of teachers had to be added to
the system, it was partially possible to control salary expenses by
controlling the mix of new teachers.  Buying large numbers of young,
inexperienced females would mean that large numbers of young, inexperienced people could be sought in the subsequent years to replace
those who would drop out of teaching or move to another district.
With both avenues of "escape" apparently cut off, and with the rate

of new entries to teaching very far down, salary control of any sort
other than negotiation with teachers' organizations has been lost.

The implications of this are not pleasant for inner-city school
systems.  They are apt to be the first to bear the brunt of salary
disputes as they will have to come to terms with the loss of control
over their salary scales before those systems on the outer fringes
of urban areas.  In time, the outer fringes will face the same problem, for if the notion of succession is correct, they are destined
to experience declining enrollment and associated problems of increasing salary expenses per student.

Table Vc presents a regression equation in which the dependent
variable is deflated salary expenditure per pupil and the independent
variables are number of students per student administrator, pupil-
teacher ratio, average training of the professional staff, and an
indicator of whether or not the district's enrollment is growing or
declining.  Each observation is a school district in a year, and
there are about fifty districts in the sample each year.  The most
interesting finding is that the per-pupil expenditures are highest
in the districts with declining enrollments, supporting our contention that declining districts become enmeshed in a cycle of hiring/
firing which increases costs.

## DOES IT ALL MATTER?

To this point, we have established the following, at least to
our satisfaction.

1. Teachers who move from one area of the metropolitan area to
   another are moving towards the fringes of the metropolitan
   area.  Moreover, the amount of movement which takes place
   has been decreasing over the past seven years.

2. Drop-out rates from teaching are highest in the outer, fast-
   growing areas of the metropolitan area (not in the central
   city where the teaching is "tough").  Drop-out rates have
   been steadily decreasing over the past seven years.

3. The percentage of experienced teachers in each ring's force
   has increased by 7 to 14%.  Almost every ring commences the
   school year with 90% of the staff carried over from the preceding year.

4. A corollary of declining growth rates and increased
   retention and survival rates is that the percentage and

number of new entries to the teaching force is declining
each year. Almost 15% of the staffs of outer rings were
new entrants in 1968. That figure is down to between 5
and 8% by 1974, ans was about 3-5% in the inner ring.
In short, it is getting harder for new graduates to enter
teaching, and the teaching force is getting older and more
experienced with each year.

5. The number and percentages of people in administrative (non-
   teaching) positions has increased since 1968 regardless of
   declines or increases in enrollment.

The foregoing points are important, in our view, only insofar
as they affect the lives of the students who enter the school system. To investigate that we must enter upon a speculative journey.

With regard to movers we must ask who can move from one district
to another? It seems probable that in a tight market the people who
are most likely to be able to move from one district to another are
those who are good. Personnel officers and recruiters who take even
modest pains to screen staff would be able to eliminate the worst
teachers and administrators from their pool of candidates, leaving
"better" people as the basic pool from which movement would take
place. This leaves the inner rings as the most likely losers of
good, experienced talent.

In addition, when we attempt to predict movement from one district to another we discover that movers tend to be less experienced,
less highly trained, and male. District attributes are also important, as people tend to move from districts characterized by low pay,
small size, high administrative presence, and high average experience.
(Table VI and VII contain the variable definitions and regression
equations).

Collectively, these things mean that districts seem to be los-
ing people that they claim they would like to keep. In order to keep
the few who leave, however, huge payments would probably have to be
made to people who would not leave the system.

A similar problem faces school systems with regard to drop-outs.
Table VIII presents the regression equations we ran to predict drop-
out behavior. The drop-outs tend to be less experienced, female,
less trained, and paid less than might be expected on the basis of
their qualifications. Further, they tend to drop out of districts
with high administrative presence, highly trained staffs, and less
experienced staffs.

Finally, we turn to administrative expansion. This problem
area may be subsumed under a more general problem of productivity
in education. Expansion of the administrative component is but one

facet of the problem, and we see a steady increase in the relative
size of that component.  This is important, for Bidwell and Kasarda
(1975) found a negative relationship between the size of the administrative component of a system and the achievement of the students
in that system.  While the sample for this study was composed of
104 Colorado School districts, Bidwell has privately reported that
it appears as if the finding will be replicated with a better sample
of students and schools in Michigan.

We also see that the pupil/staff ratio has gone steadily down
over the seven years studied in our sample.  This, combined with increases in the average experience and training of teachers means
that there has been marked improvement in the quality of services
delivered to students, or at least marked increases in the prices
of services.

Summers and Wolfe (1975) cast some doubt on the validity of
paying for experience and training.  They found that experienced
teachers did best with high achieving students, while low achievers
seemed to do better work with younger relatively inexperienced teachers.  They also found that the amount of education beyond the
bachelor's degree was unrelated to student performance, as was the
score on the National Teachers' Examination.

Hanushek (1970) had similar findings.  He found no relationship
between teachers' education, experience and students' academic achievement.  Coleman et al (1966) also found little relationship between
the training and experience of teachers and student achievement.
These four studies point to the fact that schools may be paying for
aspects of teachers' characteristics which are unrelated to productivity -- they are paying for administrative positions which do not
appear to enhance productivity, and for training and experience
which appear to be unrelated to student achievement.

In the metropolitan area, the effect of these payments is not
random.  It is the inner ring which has experienced the greatest
inflation in its salary costs, and which has experienced the upsurge
in administrative positions.  Moreover, our data probably understate
the effect of this, since it is the inner ring that many resources
have to be diverted to non-certificated people (who do not appear
in our file) such as security guards, secretaries and so forth.

Indeed, the studies show some positive aspects of the teachers'
characteristics but these are things which are not actively sought
by schools.  Hanushek found the intelligence of teachers to be positively related to achievement; the source of a teacher's training
was important in the Wolfe and Summers' study, with teachers coming
from higher ranked universities having higher achievement in their
students.  They also found that the young and relatively inexperienced teachers were the most effective with disadvantaged youths, and

yet those are precisely the teachers who are either dropping out of
teaching or who are unable to get into the schools because of decreased demand for new teachers.  Recency of education and intelligence, as well as female teacher were found to be significantly associated with student achievement by Hanushek.  These too are things
which are not sought by schools, or if sought, are not rewarded on
the pay scales.


## CONCLUSION

     Our data coupled with the studies of other researchers, lead
us to believe that there is a serious problem in the school system
of Metropolitan St. Louis.  We also believe that the problem we see
in St. Louis is typical of the nation's schools in general.  The problem is that schools are not paying for those characteristics of the
teaching force which are associated with improved pupil outcomes.
Moreover, they are not purchasing a mix of teachers and administrators which maximizes student outcomes.  Rather, the trend seems to
be towards a mix which will minimize student achievement.  We believe
that comparative studies, conducted over time, are the only way to
verify our propositions about productivity in education, and we urge
other researchers to also initiate such studies.

ANNEX
Table I.  Net Migration Flows, 1968 - 1974.

| FROM \ TO | Ring II | Ring III | Ring IV | Ring V | Ring VI |
|---|---|---|---|---|---|
| Ring I | 147 | 42 | 46 | 16 | 33 |
| Ring II | | 29 | 53 | 39 | 27 |
| Ring III | | | 40 | 16 | 2 |
| Ring IV | | | | -5 | -8 |
| Ring V | | | | | 6 |

Table II.  Survival Rates

| Entry Year | After 1 Year | After 2 Years | After 3 Years | After 4 Years | After 5 Years | After 6 Years |
|---|---|---|---|---|---|---|
| **1968-69** | | | | | | |
| Total | 60.0% | 43.3% | 38.0% | 30.1% | 27.0% | 23.9% |
| Male | 56.7% | 44.6% | 42.1% | 37.2% | 34.3% | 31.4% |
| Female | 61.1% | 42.9% | 36.7% | 28.7% | 24.6% | 21.5% |
| **1969-70** | | | | | | |
| Total | 68.4% | 55.3% | 44.0% | 37.8% | 33.7% | |
| Male | 66.5% | 55.3% | 46.4% | 42.2% | 39.4% | |
| Female | 69.1% | 55.2% | 43.1% | 36.3% | 31.7% | |
| **1970-71** | | | | | | |
| Total | 75.0% | 58.8% | 49.1% | 42.4% | | |
| Male | 70.8% | 56.4% | 47.9% | 44.4% | | |
| Female | 76.3% | 59.6% | 49.4% | 41.9% | | |
| **1971-72** | | | | | | |
| Total | 73.4% | 61.4% | 49.9% | | | |
| Male | 71.6% | 60.0% | 50.8% | | | |
| Female | 74.0% | 61.9% | 49.5% | | | |
| **1972-73** | | | | | | |
| Total | 76.8% | 65.4% | | | | |
| Male | 77.7% | 69.1% | | | | |
| Female | 76.4% | 63.9% | | | | |
| **1973-74** | | | | | | |
| Total | 81.1% | | | | | |
| Male | 79.3% | | | | | |
| Female | 82.1% | | | | | |

Table III. Drop-out and Retention Rates by Ring, 1968 – 1973.

| Year | Ring I | | | Ring II | | | Ring III | | | Ring IV | | | Ring V | | | Ring VI | | |
|------|--------|-------|-------|---------|-------|-------|----------|-------|-------|---------|-------|-------|--------|-------|-------|---------|-------|-------|
| | Number | %Drop | %Stay | Number | %Drop | %Stay | Number | %Drop | %Stay | Number | %Drop | %Stay | Number | %Drop | %Stay | Number | %Drop | %Stay |
| 1968 | 4792 | 14.4 | 84.0 | 3375 | 19.4 | 78.5 | 2372 | 16.5 | 81.4 | 1803 | 19.4 | 79.4 | 2451 | 19.8 | 77.3 | 1952 | 21.8 | 75.6 |
| 1969 | 4730 | 10.7 | 87.3 | 3537 | 19.5 | 78.9 | 2396 | 16.2 | 82.5 | 1986 | 18.8 | 79.8 | 2648 | 17.9 | 81.2 | 2088 | 19.3 | 79.7 |
| 1970 | 4778 | 8.1 | 90.6 | 3594 | 16.1 | 82.8 | 2391 | 12.1 | 86.9 | 2124 | 15.1 | 84.3 | 2833 | 16.2 | 83.3 | 2201 | 12.9 | 86.5 |
| 1971 | 4993 | 11.6 | 87.1 | 3697 | 13.0 | 85.8 | 2348 | 10.4 | 89.3 | 2244 | 12.1 | 87.6 | 3056 | 12.9 | 86.7 | 2324 | 11.9 | 87.6 |
| 1972 | 4960 | 9.2 | 89.9 | 3776 | 12.0 | 87.1 | 2345 | 12.0 | 86.9 | 2373 | 10.8 | 89.0 | 3338 | 12.1 | 87.5 | 2456 | 11.6 | 87.9 |
| 1973 | 4742 | 8.4 | 91.0 | 3754 | 12.3 | 86.3 | 2237 | 8.7 | 89.0 | 2483 | 11.0 | 88.7 | 3520 | 12.4 | 86.8 | 2480 | 9.4 | 89.9 |
| 1974 | 4659 | | | 3590 | | | 2175 | | | 2524 | | | 3698 | | | 2552 | | |

Table IV. Staff and Enrollment increases and % of New Staff, by Ring

| Year | Ring I % Change in Staff | Ring I % Change in Enrollment | Ring I % New Staff | Ring II % Change in Staff | Ring II % Change in Enrollment | Ring II % New Staff | Ring III % Change in Staff | Ring III % Change in Enrollment | Ring III % New Staff |
|------|------|------|------|------|------|------|------|------|------|
| 1968 | -- | -.6 | 12.0 | -- | 2.4 | 11.0 | -- | 3.6 | 6.3 |
| 1969 | -1.3 | -.4 | 10.5 | 4.8 | -.5 | 10.9 | 1.0 | -.8 | 7.2 |
| 1970 | 1.0 | -4.5 | 7.2 | 1.6 | 1.8 | 9.4 | -.2 | -3.0 | 5.7 |
| 1971 | 4.5 | -13.4 | 10.1 | 2.3 | -4.5 | 7.4 | -1.8 | -2.5 | 4.4 |
| 1972 | -.6 | 3.4 | 8.9 | 2.1 | -3.9 | 6.8 | -.1 | -4.2 | 4.3 |
| 1973 | -4.4 | -1.5 | 2.9 | -.6 | -3.9 | 6.0 | -4.6 | -4.7 | 3.0 |
| 1974 | -1.8 | -3.9 | 4.1 | -4.3 | -4.0 | 4.8 | -2.8 | -5.8 | 3.0 |

| Year | Ring IV % Change in Staff | Ring IV % Change in Enrollment | Ring IV % New Staff | Ring V % Change in Staff | Ring V % Change in Enrollment | Ring V % New Staff | Ring VI % Change in Staff | Ring VI % Change in Enrollment | Ring VI % New Staff |
|------|------|------|------|------|------|------|------|------|------|
| 1968 | -- | 17.3 | 14.1 | -- | 9.5 | 11.1 | -- | 15.6 | 15.4 |
| 1969 | 10.1 | 10.0 | 12.4 | 8.0 | 6.6 | 14.7 | 7.0 | 6.5 | 12.6 |
| 1970 | 7.0 | 4.3 | 11.5 | 7.0 | 5.2 | 11.4 | 5.4 | 4.5 | 9.0 |
| 1971 | 5.6 | 3.2 | 8.5 | 7.9 | 6.0 | 10.6 | 5.6 | 1.1 | 6.4 |
| 1972 | 5.7 | 2.9 | 7.8 | 9.2 | 3.6 | 10.5 | 5.6 | 1.4 | 8.7 |
| 1973 | 4.6 | 1.7 | 6.8 | 5.5 | 1.9 | 8.6 | .9 | -1.4 | 8.3 |
| 1974 | 1.7 | -.2 | 4.6 | 5.1 | 1.2 | 8.5 | 2.9 | -1.9 | 5.6 |

Table Va. Enrollment, No. of Teachers, No. of Administrators by Rings and Years.

| | Ring I | | | Ring II | | | Ring III | | | Ring IV | | | Ring V | | | Ring VI | | |
|---|---|---|---|---|---|---|---|---|---|---|---|---|---|---|---|---|---|---|
| | Enrol | #Tch | #Adm | Enrol | #Tch | #Adm | Enrol | #Tch | #Adm | Enrol | #Tch | #Adm | Enrol | #Tch | #Adm | Enrol | #Tch | #Adm |
| 1968 | 116582 | 4264 | 484 | 71803 | 2975 | 362 | 41758 | 2051 | 283 | 38603 | 1595 | 188 | 56401 | 2190 | 224 | 40527 | 1734 | 194 |
| 1969 | 116182 | 4224 | 461 | 71442 | 3077 | 401 | 41434 | 2056 | 298 | 42456 | 1742 | 207 | 60147 | 2384 | 229 | 43170 | 1843 | 216 |
| 1970 | 110934 | 4282 | 452 | 72694 | 3137 | 399 | 40200 | 2044 | 302 | 44294 | 1866 | 218 | 63304 | 2529 | 263 | 45113 | 1955 | 217 |
| 1971 | 96091 | 4455 | 491 | 69399 | 3167 | 469 | 39184 | 1999 | 301 | 45700 | 1951 | 245 | 67113 | 2698 | 309 | 45597 | 2032 | 256 |
| 1972 | 99315 | 4405 | 509 | 66705 | 3219 | 492 | 37530 | 1969 | 328 | 47029 | 2038 | 281 | 69558 | 2918 | 354 | 46226 | 2135 | 276 |
| 1973 | 97861 | 4026 | 673 | 64073 | 3184 | 500 | 35769 | 1871 | 318 | 47850 | 2140 | 288 | 70874 | 3068 | 379 | 45559 | 2147 | 284 |
| 1974 | 94068 | 3955 | 663 | 61524 | 3027 | 495 | 33682 | 1799 | 327 | 47735 | 2173 | 296 | 71698 | 3215 | 405 | 44676 | 2194 | 302 |
| Net Change(%) | -19.3 | -7.2 | +37.0 | -14.3 | +1.8 | +36.7 | -19.3 | -12.3 | +15.5 | +23.7 | +36.2 | +57.4 | 27.1 | +46.8 | +80.8 | +10.2 | +26.5 | +55.7 |
| Average Change(%) | -3.2 | -1.2 | +6.2 | -2.4 | +.3 | +6.1 | -3.2 | -2.1 | +2.6 | +4.0 | +6.0 | +9.6 | 4.5 | +7.8 | +13.5 | +1.7 | +4.4 | +9.3 |
| Pupils/Admin. '68 | 240.9 | | | 198.4 | | | 147.6 | | | 205.3 | | | 251.8 | | | 208.9 | | |
| Pupils/Admin. '74 | 141.9 | | | 124.3 | | | 103.0 | | | 161.3 | | | 177.0 | | | 147.9 | | |
| Teachers/Admin. '68 | 8.8 | | | 8.2 | | | 7.2 | | | 8.5 | | | 9.8 | | | 8.9 | | |
| Teachers/Admin. '74 | 6.0 | | | 6.1 | | | 5.5 | | | 7.3 | | | 7.9 | | | 7.3 | | |

Table Vb.  Regression Equations Predicting Size of Administrative Component from Parkinson's Law, Change in Enrollment and Change in Staff Size.

|  | (1) | | (2) | | (3) | |
|---|---|---|---|---|---|---|
| Parkinson's Law | 1.24 | (7.6)* | -- | | .86 | (5.7) |
| Change in Enrollments | 3.24 | (2.2) | 6.01 | (0.9) | -- | |
| Change in Staff Size | 0.2 | (0.1) | 11.00 | (1.6) | -- | |
| RSQ | .994 | | .789 | | .943 | |

Source of variance in size of administrative component:

    Unique to Parkinson's Law        20.5%  ((1) - (2))

    Unique to Changes in Enrollment
    and Staff Size        5.1%  ((1) - (3))

    Shared between Enrollment, Staff
    Size and Parkinson's Law    <u>73.8%</u>  ((2) + (3) - (1))

                           99.4%

*t values in parentheses

Table Vc. Salary Expense, Enrollment, No. Teaching,
No. of Administrators.

|  | Expected Sign | Coefficient | Standard Error | t-value |
|---|---|---|---|---|
| STUDSAD | − | 0.006 | 0.014 | 0.40 |
| PUPTEA | − | −16.116 | 0.854 | 18.88 |
| DVINCR | − | −32.639 | 6.598 | 4.95 |
| GRADHR | + | 7.105 | 0.382 | 18.59 |

$R^2 = .852$

$F (4,409) = 587.659$

STUDSAD:  average district ratio of students to student personnel

PUPTEA:  average district ratio of students to teachers

DVINCR:  a dummy variable with the value 1 if the district is in
a ring with increasing enrollment over the period

GRADHR:  average number of graduate hours by district

Table VI

| INDIVIDUAL LEVEL VARIABLES | |
|---|---|
| DROP | dependent variable, 1 if drop out of teaching |
| MOVE | dependent variable, 1 if move to new district |
| SEX | dummy variable for sex, female = 1 |
| SALARY | yearly salary |
| HALF | dummy variable for half-time position, half-time = 1 |
| TOTHRS | total number of hours (undergraduate plus graduate) |
| GRADHRS | numbers of graduate hours |
| DEGNONE | dummy variable for college degree, no degree = 1 |
| DEGMA | dummy variable for MA degree, MA = 1 |
| DEGSPEC | dummy variable for specialist degree, specialist degree = 1 |
| DEGPHD | dummy variable for PhD degree, PhD = 1 |
| PRESEXP | years of experience in the district |
| TOTEXP | years of experience in the profession |
| POSCENT | dummy variable for central district office personnel |
| POSSTUD | dummy variable for student related personnel |
| POSAD | dummy variable for administrator |
| POS | dummy variable for POSCENT = 1 or POSAD = 1 or POSSTUD = 1 |
| POSTEA | dummy variable for teacher |
| SALHAT | predicted salary based on TOTEXP and TOTHRS |
| SALDIFF | SALHAT-SALARY |

| SCHOOL AND DISTRICT LEVEL VARIABLES | |
|---|---|
| CLASS2 | dummy variable for district classification, "Triple-A" = 1 |
| REVSTUD | assessed value per student |
| DAPREXP | district average present experience |
| DATOTEXP | district average total experience |
| DATOTHR | district average total hours |
| DASAL | district average salary |
| STUDDIST | number of students in the district |
| TEADIST | number of teachers in the district |
| STUDTEA | pupil teacher ratio for the district |
| DSAD | number of administrators in the district |
| APD | ratio of administrators to teachers for the district |
| RELSAL | ratio of DASAL for the district to the highest DASAL of all districts that year |
| STUDGR | percent of change in number of students over the previous year |
| SSAD | number of administrators in the school |
| APS | ratio of administrators to teachers for the school |
| MINS | percent of student body in the school who are of minority groups |

Table VII.  Move Regression Results.

|  | 1968-1969 | | 1973-1974 | |
|---|---|---|---|---|
| N | 13884 | | 17458 | |
| $R^2$ | .021 | | .015 | |
| Adjusted $R^2$ | .019 | | .014 | |
| F (degrees of freedom) | 18.51 | (16,13867) | 16.59 | (16,17441) |
| CONSTANT | 0.08855 | | 0.11805 | |
| PRESEXP | -0.00183 | (9.12)* | -0.00117 | (8.98) |
| SALDIFF | 0.00704 | (5.89) | 0.07384 | (7.36) |
| STUDTEA | 0.11702 | (1.93) | -0.13854 | (4.22) |
| STUDDIST | -0.00003 | (4.88) | -0.00001 | (2.19) |
| DAPREXP | 0.00423 | (3.04) | 0.00303 | (4.27) |
| RELSAL | -0.14225 | (3.53) | -0.09702 | (4.81) |
| APD | 0.30652 | (2.83) | 0.09077 | (1.98) |
| APS | -0.00104 | ( .20) | 0.00150 | (2.05) |
| POS | 0.03562 | (2.46) | 0.00897 | (1.31) |
| TOTHRS | -0.00146 | (2.13) | -0.00102 | (2.74) |
| SEX | -0.01004 | (2.83) | -0.00804 | (4.24) |
| HALF | -0.03442 | (1.77) | 0.00900 | (1.13) |
| DEGMA | -0.00525 | (1.17) | 0.00487 | (2.06) |
| POSTEA | 0.01635 | (1.20) | -0.00765 | (1.22) |
| DEGPHD | -0.00383 | ( .17) | 0.01521 | (1.67) |
| STUDGR | -0.00026 | (1.84) | -0.00036 | (2.25) |

*
t values in parentheses

Table VIII.  Drop Regression Results.

|  | 1968-1969 | | 1973-1974 | |
|---|---|---|---|---|
| N | 16914 | | 19465 | |
| $R^2$ | .057 | | .028 | |
| Adjusted $R^2$ | .056 | | .027 | |
| F (degrees of freedom) | 60.07 | (17,16896) | 33.21 | (17,19447) |
| CONSTANT | -0.04917 | | -0.08309 | |
| TOTEXP | -0.00504 | (16.20)* | -0.00111 | (4.09) |
| SALDIFF | 0.02909 | (13.89) | 0.01668 | (12.88) |
| DEGNONE | 0.24583 | (11.44) | 0.08589 | (3.31) |
| DEGMA | -0.01860 | (1.71) | -0.01694 | (2.21) |
| STUDDIST | -0.00000 | (0.22) | -0.00001 | (0.54) |
| POS | 0.07441 | (2.60) | 0.05786 | (3.18) |
| APD | 0.46501 | (2.33) | 0.42335 | (3.74) |
| GRADHRS | -0.00102 | (4.54) | -0.00039 | (2.40) |
| DEGPHD | 0.14733 | (3.18) | 0.06144 | (2.38) |
| SEX | 0.01398 | (2.04) | 0.02352 | (4.69) |
| APS | 0.02071 | (2.03) | 0.00236 | (1.22) |
| HALF | 0.04390 | (1.33) | 0.10059 | (5.35) |
| DATOTHR | 0.01718 | (2.54) | 0.01015 | (2.41) |
| DAPREXP | -0.00573 | (2.36) | -0.00530 | (2.90) |
| STUDGR | -0.00003 | (0.98) | 0.00024 | (0.62) |
| DEGSPEC | 0.06913 | (0.83) | 0.01034 | (0.49) |
| POSTEA | 0.01173 | (0.44) | 0.02399 | (1.45) |

*
t values in parentheses

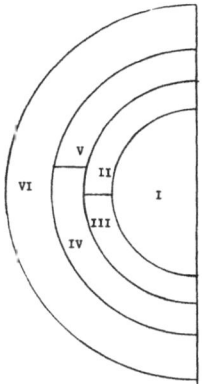

Figure I.  Division of St. Louis Metropolitan Area.

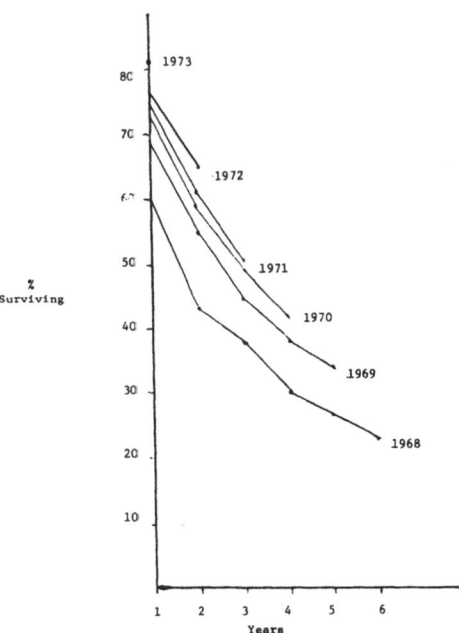

Figure 2.  Survival Rates in Teaching, 1968 - 1973.

## List of References

Becker, Howard S., "The Career of the Chicago Public School Teacher", American Journal of Sociology, Vol. LVII (March, 1952).

Bidwell, Charles E., and John D. Kasarda. "School District Organization and Student Achievement," American Sociological Review, Vol. 40, No. 1 (February, 1975) pp. 55-70.

Burgess, Ernest W, "The Growth of the City: An Introduction to a Research Project." In George Theodorson (ed.) Studies in Human Ecology. New York: Harper & Row, 1961, pp. 37-44.

Charters, W.W. "What Causes Teacher Turnover?" School Review, Vol. 64, (October 1956 (a)), pp. 294-299.

Charters, W.W. "Some Obvious Facts about the Teaching Career," Educational Administration Quarterly, Vol. 3, No. 2 (Spring, 1967) pp. 183-193.

Charters, W.W. "Some Factors Affecting Teacher Survival in School Districts," American Educational Research Journal, Vol.VII, No.1

Coleman, James S., E.Q. Campbell, C.J. Hobson, J. McPartland, A.M. Mood, F.D. Weinfield, and R.L. York. Equality of Educational Opportunity. U.S. Government Printing Office, FS5.238:38001,1966.

Gittell, Marilyn, and T. Edward Hollander. Six Urban School Districts: A Comparative Study of Institutional Response. New York: Praeger, 1968.

Greenberg, David J., and John J. McCall. Analysis of the Educational Personnel System I: Teacher Mobility in San Diego. Rand Corporation: R-1071-HEW (January, 1973).

Griffiths, Daniel E. Samuel Goldman, Warne McFarland. "Teacher Mobility in New York City," Educational Administration Quarterly Vol. 1, No. 1, 1965, pp. 15-31.

Hanushek, Eric. The Value of Teachers in Teaching. Rand Corporation: Rm6362-CC/RC (December, 1970).

Holdaway, Edward A., and Thomas A. Blowers. "Administrative Ratios and Organization Size: A Longitudinal Examination," American Sociological Review, Vol. 36 (April, 1971) pp. 278-286.

Missouri State Board of Education. Report of the Public Schools of the State of Missouri. Jefferson City, Mo.: The Board 19xx.

Missouri State Board of Education. Missouri School Directory. Jefferson City, Mo.: The Board, 19xx.

Park, Robert E., E.W. Burgess, and R.D.McKenzie. The City. Chicago: The University of Chicago Press, 1925.

Parkinson, C. Northcote. Parkinson's Law and other Studies in Administration, New York: Ballantine, 1957.

Summers, Anita A., and Barbara L. Wolfe. "Which School Resources
    Help Learning? Efficiency and Equity in Philadelphia Public
    Schools," Business Review, Philadelphia: Federal Reserve Bank
    of Philadelphia, February, 1975.

Session VI

CASE STUDIES: ASSESSMENT OF

SOCIAL IMPACT

ENVIRONMENT AND ENERGY ON THE TEXAS GULF COAST:

AN ECONOMIC EVALUATION MODEL OF ALTERNATIVE POLICIES

Kingsley E. Haynes,* Jared E. Hazleton;* W. Tom Kleeman *

The University of Texas at Austin; State Government of
Texas
U.S.A.

## INTRODUCTION

In 1973 Charnes, Haynes, Hazleton and Ryan (1975) presented a
comprehensive decision framework for evaluating competing alternative environmental and economic policies in the Texas Coastal Zone.
Such a framework had imbedded within it a series of interactive submodels. Some of these have been presented elsewhere (Haynes et al.,
1974; Haynes and Kleeman, 1975; Kleeman, Haynes and Freeland, 1975;
Phillips, White and Haynes, 1976; Charnes, Haynes and Phillips, 1976).
But a unique "rows only" ***impact evaluation procedure may be of
particular interest since we have now demonstrated its useful application not only in environmental policy evaluation but in related
evaluations of the local economic impact of alternative energy source
utilizations.

It should be pointed out with some caution that this is a short-
run impact evaluation model directed to respond to policy makers immediate evaluative needs. It does not address itself to the long-
run dynamic equilibrium adjustment process which is the more common

---

* The authors were co-principal investigators or members of the NSF
  (RANN) Texas Coastal Zone Management Research Team 1972-1976 and
  wish to express their gratitude to NSF(RANN) Grant #GI-34870X.
***
The terms "rows only" and "pass through" refer to the same basic
model. The first is a reference in terms of the calculation
technique while the latter refers to its interpretation.

basis of amost micro and macro-economic impact modelling.

We first describe the geographical and economic context of the Texas Coastal Zone and then briefly review the comprehensive Texas Coastal Zone Management Evaluation Model presented earlier.  Next, we focus on one aspect of that model - the "rows only" impact analysis.  This is discussed in terms of two applications: (1) First an environmental policy evaluation application and (2) an application in terms of the impact of a new potential energy source within the Texas Gulf Coast.


## Texas Coastal Resources

The Texas Coastal Zone, by legislative definition, comprises a tier of counties (two deep) adjacent to the Texas coast together with the adjacent bays, shallow water, and barrier island systems. Historically, the primary economic activity in this area has been based upon the extraction of oil and natural gas and the associated refining and petrochemical industries.  The availability of these fuels and the access to ports have also led to the development of steel and aluminum industries and metal fabrication industries. Major industrial complexes are located at Beaumont-Port Arthur, Houston-Texas City-Galveston, and Corpus Christi.

The major agricultural products of the coastal zone are cotton, sorghum, rice and cattle.  Much of this agricultural production requires irrigation with water drawn both from surface and artesian sources.  Networks of channels and reservoirs have been proposed by the Texas Water Development Board (1968) to increase supplies of water for both industrial and agricultural use.

It is anticipated that land-use management policies will be implemented by legislation based upon the concept of "critical environmental units."[*]  The definition of "environmental units" and "resource capability units" are being developed for both land and marine areas by the Texas General Land Office and the Bureau of Economic Geology of the University of Texas.  These units comprise broadly defined geological structures for which admissible types and intensities of use will be established based upon both surface and sub-surface geological considerations.  It will also include the consideration of biologic assemblages that are coincident or interactively dependent on these surficial conditions.  The development of these units and the establishment of associated use criteria provides a direct means

---

[*]  The existing and pending federal legislation on land use, including the much publicized Jackson Bill, refers to the need for protection and maintenance of "critical environments."

of incorporating environmental zoning considerations into the land-use planning process.  This project focussed on the development of these criteria and their application in the planning process.

Although not specified directly in the model outlined here a secondary focal point deals with the fundamental character of ports. Port cities have been of major importance in the economic development of the coastal zone.  Ports as transportation nodes and as industrial and residential centers affect bays, estuaries and littoral environments.  The influence of port cities on marine environments may manifest itself through depletion of fresh water inflows, canalization, waste inflows and recreational demands.  Construction on littoral areas often conflicts with conservation policies derived from both political and environmental consideration.

## COMPREHENSIVE FRAMEWORK

In 1973 Charnes, et al., presented a comprehensive decision framework for the NSF (RANN) supported Texas Coastal Zone Management Project.  Drawing upon the work of Charnes, Cooper and Niehaus (1972) they focussed on the design of a hierarchical goal-programming model which linked together a series multilevel sub-models with appropriate constraints into an optimizing decision structure.  The purpose of the model was to provide a normative framework within which alternative policies (goals) of an environmental land-use character could be effectively evaluated.  The model's three-level hierarchy, linking statewide, multicounty and local models allowed the explicit consideration of competing goals (policies) at different spatial scales.

The overall model is outlined in Figure 1.  The objective function to be optimized was a weighted sum of deviations from regional growth goals $g_i^{r+}(t)$, $g_i^{r-}(t)$ subject to the constraints (1)-(8) (see Figure 1).

We then examined alternative goals of minimizing unemployment $\bar{g}_{hkr}^{r+}$ or job vacancies $\bar{g}_{hkr}^{r}$ on a local basis.  The overall objective may be represented schematically as: $MIN \sum c_i^{r+}(t) + \sum c_i^{r-}(t) g_i^{r-}(t) + \sum \bar{c}_{hkr}^{r+}(t) + \sum \bar{c}_{hkr}^{r-}(t) \bar{g}_{hkr}^{r-}$ where the weights $\bar{c}_k^r(t)$, $\bar{c}_i^r(t)$, $\bar{c}_{hkr}^{r}(t)$, $\bar{c}_{hkr}^{r-}(t)$ may reflect various political and economic factors. At each spatial scale, models are linked to each other by appropriate variables and shared inputs and outputs.

Two sub-components of the model that are particularly interesting include the spatial interaction estimation procedure based on an information - theoretic extremal programming criteria, reported elsewhere, and the short term economic "pass through" impact evaluation of new technologies which will be discussed here.  The scope

Table 1

**Minimize Level 1**

$$\sum c_i^{r+}(t)g_i^{r+}(t) + c_i^{r-}(t)g_i^{r-}(t) + \bar{c}_{hKr}^{r+}\,g_{hKr}^{r+}(t) + \bar{c}_{hKr}^{r-}\,\bar{g}_{hKr}^{r-}(t)$$

| | Waste Treatment Models | Demographic Models | Water Supply Models | |
|---|---|---|---|---|
| Input - Output | $x_j^r(t) - \sum_j L_{ij} x_j^r(t) - \sum_s e_i^{sr}(t) - \sum_s e_i^{sr}(t)$ | | $+g_i^{r+}(t) - g_i^{r-}(t) - d_i^r(t)$ | (1) |
| Trade Restrict. | $E_{1S}^{rl}(t) < \sum_{s\in S} e_i^{rs}(t) < E_{1S}^{rU}(t)$ | | | (2) |
| Level 2 | $E_{2S}^{rl}(t) < \sum_{s\in S} e_i^{rs}(t) < E_{2S}^{rU}(t)$ | | | |
| Spatial Distrib. of Industry | $x_i^r(t) - \sum_{k_r \in K_r} x_{ik_r}^r(t)$ | | $= 0$ | (3) |
| Demographic Relationships | | $p_{hKr}^r(t) - p_h \lambda_{hKr}^r(t) - \mu_{hKr}^r(t)$ | $= 0$ | (4) |
| Employment | $\sum_i \sum_{k_r \in K^r} p_{ih}\, x_{ih_r}^r(t)$ | $-\lambda_{hKr}^r(t)$ | $\bar{g}_{hKr}^{r+}(t) - \bar{g}_{hKr}^{r-}\; = 0$ | (5) |
| Spatial Distrib. of Housholds | $\sum_{k_r \in K_r} p_{hk_r}^r(t) - p_{hKr}^r(t)$ | | $= 0$ | (7) |
| Land Restrict. | $\sum_i a_{ik_r} x_{ik_r}^r(t) + \sum_h b_{hk_r} p_{hk_r}^r(t)$ | | $< A_{k_r}^r$ | (6) |
| Water Restrict. | $\sum_i \sum_{k_r \in K_r} f_{ih_r} x_{ih_r}^r(t) + \sum_h \sum_{k_r \in K_r} e_{ih_r} p_{hk_r}^r(t)$ | | $-q_{K_r}(t) \quad < 0$ | (8) |
| Level 3 | | | | |

of this paper will be limited to a description of the method developed for measuring increases in cost in the private sector caused by forced internalization of additional production costs resulting from the alternative policies. Furthermore, it will demonstrate the application of the model to a regional economy.

## The Regional Economy

The region under study consisted of the thirteen counties of the Coastal Bend Council of Governments (COG). This particular region was chosen as the area for study because of the large amount of economic activity taking place near the coastline, and its relative isolation from large metropolitan complexes such as Houston. The COG is a subregion inside the larger south Texas region (number 7), which is a designation used in the ongoing Texas Interindustries Project. This project is sponsored by the Texas Governor's Office of Information Services (OIS), which has constructed a state input-output model for 1967, as well as nine regional models for the same year. Since the model for region 7 contained the required data, locations-quotient techniques were used to derive the smaller Coastal Bend subregional model (see Figure 2).

The model consists of seventy-one processing sectors that may be categories as: 'agri-ulture', 1-12, 'mining', 13; 'construction' 14-16;'manufacturing', 17-34; 'transportation and communications', 35-39; 'utilities', 40-42; 'wholesale trade', 43-49; 'retail trade', 50-60; 'FIRE' (finance, insurance, and real estate), 61 and 62; and 'services', 63-71.

Final demand consisted of seven sectors; 'households'; 'federal', 'state), and 'local' government (one sector each); 'capital formation'; 'exports'; and 'net inventory changes'. Final payments contained seven sectors: 'households'; 'federal', 'state' and 'local' government; 'depreciation'; 'impacts'; and 'residual' (which included retained earnings, profits, dividends, and savings and account balancing entries).

## Environmental Policy Evaluation

Context. The United States 1972 'Pure Water' legislation (PL-92-500) requires a national effluent standard such that major industries must apply the "Best Available Treatment" to central waste discharges by 1978, and must reduce their waste discharges to zero by 1983. To meet these standards, industries will require increased capital investments and increased operations and maintenance co-costs for central equipment for pollution. Since these costs are not distributed evenly among all industries, they will have a differential impact on a regional economy. Furthermore, since technological

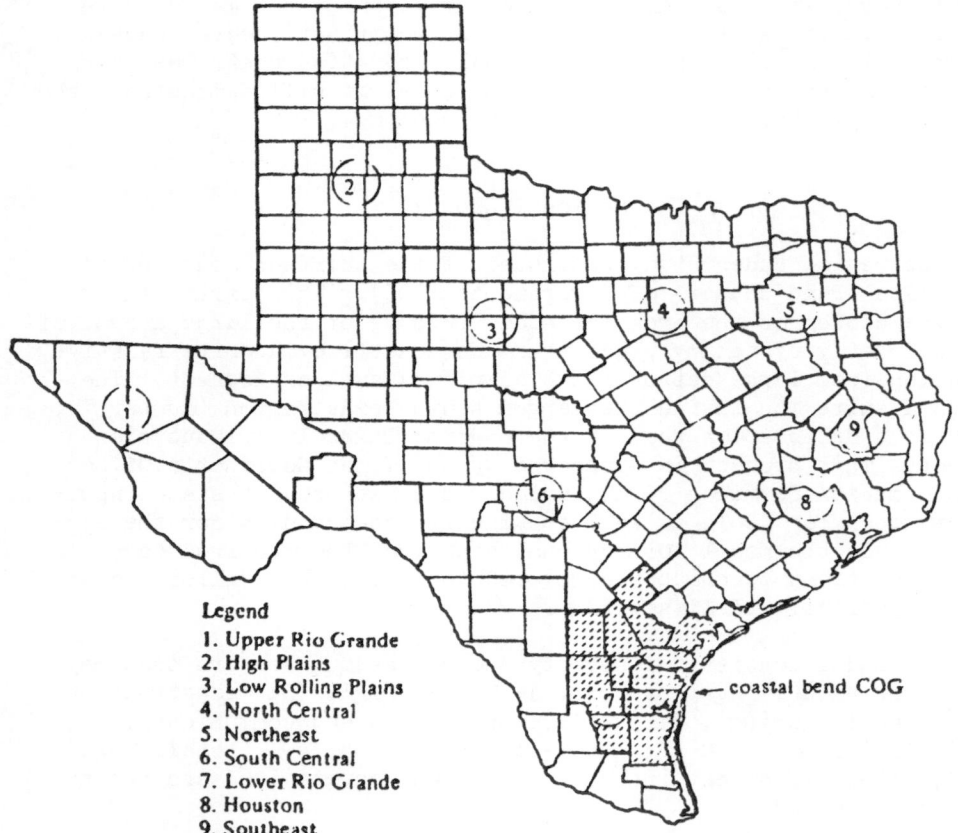

Figure 2.   Texas Input-Output Regions

coefficients in the short-run remain fixed, increased costs attributable to improved environmental quality legislation will affect the consumer in the form of increased prices.  For modelling purposes the two dates were adjusted to 1980 and 1990 respectively.

The original interest in carrying out this investigation centered around the question of whether or not environmental control policies would demonstrate inflationary tendencies.  The increased production costs resulting from the policy might bring about cost-push inflation pressures.  At the same time, a large portion of the operation and maintenance costs were accounted for by payments to households (income).  When this was coupled with no increase in output being offered to final demand, it indicated a classic case of demand-pull inflation.

When the effluent-discharge policy was first considered, it was intuitively appealing to assume that such stringent environmental protection requirements would inevitably lead to substantial increases in cost, which in turn would lead to price increases to purchasers.  The problem was to develop a method that would measure the effects of increasing prices on each sector and on the economy as a whole.  The task force studying water-needs and residuals-management developed the sets of costs to industry resulting from the implementation of the pollution-control policy (Sherman and Malina, 1974).

Capital Costs.  The first set of costs were for the additional capital outlays that would result from meeting the new standards on effluents.  The capital costs were assumed to be 'one shot' expenditures for the two target-dates, which is consistent with the generally 'lumpy' nature of capital expenditures.  Thus, the interest in capital expenditures was only with regard to its impact on the total output of the region for one time-period.  One reason for not emphasizing the potential injection of $ 11.7 million into capital investments is because $ 10 million would be outside of the processing sector.  The impact on local businesses was smaller than it would have been in more industrialized areas.  In all probability, as many jobs would be created in Houston as in the Corpus Christi area.  Second, since capital costs were being written off at the rate of 5% per year (for twenty years), this cost would show up in the depreciation sector and would be treated as an operation and maintenance cost.  The increased level of capital expenditures, and the primary sectors affected within this regional economy, are given below in terms of sales to capital formation by producing sectors.

| Sector | Number | Amount ($) |
| --- | --- | --- |
| Comment and concrete | 28 | 1448127 |
| FIRE | 62 | 248537 |
| Households | 72 | 3728048 |
| Imports | 74 | 6380702 |

Figure 3.  Sales by Sector Number

Operation and Maintenance Costs (O and M). Each of the following sectors bore the greatest impact of increased O and M costs required by the new standards: 'mining (petroleum and natural-gas extraction)',13; 'other food and kindred products', 20; 'petroleum refining and products', 26; 'cement and concrete products', 28; 'primary metals, foundries, and forgings', 29; and 'public and private
water, and sanitary service systems', 42. Each of the six sectors
affected by this hypothetical policy had payments to four sectors:
(1) 'electricity', row 41; (2) 'households', row 72; (3) 'imports',
row 77; (4) 'depreciation', row 76. The increased sales of the
three interindustry sectors are outlined in figure 4. In this way
an altered transactions table was created when the additional purchases of each sector were added to its previously projected level
of purchases. For example, previously projected sales of electricity (row 41) to petroleum and natural gas (column 13) were
$13,762,193 but have now risen by $59,951 to $13,822,145.

Obviously, these interindustry transactions will change the
technical relationships of the industries in the region. On the assumption that output levels remain the same, it will now require
more electricity, labor, imported goods and services, and payments
on depreciation capital equipment to produce that same level of output.

| Row Sales | Sector | Amount ($) | Sector | Amount ($) |
|-----------|--------|------------|--------|------------|
| 41 Electrical | 13 | 59951 | 28 | 10145 |
| services | 20 | 15629 | 29 | 5955 |
| | 26 | 38157 | 42 | 94954 |
| 72 Households | 13 | 425207 | 28 | 71953 |
| | 20 | 110851 | 29 | 42233 |
| | 26 | 270633 | 42 | 673468 |
| 77 Imports | 13 | 69943 | 28 | 11836 |
| | 20 | 18234 | 29 | 6947 |
| | 26 | 44517 | 42 | 110774 |

Figure 4.  Sales to Different Sectors

In order to determine what the changes in the technical relationships would be if total output remained the same in those sectors making the additional purchases, a new Table of Direct Requirements was derived. To do this, a new "A" matrix was derived, for
example, each total-output entry was divided back into each entry
of its respective column, $a_{ij} = X_{ij}/X_i$. Since the total output values were divided into each element of the column, for all 78 rows,
the column total would be greater than one if the environment related transactions were added but nothing else was reduced in the column. This problem was explicitly recognized by assuming that, in
the short-run, each industry would reduce its retained earnings by
an amount exactly equal to the additional payments that it had made
(retained earnings being contained in the final payments, row 78,
residuals).

With the new set of technical coefficients, it was possible to
establish what effect they would have on future dollar values of
total output. To do this, the first 71 rows and columns were extracted and subtracted from the identity matrix of similar dimensions.
The difference was then inverted, $(I-A)^{-1}$. This created what
Miernyk (1965) calls the "Table of Direct and Indirect Requirements."
The table was multiplied by the final demand vector that had been
used previously in the original set of total-output projections.
This provided a basis to make a comparison, in dollar values, between the total-output projections before and after implementation
of the policy. That is to say, total output for each industry in
the processing sector would be influenced by the transactions resulting from the public-policy change.

In order to finance these additional transactions the purchasing industries reduced their retained earnings. Obviously, most
firms will not voluntarily reduce their retained earnings for long
if they can help doing so. For this reason, it was assumed that the
costs of installation (depreciation), operation, and maintenance
would be passed on to the purchasers of each industry's products.
Given recent events in the area of energy, this seems a very sound
assumption. This is especially true since some of the industrial
sectors, as well as individual industries, in question exercise considerable market power. Furthermore, it is important to keep in mind
that this policy dealt with a nationwide effluent-control requirement. Therefore, the companies in this region were not considered
to have suffered from relative cost increases which would yield an
advantage to firms outside of the region.

To trace the method by which the costs were passed on, the
first step was to increase the residuals row back to its former level for example, where the amount of purchases had been subtracted
from the residual entry, this cell as increased to its former value.
The seven sectors in final payments were then summed into one total
final payments row, including the increased residual entries.

Next, the total output for each row was divided into each element in that sector's row in the altered Transactions Table; this produced a new matrix labelled "B". The "B" matrix was subtracted from the identity matrix, and the result was inverted, $(I-B)^{-1}$. The resulting matrix was multiplied by the new final payments vector, which included the increased residuals row. In order to avoid confusion, the reader is warned that this "B" matrix is not the same thing as the "B" matrix Leontief (1969) has discussed (see Appendix I for formal description).

As far as the authors know, this is a unique approach to tracing cost-push trends in an input-output model. Hansen and Tiebout (1963) did develop a somewhat similar technique in their rows-only approach. However, their method was used only for measuring the impact of exogenous forces on employment within a region.

Impact Results. The total increase in output for all 71 sectors resulting from the increased capital formation was only 0.02% for 1980. Listed below are the sectors with outputs greater than the average of all processors:

| Sector | 27 | 28 | 35 | 38 | 62 |
|---|---|---|---|---|---|
| Increase in output (%) | 0.17 | 2.85 | 0.07 | 0.10 | 0.25 |

The total increase in output for all 71 processing sectors in 1990 was even less, 0.01%. Below are listed all sectors with output increases greater than 0.01%

| Sector | 27 | 28 | 35 | 38 | 62 |
|---|---|---|---|---|---|
| Increase in output (%) | 0.08 | 1.10 | 0.03 | 0.04 | 0.10 |

In general, the creation of new capital for effluent control can be said to have a minor impact within this region. However, it should be noted that this model does not assume a constraint on available capital to be imported into the region to meet these environmental protection standards. If the availability of capital were determined in a free market, where all regions are trying to meet the same standards at the same time, then its scarcity and hence its costs could be a serious constraint. However, if federal funds are to play a major role in the availability of capital, as the legislation suggests, the impact of high levels of interregional competition for available environmental protection capital is not clear. Furthermore, in a policy context, a minor inflationary impact at a time of low inflationary levels is perceived differently than in a period of high inflationary pressures. Therefore, this issue is not handled explicitly within this analysis.

Operation and maintenance costs were a little more significant. Where the capital costs were assumed to be absorbed by the indivi-

dual firms, which would write them off over time, the same was not
true of operation and maintenance costs. To begin with, several of
these transactions took place within the processing sector. This
meant that some sectors would have increased operation costs due to
increased interindustry purchases. There were additional costs resulting from the increased payments to final payments sectors (households, imports, and depreciation). This would mean that the industries could absorb the additional costs, which would have appeared
as a reduction in the residual row containing retained earnings,
profits, savings, and account balancing items. The other likely
possibility was that the costs of meeting the standards would have
shown up as price increases which were passed on to consumers.

Figures 5 and 6 give the percentage price change (from a policy
of no environmental controls) as the result of a policy which contains the strong standards for effluents.

For 1980, the average price change for all 71 processing sectors was 0.04%. Listed in Figure 5 are those sectors that have price
increases of over 0.04%.

The 1990 average price increase was slightly higher, 0.07%, for
all processing sectors. Sectors with higher-than-average increases
in prices are listed in Figure 6.

In both 1980 and 1990, the sector most heavily affected was
sector 42: water and sanitation services.

The resulting new dollar values of total outputs were greater
than those projected under the no-change/policy. The differences
represent the price increases paid by purchasers. Surprisingly,
the price increases were not as great as had been expected originally.
For 1980, the only sector that had a total-output price increase of
more than one percent was water and sanitary services, which went up
by 6.93%. When all seventy-one industrial sectors were taken together the price increase averaged only 0.04%. In order to meet the
1990 zero discharge policy the prices again increased but only mod-
estly. Water and sanitary services, still the only sector with an
increase of more than one percent, showed a jump of 22.38% over its
previously projected levels. However, for all seventy-one processing sectors together the increase averaged only 0.07%.

| Sector | Price Increase (%) | Sector | Price Increase (%) |
|--------|--------------------|--------|--------------------|
| 8      | 0.06               | 41     | 0.22               |
| 20     | 0.28               | 42     | 6.93               |
| 26     | 0.15               | 63     | 0.22               |

Figure 5

| Sector | Price Increase (%) | Sector | Price Increase (%) |
|--------|--------------------|--------|---------------------|
| 8      | 0.12               | 42     | 22.38               |
| 20     | 0.57               | 52     | 0.09                |
| 26     | 0.21               | 58     | 0.14                |
| 28     | 0.48               | 59     | 0.13                |
| 29     | 0.15               | 63     | 0.70                |
| 41     | 0.40               | 66     | 0.14                |

Figure 6

## Energy Policy Evaluation

Context. The energy component was at first peripheral to our environmental research efforts, but due to the energy crisis of '74 and the identification of a possible geothermal source within the region, energy became a more significant issue (Bureau of Economic Geology, 1975). Furthermore, given the central role of energy in any assessment of the economy and the environment, the authors felt compelled to evaluate this potentially new energy source within the regional system.

The South Texas region is the poorest area of the state, and the diminishing energy reserves are not an encouraging prospect for an economically lagging area. Because of this unpromising energy and economic future, the geothermal resources of the area have been receiving increased attention.

Underlying the sparsely populated, predominantly agrarian, coastal hinterland of Corpus Christi are geopressured geothermal waters. Due to the extensive oil and gas exploration that has taken place in the area, the existence of this resource has been verified for some time. The deposits seem to run in bands along the coast, offshore, and inland as far as 75 miles (figure 7). These deposits vary in depth from 5,000 to 15,000 feet, with temperatures as high as $375^\circ F$ and pressures up to 5,000 psi at the wellhead (figure 8).

Recently two important studies have been completed regarding the economic efficacy of geopressured geothermal power plants along the U.S. Gulf Coast. These efforts have been carried out by House, Johnson, and Towse of Lawrence Livermore Laboratory (1975) and Wilson, Shepherd, and Kaufman of Dow Chemical (1974). These studies provide the best available regional estimates of the cost of building and maintaining a geothermal plant.

Figure 7.  Texas Coastal Geopressured Zone


While the Livermore report should be of interest to anyone not
familiar with the geopressured area, the level of microeconomic information available in the Dow study made it of particular value in
the present analysis.  The increased disaggregation of economic sectors by Dow made their results amenable to adaptation to the Input-
Output techniques utilized here.  Dow's team investigated six scenarios.  These included models for two locations (Hidalgo County and
the Lower Rio Grande embayment), and three production processes
(Isobutane, 1-stage flashed steam, and 2-stage flashed steam).  The
2-stage flashed steam system for the Hidalgo location was originally
developed from data obtained from the Geological Section, Oil and
Gas Division, of Dow Chemical.

The Lower Rio Grande Embayment location uses the same three
variations in power production processes and was originally developed by Paul Jones of Louisiana State University.  For the purposes
of this study, cost estimates are derived from the 2-stage Lower
Rio Grande model, as published by Dow.  The specifications of that
model are given in Figure 9.

The Lower Rio Grande 2-stage system was chosen for two reasons:
(1) there were definite cost advantages - it was nearly 4 mills per

Line of Equal Depth of 150°C Isothermal Surface
Interval 0.3 Kilometer Datum is Land Surface

Figure 8.  Depth of Occurrence of the 150° (302°F) Isothermal Surface in the South Texas Coastal Plain.

kwh cheaper than the next cheapest model, the 1-stage flashed steam; and (2) the technology for this system is at a more advanced stage of development than the Isobutane system. The Livermore report based its analysis on the use of a total flow system. Once such a system becomes operational it looks to be very promising; however, at present it is not a "state of the art" technology.

Although the geopressured sands do contain methane gas it is unclear at what level and hence for this study's purposes full saturation is not assumed. Finally, for the purposes of this analysis it was assumed that Kinetic energy potential would not be a factor in energy production.

Original Estimates of Capital Costs. The Dow report contains an extensive price breakdown on the capital components of the 2-stage model (Figure 10). The Dow cost figures represent the entire system, from the well to the lines leading out of the plant. Under these assumptions, the total capital costs came to $61,206,000. Given a kilowatt capacity of 66,500 this results in a unit capital cost of $920 per kw, hardly fortuitous for utilities that are already strapped for capital. Given the fact that nuclear plants can be constructed for $650 per kw and small-scale coal-fired plants are being constructed for approximately $600 per kw, the geopressured reservoirs would not appear attractive.

| | |
|---|---|
| Dimensions | 10 x 50 miles |
| Net Sand Thickness | 1,000 feet |
| Porosity | 18% |
| Average Water and Rock Compressibility | $9 \times 10^{-6}$ vol/vol/psi |
| Viscosity | 0.2 centipoise |
| Well Radius | 0.3' |
| Permeability | |
| 1 | 0.0275 darcy |
| 2 | .08  darcy |

Figure 9

| ITEM | COST |
|---|---|
| Wells | $24,400,000 |
| Collection and Disposal Piping | 470,000 |
| Methane Extraction | 49,000 |
| Dehydration | 112,000 |
| Cooling and Separation | 1,856,000 |
| Methane Pipeline and Compressor | 6,157,000 |
| Flash Chamber and Separator | 286,000 |
| Turbine Generator | 18,646,000 |
| Condenser | 6,274,000 |
| Cooling Tower | 2,024,000 |
| Step-up Transformer | 532,000 |
| General Site Development | 400,000 |
| TOTAL ESTIMATED CAPITAL COSTS | $61,206,000 |
| Annual Cost of ROI, Depreciation, Management and Operation | $22,646,220 |

Source: Wilson, et. al., An Analysis of the Potential Use of Geo-
        thermal Energy for Power Generation Along the Texas Gulf
        Coast, page 39.

Figure 10.   Estimated Capital and Unit Costs: Jones No. 2 66.5
                        MWE Power Plant

Reappraisal of Capital Costs. However, a closer look at the
cost figures reveals other possibilities. To begin with, the Dow
costs are for the total geothermal operation. This is analogous to
assuming that utilities which burn coal would have to acquire and
operate the coal mines and transportation systems which provide them
with fuel. While Dow's own operations span the entire process from
well-drilling to electric consumption, this is an atypical situation
with regard to the electric utilities. A more meaningful investigation would be limited to "inside-the-fence" costs. This proposition assumes that the petroleum sector would drill the well and provide the hot water for the utility. Therefore, the power plants
would need equipment only for utilizing the hot water and not for
its acquisition.

The capital costs are significantly different (Table 2) under
this limitation, since none of the cost found in the first seven
rows of the Jones cost table (wells, methane pipeline and compressor)
would be incurred by the electric utility. Thus, the total estimated capital costs for each plant would be $28,162,000. Given the

66.5 megawatt capacity, the unit capital cost is $423.50 per kw (Figure 11).

There were additional savings in annual costs. Dow figured a Return on Investment (ROI) at 20 percent and maintenance and operation costs at 8 percent. While a 20 percent ROI might be characteristic of Dow's own operations, such a return is not representative of the electric utility industry which operates, at most, on a 15 percent ROI. The 8 percent operation and maintenance cost came from the Federal Power Commission's report, "Steam-Electric Plant Construction Cost and Annual Production Expenses." The Dow report states, "It is felt this (8% rate) is a reasonable figure to apply to geothermal power plants." However, if several plants were constructed this cost could probably drop to around 6 percent of the capital investment. These two changes decreased the annual costs by 7 percent.

| | |
|---|---|
| Flash Chamber and Separator | $    286,000 |
| Turbine Generator | 18,646,000 |
| Condenser | 6,274,000 |
| Cooling Tower | 2,024,000 |
| Step-up Transformer | 532,000 |
| Site Development | 400,000 |
| TOTAL ESTIMATED CAPITAL COST | $28,162,000 |
| ROI (Straight line-15%) | $ 4,224,300 |
| Depreciation (Straight line-5%) | 1,408,100 |
| Insurance (.00275%) | 74,446 |
| Overhead (1.3%) | 366,106 |
| General and Administrative (1.6%) | 450,592 |
| Maintenance and Production (6%) | 1,689,720 |
| TOTAL ANNUAL COST | $ 8,213,264 |
| Annual Credit for Condensate | $    212,000 |
| TOTAL NET ANNUAL COST | $ 8,001,264 |
| Kw Capacity | $66,500,00 |
| Unit Capital Costs ($/kw) | $    423,50 |
| Kwh/yr at 90% Load Capacity | $5.099 \times 10^8$ |
| Unit Power Cost, (mills/kwh) | 15.7 |

Figure 11. Estimated "Inside-the-Fence" Costs: Jones No. 2 66.5 MWE Power Plant

After subtracting the annual condensate credit, the net annual cost would be $8,001,264. The unit power cost at a 90 percent load factor $(5.099 \times 10^8)$ is 15.7 mills. Given the basic assumptions concerning no credit for methane gas or Kenetic power we feel these costs are conservative.

Impact Methodology. The following procedure was used in determining the impact of geothermal development:

1) The population growth of the Coastal Bend area was projected to 1980.

2) Energy consumption was then projected based on the population growth estimates.

3) The ten-year increase in energy consumption in the area was assumed to be met by the development of geothermal plants, based on the most likely projection of population growth.

4) The number of Dow-sized plants needed to meet the increased consumption was then calculated.

5) The cost of building and operating these plants was entered into the I-O model and the results were compared with the no-new technology results.

Given the fact that both the petroleum sector and the electrical utility are highly capital intensive, employment impacts were expected to be minimal. The original interest in carrying out this investigation centered around the question of whether or not the introduction of geopressure geothermal powered electric production would demonstrate cost-push tendencies. The increased production costs resulting from the new technology might bring about such pressures (Haynes and Kleeman, 1975).

In order to determine what the changes in the technical relationships would be if total output remained the same when the utilities sector made the additional purchases, a new table of Direct Requirements was derived and the method of "rows only" pass through evaluation of the environmental policies described earlier was followed. Since total output values were divided into each element of the column, all 78 rows, this would mean the column total would be greater than one if the geothermal related transactions were added and nothing else except gas purchases were reduced in the column.

This provided a basis to make a comparison, in dollar values, between the total output projections before and after implementation of the policy. That is to say, total output for each industry in the processing sector would be influenced by the transactions resulting from the change in power generation.

A "pass-through" assumption was implicit in the treatment of

| YEAR | ZPG PROJECTION | | INTERMEDIATE PROJECTION | | CHAMBER OF COMMERCE PROJECTION | |
|------|----------------|---------|------------------------|---------|-------------------------------|---------|
| | EN (KWHx$10^6$) | GEN(MW) | EN (KWHx$10^6$) | GEN(MW) | EN (KWHx$10^6$) | GEN(MW) |
| 1970 | - | - | 4,081 | 1,682 | - | - |
| 1975 | 5,560 | 2,300 | 5,630 | 2,300 | 5,730 | 2,400 |
| 1980 | 7,050 | 2,900 | 7,270 | 3,000 | 7,580 | 3,100 |

Source: Moseley, page 65.

Figure 12. Projected Energy Consumption (KWH/CAP x POP) and Required Generating Capacity (MW) Under Alternative Growth Policies.

| YEAR | INTERMEDIATE PROJECTION | 10-YEAR INCREMENT IN ENERGY CONSUMPTION | NUMBER OF GEOTHERMAL PLANTS @ 5.099 (KWHx$10^8$) NEEDED FOR DEMAND |
|------|------------------------|------------------------------------------|--------------------------------------------------------------------|
| 1970 | 4,081 | | |
| 1980 | 7,270 | 3,189 | 6.25 |

Figure 13. Number of Geothermal Power Plants Need to Meet 1980 Demand.

increased costs.  This means that all of the additional costs incurred by the electric utility were passed on to the utility's customers.

Electric consumption rates for 1980 are given in Figure 12. The
number of geothermal plants required is given in Figure 13.

Interface with the Input-Output Model.  The previous work with
the input-output model was performed with constant 1967 dollars.
Therefore, it was necessary to reduce all of the costs in the Jones
2-stage steam model by the amount of increase in the wholesale price
index during this period, 1.557 percent.

Figure 14 shows the increased costs experienced by the petroleum sector (sector 13 in the Input-Output Model).  Since this sector performs the drilling function for the region it was assumed to
be the provider of geothermal waters for the power plants.

The payments of petroleum (column 13) for ROI, overhead and
general administrative were added to the residuals row (row 78) from
the earlier set of 1980 projects.  The column 13 depreciation payments (row 76) were increased by 5 percent of the value of the new
wells and the collection and disposal system.  Payments to insurance
(row 62) in the old 1980 model were increased by the new amount.
Finally, the petroleum sector's purchase of labor from households
to perform maintenance and production operation were added to the
previous 1980 purchases from households (row 72).

| ITEM | COSTS |
|------|-------|
| Wells | $24,400,000 |
| Collection & Disposal Piping | 470,000 |
| TOTAL COST (WITHOUT METHANE) | $24,870,000 |
| ROI (Straight Line-20% Yr.) | 5,367,600 |
| Depreciation (Straight Line-5% Yr.) | 1,341,900 |
| Overhead (1.3%) | 323,310 |
| Insurance (.00275%) | 68,392 |
| General & Administrative (1.6%) | 397,920 |
| Maintenance & Production (2%) | 497,400 |
| TOTAL ANNUAL COST | $ 7,996,522 |
| Fuel Cost/kwh: 15.7 mills/kwh | |

Figure 14.  Increased Costs to Petroleum Sector.

Estimating how the utility would pay the petroleum sector is particularly difficult. Since geopressured geothermal power plants are still hypothetical there are not many examples for guidance. In order to determine "fuel" costs for the power plant the petroleum sector was assumed to pass on its annual costs as a "fuel" charge. This figure replaced the old level of transactions between row 13 and column 41.

Since the fossil fuel plants would only account for 56 percent of the electric generation in 1980, the sale of gas to power plants (row 40 to column 41) was reduced by 44 percent.

The rest of the payments by the electric utility was treated in the same fashion as those made by petroleum, with one exception. Financing for the capital investment necessary to construct the geothermal plants was expected to come from outside of the region. Therefore, payments to imports (row 77 by column 41) reflect the cost of principle plus interest.

Impact Results. A comparison of the changes in total output is presented in Appendix II. The two sectors with the greatest negative and positive changes respectively are: 40, gas utilities and 41, electricity.

Total output for sector 40 decreased because it lost 44 percent of its previous market. The 46 percent increase in output for sector 41 represents the increase in price that resulted from the implementation of a new technology. The petroleum sector was the only economic sector that experienced an additional absolute increase in physical output (sales of geothermal water to power plants) over the previously projected 1980 levels. Increases in all other sectors represent the rise in utility rates that is implicit in the assumption that increased costs will be passed on to the consumers of electricity.

The final row in the table is the summation of all total output rows. This represents the total output of all sectors of the regional economy. In terms of inflationary tendencies the effect of geothermal plants would appear negligible. The overall price increase is approximately 1 percent. However, certain environmental considerations still must be faced (see Appendix III).

The 46 percent increase in electrical rates seems large when compared with the old projections. However, those projections were based on the assumptions of continued gas supplies and continued federal regulation and maintenance of low gas prices. These assumptions are already becoming obsolete. Given the increases in electric utility rates that have occurred during the past three years, up 30%, the geothermal option becomes increasingly competitive.

Energy Conclusions. The primary regional impact of geothermal development is reflected in the price structure of the local economy and in environmental considerations. The alternative impacts of the latter are still unclear.

When only the production side of geothermal development is examined it would seem that it is marginally competitive with alternative energy-based electric production systems. On the other side, the differential impact of geothermal energy-based systems as opposed to the present gas and petroleum-based system is evaluated via a regionalized state input-output model. The only sectors with significant changes in output levels are gas utilities and electricity, with a minor output increase in petroleum. The latter is due primarily to sales of geothermal related drilling and piping distribution services to electricity.

An alternative consideration is the cost-push impact of new technology in its initial stages before technological readjustment, when much of the new costs must be absorbed due to competitive existing technology. However, in this particular case the cost-push effects are quite small, approximately one percent. Finally, the alternative to this consideration is the impact of complete "pass-through" effects of the new technology to customers. In this case electric prices must rise by 46 percent when compared to the projected no-new technology alternative. This would seem prohibitive except that the no-new technology projection is based on 1970 electric prices which have already risen by 30 percent in the region.

From our analysis it would appear that geothermal development is a marginally viable alternative to the continued development of electric production with present technology in the South Texas Coastal Zone. The marginality of the alternative is highly dependent on the continued access of the region to reasonably priced gas and petroleum. This is an assumption with high levels of uncertainty.


CONCLUSION

Conventional input-output analysis has included no explicit consideration of social welfare, although Leontief (1969) has demonstrated its ready application to environmental problems. Given legislation aimed at obtaining a social benefit, it would seem that input-output models are useful for internalizing what were previously external costs. The inclusion of the money costs of obtaining a social benefit, for example, clean water makes the input-output model a social welfare tool. It is possible to examine a regional or national economy in terms of social goals and their likely impact on incomes, jobs, and prices. Furthermore, by using the same approach it is possible to estimate the competitive feasibility of new potential energy sources in terms of cost impacts on consumers.

Certain areas of further research should be pointed out. The method by which the prices were passed on would work only if demand, with respect to prices, were perfectly inelastic. It might well be argued that for a short period of time this is the case. Indeed, many firms will continue to purchase inputs at new prices before seeking substitutes, in order to meet the demands of customers that they cannot afford to lose. With the development of specific demand functions for each sector, it would be possible to add a dynamic element to the inherently static nature of the input-output model. Despite these further directions of research and analytic constraints, the total price increases, given the inelasticity of demand, were surprisingly small. Since this is a regional analysis, policies that can be deduced from the study are local in nature.

## REFERENCES

1. Bureau of Economic Geology, The University of Texas at Austin, "Potential Geothermal Resources of Texas" Texas Business Review Vol. 49 (September 1975), pp. 213-216.

2. Charnes, A., W.W.Cooper and R.J.Niehaus "Studies in Manpower Planning" Office of Civilian Manpower Management, Department of the Navy, Washington, D.C. (July 1972).

3. Charnes, A., K.E. Haynes, J.E. Hazleton and M.J. Ryan "A Hierarchical Goal-Programming Approach to Environmental Land Use Management: The Texas Gulf Coast" (chap. A) Proceedings of the 1973 NATO Scientific Conference on Mathematical Analysis of Decision Problems in Ecology (W.R.Lynn and A. Charnes, eds.) Heidelberg, West Germany: Springer-Verlag, Inc., 1975).

4. Charnes, A., K.E.Haynes, F. Phillips "A Generalized 'Distance' Estimation Procedure" Geographical Analysis Vol. 8 (July 1976).

5. Charnes, A., K.E.Haynes, F. Phillips and G. White "Some New Equivalencies and Dualities in the Solution of the Gravity Model of Spatial Interaction Using the Unconstrained Dual of an Extended Geometric Programming Problem" Prepared for the Commission on Quantitative Methods Symposium International Geographical Union Meetings, Moscow (July 1976).

6. Chenery, H.B. and P.G. Clark Interindustry Economics (New York: John Wiley, 1959).

7. Hansen, W.L. and C.M.Tiebout "An Intersectoral Flow Analysis of the California Economy" Review of Economics and Statistics Vol. 45 (1963) pp. 409-418.

8. Haynes, K.E. and M. Rube "Environmental Quality and Urban Popu-

lation Density" <u>Proceedings of the 1972 Meetings of the I.G.U. Commission on Quantitative Georgraphy</u> (M. Yeates, ed) (Kingston, Ont.: McGill-Queens University Press, 1974) pp. 121-135.

9. Haynes, K.E., J.E. Hazleton, T. Kleeman, F. Phillips, M. Ryan and G. White "Economics and Land Usw: Technical Assessment of Environmental Policy Impacts in the Corpus Christi Area of Texas" Final Report to <u>Establishment of Operational Guidelines for Coastal Zone Management Project</u> prepared for the National Science Foundation (RANN) and Division of Planning and Coordination, Office of the Governor of Texas; Coordinated through the Division of Natural Resources and Environment, the University of Texas at Austin, 1974.

10. Haynes, K. E. and T. Kleeman "Environmental Quality and Inflation: A Regional Perspective on the Cost-Push Impact of the 1972 Pure Water Legislation" <u>Environment and Planning A</u>, Vol. 7 (Summer 1975) pp. 567-574.

11. House, P.A., P.M. Johnson and Dr. F. Towse "Potential Power Generation and Gas Production from Gulf Coast Geopressured Reservoirs" Lawrence Livermore Laboratory, Livermore, California (May 15, 1975).

12. Isard, W. <u>Ecologic-Economic Analysis for Regional Development</u> (New York: The Free Press, 1972).

13. Kleeman, W.T., K.E. Haynes and T. Freeland "Technical Assessment of the Impact of Geothermal Development in the Corpus Christi Area of Texas" <u>Proceedings of the Second United Nations Symposium on the Development and Use of Geothermal Resources</u>, San Francisco, California (May 1975) in press.

14. Leontief, W. "Environmental Repercussions and the Economic Structure: An Input-Output Approach" <u>The Review of Economics and Statistics</u>, Vol. 52 (August 1970) pp. 262-271.

15. Miernyk, W.H. <u>The Elements of Input-Output Analysis</u> (New York: Random House, 1965).

16. Moseley, J.C., II <u>Implications of Alternative Public Policy Decisions Concerning Growth and Environment on Coastal Electric Utilities</u>, unpublished doctoral dissertation, College of Engineering, The University of Texas at Austin (May 1973).

17. Phillips, F., G. White and K.E. Haynes "Extremal Approaches to Estimating Spatial Interaction" <u>Geographical Analysis</u> Vol. 8 (April 1976).

18. Sherman, J.S., J.F. Malina Jr. "Water Needs and Residuals Man-

agement" Final Report to <u>Establishment of Operational Guidelines for Coastal Zone Management Project</u> prepared for the National Science Foundation (RANN) and Division of Planning and Coordination, Office of the Governor of Texas; Coordinated through the Division of Natural Resources and Environment, The University of Texas at Austin, 1974.

19. Texas Water Development Board <u>The Texas Water Plan Summary</u>, Austin, Texas, 1968.

20. Wilson, J.S., P. Shepherd and S. Kaufman "An Analysis of the Potential Use of Geothermal Energy for Power Generation Along the Texas Gulf Coast" <u>Report to the Governor's Energy Advisory Council</u> (USA: Prepared by Dow Chemical, October 15, 1974).


APPENDIX I

<u>Formal Description</u>. As Miernyk (1965, Chapter 7) demonstrated, production for a given sector $X_i$ goes to satisfy the demands of the autonomous sector $X_f$ (final demand) and of each non-autonomous sector $X_{in}$ (processing vestors):

$$X_i = X_{i1} + X_{i2} + \ldots + X_{in} + X_f \quad (i = 1 \ldots n) \qquad (1)$$

In his discussion of the underlying assumptions of the model, Miernyk states, that the inputs to every sector are specific functions of the output level of that sector. In other words, demand for a portion of the output of one inter-industry sector $X_i$ by another inter-industry sector $X_j$ is a function of the level of production in $X_j$.

$$X_{ij} = a_{ij} X_j \qquad (2)$$

The technical coefficients found in Miernyk's Table of Direct Requirements, or the Leontief "A" matrix, reflect the relationship between all of the sector's inputs and its ouput, or to rewrite the above equation:

$$a_{ij} = \frac{X_{ij}}{X_j} \qquad (3)$$

By using the direct requirements coefficients, it is possible to project the change in individual inter-sector transactions following exogenously determined changes in total output, e.g.,

$$Y \ (I-A)^{-1} = TO'$$

Where: Y = final demand after an exogenously induced change

$(I-A)^{-1}$ = direct and indirect requirements matrix

TO' = new total output

However, instead of tracing through a series of input changes the authors were seeking a method of tracing through price changes.

Instead of finding an "A" matrix or Table of Direct Requirements, the desire was to develop a Table of Direct Payments. This would allow the tracing of price increases that were passed on to purchasers.

Instead of dividing total output for each row into each element of its respective column, it was divided back into each element of its row:

$$b_j = \frac{X_{ji}}{X_j} \tag{5}$$

The logic used for determining the effects of exogenous changes in final demand would lead to the following steps in determining the effects of an exogenous change in final payments:

$$P \ (I-B)^{-1} = TO' \tag{6}$$

Where: $P$ = final payments after an exogenous change
$(I-B)^{-1}$ = table of direct and indirect payments
TO" = new total input

The difference between TO' and TO" for each sector, represents price changes resulting from the effluent control policy.

APPENDIX II

Comparison of Two Total Output Vectors for 1980, Old
Computed with Fossil Fuel Electric Generating Only, New
Computed with Both Fossil Fuel and Geothermal Electric Generating

| Row | Old | New | Change | Percent |
|-----|-----|-----|--------|---------|
| 1 | 8889.3779 | 8967.4118 | 78.0339 | .0088 |
| 2 | 2857.4364 | 2878.6581 | 21.2217 | .0074 |
| 3 | 6574.8423 | 6625.5155 | 50.6732 | .0077 |
| 4 | 24058.9479 | 24244.8431 | 185.8951 | .0077 |
| 5 | 37584.6900 | 37918.4634 | 333.7732 | .0089 |
| 6 | 3349.9601 | 3373.3945 | 23.4344 | .0070 |
| 7 | 57154.4954 | 57494.9120 | 340.4167 | .0060 |
| 8 | 18283.4933 | 18496.3723 | 212.8790 | .0116 |
| 9 | 1882.9382 | 1907.5201 | 24.5819 | .0131 |
| 10 | 2396.9613 | 2499.8996 | 102.9383 | .0429 |
| 11 | 9589.0460 | 9685.2505 | 96.2045 | .0100 |
| 12 | 45816.0038 | 45924.5958 | 108.5919 | .0024 |
| 13 | 5530399.0563 | 5573332.5104 | 42933.4541 | .0078 |
| 14 | 25212.4239 | 25295.9947 | 83.5708 | .0033 |
| 15 | 25137.6960 | 25199.2876 | 61.5915 | .0025 |
| 16 | 215858.0461 | 216480.2169 | 622.1708 | .0029 |
| 17 | 22892.2998 | 23048.1326 | 155.8328 | .0068 |
| 18 | 17174.5537 | 17244.7538 | 70.2002 | .0041 |
| 19 | 3433.9631 | 3455.4001 | 21.4370 | .0062 |
| 20 | 82311.1728 | 83283.7531 | 972.5803 | .0118 |

| Row | Old | New | Change | Percent |
|---|---|---|---|---|
| 21 | 15187.6825 | 10225.9159 | 38.1338 | .0037 |
| 22 | 187.6766 | 188.5937 | .4171 | .0022 |
| 23 | 15118.6864 | 15151.8149 | 33.1285 | .0022 |
| 24 | 26903.7088 | 26999.8862 | 96.1774 | .0036 |
| 25 | 313665.4205 | 321077.5827 | 7411.6622 | .0236 |
| 26 | 391926.2109 | 393037.3100 | 111.0991 | .0028 |
| 27 | 553.0171 | 565.4704 | 12.4533 | .0225 |
| 28 | 51977.3970 | 52017.4295 | 40.0324 | .0008 |
| 29 | 201312.3786 | 203078.7028 | 1766.3242 | .0088 |
| 30 | 6625.7930 | 6640.9583 | 15.1654 | .0023 |
| 31 | 12602.1729 | 12618.3291 | 16.1562 | .0013 |
| 32 | 3262.0147 | 3268.0446 | 6.0299 | .0018 |
| 33 | 6625.2079 | 6640.4275 | 15.2195 | .0023 |
| 34 | 5922.1343 | 5971.7059 | 49.5715 | .0084 |
| 35 | 94904.9779 | 95877.2718 | 972.2938 | .0102 |
| 36 | 147031.7337 | 147210.0767 | 178.3430 | .0012 |
| 37 | 24236.9809 | 24247.3353 | 10.3544 | .0004 |
| 38 | 32262.1781 | 32382.4514 | 120.2733 | .0037 |
| 39 | 87049.2125 | 87359.7471 | 310.5345 | .0036 |
| 40 | 643305.1273 | 638181.0564 | -5124.0709 | -.0080 |
| 41 | 101952.8726 | 149195.1130 | 47152.2404 | .4625 |
| 42 | 17068.4773 | 17615.8242 | 547.3469 | .0321 |
| 43 | 17733.6085 | 17772.2337 | 38.6252 | .0022 |
| 44 | 45215.1955 | 45510.8090 | 295.6135 | .0065 |

| Row | Old | New | Change | Percent |
|---|---|---|---|---|
| 45 | 4837.4986 | 4994.4861 | 156.9876 | .0325 |
| 46 | 1279.3610 | 1286.5839 | 7.2229 | .0056 |
| 47 | 58819.0042 | 59069.7934 | 250.7892 | .0043 |
| 48 | 17969.6384 | 18015.4319 | 45.7935 | .0025 |
| 49 | 141507.8343 | 142452.6283 | 944.7920 | .0067 |
| 50 | 14116.8387 | 14122.9847 | 6.1459 | .0004 |
| 51 | 5121.2945 | 5132.3515 | 11.0570 | .0022 |
| 52 | 7921.9667 | 7990.0916 | 68.1249 | .0086 |
| 53 | 57728.7070 | 58107.1415 | 378.4345 | .0066 |
| 54 | 1175992.4442 | 1184412.0683 | 8419.6240 | .0072 |
| 55 | 123241.3084 | 123803.8363 | 562.5280 | .0046 |
| 56 | 19140.5558 | 19505.8530 | 365.2972 | .0191 |
| 57 | 7883.0581 | 7936.3886 | 103.3306 | .0131 |
| 59 | 59891.2037 | 60328.7277 | 437.5240 | .0073 |
| 60 | 27205.6100 | 27462.1188 | 196.5088 | .0072 |
| 61 | 124418.1690 | 124704.0114 | 285.8424 | .0023 |
| 62 | 105008.4777 | 106170.0724 | 1161.5947 | .0111 |
| 63 | 19820.6928 | 20314.6899 | 493.9971 | .0249 |
| 64 | 43164.1166 | 43732.3182 | 568.2015 | .0132 |
| 65 | 50849.8491 | 51131.9234 | 282.0743 | .0055 |
| 66 | 7951.9117 | 8153.3851 | 201.4734 | .0253 |
| 67 | 2355.2992 | 2356.9341 | 1.6349 | .0007 |
| 68 | 28677.8066 | 28826.9710 | 149.1644 | .0052 |
| 69 | 543568.6150 | 547689.4062 | 4120.7912 | .0076 |

| Row | Old | New | Change | Percent |
|---|---|---|---|---|
| 70 | 204103.0912 | 205397.9219 | 1294.8307 | .0063 |
| 71 | 89674.1062 | 89943.6551 | 269.5489 | .0030 |
| TOTAL | 11352579.2948 | 11470699.0947 | 118119.7999 | .0104 |

APPENDIX III

Environmental Considerations. Geothermal development involves four significant environmental effects: air and noise pollution; land subsidence; and wastewater disposal. In addition to methane, geothermal waters contain various amounts of carbon dioxide, hydrogen, ammonia and hydrogen sulfide. It is possible that a market can be found for the methane. In any case, environmentally sound means must be found for dealing with these gases.

Large volumes of water would be produced by geothermal plants dependent on geopressured water-saturated sands. A number of alternatives exist for disposing of this water: (1) If the water is not too saline, it can be used for irrigation; (2) If the water is too saline, it may be profitable to desalinate most of it; and (3) If neither of these alternatives is possible, the wastewater can be reinjected into no-longer-producing oil wells, or into depleted oil wells for secondary recovery purposes.

However, some engineers believe that the wastewater must be reinjected into the geothermal wells, which may not be replenished through natural processes. If this is the case, reinjection would be necessary to extract a substantial fraction of the system's heat as well as to inhibit rapid depletion of the resource.

Land subsidence can lead to problems with surface drainage, seawater invasion of aquifers and flooding. If the wastewater is reinjected into the ground, subsidence would not be serious. This would reduce the impact of subsidence on this immediate area. If, however, the wastewater is used for irrigation the depth of the reservoir could result in a very large cone of subsidence.

ENVRIONMENTAL POLLUTION AND ITS SOCIAL IMPACT

H.J. Karpe, D. Scholz

Institute of Environmental Protection

University of Dortmund, West Germany

This paper deals with (the specific environmental conflicts of)
one of the biggest industrial agglomeration areas in West Germany,
namely the Ruhr.

Situated in this Ruhr district is the University of Dortmund
Interdisciplinary Research Institute for Environmental Protection.
It is of course understandable that one of the aims of such an institute is to help alleviate the problems of the area.  But, because of
the heavy pollution, it can be said that the research work has a
somewhat "unbalanced" theory-practice relationship.  This means that
practical and directly usable work must prevail in the interest of
the almost 6 million people living and working in this area.  The
Institute is cooperating with various faculties of the university in
basic research; but it is also elaborating upon implementable decision aids tailored for the different communities in the areas,
and many projects are initiated by local citizen groups.

The results of two such research projects illustrate the procedure and the practice-orientated goals.  The first - representing
a formalized method - applies empirical pollution indices and their
subjective evaluation to the environmental situation at community
level.  The second research project, which has only recently been
completed, is the subsequent analysis of the socio-spatial impact
of the defined pollution and the pollution indices.  One has to say
that, at least in Germany, there is only very limited knowledge on
how to define in terms of aggregated pollution factors environmental conditions and their effects on both city and population structures.  But the knowledge of these interdependent factors are necessary prerequisites for town and regional planning as well as for
the (closely) associated economic considerations.  After discussing -

within the limited time available – this complex theme "Environmental pollution and its social impact", I would like to say something on the social field in general which means the environmental awareness of engaged citizens as a latent potential for "Environmental pressure groups" – and on the general relation of citizens to their (polluted) environment. But, first of all a very short glance at the housing and industrial structure of the "Ruhr" agglomeration area as well as the socio-economic structure of the city of Dortmund to which the described research projects refer. The political representatives of the city of Dortmund – and that is quite usual – provided the initiative to the following described research projects. Of course, their interest lay in the direct application of our work and its immediate benefits. It must be emphasized however that, though the projects refer to the city of Dortmund, they can also be applied in a modified form to other big industrial cities.

The industrial area of Rhine-Westfalia – in the following referred to as the Ruhr area – is the largest and biggest industrial agglomeration of Middle Europe. Some 5.6 million people live in an area of 4,600 km$^2$. In spite of the high concentration of population, the Ruhr area is characterized by a quite inhomogenous and heterogeneous residential structure. Beside several city centres exist numerous smaller and lesser-attractive residential areas. The decentralization policy within the Ruhr area has resulted in many residential areas not having comprehensive and sufficiently supplied infrastructure potential, causing a potential social conflict in the big cities of different areas.

For more than 10 years the economic structure of the Ruhr area which was historically based on coal mining and on heavy industry is – as a consequence of important changes in demand, as for example in the energy market – in a process of basic restructuring. In the central and northern parts of the Ruhr areas the monostructure of coal mining and steel productions has had negative consequences on the economic and financial world as well as on employment and associated incomes; and as such has affected the overall development of those districts. Primarily in the years of recession of 1966/67 the structural and growth problems became obvious and resulted in difficulties on the employment market. The relatively high standard in private and in public income – as compared to the rest of Germany – disappeared. Since then it was attempted to accelerate the structural change especially by diversification in industrial production. However, the Ruhr area is still a problem area today with a high percentage of unemployment relative to the rest of Germany. It becomes obvious therefore that the quality of the environment is of basic interest for this area in order to attract new and modern industries and to slow down the present outflow of highly qualified personnel to areas with better environmental conditions (as in South Germany).

In the city of Dortmund, the large industrial location in the
east of the Ruhr area with a population of 640,000 (1972), these
positive and negative structural effects of the total region are
reflected.  Dortmund produces 30% of the beer output, and is responsible for 22% of the steel and 8% of the coal production of North-
Rhine-Westfalia.  Indices for the problems of the city are reductions
in numbers of jobs in the labour market and negative population growth
as are attempts to get away from the monostructure and effects of
the city to improve locational factors - in spite of a limited financial capacity - and to attract growth industries.  To what extent
these effects will be manifested is not known.

It is obvious that the social and economic conditions of such
an agglomeration area put an environmental stress of a very different
kind on the people living and working in that region.  The environmental stress can result from air and water pollution, from noise
and from poor standards of housing and working conditions.  There
exist information on type and degree of different pollution factors
in different institutions.  But there is an increasing difficulty in
evaluating and comparing such information as data dimensions and
structures often vary from institution to institution.  As a consequence neither planners, decision-makers nor those people affected
can include all available and relevant information in their considerations, even if they wanted to.

Therefore it is only understandable that on those administrative
levels in the community where decisions on environmental planning
are made, the demands for reproduceable and clear and concise information on environmental analyses are being very urgently made.  But
what does a planner understand by "environment", can it be expressed
by an all-embracing definition?  The present term of environment is
used differently in sciences and in politics.  The content of the
term is often unclear.  In our institute we are using, in a pragmatic way, the following definition: Environment is defined as the
sphere effective to human beings and is composed of natural factors -
that is physical, chemical and biological factors - and of cultural
factors that is in both economic-technical and in political-social
dimensions.  These factors influence - or are influenced by - human
activities.  Because of the interrelation between these parameters
the planner has to answer the questions:  how can different concur-
rent exigencies especially natural resources be evaluated and policies for these be decided as in setting priorities for water use in
industry and recreation?  On what basis can decisions on environmental impact analysis be made (a procedure that must be applied by
law for all new public investments)?

For our first research project the "Environmental Stress Model
Dortmund" - BELADO - the concept was to describe the degree of pollution and the environmental quality in the local sphere with as
few indicators as possible and to present this result in a distinct

and straightforward manner where "environmental stress" is the effect
of pollution components on the well-being or the health of human beings.  The negative effects in the working sphere however were not
included in this research work.  The following brief description outlines the procedure of the BELADO.  As already indicated, the BELADO
is a formalized method for the evaluation and consequent representation of spatially different pollution factors.  The model - developed
as an application of the "utility-value-method" - consists of the
following steps:

1.    Characterization of the pollution situation by selecting rele-
      vant pollution indicators.

2.    Definition of measurement procedures and definition of spatial
      differentiation of the indicators.

3.    Transformation of differentiation indicators to a common scal-
      ing base.

4.    Weighting of the indicators according to their relative influ-
      ence on the environmental stress as compared to the other in-
      dicators.

5.    Aggregation of the transformed and weighted values to partial
      and total stress indicators.

6.    Transformation of resulting model (single indicators, partial
      indicators and overall aggregated value) into maps.

      The selection of pollution indicators was the subject of a num-
ber of criteria like the representativity and relevance of pollution
factors, reliability of the measurement procedure, spatial differentiation of pollution and only those factors were included that differentiated in the analysed region.  Since in this trial model only
existing data material could be included, a preselection of the relative few factors that complied with the described criteria was
made.  According to the original definition of "environment" the factors were set into a hierarchy.  On the one side the "natural" pollutions, like air and noise, and these were sub-divided into sulphur
dioxide ($SO_2$) and dust as well as in day-noise and in night noise,
on the other side the block of "socio-economic" stress factors like
the lack of green areas, pool public transport systems and inadequate
sport and recreational facilities (sport fields, playgrounds etc.).
As to the selection of the factors in the socio-economic sphere it
must be stated, that the research groups pragmatically only used
those factors that were relatively easily accessible and had a certain relation to the selected physical environmental factors.  This
was necessary in order not to delay the work and time-table of the
project - a very comprehensive representation of this field was thus
not possible.  The principal character of the factors was more important.

In the phase of transforming different pollution factors to a
uniform scale an interval of 0 to 1 was selected. Every environmental situation was compared with two ideally-typical situations:
one with a pollution-free and one with a pollution-saturated situation. The scaling value of "zero" indicates there is no pollution
and no pollution-stress, whereas the factor "one" means that a predetermined maximum pollution standard is reached or has been exceeded. The transformation of a measured pollution factor to a value
on the linear scale between 0 and 1 was made possible by applying
specific scaling functions. Since the present knowlege of pollution
impacts is still relatively incomplete, some simplifying assumption
had to be made. The scaling functions therefore cannot give a complete picture of the true impacts. By the transformation to a uniform scale the necessary preconditions for the aggregation was created. The subsequent step after the scaling consisted in the weighting of the pollution factors. This is where subjective values play
an important role. Therefore it was necessary to document precisely
who placed the weights, on which information of pollution impact the
judgement was based and with what justification the weighting was
performed. For this procedure the Delphi-Method was used in which
competent experts could describe their valuation in form of weights.
Scaling and weighting of the pollution factors were the methodological preconditions for the next step: the aggregation of pollution
factors.

In view of the many known and many assumed interrelations (syn-
ergisms) of pollution factors we know that every aggregation is very
problematic. But this type of aggregation has the big advantage
that this formalized procedure is reproducible, controllable and -
if necessary - revisable.

The described work steps reproduce in very general and simplified terms the logic of the "stress" model. The goal had been to
define and to describe the environmental pollution of a municipality
and to evaluate the situation and to present the results in a carto-
graphic form. Now, what are the direct or indirect consequences
and where are the limits of such a theoretical approach?

The results show that the selected method is a possible way for
a documentation of the environmental situation. But there are of
course restrictions in the use of pollution maps as a unique decision instrument. The results of BELADO can neither replace the expertise of the planner nor political decisions. The main value of
the model lies in the use of existing data material and in the transformation of very difficult and complex single data to understandable
and clear information. The presentation in map form is very often
much more useful than in the form of numbers and in tables. The
maps clearly indicate areas of heavy pollution impact and areas in

which several components exceed given standards.  Also, the origins
of pollution can be traced. It can be analysed where necessary pre-
cautions have to be taken and which points have to be considered in
locating new industries and in making environmental impact analyses.
In spite of all immanent weaknesses the BELADO proved to be a real
though only partial decision aid.  Considering the present stage of
scientific knowledge - especially in the field of planning and decision - from a pragmatic point of view that is quite an advance.

A few remarks to the follow-up project of BELADO: "Environmental Pollution and its Social Impact", our so-called "UMSOZ" project.
The main aspect of this empirical work was to find out what the
spatial-social reactions to pollution were, what expectations exist
in the population concerning environmental quality and the reactions
of the population concerning pollution.  The BELADO does not give
any answers to these questions since it refers strictly to the area
of the city of Dortmund and since it leaves open how many people are
affected in the different parts of the city.  But the answer to this
question is of especial importance as this facilitates recognition
and possible reduction of conflict potentials.  There are some indications that pollution is distributed through the various social
groups in a very inproportional manner and that especially lower
income groups are suffering from a heavier pollution burden.  This
can lead to a "ghetto-like" situation in areas of heavy pollution,
the emigration of those of higher social status (doctors, pharma-
cists, managers etc.) to better areas and to the immigration of
foreign workers (especially in Dortmund).  Especially at the beginning of the process of structural changes it can come to deep differences of opinion between the political-administrative organs and the
population of the area concerned.  On the one hand - and that generally is the rule - the process of environmental degradation brought
about by man himself becomes so frequent and commonplace that people's
attitudes to their environment and indeed, to community life, tend
to be rather lackadaisical.

It can be shown that low income groups in areas of high pollution are very often insensitive to an environment of low quality.
On the other hand, in Dortmund one notices that areas having a relatively good environmental quality show strong reactions if there
is only a minor degradation of environmental quality and that immediately citizen action groups are formed which very often have a
strong influence on those responsible for making political decisions.

Finally, the bad environmental quality leads to a migration of
qualified hand-workers to other industrial areas having a lesser
degree of pollution which only worsens the structural crises of the
Ruhr area.

In analogy to advanced industrial - sociological investigations
about satisfaction on working conditions, workers motivation and

willingness to accept conditions which conflict with their ideals
the UMSOZ-Project analysed these factors to discover the importance
the state of the natural environment has for the "social situation",
what reactions occur and to find what their subjective feelings on
their environment were.  The work is not completely finished but the
important thing was to analyse comprehensively the effects of pollution in an urban agglomeration area.  The difficulty in defining
and presenting indicators as well as the problem of the different
interdependent factors affecting the socio-spatial elements in the
field of the environment only shows how much research work is still
required in this field.  We are only at the beginning of developing
positive management of our resources and environment, in analysing
relevant factors, and in carrying out an efficient analysis of environmental impacts.  The government in West Germany has defined
quite an impressive number of laws which embrace a variety of subjects concerning the environment aimed at protecting its citizens.
But how aware is the citizen himself of the importance of a clean
environment?

   Two surveys involving 60,000 questionnaires were carried out in
Dortmund to find out the views of the citizen on his environment.
The selection of pollution factors was limited to those that were
physically directly perceivable like noise by day and by night,
odours and dust.  Apart from a subjective judgement of damage, the
questionnaires also asked after the suspected origins of pollution.
The results show that questionnaires and surveys are an important
aid as far as measurable and identifiable pollution factors are concerned.  But the concrete analysis of social effects is one thing,
and to inform the citizens about general consequences of pollution
factors that are not directly known is another - as in those factors
not normally detected by the population that are poisonous substances taken up in the food chain, and other not directly tasteable substances like dust in the air that enter directly into the lungs and
which can also carry cancerogenic (like benzopyrene) particles.

   The basis for a long-term recognition and minimization of environmental pollutants and support in the reduction of conflicts is
only possible with the cooperation of the populace.  The consequences
of a lack of information within the population concerned can be big
losses in the credibility of the political administration.  The politicians and also the scientists should be aware of their strong and
common responsibilities.  In a general questionnaire survey that was
not limited to the city of Dortmund in 1972 the Institute for Applied
Social Sciences in Bonn had tested the environmental consciences of
those citizens who were familiar with the problem of environmental
planning.  90% of those persons asked felt that environmental values
had been previously underestimated in their assessment for an adequate standard of living.

   It became quite obvious as a result of this survey that the

environmental issue has become a subject of increasing importance
for citizen action-groups.  This survey also produced quite a hope-
ful perspective:  The environment was not seen as a problem with
endless dimensions but as a task which, though difficult, could be
overcome in the not too distant future.  This should be an incentive
for joint action of those concerned in trying to get closer to the
goal of a human environment that could be accepted and lived in and
enjoyed by all.

ANNEX

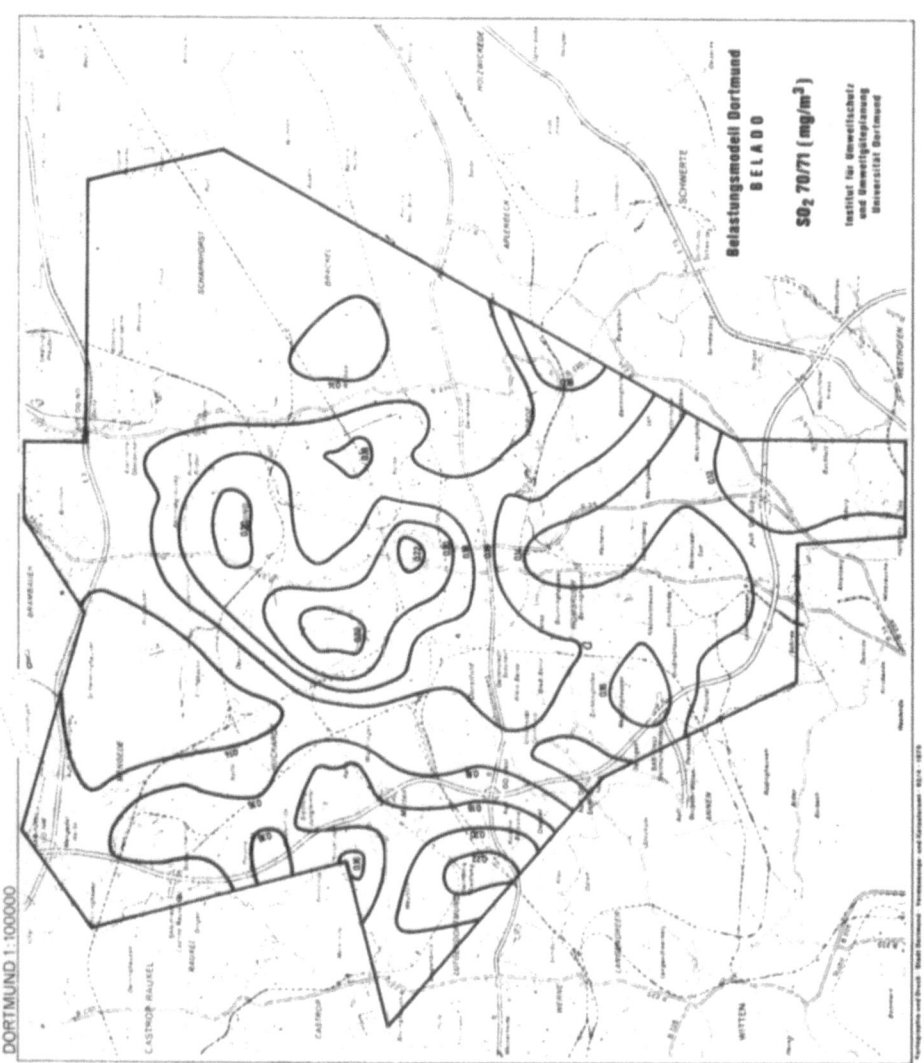

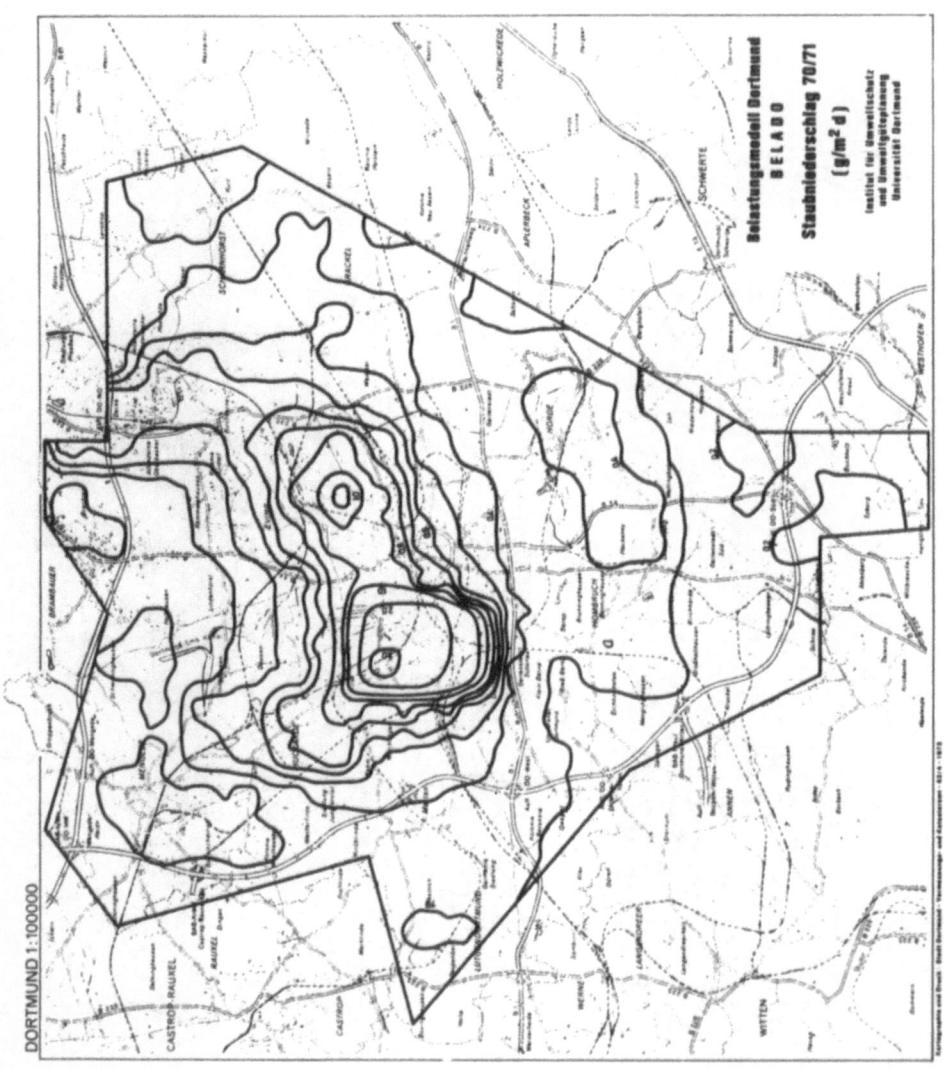

# SOCIAL PERCEPTION OF INDUSTRIAL ODORS

D. Agrafiotis, A. Baumerder, J. Brenot[1], F. de Lavergne

[1] Commissariat à l'Energie Atomique, Institute de
Recherche et d'Information Socio-économique
FRANCE

## INTRODUCTION

Increasing concerns of the public about environment quality
have stimulated local authorities and governments to look for compromise solution between economic and industrial development on the
one hand and environmental safety on the other.  For air pollution
and specially odor pollution, information relating to measurement,
control and social perception of odors would help to a better understanding of the problem and accordingly to define a policy.  At the
moment, there is no precise knowledge of what levels of odors are
acceptable by the public; so, how to establish emission standards
for industrial facilities?

The study relates quantitative odor exposure data to affected
community reaction.  Sensory techniques of odor measurement were
used to obtain quantitative data and an attitude survey was designed
for social perception. Without ignoring the sensory technique importance, this paper is focused on the survey investigation and develops a theoretical model of social perception.

Unlike other types of pollution ($So_2$, water, ...), "objective"
methods for odor evaluation are not sufficient.  Odorous chemical
compounds must be perceivable to carry away an odor problem.  The
personal nature of odor evaluation appears in the interaction of
human nose with chemical substances; this chemical and physiological phenomenon is intimately related to odor-affected individual's
psychology which is structured and influenced by social environment.
This chain shows the plurality of contents of environmental odor
evaluation.

Objective odor evaluation, here by sensory techniques, is based on a theoretical scientific background, needs instruments, provides concentration measurements or levels.  Social odor evaluation
requires an approach of social psychology, uses attitude survey,
expresses reactions, discloses annoyance and potential action.

A technocratic policy would be supported by objective odor evaluation, but it is rather dangerous to neglect or under-estimate
social reactions.  So, to combine physics and social rationalities,
it is necessary to have results of both evaluations because they
are complementary in order to make a decision.  This recommendation
underlines the possibility to conduct such a policy for a community,
each odor problem being considered as unique because of odor-type
and socio-economic structure, and the extreme difficulty to elaborate national odor control standards.

## THE ODOR AFFECTED COMMUNITY

The study concerned a small city of Brittany (France).  Offensive smell of a big industrial plant devoted to protein extraction
from meat scraps (fish, poultry, ...) affected the community.  Malodorous compounds were caused by natural and bacterial decomposition
of animal tissues and by the cooking operation.  Around, sewers,
used water treatment plant, paint unity and the fish harbour have a
little relative importance.  The factory is outside town and 2.5 km
far from city center and residential area.  The windward is essentially country; see Fig. 1.

Many studies (1) have already been made either with an objective odor evaluation to develop pattern odor intensities, or with
survey investigation to precise the extent of odor annoyance for
oil refineries, pulp mills, sulfate cellulose industries.  Few of
them relate an objective evaluation of odor and a social one (2),
and to our knowledge none treats this particular malodor.

Intensity and areal extent of odors were evaluated by sensory
techniques in the plant surroundings for some typical weather conditions.  Application of an atmospheric dilution mathematical model
permits to define the odor effected area; information on weather
time-series allows odor frequency for the year to be estimated for
each location in the area.  Complete results for the community are
presented in (5).  As a summary, one would expect a 1.5% odor frequency in A-zone, 2% in B-zone and about 12% for a point located in
C-zone which is in close neighbourhood of the industrial plant -
see Fig. 1.

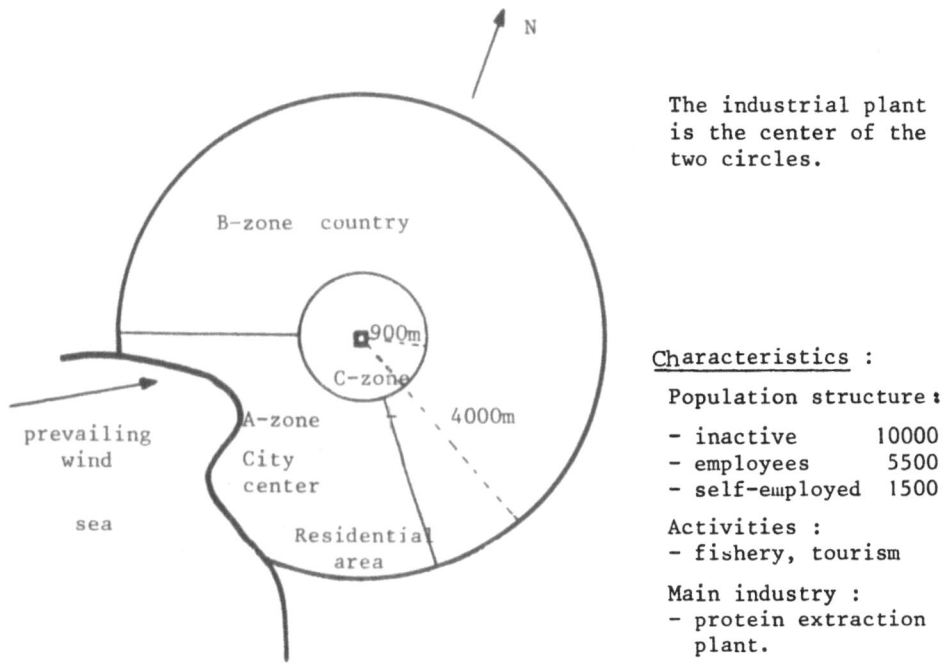

The industrial plant
is the center of the
two circles.

Characteristics :

Population structure :

- inactive        10000
- employees        5500
- self-employed    1500

Activities :
- fishery, tourism

Main industry :
- protein extraction
  plant.

Figure 1.  Objective Olfactory Evaluation (3, 4)


SOCIAL ODOR-EVALUATION

The current concern of the public with environmental pollution
problem leads sociologists to approach a pollution problem globally:
technical constraints, ecological and economic considerations, social preferences are studied simultaneously.

A scheme of social perception (7) applied to odor-pollution
will show how we proceed and will be an useful guide for the types
of reaction to offensive smells which are searched through an attitude survey.

We mean by social perception the numerous representations of
the real world and their organization.  Some of these which appear
in groups of individuals, help to characterize these social groups.
In that sense, "objective evaluation" using instruments, measurement

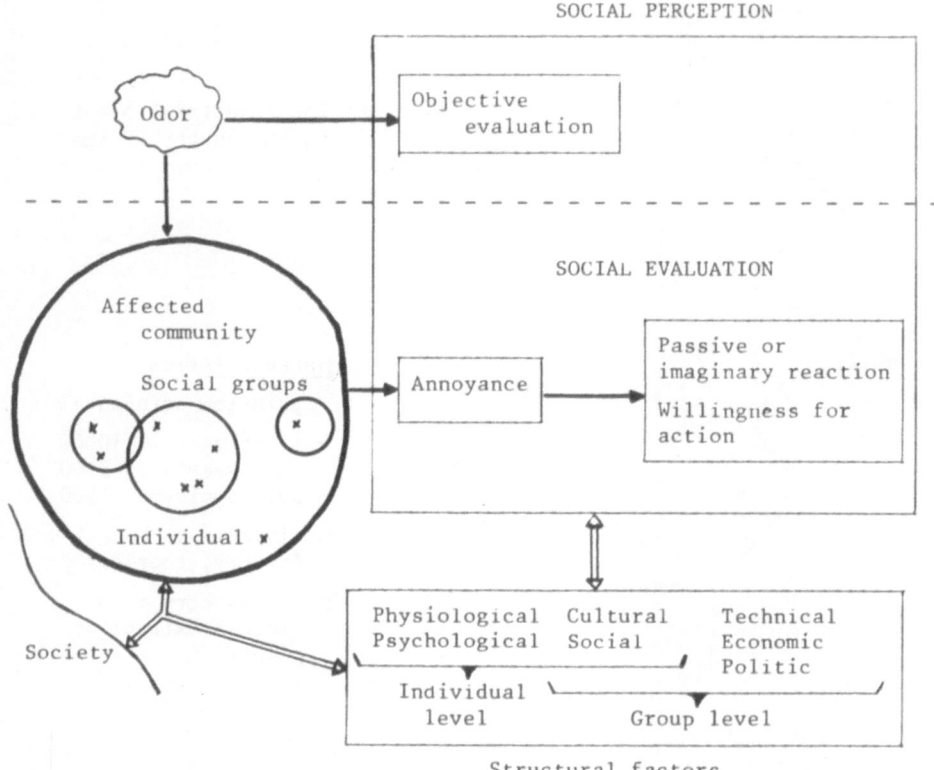

Figure 2

and calculations can be considered as a discourse of scientists. Obviously, this discourse is more or less based on and organized according to scientific standards. "Social evaluation" begins with expressions to offensive odors, rarely pleasure, currently annoyance which is the extent to which people report being bothered, disturbed or irritated (Jonsson (6) ). Following annoyance, there are passive attitudes which often come from the fact that people refuse to face the problem and give some "rational" justifications for no-acting and there is also willingness to take action either individual to modify self-environment or collective through complaints. Our purpose is to find structural factors which influence and organize social perception. Some of these interfere with the individual level: physiological, psychological, cultural and social; as for the group, it is determined by cultural, social, technical, economic and political factors. All these factors are guiding the design of

our questionnaire for the attitude survey.  The questionnaire was
elaborated to point out the role of these structural factors on relationship between objective evaluation and annoyance on the one
hand, and between annoyance and desire of action on the other.

The questionnaire structure was the following:

- Part 1 reported frequency, intensity, types of odors in order to
compare with the objective evaluation;
- Part 2 was concerned by annoyance, mainly interference with everyday activities (rest, entertainment,...) and physical symptoms of
physiological changes (nausea, vertigo,...);
- **Part** 3 was devoted to willingness of action either individual or
collective and also to appreciate denying attitudes;
- Part 4 emphasized the structural factors: physiological condition;
insertion in the community (job, civil status, club membership, new
media interest); opinions about the odor pollutant plant; socio-
economic conception for their environment (industrial growth and
development or "return to the nature").


RESULTS

The survey covered about 200 persons, i.e. 1% of town population,
interviewed personally.  We did not succeed to respect exactly some
social and demographic percentages defining the community in the survey.  Indeed, workers are few whereas other employees, independent
jobs with specially retail dealers are above quotas; however, it is
not really troublesome if we consider economical and political importance of these groups in the community.

88% of the respondents perceived odors; 65% were odor-bothered
and among those bothered 18% and 13% proclaimed respectively to be
disturbed and irritated.  Almost all of these respondents attributed
odors to the protein-extraction plant.  Despite the complaints, 70%
found that, in spite of all that the air-quality is good and recognized an economic advantage in having the plant in the area.

The results of this part of the survey reveal also individual
difficulty to give precise evaluation of odor frequency and time
length of odor "puffs".  The odor phenomenon seems to be perceived
as a whole; it is or it is not.  Physical effects such as health
diseases, disturbing changes in every day activities were not reported.  Women were more bothered than men; this is a well known physiological fact.  When home is odor-affected we found that housewives
were more discomforted than working women; it can be justified by
a greater home-investment of housewives.  Comparisons between the
three different zones A, B, C of the area indicated no concordance
of objective odor evaluation with expressed inconvenience.  Interviewees who have residences in plant neighbourhood (C-zone) or in

B-zone are in general not as affected by odor whereas A-zone residents undergo severe and major annoyance. Table 1 shows this opposition between urban area and country which was already observed in
a previous interview-study (8) concerning a sulfate cellulose factory.

Table 1 shows that annoyance results from conditions which do
not only depend upon emission of the factory; odors seem to act as
a revealer of some social discomfort and it is not surprising that
urban area residents are more concerned by quality of life. Other
structural factors such as age, health condition, job, social group
membership do not interfere with annoyance.

After these results referring to annoyance and its relation
with objective evaluation, we had to precise how community members
thought to bring the odor-problem out and help to resolve it. This
willingness to take action is more or less shared by the individuals,
some of them are active, others are frankly passive. Respondents
were invited in the questionnaire to say if they agreed with either

|  | % of the Zone population expressing major annoyance | Average number of odorant days in any point of the Zone |
|---|---|---|
| A-ZONE <br> . City center <br> . Residential area | 38 <br> 42 | 1.5 day for 100 days |
| B-ZONE <br> . Country | 14 | 2 days for 100 days |
| C-ZONE <br> . Plant | 18 | 12 days for 100 days |

Table 1

direct or indirect forms of individual or collective action.  At
first, 80% of the respondents were convinced that there is a solution for reducing the odor.  Authorities such as officials of town
council, of government or odor specialists were deemed credible by
40%; about 50% believed in the validity of more spontaneous actions
to push on the solving of the problem (citizen committee, complaints,
sign petitions, attend a meeting); less than 20% considered direct
individual action: going in justice court; giving money; leaving the
area.  Moreover, interviewees' responses have shown how they ranked
the proposed forms of action according to an "involvement degree"
as we call it.

Table 2 gives the hierarchy:

| Forms of action | Involvement degree | Level |
|---|---|---|
| Go in justice court<br>Ask for money | very strong | institutional |
| Meeting<br>Give money<br>Leave the area | strong | political<br><br>escaping<br>attitude |
| Petition<br>Committee<br>Vote for an air-quality<br>Concerned candidate<br><br>Government intervention<br>City            " | average | political |
| Odor-specialists<br><br>Enforce the air-pollutant<br>plant managers | weak | no concern<br><br>delegation |

Table 2

Respondents expressing major annoyance (disturbed and irritated by odor) were extremely involved in the problem and chose any form of action either institutional or political.  Women had specific attitudes; they trusted city council efficiency, specialist compet- ence and were cautious against non-institutional action such as at- tending a meeting and signing a petition.

We were interested to see whether social group membership could explain why respondents preferred or were against some forms of odor reducing action.  The French usual nomenclature was retained: firstly inactive, secondly farmers, craftsmen, retail dealers, liberal pro- fessions which are indepent jobs; thirdly, employees differed from the others because their answers were rather "yes" to proposed ac- tions while the others said rather "no" or refused to give their opinions.  A specific attitude of liberal professions emerged: no committee membership, no petition signing, no meeting attending. Executives as opposed to other employees did not assign the odor- problem to specialists and to authorities and as big city tax-payers were against a financial contribution of the city.

The other employees presented a global social behaviour; they attributed some efficiency to government, city council and to elec- ted representatives; they respected specialist knowledge; they were convinced by collective actions utility - in short, odor-reducing actions depended upon the annoyance degree and the interviewees got more involved specially when they were men, active, employees.


CONCLUSIONS

This study shows the complex aspects of an odor-pollution prob- lem.  If we limit odor perception to objective odor evaluation, we can be satisfied by having eliminated the unavoidable bias which is injected by the respondents to any attitude survey.  We have remarked that major odor annoyance was not associated with high level given by an objective odor evaluation method in the odor-affected area. Also, we can expect that any malodorous plant will act as a revealer of some social discomfort in an urbanized area.  To keep close to social reality, it is useful to conduct an attitude survey simul- taneously with "objective" odor measurements.  If annoyance inten- sity naturally does not interfere with social group membership, it is quite different for the kinds of odor-reducing action which will be considered.  Some social groups more readily than others appeal to some institutional or political means to push on the solving of the problem; thus, any decision-maker must realize it.

For such an air-pollution, we are convinced that an odor- reducing solution at a local level can be found for each affected community.  This solution, to harmonize the social system with its environment, must be based on a collaboration between the managers

of the odor-polluting plant on the one hand and on the city council
regarding to citizen complaints and on government officials on the
other.

## ACKNOWLEDGEMENTS

We wish to express our sincere appreciations to R. Haulet for
suggesting this research and for his support.  This paper is based
mainly on a study performed under the French Air-Quality Program.

Questionnaire I

Non Oui Autre

Existe-t-il, actuellement, une (ou des) odeur(s) dans
votre région?
Vous, personnellement, la (ou les) sentez-vous?
      . dans votre lieu de travail ...................
      . dans votre quartier .......................
      . ailleurs ..................................

Cochez la case de la proposition qui correspond le plus
à ce que vous pensez?
      . vous la ressentez,mais elle vous laisse indif-
        férent ou elle ne vous déplait pas ...........
      . elle vous incommode ........................
      . elle vous incommode beaucoup ...............
      . elle vous est insupportable ................

Combien de fois par mois, en moyenne, la sentez-vous?
      . entre 0 et 2 fois par mois .................
      . entre 3 et 6 fois..........................
      . entre 7 et 14 fois ........................
      . plus de 14 fois par mois...................

Quand vous la sentez, combien de temps cela dure-t-il
en moyenne?
      . moins de 2 heures..........................
      . entre 2 et 6 heures........................
      . entre 6 et 12 heures ......................
      . entre 12 et 24 heures .....................
      . plus d'une journée ........................

Choisissez le (ou les) types(s) d'odeur le (ou les)
plus proche(s) de celle(s) que vous sentez
      . peinture ...................................
      . oeuf pourri ................................
      . poisson ....................................
      . caoutchouc brulé ...........................
      . chair brulée ...............................
      . ordures ....................................

---

Les gens, autour de votre domicile, se plaignent-ils
des odeurs?
Vous-meme, vous arrive-t-il de parler de l'odeur pré-
sente avec vos voisins, vos amis, votre famille ou à
votre lieu de travail?
Est-ce que l'odeur a diminué depuis un an?...........
Pensez-vous que malgré l'odeur la qualité de l'air
est bonne? ...............................................
Cochez, parmi ces divers problèmes, les quatre qui, lo-
calement,vous préoccupent le plus, à l'heure actuelle
      . bruit ......................................
      . odeur ......................................

Non Oui Autre

. béton ............................................
. pollution atmosphérique .......................
. publicité ....................................
. tourisme ......................................
. pollution des rivières .......................
. pollution des océans .........................
. déchets radioactifs ..........................
. engraise et pesticides .......................
. circulation ...;.............................
. ordures ......................................

---

Questionnaire II

L'odeur vous empeche-t-elle :
        . de dormir? ...................................
        . d'apprécier vos repas? ......................
        . de recevoir vos amis? .......................
        . de profiter pleinement de vos loisirs?.......
        . de travailler? .............................

L'odeur vous rend-elle malade? .....................
L'odeur vous donne-t-elle mal à la tete? ...........
L'odeur est-elle une cause de fatigue? .............
L'odeur vous rend-elle nerveux? ....................
L'odeur vous fait-elle éternuer? ...................
L'odeur vous oblige-t-elle à contourner certains endroits?
Vous-meme, y pensez-vous souvent? .................

Quelles sont les sources d'odeur genantes dans votre
region?  (cochez la case correspondante)
        . usine du Poteau Vert .......................
        . usine de peinture ..........................
        . décharge d'ordures .........................
        . égouts .....................................
        . porcherie ..................................
        . usine de traitement des eaux usées .........

---

Pensez-vous qu'on puisse agir contre les odeurs?...
Le pollueur devrait-il régler lui meme son problème odeur?
La commune devrait-elle y contribuer financièrement? .
En cas de refus du pollueur, pensez-vous qu'on puisse
l'obliger à agir? ..............................
Un Comité ou une association contre les odeurs serait-
il efficace? ....................................
La municipalité peut-elle avoir une action efficace?
Est-ce au Gouvernement de régler le problème odeur?
Pensez-vous que le pollueur devrait vous dédommager
financièrement de cette gene? ....................
S'il n'y a aucune amélioration, envisagez-vous de
changer de quartier? .............................

Non Oui Autre

Ce problème est-il uniquement l'affaire des
spécialistes ou des administrations? ..............
Voteriez-vous de préférence pour un dandidat qui,
entre autre, proposerait de combattre la pollution odorante? .....................................
Accepteriez-vous de signer une pétition contre les
odeurs? ............................................
Seriez-vous pret à donner de l'argent pour que cela
sente moins? .......................................
Seriez-vous pret à porter le problème en justice?...
Participeriez-vous à une manifestation publique
contre la pollution odorante? ......................
Pensez-vous que l'usine du "Poteau Vert" facilite le
développement économique de la région?..............
L'implantation de l'usine était-elle indispensable
pour éviter le chomage dans la région? .............

Questionnaire III

Etes-vous souvent fatigué?.........................
Avez-vous souvent l'impression d'etre triste et deprimé?
Etes-vous anxieux? ................................
Passez-vous pour etre nerveux? ...................
Etes-vous agacé par la critique des autres? .....
Etes-vous incompris des autres? .................
Etes-vous souvent malade? .......................
Consultez-vous souvent votre médecin? ...........
Utilisez-vous souvent des médicaments? ..........
Avez-vous beaucoup de mal à vous endormir ou à
rester endormi? ...................................
Avez-vous souvent des maladies des voies respiratoires?
Avez-vous souvent d'appétit?
Souffrez-vous souvent de maux de tete?............
Avez-vous souvent le nez bouché? .................
Aimez-vous:
    . regarder la télévision? ...................
    . les voyages? ..............................
    . la jeunesse actuelle ?....................
    . le cinéma?................................
    . la pop-musique? ..........................

Lisez-vous régulièrement un journal? .............
Suivez-vous de près l'actualité politique? .......
Suivez-vous de près l'évolution des problèmes locaux?
Faites-vous partie d'une association? .............
    . politique ................................
    . syndicale ................................
    . d'intéret local ..........................
    . de consommateurs .........................

Non Oui Autre

. de défense de l'invironnement .......... ...
. sportive ..................................
. religieuse ................................
. culturelle ................................

Vous considérez-vous comme un membre actif? .......
Avez-vous des responsabilités au sein de cette association?
Avez-vous des responsabilités importantes dans votre travail?
Vers quel type d'orientation voudriez-vous qu'on accentue la
politique régionale? (cochez la case correspondante).
    . une région peu développée (peu d'emplois, faibles revenus)
      mais encore naturelle ......................
    . une région en expansion (beaucoup d'industries, nombreux
      emplois), mais comportant quelques inconvénients ou
      nuisances;

Etre-vous satisfait de l'endroit où vous habitez? .........
Ou préféreriez-vous habiter, si vous en aviez la possibilité?
(cochez les cases correspondantes)
    . à l'intérieur d'une ville?..........................
    . à l'extérieur d'une ville?..........................
    . dans un pavillon? ..................................
    . dans un appartement? ...............................
Accordez-vous de l'importance à la beauté du cadre dans
lequel vous vivez?....................................
Ressentez-vous le besoin d'accorder plus de temps à vos loisirs?
Aimeriez-vous avoir plus d'espaces verts dans la ville? .....
Souhaitez-vous qu'on développe les équipments collectifs
tels que les hopitaux, les écoles, les crèches? .............
Pensez-vous que la voiture soit un instrument de liberté?..

Questionnaire IV

Age:
Sexe  . masculin
      . féminin
Habitez-vous:
    . une cité .........................................
    . un pavillon ......................................
    . un appartement ...................................
Nombre de pièces (hormis cuisine, salle de bains, WC)...
Combien de personnes habitent avec vous (en vous comptant)
Etes-vous propriétaire? .................................
Etes-vous locataire?....................................
Habitez-vous cette région en permanence? ...............
Depuis combien de temps résidez-vous dans la région? ...
Où habitiez-vous précédemment? ..........................
Placez vos lieux d'habitation (en mettant H) et de travail
(en mettant T) sur les cartes jointes en annexe.
Etes-vous marié? ........................................
Votre conjoint travaillet-il? ..........................

Nombre total d'enfants que vous avez en ..............
Possédez-vous une voiture (ou en avez-vous l'utilisation permanente)? ........................................

Profession
    . inactif ........................................
    . non salarié ........
          . agriculteur exploitant...................
          . patron de l'industrie ou du commerce .....
          . profession libérale, cadre supérieur .....
          . autre ...................................

    . salarié
        . salarié agricole ..........................
        . cadre supérieur ..........................
        . cadre moyen ..............................
        . employé ..................................
        . ouvrier ..................................
        . personnel de service .....................
        . autre ....................................

Avez-vous des diplomes ? ...........................
    . CEP ..........................................
    . BEPC ou CAP ou BEP .........................
    . baccalauréat ...............................
    . études supérieures ........................

REFERENCES

1.    Jonsonn, E.  Annoyance reactions to environmental odors, p.
      329 in Human Responses to environmental odors, Ed. Turk A.,
      Johnston, J.W., Mouldton, D.G., Academic Press (1974).

2.    Jonsonn, E., Deane, M. and Cederlof, R., Community reactions
      to odors pulp mills.  A pilot study in Eureka, California.
      Paper presented at the Karolinska Institute Symposium on En-
      vironmental Health, 3rd. Stockholm, June 1970.

3.    Sigli, P., Thal, M.F., Zettwoog, P., Les bases de l'olfacto-
      métrie industrielle. Pollution Atmosphérique No. 63, (1974).

4.    Turk, A., Johnston, J.W., Moulton, D.G. Editors, Human Respon-
      ses to environmental odors, Academic Press (1974).

5.    Haulet, R., Mesure du niveau de nuisance par les odeurs et du
      niveau de détriment ressenti par les populations, CEA/D.Pr/
      STEPPA (1975).

6.    Jonsonn, E., Annoyance reactions to external environmental
      factors in different sociological groups.  Acta sociologica
      7, fasc 4 (1964).

7.    de Lavergne, F., Nuisances ressenties et moyens d'action en-
      visagés par une population soumise à une pollution physique:
      cas des odeurs industrielles.  Cahier de l'Association Fran-
      çaise de Science Economique (to be published).

8.    Tapia, C., Les attitudes à l'égard de la pollution atmosphé-
      rique dans la région bordelaise Etude psychosociologique.
      CEA/D.Pr/STEPPA (1974).

IMPACT ASSESSMENT AND PARTICIPATION:

CASE STUDIES ON NUCLEAR POWER SITING IN WEST GERMANY

Volkmar J. Hartje, Meinolf Dierkes

International Institute for Environment and Society
Science Center Berlin
West Germany

## IMPACT ASSESSMENT AND PARTICIPATION

Greater interdependency between ecological and socio-economic systems, increasing number of interventions and the cumulative effects of past interventions, have resulted in a higher likelihood of secondary, unintended impacts of public intervention. Due to differing values, goals and perceptions the assessment of these impacts is not only a question of information and valid projection, but also one of political evaluation and of the integration of these effects into policy formulation and implementation. Impact assessment then, comprises two major concerns: the development of techniques and methods to implement the concept (how to assess); and the evaluation of these impacts in the political processes (who assesses). The second concern gains its importance from the fact that governmental bureaucracies control the selection and weight of the criteria that determines the result of the impact assessment process. This has been clearly demonstrated in the context of nuclear power plant siting in all western industrialized countries.[1] Citizen groups are increasingly questioning the validity of assessment data produced by power plant companies and governmental authorities. The right of these institutions to decide unilaterally on specific locations has been challenged. Where opposing values, different perceptions of risk and a high degree of uncertainty characterize the impact assessment process, citizen participation in decision-making becomes a significant aspect.

---

The demands for more citizen participation have so far been attributed by a number of researchers to a general legitimacy crisis of parliamentary democracies in western industrialized countries.

Various factors are supposed to have contributed to this legitimacy
crisis.  Among these are:

1)  Parliaments have lost many of their control powers and have
    delegated discretionary policy-making functions to the ex-
    ecutive branch and to various government agencies.[2]  The
    loss of control function of parliaments vis-à-vis govern-
    ment bureaucracies undermines traditional participation
    channels, basically the election process and voting beha-
    vior.  Citizens who have been frustrated by their inability
    to affect policy concerning their daily lives or to express
    their concern effectively through these channels have in-
    tensified their demands for increased direct participation
    in the assessment and decision-making process.[3]

2)  The centralisation of decision-making power on these issues
    into a network of a few governmental agencies, which cannot
    be easily influenced by the individual, is yet another fac-
    tor contributing to demands for more citizen participation.
    This development towards centralization is at the same time
    linked to the functional decentralisation of decision-making
    in problem areas that are cross-functional in character like
    environmental and energy policy.  The functional decentra-
    lisation is the result of past assignment of competence to
    deal with the problem areas.  This creates an additional
    problem.  The large number of factors involved in such de-
    cisions, the different degrees of competence, and the com-
    plexity of interaction all tend to further minimise the
    ability of individuals or local groups to influence inter-
    governmental decisions since the mechanism appears to be
    less and less transparent.

The hypothesis that the tendency towards bureaucratisation causes a
legitimacy crisis in industrialized countries - citizen groups being one expression of this crisis - tends to overlook the loyalty
achieved by these bureaucracies in serving their specific constituencies.

    Governmental bureaucracies are competing for resources in order
to survive and to grow.  To support their claims for resources,
these bureaucracies need the backing of organized interest groups
as constituencies, as a basis of legitimacy of their claims.  The
bureaucracies have to be responsive to the demands of these constituencies in order to maintain their continuing support and, thereby,
they make the overall system legitimate for these constituencies.

    A political system that relies heavily on functionally decen-
tralized governmental bureaucracies does not only specialize with
respect to policy making but with respect to loyalty provision as
well, reflecting the pluralistic structure of a society composed of

competing interest groups.  It, thus, serves several purposes: besides reducing the complexity of policy-making by delegation, it
allows bureaucracies to pursue their organizational goals of survival and growth, and at the same time, it integrates interest groups
into the overall system by mediating their demands.  However, such
a system has a number of shortcomings with respect to its ability
to achieve loyalty of groups to the overall political system.  It
can only fulfill this function if the competing demands of the various agencies and their constituencies can be met.  The likelihood
of such a demand satisfaction is higher in a situation of growing
resources because these demands tend to grow for redistributional
reasons or for aspiration levels based on past growth rates.  With
lower rates of economic growth, these competing demands will put the
overall system under considerable stress.  The major deficiency,
however, is that it only accounts for demands of new constituencies
if they have organizational backing.  Thus, the system neglects
those 'constituencies' that are unable to achieve organizational
status.  The open question about the demands articulated by citizen
action groups is whether they are an indicator for new demands, in
the process of organizing themselves, or whether they are expressing
doubts about the overall validity of the delegation of specialized
functions to governmental agencies.

    Any impact assessment of public policies that takes place in
such a decision making environment will always be evaluated by agency-
specific criteria that tend to reflect the interests of its constituency.  Groups who are affected by agency decisions but are not part
of its constituency usually do not have the resources to change the
evaluation criteria in their favor.  Proposed solutions to increase
the ability of these 'not-yet' or 'not-at-all' constituencies to have
an impact on agency policy are to change organizational design (e.g.
the division of the Atomic Energy Commission in the United States)
or to prescribe specific decision making rules (e.g. the Environmental Impact Statements).

THE NEED OF NEW DECISION-MAKING STRUCTURES: NUCLEAR POWER PLANT
SITING AS A CENTRAL CASE

    The growth in size and number of citizen groups and their grea-
ter political visibility in industrial countries are key indicators
for the strength of citizens' demands for direct involvement in decisions significantly affecting the social, environmental and economic aspects of their lives.  Most countries possess no adequate,
organised process formally allowing citizens to participate meaning-
fully in these decisions or even providing them with adequate information on these issues.  Decision-makers are increasingly perceived
to be withholding valuable information with respect to the total
environmental and social impact of their decisions.

Although citizens in some countries do have the legal right to
require this information, the actual response of decision-makers towards these demands many times weakens the position of citizens.[4]
There is often no formal structure through which to force officials
to provide a more comprehensive evaluation procedure or to integrate
citizens' values, goals and perceptions into the decision-making
criteria.  Assessment processes comparable to the Environmental Impact Statement in the United States,[5] which, at least in principle,
allow for more involvement of citizens, do not exist in Europe. The
procedure adopted by the US Atomic Energy Commission, for example,
has been severely criticised for the lack of opportunity for mean-
ingfull participation.[6]  The actual procedure does not balance the
relative positions of proponents and opponents in siting decisions.
With respect to information, time, knowledge, personnel and financial resources, the proponents are still in an advantageous position.

The inadequate decision-making process has therefore compelled
citizens concerned about environmental impact in Europe to rely on
a variety of political actions ranging from demonstrations and public pressure to active interference with the construction and operation of disputed projects.  Nuclear power plant siting provides an
appropriate example of this problem.  The objective of governments
to secure energy supplies conflicts sharply with the reluctance of
citizen groups to accept the political validity of further economic
growth, increasing energy consumption and the environmental impacts
of power plants on the local level.

THE FIRST STEP IN DEVELOPING A NEW DECISION-MAKING STRUCTURE:
UNDERSTANDING THE DEMAND FOR MORE CITIZEN INVOLVEMENT

A proposal of a structure allowing for more citizen involvement
in nuclear power plant siting cannot be discussed in an abstract way.
It has to be placed in the social and cultural framework of a particular society.  The current possibilities and reasons for citizens'
involvement as well as an understanding of the kind of information
necessary for their effective involvement must be examined.  Intensive case studies seem to be a necessary contribution to develop such
a concept of citizen participation in the process of technology and
environmental impact assessment.  This paper attempts to do so by
reporting of a project studying the involvement of citizen groups in
nuclear power plant siting in West Germany.[7]  The conclusions are
therefore basically bound to the social and cultural situation of
this country and its history of citizens movements.  However, the
authors are convinced that results from these studies could be used
in a broader perspective: they present a basis for discussion about
the development of structures designed to integrate public participation into the decision-making process in other European countries.

The Current Situation of Potential Citizen Involvement:
The Nuclear Power Plant Siting Process in West Germany

Since West Germany has no equivalent of the United States National Environmental Policy Act as a general framework for citizen's participation in decisions with major environmental considerations, the provision for public participation is restricted to particular laws regulating the various types of decision-making with environmental impacts. Nuclear power plant licensing is regulated by the Federal Department of the Interior under the terms of the Nuclear Energy Law (Atomgesetz).[8] This law has basically two objectives: to sponsor the development of peaceful use of nuclear power plant and to protect the population from the specific risks of nuclear energy. The German Supreme Court for Public Law (Bundesverwaltungsgericht) has ruled that the protection objective must be accorded priority over that of development.[9] Executive powers lie with state agencies under the direction of the Federal Department of the Interior.

The licensing responsibility of the nuclear power authority is limited to the specific risks and hazards resulting from the use of nuclear fuel as energy source and not to the general environmental impacts. The authority has the responsibity to cooperate with other governmental agencies affected by the siting decision, but the violation of this obligation has no bearing on the legality of an issued licence. Its purpose is to speed up the licensing process and avoid proliferation of controls. The licences according to land use, water pollution and construction safety regulations are issued separately by a number of state and local authorities. (Table 1 shows the actors in the complete licensing process.)

Having received the application for the construction and operation of a nuclear power plant, the state licensing authority has to make the project public with special reference to the possibility of objections and the availability of the documents. Objections have to be filed by individuals within a month. Only those objections referring to the specific nuclear aspects of the power plant are discussed in a public hearing. Other environmental or social considerations cannot - according to law - be brought up at the public hearing. After this hearing and after consulting safety experts the licensing authority reaches its decision. The decision becomes effective 14 days after it has been sent to the applicant and to the objectors.

Anybody who considers the decisions as being unbalanced with respect to the objections made during the hearing can go to court. The suit must be based on arguments already made during the hearing. Individuals are allowed to take legal action only if one could reasonably assume an interference with individual rights (life, health, property). The courts define the limits of possible interference of rights by nuclear power plants as being up to 60 miles distance.

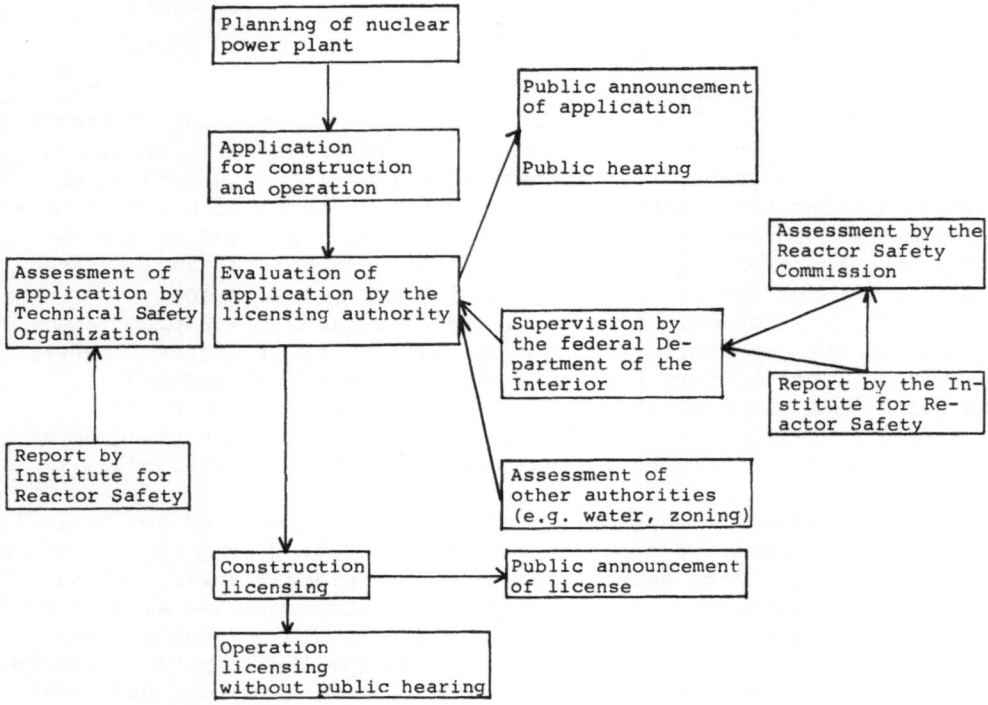

Table 1.   Procedure of Nuclear Power Plant Licensing in West Germany

Source: Bundesministerium für Forschung und Technologie (Ed).

Viertes Atomprogramm der Bundesrepublik Deutschland für Jahre 1973-76, p. 80.

Although the courts took a generous position with respect to the admittance of legal actions they were restrictive with respect to their validity: the burden of proof lies with the plaintiff and, as yet, no nuclear power plant rulings have been overturned. The effect of these decisions has been to undermine the above-mentioned ruling of the Supreme Court.

### The Projected Development of Nuclear Energy in West Germany as a Framework

The nuclear energy program is a joint program of the Federal Government and the states in West Germany.[10] Its main purpose is to secure energy availability by the development of nuclear power plants. The fourth nuclear energy program (1973-76) costs about 2.4 billion dollars. This committment to nuclear energy is based

on the assumption that nuclear energy will constitute 15% of primary
energy sources by 1985.  The projections of energy consumption, made
in 1974 during the recession and the oil embargo, indicate an increase of energy consumption of 100% by 1985.  The major four parties in West Germany generally support the nuclear energy program.
One can consider the nuclear power industry and the utilities as
the constituency of this program.


## DESCRIPTION OF THE CASE-STUDIES

The two siting cases to be discussed here differ by the character of their region and by the timing of the siting decision:

a.   Ludwigshafen

Ludwigshafen is an industrial town in South-western Germany,
with a population of 180,000 with an industrial structure dominated by a chemical company.  Ludwigshafen is part of the
Rhine corridor being the major industrial area in West Germany.
It is located in a metropolitan area with a population of 1.1
million 60 miles south of Frankfurt.  A chemical company applied
for a licence to construct a plant in 1969.

b.   Wyhl

Wyhl is a small town in an agricultural county which has a population of 120,000, located in the upper Rhine valley, west of
the Black Forest.  The agriculture consists predominantly of
fruit and wine growing.  The State government is planning a
number of projects to industrialise the area which is considered
to be strategically valuable for its central location within
the Common Market, its good transportation network and its proximity to Basel and the southern end of industrial Rhine corridor
around Ludwigshafen/Heidelberg.  Wyhl was the second choice of
a state-owned utility company after a first site in the same
area, in Breisach, failed due to public pressure.  The selection was made public in the summer of 1973.

The public reactions to both announcements varied between acceptance by the traditional political groups and the rejection by citizen groups with little influence in the general population in Ludwigshafen, and a resistance from local citizen groups with widespread
political support in the population of Wyhl.


## THE FORMATION OF CITIZEN GROUPS

In both locations, citizen action groups grew out of a reaction
against developments perceived as having a negative impact on the
local environment: in Ludwigshafen the earlier planning of an airport

caused the formation of environmental groups who subsequently took
part in the discussion about the nuclear power plant.  In Wyhl, the
announcement of the nuclear power plant plans instigated citizens
to join together in action groups.  The formation process started
with loose, informal meetings and led to the creation of an association as a legal device to avoid liability claims on the individuals.  The objectives during the formation process differ in both
locations.  In the metropolitan area, the purpose was to develop a
broad environmental awareness and specifically to increase the concern about health hazards by distributing information about the problem.  In the agricultural county, the objective was very concrete,
to prevent or postpone the construction of the nuclear power plant.
Some of the motives of the individuals involved were similar:  in
Ludwigshafen, the perceived general structural deficiencies in the
public administrative decision-making process were at the root of
citizen dissatisfaction.  In Wyhl, the indignation about the secrecy
of planning in a democratic state was a significant complement to
the obvious environmental considerations.


## THE FUNCTIONING OF CITIZEN ACTION GROUPS

     The actual information policy of the licensing authority and the
applicant led to a "planning shock" expressed as a feeling of deliberate exclusion from information.  The complexity of the technology
and the licensing process at first increased the feeling of infer-
iority experienced by the laymen.  The initial step of citizen groups
was to bridge the information gap by self-education and by cooperation with independent predominantly university scientists.  This
information was disseminated to the general public.  In the rural
county, the process was facilitated by the experience gained during
the conflict over the first site in Breisach.

     The structure of both groups generally reflected the social
stratification of the general population, with a slight overrepres-
entation of economically independent groups (doctors, teachers, etc.).
The citizen groups considered themselves to be non-partisan.  Although one major opponent was the state government supported by one
party, the membership was by and large cross-partisan and comprised
a broad spectrum of political opinion.  The internal structure of
the action groups was not hierarchical: the experts did not automatically assume leadership.

     The argumentation of both groups was twofold: impact-related
and process-related.  The impact-related arguments generally tried
to emphasize environmental criteria by taking into account long term
effects and impacts on other media, by increasing the perception of
possible risks and by suggesting alternative sources of energy as
well as concepts of a low growth economy.  The process-related arguments denounced the lack of information, the biases in the report

and the conflicts of interests of the decision-makers.

In Wyhl, the environmental concern of the action groups centered on the effect of the cooling towers on the wine-growing agriculture. The water vapors from the towers were expected to increase the frequency of inversion situations and,thus, reduce the quality of the local wine.  Since a report of the authority, based on a study conducted by consultants, neglected these impacts which were discovered by the citizens themselves, the credibility of most reports put forward by the authority was lost.  These impacts were most important since they severely affected the income of a large portion of the population.  Other nuclear power issues, like radiation, accident risks, heat waste were still important, but secondary to the impact of a changing local climate.  In addition to these issues the citizens protested against the prospective industrialization of the upper Rhine valley, of which the nuclear power plant was a part. The citizens feared that the industrialization would affect the local lifestyle negatively and that this process would take place without local control.

In Ludwigshafen, the arguments of the citizen groups against the power plants were broader and correspond more to the typical general objections.  Safety had priority, particularly since the site was located within a metropolitan area.  The impacts of the heat waste on the ecology of the Rhine was another major concern next to the economics of nuclear energy in general.  The exclusion from assessments of the costs necessitated by safety and radioactive waste disposal were considered a short-sighted basis for decision-making.  The dominance of the growth objective as a decision-making criteria was seen to restrict the investigation of possible alternatives to nuclear energy.

The specific circumstances led to two different strategies that were pursued by both citizen action groups.  In Wyhl, the objective was to increase environmental consideration by lengthening the time for decision-making.  Thus, the group was asked for more, unbiased consulting and in-depth studies on environmental impacts.  In the beginning the style remained objective and argumentative to achieve credibility with the licensing authority.  This group was willing to cooperate and to compromise. They tried to establish an image of being middle of the road, serious, concerned citizens protesting against inconsiderate government.  The actions were directed at two audiences: at the general public for support, and at the licensing authority to influence its decision.  The action groups organized demonstrations, collected signatures, established a measurement station, supported the occupation of an industrial site in a neighbouring French town and expressed objections in the hearings.

In Ludwigshafen, the main objective was to raise concern about the nuclear power plant by bringing the issue to the local agenda.

The actions consisted of writing letters to newspapers, organizing public discussions and distributing pamphlets. The citizen groups considered themselves as a lobby for future generations and remained uncompromising of the very general nature of their objection to the nuclear power.


## PERCEPTION OF THE CITIZEN ACTION GROUPS BY APPLICANTS AND THE AUTHORITY

The citizen action groups were generally viewed negatively by their opponents (applicant, licensing authority, producers of the nuclear power plant). The opponents expressed strong doubts about the legitimacy of the action groups and the validity of the response in the general public. They suspected, partly, communist or anarchist origins in the opposition to their plans and complained about the ignorance among the citizen groups with respect to the "real" issues. This misperception on the part of the opponents was based on inadequate information and on elitist understanding of democracy as well as an equating of representative democracy with democracy as such. The exclusiveness of representative democracy models ruled out a bargaining element of law. The administrative decision-makers believed they were serving the public interest. Their training in the German legal traditions, which is based on a Gesellschaft-type understanding of society, did not allow for the interference by citizen action groups with the proper decision-making process.

The course of the hearing process in Wyhl illustrates the perceptions of the licensing authority quite clearly. Unlike the Environmental Impact Statement hearings in the United States at which the opponents have substantial rights, the procedure and the agenda were set by the licensing authority. The rules were clearly disadvantageous to the interests of the citizen groups and, thus were brought up as a point of discussion without being considered by the licensing authority. This exacerbated the conflicts during the hearing, and the session was broken off.


## BACKING OF THE CITIZEN GROUPS BY THE GENERAL POPULATION

The difference in both locations, Ludwigshafen and Wyhl's county (Emmendingen) are reflected in the response of the general population. A survey, taken in 1974, showed stronger opposition against nuclear power plants, stronger support for environmental objectives and stronger support for the citizen groups in Emmendingen than in Ludwigshafen. The level of knowledge about the planned project was higher in Wyhl (99%) than in Ludwigshafen (72%) (Table 2). Whereas 20% opposed the project in Ludwigshaften, 60% opposed it in Emmendingen, 35% of the population supported the project in Ludwigshafen, but only 14% in Emmendingen (Table 3). While 37% supported environmental objectives in Ludwigshafen, 54% were in favor of environmental objectives in Emmendingen. The respective percentage in favor of further economic growth as the main objective growth are 52% in Ludwigshafen and 35% in Emmendinger (Table 4). The knowledge about the citizen groups is higher in Emmendingen (68%) than in Ludwigshafen (45%) (Table 5). The attitudes towards citizen groups were more positive than those for the traditional political system in Emmendingen (63% versus 14%). In Ludwigshafen "only" 47% trusted the citizen groups and 16% the traditional political system (Table 6). In Emmendingen, 54% of the population agreed with the objectives of the citizen groups, in Ludwigshafen only 26% agreed (Table 7). The citizen groups in Emmendingen seemed to be more effective in reaching the population since 75% of the population received information from them compared to 23% in Ludwigshafen (Table 8). Despite the visibility of the actions of the citizen groups and the general support for their objectives, the population in both locations was sceptical about the success of the opposition to the projects: in Emmendingen, 74% believed it that the power plant would be built despite the efforts of the action group and the widely shared concern of the citizens, in Ludwigshafen 86% believed it (Table 9). How strong the effect of the citizen groups has been on mobilizing the population and how much they were reflecting attitudes that existed already remain unanswered questions.

| Area | yes % | do not know % | Total |
|---|---|---|---|
| Ludwigshafen | 72 | 28 | 213 |
| Emmendingen | 99 | 1 | 139 |
| Freiburg[1] | 82 | 18 | 184 |
| Total | 82.3 | 17.7 | 536 |

Question: There is a plan to build a nuclear power plant here in the area. Do you know where?

Source: Battelle Institut, Frankfurt, Bürgerinitiativen in Bereich von Kernkraftwerken, Bericht für das Bundesministerium für Forschung und Technologie, Bonn 1975, p. 257.

[1] Freiburg was included as neighbouring town to Emmendingen as a control area.

Table 2. Knowledge about the planned nuclear power plant

562                                                V. J. HARTJE AND M. DIERKES

| Area | against % | nothing against % | split opinion % | indif-ferent % | no comment % | Total % |
|---|---|---|---|---|---|---|
| Ludwigshafen | 20.2 | 34.6 | 28.6 | 13.6 | 2.8 | 213 |
| Emmendingen | 60.4 | 14.4 | 16.5 | 10.1 | 1.4 | 139 |
| Freiburg | 32.6 | 23.3 | 24.0 | 18.5 | 6.0 | 184 |
| Total | 34.9 | 25.6 | 25.7 | 14.4 | 3.5 | 536 |

Question:   What is your opinion of this plan?
Source:     Battelle Institut, op. cit., p. 258

Table 3.  Opinion towards the nuclear power plants

| Area | in favor of growth (A) % | in favor of environment(B) & | do not know % | no comment % | Total |
|---|---|---|---|---|---|
| Ludwigshafen | 52 | 37 | 4 | 4 | 213 |
| Emmendingen | 35 | 54 | 11 | 0 | 139 |
| Freiburg | 34 | 55 | 7 | 4 | 184 |
| Total | 41 | 48 | 7 | 4 | 536 |

Question: Which statement do you agree with?
        A: We need more electrical energy and new energy sources. Only by new energy we can secure our economy and keep our jobs, wages and our living standard.

        B: We have to question economic growth. Now, environmental protection is more important. What is the use of more growth if our ecological base is endangered.

Source: Battelle Institut, op. cit., p. 238.

Table 4.  Attitudes about growth and environment

| Area | yes % | no % | do not know % | Total |
|------|-------|------|---------------|-------|
| Ludwigshafen | 45 | 14 | 41 | 213 |
| Emmendingen | 68 | 14 | 18 | 139 |
| Freiburg | 58 | 8 | 34 | 184 |
| Total | 55 | 12 | 33 | 536 |

Question: Is there a citizen group against the nuclear power plant in your town?

Source: Battelle Institut, op. cit., p. 269

Table 5. Knowledge of citizen groups

| Area | Opinion A (parties) % | Opinion B (citizen groups) % | split opinion % | do not know % | Total % |
|------|----------------------|------------------------------|-----------------|---------------|---------|
| Ludwigshafen | 16 | 47 | 32 | 5 | 213 |
| Emmendingen | 14 | 63 | 16 | 7 | 139 |
| Freiburg | 9 | 72 | 14 | 5 | 184 |
| Total | 13 | 60 | 21 | 6 | 536 |

Question: Which statement do you agree with?
A: We do not need citizen groups. We can vote and express our opinions within a party.

B: The government and the parties do not listen to us anyway. Thus, we need the citizen groups to represent our interests.

Source: Battelle Institut, op. cit., p. 240

Table 6. Opinions about citizen groups

| Area | yes % | no % | split opinion % | do not know % | Total % |
|---|---|---|---|---|---|
| Ludwigshafen | 26 | 41 | 11 | 22 | 96 |
| Emmendingen | 54 | 13 | 23 | 10 | 94 |
| Freiburg | 36 | 29 | 23 | 12 | 108 |
| Total | 39 | 27 | 20 | 14 | 298 |

Question:   Do you agree with these objectives?
Source:     Battelle Institut, op. cit., p. 275

Table 7.   Agreement with the objectives of citizen groups

| Area | yes % | no % | no answer % | Total |
|---|---|---|---|---|
| Ludwigshafen | 23 | 76 | 1 | 96 |
| Emmendingen | 75 | 24 | 1 | 94 |
| Freiburg | 56 | 43 | 1 | 108 |
| Total | 51 | 48 | 1 | 298 |

Question: Have you received information about the planned nuclear power plant from the citizen groups?
Source:     Battelle Institut, op. cit., p. 273

Table 8.   Communication between citizen groups and population

| Area | will be built & | will not be built & | open question & | do not know & | Total |
|------|------|------|------|------|------|
| Ludwigshafen | 86 | 1 | 9 | 4 | 213 |
| Emmendingen | 74 | 7 | 6 | 13 | 139 |
| Freiburg | 71 | 4 | 12 | 13 | 184 |
| Total | 78 | 4 | 9 | 9 | 536 |

Question: Do you believe that the nuclear power plant will be built
          or not?

Source:   Battelle Institut, op. cit., p. 280

Table 9. Expectation towards construction of nuclear power plant

## SUBSEQUENT DEVELOPMENT

In January 1975, the supervising governmental agency announced
the issue of the construction licence. The reactions of citizen
groups towards the decision-making process and the decision itself
culminated in the occupation of the site in February 1975. Since
the groups had considerable support for this action among the general population, threats of legal actions by the construction company and the state government only aggravated the confrontation.
The issue became nationally known by a favorable TV report in February 1975 and the opposition party to the state government started
to support the citizen groups. State government was thus forced to
step down, and negotiate with the citizen groups. As a result, in
January 1976, it was decided to postpone the beginning of the construction for a year.

## CONCLUSIONS

Under the given decision-making system, the given interests of
the bureaucracies and their constituencies involved, the outcome of
impact assessment will be predictably in favor of the siting of the
nuclear power plants, assuming the compliance with existing safety

regulations.  The criteria of evaluation, used by these actors
systematically neglect other adverse impacts.  Those adversely affected by these impacts are only able and willing to organize themselves for a political conflict if primary areas of interest, particularly if economic interests are affected.  Since the adverse impact
is felt locally, and since the organization costs are relatively low
locally, the opposition against the general nuclear energy policy
is focusing on the siting of individual nuclear power plants.  There
are some indicators for an increased willingness to organize locally
to defend interests beyond primary economic interests. However, it
remains an open question whether these efforts will have a similar
impact on governmental decisions as in the siting case in Wyhl since
the organizational strength and ability to endure conflicts is likely
to be weaker.

If the environmental concerns should have more weight in the
assessment process it becomes necessary to change the decision making rules and structures.  A number of proposals have already been
made which are designed to improve the participatory character of
the siting process.  Among these are:

- To permit citizen groups to take legal actions on behalf of their
  members,[11]

- to place the burden of proof on the applicant,

- to fund the citizen groups for assessment reports for the environmental issues involved,[12]

- to make the documents more easily available,

- to structure the hearing towards more balance between applicant
  and opponents,[13]

- to make environmental criteria stricter in the licensing process
  and to include criteria for the nuclear waste cycle.

It remains an open question whether the implementation of these
provisions will lead towards an actual moratorium of nuclear power
plant construction.  This could only result in a re-evaluation of
energy policy at the federal level.


NOTES

1.    W.P.Lange u.a. (Hrsg.): Zur Rolle und Funktion der Bürgerinitia-
      tiven in der Bundesrepublik und West Berlin,Bericht einer For-
      schungsgruppe an der FU Berlin, Zeitschrift für Parlamentsfra-
      gen, Juni 1973.

      Richard Lewis, The Nuclear Power Rebellion: Citizen vs. the
      Atomic Industrial Establishment, New York 1972.

D. Nelkin, Nuclear Power and its Critics, Ithaca 1971.

2.  Th. Ellwein, A. Görlitz, Parlament und Verwaltung, Stuttgart 1967.

3.  C.Offe, Strukturprobleme des kapitalistischen Staates, Frankfurt 1972, p. 124.

4.  Ebbin, Kasper, Citizen Groups and the Nuclear Power Controversy: Use of Scientific and Technological Information, Cambridge, Mass. 1974, p. 262.

5.  National Environmental Policy Act of 1969 (NEPA), 42 USC § 4321-4347 (June, 1970).

6.  S. Ebbin, R. Kasper, lcc. cit., p. 254

7.  The Case Studies are based on:
    Battelle Institut, Frankfurt am Main, Bürgerinitiativen im Bereich von Kernkraftwerken, Bericht für das Bundesministerium für Forschung und Technologie, Bonn 1975.

8.  Atomgesetz von 23.12.1959, revised 19.12.1975 (BGB/I p.814).

9.  BVerwG, Deutsches Verwaltungsblatt 1972, p. 678.

10. Bundesminister für Forschung und Technologie (Hrsg.) Viertes Atomprogramm der Bundesrepublik Deutschland, 1973-76.

11. E. Rehbinder, Möglichkeit, Notwendigkeit und Zweckmäßigkeit, eine Klagebefugnis für private Umweltorganisationen, Frankfurt 1974, (Manuskript).

12. Evangelische Akademie Baden (Hrsg.) Wiedenfelser Entwurf zur Umgestaltung des Genehmigungsverfahren im Umweltschutz, Karlsruhe 1973 (Manuskript).

13. Battelle Institut, op. cit., p. 317.

# ENVIRONMENTAL PLANNING AND SOCIAL RESPONSE AT THE STRATEGY LEVEL

F.E. Joyce, C.W. Sinclair

University of Aston

Birmingham, United Kingdom

## PREFACE

Some of the work described in the paper is being undertaken in conjunction with the West Midlands County Council. It is funded by the Science Research Council and the West Midlands County through the mechanism of the University of Aston's interdisciplinary higher degree scheme. The work is supervised in the University by Mr. F. E. Joyce, Head of the Joint Unit for Research on the Urban Environment and on behalf of the County Council by Mr. M. Burns of the County Planning department. Special acknowledgement is due to the help and assistance of the County Council in the design of the Strategic Choice Game. However it should be noted that the ideas and proposals discussed in the body of the text in no way commit the Council to adopting the techniques discussed.

--------

"Many advanced states are experiencing ... a gap between public opinion and the machinery of government. A paradox of the modern technological society is revealed: society creates problems so complex that they can be handled only by those with specialist skills and intricate knowledge, and at the same time it produces people who are in general more highly educated and inquiring than previous generations. It centralises decision-making but spreads the desire to make decisions. How can democracy, in this predicament, satisfy both the need for greater efficiency and the need for wider participation?" So wrote the London Times in May 1968. This is a dilemma which besets many aspects of governance in a technological society but none more acutely than the field of environmental planning. Planning at both the urban and regional scale has seen

569

an increasing concern with the development of complex analytical and predictive techniques, used in both explanatory and normative modes, and at the same time a growing awareness on the part of practitioners and the general public that the people affected by both the environmental planning process and its output should be consulted and involved in the decision making process.

In the U.K. the 1968 Town and Country Planning Act was the first major piece of British legislation which attemtpted to come to terms with these problems. It made provision for full public involvement in the planning process (a feature hailed as a legislative landmark by some (Cullingworth 1972) ). It laid down a general pattern of procedure for the preparation of structure plans concerned with broad strategy and for local plans operating within the framework and explicitly emphasised the need for adequate publicity and public participation. The requirements at key stages similar in both types of plan, are in outline as follows:-

a. Publication of a <u>report of survey</u>, followed by a period of publicity on issues arising and opportunity for public representations.

b. Publication of <u>matters proposed to be in the plan</u>, i.e. a draft plan. This involves similar opportunities for public comment, and a statement on changes which are made because of these comments.

c. Formal <u>placing of the plan on deposit</u> and invitation of written objections. A copy going to the Secretary of State for the Environment.

d. The <u>examination in public</u>,carried out by a panel appointed by the Secretary of State, who reports to the Secretary having taken written and verbal evidence from interested parties[1].

e. <u>Publication of proposed modifications</u> to the plan and consideration of resulting objections.

f. Publication of the <u>adopted version of the plan</u>.

The aspirations and good intentions of the legislation, however, were and are not matched by the availability of suitable methods and techniques for communication with the public at large about planning issues or indeed for the assessment and evaluation of social response to them. This problem is particularly critical at the structure plan level as structure plans, in British planning practice, are

---

[1] The procedures are somewhat different for a local plan in this regard as the report is to the appropriate planning authority.

essentially documents, accompanied by appropriate diagrammatic illustration, intended in part to translate national and regional economic and social policies into a local context. In so doing they
attempt to provide a framework for the preparation and implementation of local plans. The latter may take a number of forms from
'Action Area' plans concerned with comprehensive renewal or rede-
velopment over a short to medium term time horizon, to 'district'
plans concerned with local planning policies for a sector of the
city, to 'subject' plans which may set out to tackle specific topics such as the provision of leisure and recreational facilities.

Without doubt, the larger the scale of the planning role and
the higher the level of abstraction of the planning policy, then the
greater the difficulties of both communication and the eliciting
of public response become. The objectives of this paper are to
discuss these problems and to suggest, albeit tentatively, some
methods and techniques which may help in overcoming some of the
difficulties. The discussion reflects the authors' involvement in a
case study research situation in the context of the West Midlands
Metropolitan County Council's structure planning activities. These
activities are currently providing the authors with a framework and
set of substantitive issues within which to explore and develop techniques of eliciting social response to strategic planning issues.

## THE INSTITUTIONAL AND SOCIO-ECONOMIC CONTEXT OF THE WORK IN THE WEST MIDLANDS

The reorganization of local government in England and Wales in
1974 created a two-tier system of administration. In the major metropolitan areas we now have 'Metropolitan County Councils' who are
broadly responsible for strategic planning and strategic services
and a second tier of 'district' authorities responsible for local
planning and local services. The West Midlands Metropolitan County
is one of the largest of the new authorities containing a population
of 2,700,000. Fig. 1 shows the county and the constituent district
authorities, and Table I indicates their area and population. It
is a basically industrial and manufacturing area with the largest
concentration of employment in the metal manufacturing and vehicle
construction industries.

The West Midlands Metropolitan County Council (WMMCC) is currently preparing a structure plan for its administrative area. The
plan will have to tackle a range of planning issues and of major
importance in this context will be the County's industrial and employment base. In 1971 an economic appraisal of the wider region
of which the County is a part identified several major weaknesses
in the regional economy stemming from overdependence on the traditional industries of metal manufacture and metal goods. It high-
lighted the need for a more positive economic policy, selective

introduction of industries exploiting new technological developments
and the stimulation of the service sector of the economy.  In the
transportation sector the County faces the problem familiar to many
major conurbations - inability to cater for the demands of the private car and an awareness of the need to regenerate and expand public transport provision.  In this regard it faces a particular dilemma in that undue restraint of the private car is a particularly
sensitive issue in an area dependent for employment opportunities
on vehicle manufacturing industries.  In the housing and population
sector, the plan must tackle the fundamental social and environmental problems created by emmigration from inner city areas and peripheral pressure for new development in protected 'green belt' areas.

These brief comments are mainly intended to give an impression
of the kinds of issue which have to be communicated to the general
public at the various stages in the plan preparation process.  In
the section below we discuss the problems of communication and evaluation of such issues.

## SOCIAL RESPONSE AND PUBLIC PARTICIPATION

The analytical definition of strategic problems poses fundamental problems for the design of a public participation exercise.  In
fact the early optimism of British planners concerning participation
was, in part, based on the American experience which was concerned
primarily with citizen involvement at a city level and the organisation of the public at a neighbourhood level.  The nature of strategic considerations aggravates such problems as the specification
of groups with different needs, and different influence, the possibility of consensus, the determination of weights for interest
groups and the decisions as to which groups should be consulted. In
particular, strategic issues lack the immediacy of local issues; and
the individual perception of concrete problems.  Briefly, strategic
planning problems differ in that they are abstract, complex and
interrelated.  There are constraints on the information available
to planners about problems and their definition, so that issues presented to the public are never clear-cut.  There are constrains on
the extent to which an authority can deal with 'problems', firstly
because of its narrow legal responsibilities, and secondly, because
of its limited resources, so that the depth of concern about certain
problem areas will not reflect the public's concern with them.  The
issue of techniques presents equally complex problems.  For example,
with regard to the timing of surveys, if it is to be administered
at the 'report of survey' or 'definition of objectives' stage then
different techniques of presentation will be required than at assessment or evaluation stages.  Similarly, the survey designer will face
the problem that at the 'problem identification' stage the issues
will be only vaguely defined and unlikely to relate to individual
perception of problems.  Measurement problems raise another set of

Estimated Area                                                April 1974

|                  | Population | Acres  | Sq. Km. |
|------------------|------------|--------|---------|
| W. Midlands Cty. | 2793286    | 214146 | 866.6   |
| Birmingham       | 1097942    | 59059  | 239.0   |
| Coventry         | 336746     | 23673  | 95.8    |
| Dudley           | 293926     | 20856  | 84.4    |
| Sandwell         | 330161     | 21845  | 88.4    |
| Solihull         | 192068     | 44826  | 181.4   |
| Walsail          | 273333     | 25378  | 102.7   |
| Wolverhampton    | 269110     | 18509  | 74.9    |

Source: County Surveyor

Table 1

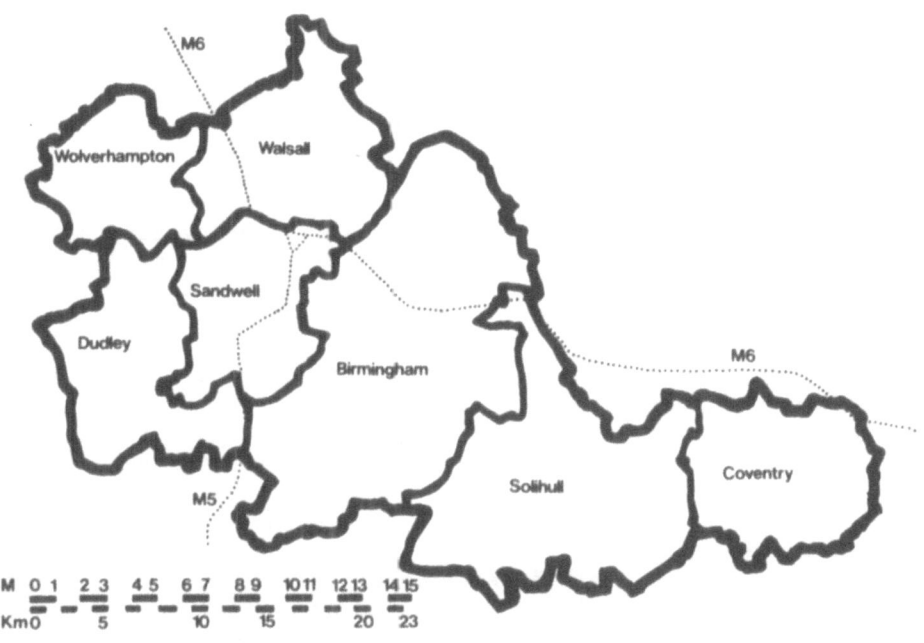

Figure 1.  Sketch map of the West Midlands County showing district
           boundaries.

questions, for example:

- should it be a general sample, or structured regarding main issues, or particular social groups?

- what are the aims, is it a general survey or a survey of preferences regarding choices?

- are the issues made comprehensible?

- does the instrument take into account the interrelations between issues/problems?

- should the instrument force people to trade off between choices?

- should the choices be based on budgetary constraints etc?

- should people make judgements for themselves or for the community?

Finally theoretical and methodological constraints influence the assessment and evaluation of the output of such surveys. Preference selection, ranking, majority voting procedures must all take place within the framework of sound and equitable principles of evaluation. Thus variations in intensity of individual response, and baseline positions with regard to the balance of advantages and disadvantage in environmental terms and the nature and incidence of change must all in some way be reflected in the methodology.

METHODOLOGICAL PROBLEMS IN ELICITING SOCIAL RESPONSE TO STRATEGIC
PLANNING PROBLEMS

All environments can be seen as complex patterns and relationships of components and systems in interaction. Descriptions of environments or parts of environments (e.g. for planning purposes, or for presentation in interview situations) depend upon how the environment is being looked at, and by whom, i.e. upon the conceptual and methodological basis of description.

Although there is a growing literature on how people perceive and attach meaning to environmental stimuli, and which stimuli they see as significant, this literature is largely concerned with small scale, immediate environments rather than more abstract strategic environments. Thus, when one attempts to consider complex abstract environments and environmental processes within the terms of available methodologies they prove very inflexible. What is required is a systematic methodology which is equally applicable to a large range of settings. The difficulties of developing such a methodology can be discerned by considering the limitations and inconsistencies of some of the present approaches. In psychology for example it is generally accepted that between 1920-60 psychology was dominated by the conceptualisation of behaviour as 'mechanistic' rather than cognitive, by the mechanistic behaviourism of Watson and the neo-behaviourism of Hull. Its goal was to 'predict, given the stimulus, what reaction will take place; or given the reaction, state what the stimulus is that causes the reaction' (Watson 1924). The general trend in the contemporary study of motivation is to question the validity of the S-R (Stimulus-Response) formula, and to replace it with a cognitive model which presents the formula S-C-R (S-Cognition-R) thus including higher mental processes. The antecedent stimulus is seen as a source of information rather than stimulation, and is then imbued with meaning which then determines behavioural response. A major concern of environmental psychologists in this context has been to examine the operation of the cognitive systems which bridge the gap between stimulus and response - e.g. work on 'cognitive persuasions' (Holland 1972). Our concern and the concern of environmental planners in general, has a different focus, that is to examine different stimuli (especially displays and information, in interview situations), and different response types (e.g. preferences, judgements etc.) as well as the properties of cognitive processes.

The work, therefore, faces a number of methodological problems:

   (i) problems of definition. That is, which parameter of social response do we select as representing the general system - environmental preference, environmental judgement, environmental behaviour etc.
  (ii) problems of content. In particular, how can one differentiate the extremely complex subject matter broadly described as the environment.
 (iii) problems of presentation. That is, what type of information should one give as being representative of the stimulus (subject-matter).
  (iv) problems of technique. Obviously the nature of evidence is influenced by the techniques employed e.g. questionnaire, laboratory studies, participant observation.
   (v) problems of context. In particular, the operational context of the research will have implications for the reliability of data and its generalisability for theory construction.

Any methodology concerned with social response must therefore resolve, or at least in the short term circumvent some of these problems.


TOWARDS A METHODOLOGY

Methodological developments in environmental psychology have recognised that people perceive, structure and take meaning from their environment and have also accepted the concept of a 'stimulus-

response' model which takes account of the connection between individuals and milieu ('behaviour setting' (Barber 1968) ), allowing
for the perception of both goals and the paths to them.  They have
recognized that people continuously make adaptations to their environments in pursuit of such goals and hypothesised the existence of
a 'cognitive map' of individual environments within the framework of
which the individual is in the process of translating afferent patterns into efferent patterns - a product and process of the stimulus and internal evaluations.

     In the work outlined here, however, with its concern with com-
plex, non-immediate stimuli, the respondent may not be in possession
of a fully formed cognitive map nor be aware of 'goals' or 'objectives' with respect to these stimuli.  The problems of definition,
context and presentation and associated techniques are thus of critical significance and are only to be resolved by the development
of more responsive conceptual structures.

     With regard to the problem of definition we note that a number
of conceptual frameworks developed by psychologists concerning 'attitude' may prove useful.  In particular the commonly accepted formulation of attitude as consisting of three components - cognitive,
behavioural and affective (Upshaw 1968), has direct relevance in any
attempt to classify responses.  The problem of content can perhaps
be resolved by conceptualising the stimuli with respect to a number
of dimensions.  We may for example classify content with regard to
unidimensionality/multi-dimensionality, simplicity/complexity or
specificity/abstraction.

     With regard to presentation we must recognise the conceptual
difficulties mentioned above and note that if the respondent is
likely to have an 'internal representation' of the stimuli under consideration it may be possible to elicit a response with a minimum of
presentational material.  However this may not be the case with respect to 'strategic environmental stimuli' where extensive information
may be required and indeed the process may have to be didactic (and
therefore iterative) in character.

     The problems of technique are a conglomerate of the other meth-
odological problems, including the operational context of the research.  A useful methodological distinction through which one can
approach the problems of choice of technique, is the distinction
between those techniques which elicit a response via external representations i.e. by asking the respondent to draw pictures, build
a model or verbalise descriptions of his environment, and those which
infer representations from the respondent's behaviour in natural
environments, and inferences drawn from specific tasks given to the
respondent.  However, a primary concern of both types of technique
is to elicit information in situations in which people respond in a
meaningful way to the stimuli.  In this context 'priority evaluation

techniques' have seen growing use in environmental planning.  Since
1972 they have been used in the development of structure plans by
Merseyside, Glamorgan, Nottinghamshire, South Yorkshire and Hert-
fordshire County Councils in the U.K.  Such techniques seek to determine how people evaluate their present environments, determine
how people evaluate their present environments, determine preferred
improvements from this base line and by limiting the number of improvements expose 'trade offs.'  The most advanced development of
these techniques to date are to be seen in the 'Priority Evaluator
Technique' (P.E.T.) developed by Social, Community, Planning Research
(Hoinville 1971).

However such techniques, together with more conventional attitude scaling techniques are found wanting with respect to the primary presentational criteria i.e. the extent to which the techniques
are capable of presenting abstract, complex and interrelated issues.
They are also inadequate (though P.E.T. to a lesser extent) with respect to secondary criteria which reflect the need for the techniques
to perform as a learning instrument in some circumstances.

## PRIORITY EVALUATION AND STRATEGIC CHOICE

The growing use of P.E.T.'s in British planning practice merits
further analysis of their suitability to structure/strategic planning issues in the knowledge that this analysis will also provide
some prescriptive response to the methodological difficulties outlined above.

The general problems associated with the application of such
techniques at the strategic planning level can be summarised briefly:

(i)     The presentation of attributes: The stimuli attributes
        must inevitably be reduced and simplified and thus assump-
        tions must be made concerning experience levels of respon-
        dents, perception of attributes and the true relationships
        between the attributes and the symbols used to present them.

        At the strategic level these assumptions become critical.
        To be more specific:
        - P.E.T.'s deal with general, non-specific attributes, where-
          as specific problems/issues facing the authority should
          be the focus.

        - Individual preferences on non-specific attributes are in-
          adequate.  Strategic issues require judgements for the
          community.

        - P.E.T.'s present attributes which are not interrelated
          either in conception or presentation.  The nature of
          strategic planning is such that the interrelation between
          variables and systems is of paramount importance, and must

be mirrored in any technique which provides a meaningful 'choice'.

- P.E.T's offer choices across attributes which are immediately comprehensible to the respondent (e.g. more, or less traffic) and where his present situation regarding the attributes is crucial in developing 'community values'. Strategic issues are not, however, immediately comprehensible to the respondent, therefore, a learning situation must be created within which the respondent is aided in his choice between strategic elements.

- Arguably the respondents' present position regarding the attributes should not be the basis for analysis, rather he should be asked for his 'judgements on behalf of the community', based on information given to him which presents the complexity of strategic problems.

(ii)      The scaling of the attributes: It is usually assumed that the attribute can be represented pictorially or otherwise, by values on a linear scale, but for strategic elements and their interdependence make such scaling inadequate, firstly as a measure of intensity of preference, and second ly, as a measure of the relative importance of attributes.

In summary then a number of requirements for techniques of eliciting strategic responses can be specified.

  (i) They must present problem/policy specific issues.
 (ii) The choices should be presented in a way that makes them as realistic as possible as opposed to utopian.
(iii) They should allow for the fact that information dissemination will be an integral part of the process.
 (iv) Consistency of choice and feedback by an iterative mechanism should be incorporated.
  (v) Respondents should be asked to provide 'ethical' (or community) judgements (Harsanyi 1955).
 (vi) Respondents' present position should not form the only basis for assessment. Analysis should also show comparative responses between different social and spatial groups and provide opportunities for appropriate social 'weights'.

     The West Midlands County Council and the authors have designed a prototype survey instrument at present labelled a 'strategic choices game' (S.C.G.) which attempts to meet these requirements. It utilises a simple model of the West Midlands County together with choice cards to represent the stimuli. It is hoped that the basic components of the structure plan can be presented in kit form to the respondent under controlled conditions.  The respondent will be asked to make a succession of choice/judgements, each governed by the preceding choices which act as constraints, i.e. it is of the form of an 'unfolding technique.'  In some cases the respondent will trade-off some units of one good against units of an alternative good.  Wherever possible choice and decisions will be shown on maps. Iteration will be encouraged.  The respondent will therefore be asked to proceed through a series of planning choices, asked to allocate counters in a preference stage, and to allocate money to specific projects.  Throughout the 'game', figures (e.g. population estimates) and project costs, and implications of one choice for another, will be presented by the interviewer to the respondent in a standardised form with the data taking as realistic and accurate a form as possible.

It should be noted that at this stage the aim of the S.C.G. is neither to examine people's broad value systems (or internal representations) nor to obtain support for specific strategies.  Rather, it is to examine how people perceive and evaluate planning problems and to recreate specific planning choices in a situation which balances a requirement for reality on the one hand, with the need to retain the involvement of the respondent on the other.

It is clear that, in games such as S.C.G., where decisions/ judgements are based on internal representations and on information given to the respondent during the game, the behavioural output is inevitable constrained, i.e. it does not totally reflect the respondent's representation of the environment - it is in a sense a quasi-representation. Our most important priority is therefore to test hypotheses regarding the presence, and structure, of internal representations of strategic environments/environmental processes in respondents, and subsequent to administering S.C.G., to test hypotheses regarding the effect of S.C.G. (i.e. its form and information content) on those internal representations.  The interaction between internal cognitive processes and survey techniques themselves (in this case S.C.G. is the independent variable) is therefore considered to be of major importance in research into the use of social survey techniques in eliciting response to strategic environmental issues.

Finally it should be noted that even if successfully developed S.C.G.'s will only have a minor role with respect to the overall process of public participation in environmental planning. S.C.G.'s will undoubtedly require supplementation with techniques and processes handling issues with greater immediacy and occurring at different stages in the planning process, but hopefully it will, at least, make a small contribution to the resolution of the dilemma of technological complexity and democratic choice[2].

REFERENCES

Barker, R.G. 'Ecological Psychology' (Stanford University Press, 1968).

Cullingworth, J.B. (1972) 'Town and Country Planning Britain'.

Harsanyi, J.C. 'Cardinal Welfare, Individualistic Ethics and Inter-personal Comparisons' J.Pol.Econ., 63, 1955.

Hoinville, G. 'Evaluating community preferences' in Environment and Planning' 3(1971) 33-50.

Holland, V. 'Cognitive Persuasions, assumptions and presumptions' in Proceedings of the edra 3/ar 8 conference, University of California, January 1972.

Upshaw, H.S.  Attitude Measurement in Blalock H.M. and Blalock A.B. 'Methodology in Social Research' 1968.

Watson, J.B. (1924) 'Behaviourism', People's Institute, N.Y.

---

[2]  The appendix to this paper, entitled "Comments on the Form and Content of the S.C.G.", includes extracts from the interviewer's script and outlines the form of the S.C.G.  The appendix, and any other information regarding the research, are available from the authors.

REGISTER OF NAMES


| Agrafiotis, D. | | France |
|---|---|---|
| Agarwal, M. | State University of New York at Buffalo,    New York | U.S.A. |
| Anderson, B. | Washington University, St. Louis | U.S.A. |
| Baumerder, A. | | France |
| Bayraktar, B.A., Dr. | NATO Scientific Affairs Division, B 1110 Bruxelles | Turkey |
| Bowen, K.C. | Defence Operational Analysis Est. Ministry of Defence, Parvis Road, West Byfleet, Surry, KT 14 6LY | U.K. |
| Brenot, J. | Commissariat à l'Energie Atomique, Dept. de Protection, F 92260 Fontenay-aux-Roses | France |
| Buehring, F. | Osaka University, Osaka | Japan |
| Buehring, W. | Osaka University, Osaka | Japan |
| Burkhardt, D., Dr. | IABG, Einsteinstraße, D 8012 Ottobrunn | West Germany |
| Cooper, D., Dr. | Department of Mathematics, Royal Holloway College, Englefield Green Surrey TW 20 OEX | U.K. |
| Cropley, A.J.,Prof. | Unesco Institut für Pädogogik, Feldbrunnenstraße 70, D 2000 Hamburg 13 | West Germany |

Cruon, R.                    Ministère de la Qualitè de la
                             vie Environment,
                             14 bis, Rue de Cirque,
                             F 51000 Chalons-Sur-Marne          France

Davoll, D., Dr.              The Conservation Society,
                             12 London Street, Chertsey,
                             Surrey KT 16 8AA                   U.K.

Dennis, R.                   Osaka University, Osaka            Japan

Despontin, M.                Vrije Universiteit Brussel,
                             Centrum voor Sociologie,
                             Campus Oefenplein, Pleinlaan 2,
                             B 1050 Brussel                     Belgium

Dierkes, M.                  International Institute for
                             Environment and Society,           West
                             Science Center, Berlin             Germany

Dogrusöz,H., Prof. Dr.  Department of Mathematics,
                             Middle East Technical University,
                             Ankara                             Turkey

Emmett, B.                   Office of the Science Advisor,
                             Environment of Canada,
                             Ottawa K1A OH3., Ontario           Canada

Fenker, R., Prof.            Psychology Department,
                             Texas Christian University,
                             Forth Worth, Texas 76129           U.S.A.

Ferguson, N.                 Plessey Radar Research Centre,
                             Southleigh Park House,
                             Eastleigh Road, Havant, Hants      U.K.

Fischer, D.W.                University of Waterloo,
                             Ontario                            Canada

Foell, W.K. Prof.            IIASA,
                             A 2361 Laxenburg-Schloß            U.S.A.

Freckleton, S.               Plessey Radar Research Centre,
                             Southleigh Park House,
                             Eastleigh Road, Havant, Hants      U.K.

Gülec, K. Dr.                Sanayi ve Teknoloji Bakanligi,
                             Bilim ve Teknoloji Dairesi Baskanligi
                             Yardimcisi Tandogan, Ankara        Turkey

Gürkaynak, M.                Technical University, Ankara       Turkey

Harris, J.I. Ms.             Defence Operational Analysis Est.
                             Ministry of Defence,
                             Parvis Road, West Byfleet,
                             Surrey, KT 14 6 LY                 U.K.

| | | |
|---|---|---|
| Hartje, V. | International Institute for Environment and Society, Science Center, Steinplatz, D 1000 Berlin 12 | West Germany |
| Haynes, K. Prof. | LBJ School of Public Affairs University of Texas, Austin, Texas | U.S.A. |
| Hazleton, J. E. | The University of Texas, Austin, Texas | U.S.A. |
| Ito, K | Osaka University, Osaka | Japan |
| Ittelson, W.F.Prof.Dr. | Department of Psychology, The University of Arizona, Tucson, Ariz. 85721 | U.S.A. |
| Johri, H.P. Dr. | Bureau of Management Consulting, 365 Laurier Av. West, Ottawa/Ontario | Canada |
| Joyce, F.E.,Dr. | Joint Unit for Research on the Urban Environment, University of Aston in Birmingham, Gosta Green, Birmingham B4 7ET | U.K. |
| Karayalçin, I.I., Prof. Dr. | School of Mechanical Engineering, Technical University of Istanbul, Gümüssuyu, Istanbul | Turkey |
| Karpe, W.A., Dr. | Universität Dortmund, Haus Weicken, Rosemeyerstr. 6, D 4600 Dortmund | West Germany |
| Keeney, R. | Woodward Clyde Consultants, San Francisco, California | U.S.A. |
| Keith, R.W. | University of Waterloo, Ontario | Canada |
| Keles, R.,Prof.Dr. | Faculty of Political Sciences, Ankara University, Ankara | Turkey |
| Kerpel, A. de | Vrije Universiteit Brussel, Centrum voor Sociologie, Campus Oefenplein, Pleinlaan 2 V 1050 Brussel | Belgium |
| Klaus, J. Prof. Dr. | Volkswirtschaftliches Institut Universität Erlangen-Nürnberg, Hauptmarkt 2/II, D 8500 Nürnberg | West Germany |
| Kleeman, W.T. | State Government of Texas, Texas | U.S.A. |

| | | |
|---|---|---|
| Knapper, C.K. Prof. | Psychology Department, University of Regina, Regina, Saskatchewan | Canada |
| Lapillonne, B. | Woodward Clyde Consultants, San Francisco, California | U.S.A. |
| Lavergne, F. de | | France |
| Le Compte, W.A. | University Beytepe, Ankara | Turkey |
| Lesthaeghe, R. | Vrije Universiteit, Brussel | Belgium |
| Liebermeister, U.,Dr. | IABG, Einsteinstraße D 8011 Ottobrunn | West Germany |
| Mahajan, V., Dr. | Department of Management Systems, State University of New York, Buffalo, New York 14214 | U.S.A. |
| Mark, J. | Washington University, St. Louis, | U.S.A. |
| McDowell, M.R.C.,Prof. | Department of Mathematics, Royal Holloway College, Englefield Green, Surrey TW 20 OEX | U.K. |
| Milne, J.D. | Government of Canada, Ottawa | Canada |
| Moncaster, M.E. | The Home Office, London | U.K. |
| Perron, L.E. Ms. | Office of the Science Advisor, Environment Canada, Ottawa | Canada |
| Ricci, P. Dr. | Dept. of Georgraphy Regional Planning University of Ottawa, Ottawa | Canada |
| Roberts, P.C. | Department of the Environment, 2 Marsham Street, London SW1 | U.K. |
| Rupp, E.,Dr. | AGF-Programmleitung, c/o DFVLR Linder Höhe D 5000 Köln 90 | West Germany |
| Scholz, D. | Institute of Environmental Protection, University of Dortmund, | West Germany |
| Sinclair, C.W.,Dr. | JURUE, University of Aston in Birmingham, Gosta Green, Birmingham | U.K. |

| | | |
|---|---|---|
| Solem, E.,Dr. | Directorate of Strategie Analysis ORAE<br>Department of National Defence<br>Ottawa, Ontario | Canada |
| Theys, J. | Ministère de la Qualité de la<br>vie Environment,<br>14 Boulevard du Général Leclerc<br>F 92521 Neuilly-Sur-Seine | France |
| Timmers, H. | Institute for Perception TNO<br>Kampweg 5,<br>Soesterberg, | Netherlands |
| Wagenaar, W.A. | Institute for Perception TNO<br>Kampweg 5,<br>Soesterberg | Netherlands |
| Weber, H.P.,Dr. | RegDir, Bundesverkehrsministerium<br>Kennedyallee 72,<br>D 5300 Bonn-Bad Godesberg | West<br>Germany |
| Wijewickrema, S. | Vrije Universiteit,<br>Brussel | Belgium |
| Winkel, G.H., Dr. | Environmental Psychology Program,<br>CUNY Graduate Center,<br>33 W 42 Street, New York | U.S.A. |
| Wolf, Ch.P., Dr. | Environmental Psychology Program<br>CUNY Graduate Center,<br>33 W 42 Street, New York | U.S.A. |
| Wright, R.E. | Government of Canada,<br>Ottawa, Ontario | Canada |

INDEX

Accidents, economics of, 250
  environemntal causes of, 264
  rail (*See under* Rail)
  road traffic (*See under* Road
    traffic)
Air pollution
  social impact of, 527
Analysis
  for decision making, 173
  levels of, 165
Analytic forecasting, 71
  cross-impact analysis, 75
  Delphi technique, 74
  methods and techniques, 74
  morphological analysis, 77
Anthropology, 382
Assessment, 197
  definition of, 135
  risks, 136
  systemic distortion in, 3

Behaviour
  game theory and, 169, 171,
    174
  in social systems, 166
Behaviour orientated models, 84
Behaviour settings, 225
  assessment of, 228
    single situation, 228
    single variable, 229
  backstage, 226
  ecological approach, 227
  homology, 229
  level of manning, 227
  occupancy time, 227
  organization, 226

Behaviour settings (cont'd)
  population size and, 230, 231
  in rehabilitation hospital,
    232, 233
  territorial range of, 227, 234
  total institutions, 232
  units, 226
BELADO, 527
Belgium
  demo-economic policy in, 29
  population of, 36
Bureaucracy, 552

Canada
  distribution of environmental
    quality in, 361
  public policy in, 377
Canadian railway system
  government regulation of, 242
  safety on, 242
    conceptual model of, 248
    economic system of, 250
    level of risk, 251, 252, 253
    physical system of, 248
    policy alternatives, 257
    subsidies and, 257
    taxation and, 257
    value system, 251
Capitalism
  definition of, 50
  environemntal deterioration
    and, 54
  regional inequalities and, 52
  urban growth and, 49, 51
Carbon dioxide emission, 245

Carrying capacity, 377
  concept of, 377
  in ecology, 379, 381
  geographers and, 383
  legislated into action, 390
  model, 378
  policy making and, 385
  population growth and, 390
Cause-effect relationships, 138
  measures, 139
  natural law and, 141
Change
  secondary effects of, 123
Check lists
  application of, 156
  characteristics of, 157
  disaggregation of, 150
  for environment, 145, 150,
      158
  impact trees and, 133
  magnifying glass, 152
  specialized, 156, 158
  water, 149
'Chicken games', 169
Cities, (*See* Urban areas)
Citizen action groups
  backing by population, 560
  formation of, 557
  functioning of, 558
  perception of, 560
Coal mining in Ruhr, 526
Cocoyoc declaration, 55
Committees
  drawbacks of operation, 75
Conflict
  analysis studies of social
      systems and, 23
  decision making and, 169
  decision theory and, 161, 162
  game theory and, 161, 168,
      174
  in industrial production, 25
  research into, 160
    relevance of, 164
Conflict transposition in
      social systems, 22
Cost of living, 113
Crises, prevention of, 23
Cross-impact analysis, 75

Data, information content of,
      170
Decision analysis
  of energy/environment system,
      197
Decision making, 9, 10, 163, 166
  analysis of, 173
  assessment of, 197
  centralization of, 552
  in conflict, 161, 162, 169
  information and, 170
  need for new structures, 553
  public concern in, 569, 570
  public participation in, 554,
      572, 577, 578, 579
  public reaction to, 553, 554
Decision trees, 77
Defensible space, 435
Delphi technique, 74
Demo-economic models
  principles of, 34
  in Belgium, 29
Demographic forecasting, 77
Developing countries
  demand for technology, 24
  housing in, 58
  urban growth in, 49
  urban planning in, 57
  urban systems in, 57
Development, difference between
      two courses, 136
Drinking water, 206, 210
  economics of, 211, 212
Dynamic investment planning, 87

Earth resources, 5
Ecology
  behavioural, 227
  carrying capacity and, 379, 331
  conflict with economics, 7
Economic development
  forecast of, 8
  rate of, stationary populations
      and, 36
  social welfare and, 56
Economic policy, problems in, 42
Economic strategy, multiple
      criteria decision in,
      40

Economics, 177
  assumptions of, 9
  conflict with ecology, 7
  decline in population and, 31
Economy, collapse of, 8
Education
  administration in, 476
  costs of, 474, 477
  staff-pupil ratio, 477
Effect, definition of, 136
  pattern of, 137
  union set, 143
Effluent control, cost of, 502
Electrical energy, 513
  impact on environment, 195
Energy
  capital costs of, 507
  decision making in, 552
  future demand for, 191
  impact on environment, 510
  policy evaluation, 504
Energy/environmental systems
  alternative policies, 493
  appraisal of models, 188
  assessment of, 335
  decision analyses in, 197,
    352
  format of research in, 184
  impact of, 510
  implementation of research,
    199
  models, 192
  options and strategies, 196
  policy assessment of, 183,
    504
  policy framework, 495
  regional, 186
  scenarios, 194, 336, 343
    building, 189
  structure of, 187
  in Texas Coastal Zone, 493
Energy use, primary, 343
Environment
  assessment of, 395
  changes in, adoption over
    time, 444
  checklists for, 145, 150, 154
  constraints and control in
    policies, 404

Environment (cont'd)
  definition of, 362
  definition of issues, 399
  degradation of, 6
    socialism and, 60
    under capitalism, 54
  diffusion of innovation, 442
  distribution of direct
    services, 367
  distribution of indirect
    services, 364
  energy impact on, 194
  evaluation, long-term, 400
  impact of electrical energy on,
    195
  impact trees and, 142
  improvement of, capital costs,
    499
  management of, 363
    distribution considerations,
    371
  models, 396
  pollution damage, 420
  public opinion of, 531
  public policy on
    adoption analysis, 447
    diffusion analysis, 441
    dimension of innovation, 451
    innovation index, 448
    multi-dimensional scaling
      and, 446
    social, economic and
      political factors, 447
  quality of, 361
    application of models, 413
    assumptions, 405
    economics of pollution, 426,
      429
    evaluation models, 415
    management strategies, 431
    models, 396
    pollution damage, 420
    residual generation
      discharge, 411
    scenarios, 403
    simulation models, 405
    social economic modules, 409
    structural analysis, 401
  quality change, 395

Environment (cont'd)
  social reaction to, 530, 574
  stress factors, 528
Environmental impacts, 344
Environmental planning (*See under* Planning)
Environemntal psychology
  methodological developments, 575
Experience and exponential growth, 110

Fertility, decline in, 29
Food prices, 365, 372, 373
  prediction of, 113
Forecasting
  (*See also* Planning *and under* Subjects)
  analytic, 71
  basis of, 73
  complexity in, 73
  cross-impact analysis, 75
    decision and relevance trees, 77
  Delphi technique, 74
  demographic, 77
  discounting in, 127
  games and gaming, 79
  mathematical models, 76
  measurement and values, 73
  methods and techniques, 74
  models, 126
    under conditions of interaction, 123
  morphological analysis, 77
  paradox and, 125
  problems of, 71
  public involvement in, 81
  scenarios, 79
  systems analysis, 78, 80
  trend extrapolation, 76
  uncertainity in, 72, 81
Forecasts, official, 13
France
  energy/environment system, 189
  environmental quality change in, 395
Freight transport, 96

Future
  problems of, 71
  prospects for, 14
Fuzziness, 174
  concept and value, 171
Fuzzy sets, 162

Games and gaming, 79
Games theory, 174
  behaviour and, 169, 171, 174
  conflict and, 161
  importance of, 166
  issues in, 168
  new types of, 167
  research into, 173
Generation processes, 411
Geographers, carrying capacity and, 383
Geothermal energy, 504
  impact of, 510, 514, 523
German Democratic Republic
  alternative energy/environment futures, 335
  energy/environment systems in, 184
  environmental impacts in, 184
  health and safety in, 347
  primary energies in use, 343
Germany
  nuclear power siting in
    social impact of, 551
  traffic forecasting in, 83
Government
  intervention by, 34
  planning of, 14
  public opinion and, 569
Government agencies, power of, 552
Gross national product, 38
Growth, 379, 406
  direct presentation of processes, 114
  exponential, 103
    factor of experience, 110
    graphical presentation of, 110
    inverse statistics, 118
    numerical presentation of, 105

Growth (cont'd)
  intuitive prediction of, 103
  inverse statistics, 118
  misperception phenomenon, 104
  as natural process, 7
  promoting mechanisms, 7
  time factor, 113
    of towns and cities, 49
      center-periphery
        relations, 51

High speed transport, 88
Hospitals, treatment
        environment, 232, 233
Housing
  accessability of services,
    435, 439
  conditions, 53
  control of, environment and,
    437
  defensible space, 435
  economics of, 439
  personal security in, 438
  planning change in, 433
    approach to, 436
    impact of, 438
  privacy in, 436, 439
  problems of, 56
  quality, 371
  repairs, 436, 439
  residents control of, 434
  security in, 435
  social inequalities in, 58
    in Turkey, 62
Human activity, 6

Impact trees
  basic principle, 139
  environment and, 142
  graphical compilation of
      effects, 139
  levels of abstraction, 145
  measure and, 142
  mental associations, 144
  modularity of, 141
  universal checklists in, 133
Income, distribution of, 52, 374
    374
Industrial activity, 337

Industrial development, 11
  on Texas coast, 501
Industrial odors,
  community affected by, 536
  evaluation of, 537
  public reaction to, 539
  social class and, 542
  social perception of, 535
Industrial production, conflict
    in, 25
Industrial Revolution, 17
Industrial safety, 347
Industry
  economics of, 501
  in Ruhr, 536
  water use by, 206, 210
    economics of, 211
  working conditions in, 530
Information
  decision making and, 170
  definition of, 170
Interaction, prediction and, 123
Interventionism, rise of, 55

Land
  carrying capacity of, 382
  ownership of, 63
  price of, 62
  private ownership of, 53, 60
  public ownership of, 56
  urban, in service of society,
    60
Land prices, 53
Land speculation, 63
  urban, 53
Land use in Wisconsin, 348
Language, 160, 172
  vagueness of, 172
Level of manning in behaviour
    setting, 227
Local government in U.K., 571

Magnifying glass technique, 152,
    153, 155
Market economies
  cities in, 50
  planning models, 76
  pollution and, 54
Measurement, 165

Measures, 139
  space of, 151
  union sets of, 143
Medical sociology, 27
Mesarovic-Pestel strategy for
        survival, 78
Misperception phenomenon, 104
Models, 3 (See also under
        types)
Morphological analysis, 77
Multidimensional scaling, 446

Natural gas, 504
Neckar River, 203
  basin of, 205
New towns, 56
Nuclear energy
  development in W. Germany, 55
        556
  public involvement in
        decision making, 554
Nuclear power siting
  citizen action groups and,
        557
  decision making and, 553
  environmental factors, 559
  social impact of, 551

Occupancy time in behaviour
        settings, 227
Occupancy time index, 232
Oil, 504, 512, 513
Operational research, 160
  definition of, 166
  place of, 167
Overmanning, 230

People, problems with, 163
Perception, 160
Personal autonomy, 10
Petrol, 366
  price of, 373
Petroleum development, 318
  actor analysis framework, 321
        321
  infrastructure support
        systems, 322
  programme, 320
Planification process, 397

Planners
  action of, 5
  attitudes of, 9, 11
  personal interests of, 12
  role of, 13
Planning, 17 (See also
        Forecasting)
  importance of, 3
  presentation of attributes, 577
  priority evaluation techniques,
        577
  public participation in, 554,
        570, 577, 578, 579
  scaling of attributes, 577
  social response to, 574
    public participation in, 572
    strategy level, 569
  systematic distortion in, 3
  technology and social systems
        and, 19
  time scale of, 9
Police
  road transport and, 284
Policy issues in energy/
        environmental system, 190
Policy making
  carrying capacity and, 385
  effects, 384
Policy-sensitive models, 83
Political progress, 81
Pollutants, 400, 412
  amount of, 430
  categories of, 429
  damage from, 425
  ranking of, 427, 428
Pollution, 6, 191, 367, 401
  of affluence and poverty, 57
  control of, 372, 401
    costs of, 502
  damage by, 415, 423, 425
  damage to receptor, 420
  delayed action of, 128
  economics of, 128, 426, 429,
        497, 498
  growth of, 104
  indicators of, 528
  natural, 528
  odor, 536
  pace of, capitalism and, 54

Pollution (cont'd)
  prevention of, 497
  public attitude to, 531
  social impact of, 525
    evaluation of, 528
  socialism and, 60
  survival and, 128
Population
  decline of, 29
    economics and, 31
  demo-economics and, 34
  growth, 6, 390
    propensity for, 6
  movement in urban areas, 471
  pollution and movement of,
    530
  stable, 32
  stationary, 31
    economic growth rate and,
    36
  of Turkey, 61
Poverty, 57, 58
Power plants, heat discharge
    from, 350
Prediction, (*See* Forecasting
    *and* Planning)
Priority evaluation technique,
    577
Privacy in housing, 436
Problems
  well-structured and ill-
    structured, 73
Problem solving, 72
Profitopolis, 55
Programme policy budgetary
    systems, 80
Progress, concept of, 12
Public education
  change in urban system and,
    469
Public involvement
  in forecasting, 81
  in planning, 554, 570, 577,
    578, 579
Public opinion, government
    and, 569

Quality of life, 384

Rail safety, 241-261
  conceptual model of, 247
    control group and, 251
  definitions of, 245
  economic system and, 250
  government regulations and, 242
  level of risk, 251, 252, 253,
    258
  physical system of, 248
  policy alternatives for, 257
  subsidies and, 257
  taxation and, 257
  value system, 251
Receptors, 419
  damages to, 420, 421
Recreation
  water resources and, 210, 221
    economics of, 211, 212
Regional development, 59
Regional inequalities, 52
Relevance trees, 77
Resources
  depletion of, 6, 127
  regulation of use of, 7
Rhone-Alps
  energy/environmental future of,
    335
  environmental impacts in, 344
  heat from power plants in, 350
  primary energy use in, 343
Rich and poor countries,
    competition between, 8
Risk, acceptable levels of, 254
Road schemes
  benefits in time saved, 281
  cost benefits of, 282
Road traffic accidents, 303
  attitudes towards, 267
  causes, 263, 264, 274
    research into, 286
  driver error in, 286, 287
  economics of, 282, 288, 289,
    299
  environmental causes, 264
  health service and, 287
  legal aspects of, 283
  police patrols and, 286
  risk of, 303

Social systems (cont'd)
  technical progress and, 17
    planning in, 19
  transformation of, 20
Social welfare, 56
Socialism
  environmental degradation
        and, 60
  urban growth and, 49
Society
  control systems, 10
  ethical factors, 13
  future of, 8, 9
Species, survival of, 5
Standstill principle, 402
Starvation, 8
Statistics
  inverse, 118
Steel production, 526
Strategic choices game, 578
Stress factors, 528
Sulphur dioxide emissions, 346
Survival, 5, 124
  Mesarovic-Pestel strategy
        for, 78
  pollution and, 128
Systems analysis, 78, 80

Taxation
  rail safety and, 257
  road safety and, 294
Technical progress, social
        systems and, 17
Technological systems
  assessment
    actors, 326, 330
    actor analysis framework,
        321
    actor classification
        framework, 323
    actor issues, 327
    allied supporting actors,
        324
    core actors, 324
    decision framework, 330
    decision issues, 327
    evaluation, 317
    issues analysis framework,
        326, 328

Technological systems (cont'd)
  assessment (cont'd)
    rivals and adversaries, 325
    strategy analysis framework,
        329
Technology
  cost of innovation in, 18
  course of, 6
  demand for in developing
        countries, 24
  new
    introduction of, 18
    rule of conduct, 18
  progress in, 21
  social systems and, 19, 21
Territorial range, 234
  in behaviour settings, 227
Texas Coastal Zone
  capital costs of, 500
  energy policy evaluation in,
        504
  energy/environmental system
        in, 493
  industrial development in, 501
  ports, 495
  regional economy, 497
  resources, 494
Third world
  demand for technology, 24
  housing in, 58
  urban growth in, 49
  urban planning in, 57
TORSCA, 447, 454
Towns (*See* Urban areas)
Traffic, 27  (*See also under*
        Road *and* Rail)
Traffic development, 83
Traffic forecasting, 83
  goods, 84, 86, 96
  long distance road transport,
        97
  models, 84, 88, 91
  passengers, 86, 125
  research, 86
  supply models, 91
Transport (*See also under*
        Rail *and* Road)
  economics of, 281, 282, 373

Road traffic accidents
  (cont'd)
  speed and, 306
  in U.K., 288
Road transport
  capital expenditure on, 290
  control of, 279-298
    division of responsibility,
      291
    financial, 294
    health service involvement
      in, 287
    insurance and, 287
    legal constraints, 282, 291
    parking, 295
    pressure groups in, 293
    research into, 296
    role of courts, 286
    role of local government,
      292
    role of police, 284, 291,
      293
    speed limits, 292
    sub-optimization, 295
    taxation, 294
  costs of, 98, 287
  driver behaviour, 264, 299
    age and experience, 265
    attitudes and opinions,
      266
    bad, 271
    culture and, 266
    effect of communication, 3
      309
    gap acceptance, 307, 312
    sex and intelligence and,
      265
    social interaction in,
      268, 273
    social psychology and,
      268, 272
    study of, 269
    variables, 265, 272
  effect of radar speed checks,
    308
  improvement of raods, 282
  interactions in, 279
  long distance, 97

Road transport (cont'd)
  policing, 299-315
    activity measurement, 303
    effect of communication, 309
    effect on gap acceptance
      behaviour, 312
    effect on speed, 308
    research and design
      procedure, 302
    resource allocation, 301
    rural areas in, 300
    vehicle defects, 304
  social psychology of, 263-278
  speed limits, 300
  traffic flow, 304
  user expenditure, 290
  vehicle speeds, 305
    accidents and, 305
Rotary clubs, 230, 231
Ruhr
  industry in, 526
  social impact of pollution in,
    525

Safety
  cost of, 250
  definition of, 245
  in transportation, 243
Scenarios, 79
  writing of, 196
School teachers
  becoming administrators, 473
  drop-out patterns, 472, 475,
    476
  mobility in urban areas, 475
Slum clearance, 53
Smells
  social perception of, 535
Social economic module, 409
Social needs, 405
Social sciences, emphasis on, 15
Social systems
  behavioural factors, 152
  controlled conflict
    transposition, 22
  development and reform, 20
  as instruments of technical
    progress, 21

Transport (cont'd)
  energy/environmental systems
    and, 190
  income elasticities of
    demand, 366
  safety and, 243
    conceptual model, 247
    control group, 259
    definitions, 245
    level of risk, 259
Trend extrapolation, 76
Turkey
  land ownership in, 63
  land prices in, 62
  population of, 61
  socio-economic systems, 49
  urbanization, 49, 61

UMSOZ project, 530
Uncertainty, 72, 81
Unemployment, 526
Union sets, 143, 144
United Kingdom
  road traffic control, 279
  transport policy, 281
United States
  environment and public
    policy in, 441
Universe, 137
  division of, 145
  meaning and purpose of, 5
  scanning of, 143
  span of, 137
Urban areas
  center-periphery relations,
    51
  change in
    public education and, 469
  conflict with villages, 59
  energy/environment systems
    in, 190
  growth, 49
    in market economies, 50
  housing problems, 52, 56
  income distribution in, 52
  land in service of society
    in, 60
  land speculation in, 53
  odor affected communities
    in, 536

Urban areas (cont'd)
  optimum size of, 51, 56
  planning of, 63
  population movement in, 471
  poverty in, 57
  private ownership of land in,
    60
  public ownership of land in,
    56
  rise of interventionism, 55
  satellite, 56
  school teachers and, 469
  in socialist countries, 58
  structure of, 50, 59
  teacher mobility in, 469, 475,
    476
Urban renewal, 53

War precursors of, 160
Waste disposal, capital costs
  of, 499
Waste water treatment, 203
Water
  checklist for, 149
  cost of, 503
  drinking
    economics of, 206, 210, 211,
    212
  industrial use of, 210
    economics of, 211
  population growth and, 379
  quality management
    assessment of, 203-244
    complex evaluation method,
    217
    cost-benefit analysis,
    206, 208, 210
    economics of, 216
    environmental benefits, 214
    evaluation, 204
    goal structure, 208
    participating experts, 216
    non-economic benefits, 207
    participating experts, 216
    recreation and, 210, 221
      economics of, 211, 212
    social impact of pollution,
    527
    treatment of, 497
    utilization of, 208

West Dallas housing project,
       434
Wisconsin
   decision making in, 189
   energy/environment systems
          in, 192, 193, 335
   health and safety in, 347

Wisconsin (cont'd)
   heat from power plants in, 350
   land use in, 348
   primary energy use in, 343
Working conditions, 530
Work places, population size,
       230